MINE VENTILATION AND AIR CONDITIONING

MINE VENTILATION AND AIR CONDITIONING

THIRD EDITION

Howard L. Hartman
The University of Alabama (Emeritus)

Jan M. Mutmansky
The Pennsylvania State University

Raja V. Ramani
The Pennsylvania State University

Y. J. Wang
West Virginia University

A WILEY-INTERSCIENCE PUBLICATION

JOHN WILEY & SONS, INC.
New York • Chichester • Weinheim • Brisbane • Singapore • Toronto

Library of Congress Cataloging in Publication Data:
Mine ventilation and air conditioning / Howard L. Hartman ... [et
    al.].—3rd ed.
        p. cm.
    Rev. ed. of: Mine ventilation and air conditioning / Howard L.
Hartman. 2nd ed. 1991.
    "A Wiley-Interscience publication."
    Includes bibliographical references and index.
    ISBN 0-471-11635-1 (cloth : alk. paper)
    1. Mine ventilation. 2. Air conditioning. I. Hartman, Howard L.
TN301.M554   1997
622'.42—dc21   97-547

17 16

To our wives,

Bonnie, Diane, Geetha, and Janet

*whose support, understanding,
and patience on the home front enabled
us to write this book.*

CONTENTS

That this book has enjoyed a modicum of success and evolved into a third edition was hardly anticipated when the original work was published in 1961. The second edition, issued in 1982, was a group effort by 25 contributors and modernized and expanded the coverage. When it was decided to revise the book again, four of us agreed to undertake the task.

Our objectives in the third edition are largely the same as before: (1) to present an integrated engineering design approach to mine ventilation and air conditioning, (2) to advance an understanding of comprehensive environmental control of the mine atmosphere, and (3) to advocate total mine air conditioning through simultaneous control of the quality, quantity, and temperature–humidity of the underground atmospheric environment.

What may differ in this revision is the emphasis on an undergraduate treatment of the subject matter. We have intentionally restricted the scope and level of the coverage so that students can cover the material in one semester, assuming a background in the basic and engineering sciences and introductory mining engineering courses.

We have also directed the book to practitioners of mine ventilation in the field. It should provide adequate depth and breadth to those who design or operate mines, with responsibility for environmental engineering and especially for the health and safety of miners who rely on the underground atmosphere for survival.

To be responsive to current trends, we have again made use of dual mathematical units (English and SI) throughout.

We are indebted to our colleagues who contributed to the second edition; with their permission, we have drawn liberally in this revision from their earlier work. Additionally, sources in the current literature, manufacturers, and practicing ventilation engineers have generously provided state-of-the-art knowledge for our efforts.

To my three coauthors in this endeavor—all Penn State associates or former students of mine—I give warm thanks. To them belongs any credit for the lasting contribution this volume may make to our profession. In acknowledgment of our common educational roots, we have assigned all royalties from this book (as we also did with the second edition) to endow a mining engineering scholarship at our alma mater.

Sacramento, California   HOWARD L. HARTMAN

■■■■ ACKNOWLEDGMENTS

The following persons authored chapters that appeared in the Second Edition of this book (1982) and gave permission to the present authors to draw on that material as appropriate. Their contributions to the Third Edition are gratefully acknowledged.

James L. Banfield, Jr. (deceased)
Formerly of the
Mine Safety and Health
Administration

H. Douglas Dahl
Eastern Associated Coal Company

Rodolfo V. de la Cruz
University of Wisconsin at
Madison

Ralph K. Foster
Retired, Formerly of the
Mine Safety and Health
Administration

Y. S. Kim
Private Consultant

Richard J. Kline
Mine Safety and Health
Administration

Thomas Novak
The University of Alabama

Richard L. Sanford
The University of Alabama

Stanley C. Suboleski
A. T. Massey Coal Company

Peter M. Turcic
Mine Safety and Health
Administration

Floyd C. Bossard
Floyd C. Bossard & Associates, Inc.

Robert W. Dalzell
Retired, Formerly of the
Mine Safety and Health
Administration

C. Frederick Eben
Retired, Formerly of
Bethlehem Steel Corporation

Bruce R. Johnson
Zephyrus Mining Consultants Inc.

John D. Kalasky (deceased)
Formerly of Island Creek Coal
Company

Edward J. Miller
Mine Safety and Health
Administration

Thomas J. O'Neil
Cleveland Cliffs Inc.

Madan M. Singh
Engineers International, Inc.

Pramod C. Thakur
Consol Inc.

Richard W. Walli
Private Consultant

Edwin B. Wilson
Bethlehem Steel Corporation

The present authors also extend thanks to John E. Urosek, Chief, Ventilation Division, Mine Safety and Health Administration, and his colleagues who read and provided technical advice on Chapter 15, Control of Mine Fires and Explosions.

Mathematical symbols associated with the literature and practice of mine ventilation have evolved with time. Originally, they were based on American National Standards Institute (ANSI) codes, but these standards have fallen into disuse over the years. The symbols employed in this book are representative of the ones customarily employed in mine ventilation in the United States and have been adopted because of their clarity, consistency, and recognizability.

Symbol Letters

A	Area, ft² (m²)
a	Radius, ft (m)
B	Characteristic gas flow constant of coal seam, atm^{-1}
B_g	concentration of gas in air, % or fraction
b	Modified seam characteristic (dimensionless); width, ft (m)
C	Cost, $; concentration of tracer gas, %; concentration of methane in coal, ft³/ft³ (m³/m³)
C_a	Anemometer correction factor, ft/min (fpm) (m/s)
C_c	Coefficient of contraction (dimensionless); Cunningham correction factor (dimensionless)
C_r	Thermal conductance of rock, Btu/h·ft²·°F (W·m²·°C)
CF	Altimeter correction factor (dimensionless)
c	Specific heat, Btu/lb·°F (kJ/kg·°C); unit cost, $/ft³ ($/m³)
D	Diameter, ft (m); coefficient of diffusion or diffusivity, ft²/h (m²/s)
DR	Altimeter density ratio (dimensionless)
d	Depth (or height), ft (m)
E	Exposure concentration, mg/m³; excavation cost, $/ft ($/m) of advance
E_c	Effective combustible parameter, %
E_i	Effective inert parameter, %
F	Sensible-heat factor (dimensionless); force, lb (N)
f	Coefficient of friction (dimensionless)
G	Weight flow rate, lb/h (kg/s)
G_w	Liquid weight flow rate, lb/h (kg/s)
g	Acceleration due to gravity = 32.174 ft/s² (9.807 m/s²)

H	Difference in head, inches (in.) water (mm or Pa)
H_f	Friction head loss, in. water (mm or Pa)
H_l	Head loss, in. water (mm or Pa)
H_m	Mechanical ventilation head, in. water (mm or Pa)
H_n	Natural ventilation head, in. water (mm or Pa)
H_s	Static head, in. water (mm or Pa)
H_t	Total head, in. water (mm or Pa)
H_v	Velocity head, in. water (mm or Pa)
H_x	Shock head loss, in. water (mm or Pa)
H_z	Elevation or potential head, in. water (mm or Pa)
h	Enthalpy, Btu/lb (kJ/kg)
h_c	Cooling power, Btu/h·ft^2 (W/m^2)
h_{corr}	Enthalpy correction, Btu/lb (kJ/kg)
h_{SCP}	Specific cooling power, Btu/h·ft^2 (W/m^2)
I	Internal energy, Btu/lb (kJ/kg); inhalation rate, m^3/h
ICO	Index of carbon monoxide or Graham's ratio (dimensionless)
i	Sound intensity, dB
K	Friction factor, lb·min^2/ft^4 (kg/m^3)
K_c	Compression friction factor, lb^2·in.·min^2/ft^7 (kg^2/s^2·m^4)
k	Conductivity, Btu/h·ft·°F (W/m·°C); constant of proportionality (dimensionless); permeability of coal, millidarcies (m^2)
L	Length (or distance), ft (m)
L_e	Equivalent length due to shock loss, ft (m)
L_p	Sound pressure level, dB
L_w	Sound power level, dB
M	Metabolism rate, Btu/h
M_g	Mass of tracer gas, lb (kg)
M_t	Volume of gas desorbed in given time, ft^3/h (m^3/s)
m	Molecular weight (dimensionless); radius ratio $= r/b$ (dimensionless)
N	Number or ratio (dimensionless)
N_a	Number of airways in parallel
N_b	Number of branches in network (dimensionless)
N_l	Layering number (dimensionless)
N_m	Network degree or number of meshes in network (dimensionless)
N_n	Number of nodes in a network (dimensionless)
N_{Re}	Reynolds number (dimensionless)
NVE	Natural ventilation energy, ft (m)
NVP	Natural ventilation pressure, in. water (Pa)
n	Rotational speed, r/min (rpm); process index (dimensionless)
O	Perimeter, ft (m)
O_e	Equivalent orifice, ft^2 (m^2)
P	Probability (dimensionless)
P_a	Air power, hp (kW)
P_i	Electrical input power, hp (kW)

P_m	Mechanical power, bhp (kW)
P_w	Sound power, W
p	Root-mean-square sound pressure, microbar; pressure, in. Hg or lb/in.^{2} (psi) (mm Hg or Pa)
p_a	Partial pressure of dry air, in. Hg or psi (mm Hg or Pa)
p_b	Barometric pressure, in. Hg or psi (mm Hg or Pa)
p_r	Reference sound pressure, microbar
p_s	Saturation vapor pressure, in. Hg or psi (mm Hg or Pa)
p_v	Partial pressure of water vapor, in. Hg or psi (mm Hg or Pa)
Q	Quantity of flow, ft^3/min (cfm) (m^3/s)
Q_g	Gas inflow, cfm (m^3/s)
Q_w	Liquid volume flow rate, gal/min (gpm) (m^3/s or L/s)
$\dot{q}$	Input of heat energy, Btu/lb (J/kg)
q	Rate of change in heat content, Btu/h (W)
q_L	Rate of change in latent-heat content, Btu/h (W)
q_R	Amount of refrigeration, tons
q_S	Rate of change in sensible-heat content, Btu/h (W)
q_c	Rate of heat transfer by convection, Btu/h (W)
q_e	Rate of heat transfer by evaporation, Btu/h (W)
q_r	Rate of heat transfer by radiation, Btu/h (W)
q_s	Rate of heat transfer by storage, Btu/h (W)
R	Gas constant = 1545/molecular weight, ft·lb/lb mass·°R (J/kg·K); mine or airway resistance, in.·min^2/ft^6 (N·s^2/m^8)
R_a	Anemometer reading, fpm (m/s)
R_{eq}	Equivalent resistance, in.·min^2/ft^6 (N·s^2/m^8)
R_h	Hydraulic radius = A/O, ft (m)
R_{mc}	Ratio of methane to total combustible in atmosphere (dimensionless)
RF	Recirculation factor, % or fraction
r	Radius, ft (m)
r_e	Hydraulic radius of mine opening modified for roughness, ft (m)
S	Rubbing surface area = $O·L$, ft^2 (m^2)
s	Entropy, Btu/lb·°F (kJ/kg·°C); specific gravity (dimensionless)
T	Absolute temperature, °R (K)
TR	Trickett's ratio (dimensionless)
TLV	Threshold limit value, %, ppm, mg/m^3, etc.
t	Temperature, °F (°C)
t_d	Dry-bulb temperature, °F (°C)
t_{dp}	Dew-point temperature, °F (°C)
t_e	Effective temperature, °F (°C)
t_g	Globe thermometer reading, °F (°C)
t_0	Rock temperature at surface, °F (°C)
t_r	Virgin-rock temperature, °F (°C)
t_w	Wet-bulb temperature, °F (°C)
t_z	Rock temperature at depth Z, °F (°C)

U	Relative velocity, ft/s (fps) (m/s)
V	Average velocity of flow, ft/min (fpm) (m/s)
V_c	Critical velocity, fpm (m/s)
V_t	Particle settling velocity, fpm (m/s)
v	Specific volume, ft³/lb (m³/kg)
W	Specific humidity, grains/lb or lb/lb (kg/kg)
$\dot{W}$	Input of mechanical energy, ft·lb/lb (J/kg)
W_k	Rate of doing work, Btu/h (W); air quantity flowing in chord k of a network, cfm (m³/s)
w	Specific weight, lb/ft³ (kg/m³)
X	Shock-loss factor (dimensionless)
x	Constant of proportionality (dimensionless); concentration of contaminant, % or fraction; linear distance, ft (m)
Y	Volume, ft³ (m³)
y	Expansion factor (dimensionless)
Z	Elevation above datum, usually sea level, or potential energy, ft (m)
z	Contraction factor (dimensionless)

Greek Letters

α (alpha)	Thermal diffusivity, ft²/h (m²/s)
γ (gamma)	Specific-heat ratio at constant pressure and volume = c_p/c_v (dimensionless)
ϵ (epsilon)	Thermal emissivity, Btu/h·ft²·°F (W/m²·°C); Goch–Patterson heat-flow term (dimensionless)
λ (lambda)	Mean free path of gas molecules, m
η (eta)	Efficiency, %
θ (theta)	Angle, degrees
μ (mu)	Absolute viscosity, lb·s/ft² (Pa·s); degree of saturation, %
ν (nu)	Kinematic viscosity, ft²/s (m²/s)
ρ (rho)	Mass density, lb·s²/ft⁴ (kg/m³)
τ (tau)	Time; s, min, h, or year
ϕ (phi)	Relative humidity, %; pseudoporosity, ratio of volume of gas adsorbed on coal surface per atmosphere per volume of coal (dimensonless)
ω (omega)	Goch–Patterson heat-flow term (dimensionless); angular velocity, radians/s

The symbols listed here alphabetically are typical, but not standard, symbols for use on mine ventilation maps. Because standards do not exist, the reader should be careful in interpreting symbols on any given ventilation map. Variations in practice are common. First, many companies use color-coded maps to help in identifying ventilation airflows. For example, intake air can be denoted by blue arrows, return air by red arrows, escapeways by green arrows, and belt air by yellow arrows. The color scheme differs from mine to mine. Second, the style of the air directional arrows differs from mine to mine with different types of arrows used to denote intake and return airstreams.

Symbol	Description
	Airflow (intake)
	Airflow (return)
	Airlock; a double-door system to allow equipment to pass through without disrupting the ventilation circuit
	Auxiliary fan and vent pipe or tubing (flow direction may be indicated by an arrow)
	Brattice (also called a *line brattice*); a curtain of plastic or plastic-covered fabric hung from the roof to direct air to or from a working face
	Box check; a stopping with a hole in it to allow a conveyor or other equipment to pass through while limiting the airflow quantity
	Check curtain; a barrier of plastic or plastic-covered fabric hung across an opening from the roof to block the flow of air
	Door
	Escapeway with direction of escape in the direction of airflow
	Escapeway with the direction of escape in the direction opposite to the airflow direction

Fan (flow direction may be indicated by an arrow)

Fire door (normally open)

Main fan (the dotted lines show the location of the weak wall)

Overcast or air crossing; an area where roof material is taken to allow one airflow to pass over another without mixing (the parallel lines indicate the airway that goes straight through the overcast); may also be constructed as an undercast or sidecast crossing

Overcast with a built-in regulator

Pipe overcast; a method of using pipes to pass a small quantity of return air through an intake airflow without mixing the two airflows; generally used for taking belt air directly to the return in a coal mine

Regulator

Seal

Self-contained self-rescuer cache location

Shaft with a downcast flow of air (alternately, this symbol may represent an undercast)

Shaft with an upcast flow of air (note that this symbol could also represent a gas well or a borehole location on some mine maps)

Stopping (permanent); an impermeable stopping made of masonry, steel, or other flame-resistant material to block the flow of air through an opening

Stopping (temporary); a quickly erected and movable stopping normally made of brattice material to temporarily block the flow of air through an opening

Stopping with small door to allow the passage of personnel

MINE VENTILATION AND AIR CONDITIONING

PART I

INTRODUCTION

CHAPTER 1

Environmental Control of the Mine Atmosphere

1.1 PURPOSE AND IMPORTANCE

In the vacuum of outer space, human astronauts rely on the artificial atmosphere of a spacecraft for their life support system. While differing in locale and mission, human miners are no less dependent on an artificial atmosphere to sustain them in underground mines where the air may be stagnant and contaminated.

It is evident that both miners and astronauts confront a hostile environment, and that both groups must depend on a ventilation–air conditioning system to supply them adequate air for breathing.

Under even normal circumstances, excavation in the earth—like exploration in space—can be fraught with a variety of environmental problems and hazards. While ground support is an obvious and compelling need, the most vital aspect of the mine environment to control is the atmosphere of the workplace.

To the mining engineer, ventilation is the most versatile atmospheric control tool. It is the process relied on to accomplish most environmental control underground. *Mine ventilation* is essentially the application of the principles of fluid dynamics to the flow of air in mine openings. As the primary means of quantity control, ventilation is responsible for the circulation of air, in both amount and direction, throughout the mine. It is one of the constituent processes of *total mine air conditioning,* the simultaneous control within prescribed limits of the quality, quantity, and temperature–humidity of mine air (Anon., 1993).

Increasingly, in underground mining, environmental objectives require that we condition air to meet quality and temperature–humidity standards as well as quantity criteria. In recent years, these standards have been raised substantially. Although threshold limits are based on human safety and tolerance, increasing concern is being expressed for standards of human comfort as well. The provision of a comfortable work environment is both cost-effective and humanitarian. Worker productivity and job satisfaction correlate closely with environmental quality. Further, excessive accident rates and workers' compensation rates are a consequence of unsatisfactory as well as unsafe environmental conditions. No mining company today can afford to be lax in its environmental and air-control practices.

Historical Perspectives and Natural Constraints

The importance of mine ventilation and air conditioning has not just newly been recognized. From the onset of underground mining in Paleolithic times, perhaps as early as 40,000 BCE (B.C.) (Gregory, 1980, p. 50), miners confronted oxygen deficiency, toxic gases, harmful dusts, and debilitating heat. As miners became more skilled, by the first millennium BCE, they learned to course the air through multiple openings or circuits to provide fresh air to the working face (Lacy and Lacy, 1992, p. 5) and to use fire-induced air currents (McPherson, 1993, p. 2).

By the Middle Ages, mine ventilation enjoyed the status of a mining art. In the most celebrated early mining treatise, Georgius Agricola (1556, p. 200), a respected German scholar and scientist, decried the evils of the foul atmospheric environments in which miners had to work and pictured their still-primitive efforts to combat these conditions:

> I will now speak of ventilating machines. If a shaft is very deep and no tunnel reaches to it, or no drift from another shaft connects with it, or when a tunnel is of great length and no shaft reaches to it, then the air does not replenish itself. In such a case it weighs heavily on the miners, causing them to breathe with difficulty, and sometimes they are even suffocated, and burning lamps are also extinguished. There is, therefore, a necessity for machines which the Greeks call πνευματικαί and the Latins, *spiritales*—although they do not give forth any sound—which enable the miners to breathe easily and carry on their work.*

Figures 1.1a–c, taken from Agricola's book, portray some of these early "ventilating machines." A contemporary history of mine ventilation is presented by McPherson (1993, pp. 1–7).

Technology has vastly improved mine ventilation, although environmental challenges underground still abound. Depth, the most serious natural constraint, sets the ultimate limit, specifically through rock pressure and rock temperature. Not only do rock pressures rise inexorably with depth but temperatures do also, with subsequent deterioration of the atmosphere. According to Spalding (1949, p. 238):

> Of all the factors which affect mining operations, high rock temperature is the one most often likely to limit the depth to which those operations can be extended. The science of ventilation is therefore rapidly becoming the most important branch of deep mining.

* By permission from Dover Publications, Inc.

At great depths, ventilation requirements and costs eventually climb to unsustainable levels. To preserve mine atmospheric quality under these intense heat conditions, ventilation at great depths must be supplemented by air conditioning.

Although heat generated by depth imposes the ultimate limit, the mine and its atmosphere have other detrimental conditions to withstand. These consist usually of airborne contaminants such as gases and dusts. As mines expand in size, complexity, manpower, and mechanization, demands on the ventilation–air conditioning system to maintain more stringent standards of environmental quality likewise rise. Fortunately, advances in mining science and technology tend to keep pace with worsening hazards underground. The struggle, however, is a continuous one reflected in both human safety and operating costs.

1.2 CONTROL PROCESSES

Lest confusion arise in the mind of the reader, it is well to clarify some of the terms related to environmental control of the mine atmosphere. Used alone, in mining parlance, *air conditioning* denotes only the function of temperature–humidity control, generally cooling or heating. To signify *total* mine air conditioning and all the functions of environmental control it entails, the qualifier term "total" should be used.

To reiterate, the functions encompassed by total air conditioning are (1) *quality control,* (2) *quantity control,* and (3) *temperature–humidity control* of the atmosphere. To accomplish these objectives, individual conditioning processes are employed; in mining, they consist of the following:

1. Quality control (purifying air and removing contaminants)
   a. Gas control—vapors and gaseous matter, including radiation
   b. Dust control—particulate matter
2. Quantity control (regulating magnitude and direction of airflow)
   a. Ventilation
   b. Auxiliary or face ventilation
   c. Local exhaust
3. Temperature–humidity control (controlling latent and sensible heat)
   a. Cooling
   b. Heating
   c. Humidification
   d. Dehumidification

Control processes may be applied individually or jointly. If the objective is total air conditioning of the mine, then all three goals must be met, and multiple processes may be applied simultaneously. Several processes can serve more than one function; for example, ventilation, the most common

A—Sills. B—Pointed stakes. C—Cross-beams. D—Upright Planks.
E—Hollows. F—Winds. G—Covering disc. H—Shafts. I—Machine
WITHOUT A COVERING.
(a)

A—Tunnel. B—Pipe. C—Nozzle of double bellows.
(b)

A – Drum.   B—Box shaped casing.   C—Blow-hole.   D—Second hole
E –Conduit.   F—Axle.   G—Lever of axle   H—Rods.

(c)

FIGURE 1.1 Mine ventilation machines of Agricola's day: (*a*) deflectors; (*b*) bellows; (*c*) fans. [Parts (*a*)–(*c*) from Agricola, 1556. By permission from Dover Publications, Inc., copyright 1950.]

one in mining, performs mainly quantity control but may serve also for quality control and temperature–humidity control.

In coping with atmospheric environmental hazards in mining, certain engineering principles are fundamental and applicable to control of any contaminant. These contaminants consist mainly of gases and dusts but include heat and humidity as well. In order of preference of their application, *engineering control principles* consist of the following (Hartman, 1968):

1. Prevention or avoidance
2. Removal or elimination
3. Suppression or absorption
4. Containment or isolation
5. Dilution or reduction

For example, if quality control of a dust hazard is the objective, then these five steps should be evaluated and, as appropriate, applied in the order given. Ventilation, a dilution measure, may be the ultimate solution, but it should be employed in conjunction with prevention, removal, suppression (by water), and containment (by suitable enclosure of the source).

In addition to engineering control, there are other measures at the disposal of mine officials responsible for safety, ventilation, and air conditioning. *Medical control principles* consist of education, physical examinations, lung x-rays, personal protective devices, prophylaxis, and therapy. Last are *legal control principles,* which consist of statutory and regulatory provisions and workers' compensation laws. All are resources to employ in combating environmental hazards.

Up to this point, the stated or implied reason for air conditioning or other environmental control processes is the preservation or enhancement of human life. Conditioning that controls the atmosphere that human beings breathe is termed *comfort air conditioning.* Nearly all mine air conditioning systems are of this type. On occasion, however, *product air conditioning* is employed when the objective is preservation of the plant or quality of the product. Examples are air temperature or moisture reduction to prevent slaking of coal mine roofs, absorption of water by drying to preserve deliquescent minerals, heating of water lines in downcast shafts during winter, and dehumidification of air in wet upcast shafts.

1.3   COORDINATION OF MINING AND VENTILATION SYSTEMS

Notwithstanding its criticality to the life support process, environmental control in mines poses a paradox. It does in all industry. On the one hand, environmental control is essential to the preservation of human life and necessary for the conduct of underground operations. On the other hand, it is ancillary to the primary objective of mining: the production of ore, rock, or coal from a mineral deposit. The paradox is that environmental control contributes nothing to production directly and yet makes the production cycle possible.

As stated previously, the two most vital environmental control measures in mining are (1) ventilation and air conditioning and (2) roof support and ground control. Ideally, they should be performed with minimum interference and cost to the production operation. Realistically, they are essential and must carry top priority in the entire mining system.

How is the dilemma resolved? The answer is that *environmental control measures are auxiliary operations that are programmed into and performed as an integral part of the production cycle.* Thus they receive the attention they must but are managed to optimize the productivity of the overall mining operation. The matter receives more detailed consideration in Chapters 13 and 14.

There is also a unique interdependency between the mine production system and the environmental control system (Hartman, 1973). In mine ventilation, air is coursed through the mine workings and openings themselves (and for auxiliary purposes through vent tubing and ducts). Quite clearly, joint optimization of the two systems can lead to the most satisfactory environment and to the most cost-effective mining.

Through the centuries, recalling Agricola's discourse, and into more recent years, ventilation systems and mining systems have tended to evolve together. Probably the prime example of the evolution is in room and pillar mining of bedded or tabular mineral deposits, especially coal. Here constraints of the ventilation system have imposed numerous changes in the configuration of the mining plan in response to technological progress in machines and methods. For example, ventilation requirements have led to the provision of special underground openings, called *bleeders.* And in all mining systems, one finds concessions to ventilation needs in the form of air shafts, escapeways, multiple openings, and various airflow control devices.

Much of the joint evolution of mining and ventilation systems, however, has been fortuitous and unplanned. In spite of the progress that has been made, the safety record in underground mining remains one of the poorest of all U.S. industries. There is now reason to believe that the situation may be changing (Hartman, 1982). Four highly important developments in the last three decades, two technological and two nontechnical, are responsible:

1. The high-speed, electronic *digital computer,* permitting advanced solutions to ventilation circuits and networks heretofore unsolvable

2. The *systems approach,* which optimizes complex industrial operations, permitting personnel, materials, and methods to be coordinated in the most efficient way

3. Extensive federal *legislation,* embodying a strict code of regulations to improve the safety of mining operations

4. The advent of *socioengineering,* the applying of technology with full consideration of the social, political, economic, and environmental consequences as well as the technical benefits

Their application is considered as appropriate in succeeding chapters; the literature now is replete with examples. It is worth noting that none of these developments existed (or certainly not as state-of-the-art technology) at the time the first edition of this book appeared in 1961. This progress alone necessitates and justifies the current revision.

1.4   FOUNDATIONS OF MINE VENTILATION AND AIR CONDITIONING

The foundations of the field of ventilation and air conditioning are laid on many disciplines. Without being identified solely with any one, they apply basic concepts of physical chemistry, thermodynamics, fluid mechanics, and mechanical design to control of the physical, chemical, and thermal properties of air.

Mining engineers, in coping with the more specialized field of mine ventilation and air conditioning, further draw from their knowledge of mining methods in designing underground conditioning systems. They are more limited in their practice than mechanical engineers in dealing with industrial air conditioning, however; as just described, their major conduits for airflow must coincide with the openings driven in the rock for mining purposes. Understanding of mine ventilation and air conditioning thus requires a knowledge of mining technology as well as basic science.

The goal of this book is to instruct mining engineers in the principles and practices of ventilation and air conditioning applicable to the underground atmosphere and the unique environmental conditions found in mines. If they employ the latest available technology, mining's myriad challenges of depth, size, complexity, manpower, and mechanization can be adequately met today.

Since total mine air conditioning is conveniently divided into three subareas or functions, the theory and practice of each are presented in separate parts of this volume. For mining engineers who need to develop competence in the design of mine conditioning systems, *the application of basic theory to practical design is of the utmost importance.* To this aim, the book is dedicated.

Accuracy of Calculations

In mine ventilation and air conditioning calculations, it is a general rule to express numerical values to three or preferably four significant figures. This is because the precision of most ventilation and air conditioning measurements does not exceed that range. Thus there is no need for a high degree of accuracy in the bulk of mine ventilation and air conditioning work. This is not to say, however, that the speed, capacity, and precision advantages of digital or other types of computers are not utilized; complex ventilation networks can be solved only by programmable computers. For routine, basic calculations, pocket calculators are generally employed and are entirely adequate. Although answers may be rounded off to four significant figures, intermediate steps should be carried to the capacity of the calculator.

Mathematical Units

It is a reflection of the times and the complexities of socio-politico-technological change that this third edition retains the practice of employing both English and SI mathematical units adopted in the second edition (the first edition adhered strictly to English units). The authors considered alternatives but in the end elected to make no significant change. While virtually all the sciences have adopted SI usage, some technologies, especially the American mining industry, remain attached to English units.

Even with dual units, English usage still receives some preference in this book. English units generally appear first with SI in parenthesis. Formulas are stated in both units if necessary, with English first. Examples and problems also employ both units. If, for space reasons, illustrations and tables lack dual units, applicable conversion factors appear in the captions or footnotes.

For a more detailed explanation of unit usage and conversion practice as well as a table of conversion factors, see Appendix B.

Mathematical Symbols

Symbols generally employed in scientific notation as well as those peculiar to the fields of air conditioning and mine ventilation have been standardized and adopted throughout this book. They are listed in the front matter of this book preceding the list of Map Symbols.

Map Symbols

Standardized symbols for fans and control devices are customarily employed in mine ventilation mapping and surveying. Those adopted in this book appear in the front matter following the list of Mathematical Symbols.

CHAPTER 2

Properties and Behavior of Air

2.1 NATURE AND COMPOSITION OF AIR

The fluid substance of chief concern in the mine environment is air. *Air* is a gaseous mixture, existing as a vapor, that constitutes the natural atmosphere at the surface of the earth. Thermodynamically, it may be thought of as a mechanical mixture of dry air and water vapor, whose behavior is complicated by changes of state in the water vapor. Chemically, the composition of so-called dry air at sea level is as follows (where vol% and wt% are percents by volume and by weight, respectively) (Bolz and Tuve, 1973):

Gas	Vol%	Wt%
Nitrogen	78.09	75.55
Oxygen	20.95	23.13
Carbon dioxide	0.03	0.05
Argon, other rare gases	0.93	1.27

For calculations involving quality control, it is customary to assume dry air and compute problems on a volume basis, taking the composition approximately as

Oxygen	21%
Nitrogen and "inert" gases	79%

The various rare gases are grouped with nitrogen because they are chemically and physically inert insofar as air conditioning is concerned. For problems involving carbon dioxide, use 0.03% or the actual content by volume.

It must be borne in mind, however, that *dry* air does not exist in normal atmospheres. It is a hypothetical term, one we assume in quality control or use as a convenience in psychrometric calculations. *Saturated* air, which is air containing all the water vapor possible at the existing conditions of temperature and pressure, is more than occasionally encountered; and even *supersaturated* air (fog) is not uncommon. The usual situation confronted in mine air conditioning is *moist* air, or *normal* air, which is a mixture of dry air and water vapor, varying from 0.1 to 4 vol% (usually well over 1% in mines). This is a "normal" atmosphere, the basis for ventilation and air conditioning calculations. So-called "standard" air is a misnomer; generally, normal air is meant. (On the other hand, *environmental standard* and *standard conditions* are appropriate terms.)

2.2 PROPERTIES OF AIR

With regard to *chemical properties,* air is colorless, odorless, and tasteless and supports combustion and life. These are important in quality control. Other properties of air may be classified as physical or psychrometric. *Physical properties* consist of those of the fluid, both at rest or in motion; quantity control (ventilation) is concerned principally with the dynamic properties. *Psychrometric properties* relate to the thermodynamic behavior of air–water-vapor mixtures and are of particular importance in temperature–humidity control (air conditioning). For convenience, all the properties of air used in this text, together with their customary units, appear in the List of Mathematical Symbols in the frontmatter.

The following table of general air constants is useful in ventilation and air conditioning work (Bolz and Tuve, 1973):

Molecular weight m	28.97
Specific gravity s	1
Gas constant R	53.35 ft·lb/lb mass·°R (287.045 J/kg·K)
Specific weight w at standard conditions (at sea level, 29.92 in. Hg and 70°F)	0.0750 lb/ft³ (1.2014 kg/m³)
Standard barometric pressure p_b (at sea level)	29.92 in. Hg or 14.696 psi (760 mm Hg or 101.33 kPa)
Specific heat at constant pressure c_p	0.2403 Btu/lb·°F (1.006 kJ/kg·°C)
Specific heat at constant volume c_v	0.1714 Btu/lb·°F (0.717 kJ/kg·°C)
Ratio of specific heats at constant pressure and volume γ (for any diatomic gas)	1.402

Psychrometric Properties

The science of the evaluation of the psychrometric properties (heat and humidity) of air under given conditions and during different temperature–hu-

midity control processes is termed *psychrometry*. It concerns the thermodynamics of air–water-vapor mixtures. Because normal air is a mixture of dry
air and water vapor, both of which behave as nearly perfect gases, and
because of changes of state involved with the water vapor, tracking the
behavior of the mixture during air conditioning processes requires the application of thermodynamic principles.

The determination of certain psychrometric properties of air at given conditions, termed the *state point,* is prerequisite to solving problems involving
air conditioning processes. Two aids are available in finding these properties:
psychrometric tables and psychrometric charts. Their use is demonstrated
in the sections that follow. For a more complete discussion, refer to a current
air conditioning text (e.g., McQuiston and Parker, 1982).

At a known barometric pressure, any two psychrometric properties of air
fix the state point; those easiest to measure and to employ are the dry-
bulb and wet-bulb temperatures. (As many other psychrometric properties
as desired may then be determined, although the one property essential to
all process calculations is enthalpy.)

Barometric pressure p_b: atmosphere pressure as read by a barometer; in
in. Hg or psi (mm Hg or Pa).

Dry-bulb temperature t_d: temperature indicated by a conventional dry
thermometer, a measure of the sensible heat content of the air; in °F (°C).

Wet-bulb temperature t_w: temperature at which water evaporating into
air can bring the air to saturation adiabatically at that temperature, a measure of the evaporating capacity of the air; indicated by a thermometer
with a wetted wick; in °F (°C).

Mathematically, the barometric pressure equals the sum of the partial pressures of dry air p_a and water vapor p_v:

$$p_b = p_a + p_v \tag{2.1}$$

Psychrometric Tables

Psychrometric tables have been prepared to summarize the psychrometric
properties of air under dry and saturated conditions, and these are repro-
duced in Appendix Table A.2. (So-called steam tables are abbreviated psychrometric tables at high temperatures.) Using these tables, needed psychrometric properties of air at dry or saturated conditions can be read. The
properties at the given state point (usually moist air, neither dry nor saturated) can then be calculated. Formulas used in these calculations are
as follows; for derivations, see the American Society of Heating, Refrigerat-

ing & Air Conditioning *Handbook of Fundamentals* (Anon., 1993): Given: t_d, t_w, p_b. From the tables, read properties of saturated air:

1. Saturation vapor pressure at t_d, p_s in in. Hg (Pa)
2. Saturation vapor pressure at t_w, p'_s in in. Hg (Pa)
3. Saturation specific humidity at t_d, W_s in lb/lb (kg/kg)

If steam tables are not available, calculate saturation vapor pressures by the relation

$$p_s = 0.18079 \exp\left(\frac{17.27t_d - 552.64}{t_d + 395.14}\right) \text{ in. Hg} \tag{2.2}$$

$$p_s = 0.6105 \exp\left(\frac{17.27t_d}{t_d + 237.3}\right) \text{ kPa} \tag{2.2a*}$$

To compute p'_s, substitute t_w for t_d. Then use Eq. 2.5 to calculate W_s, substituting p_s for p_v. *Find:* All properties at the state point.

1. *Vapor pressure*—partial pressure of water vapor in air (related to barometric pressure and partial pressure of dry air by Eq. 2.1):

$$p_v = p'_s - \frac{(p_b - p'_s)(t_d - t_w)}{2800 - 1.3t_w} \text{ in. Hg} \tag{2.3}$$

$$p_v = p'_s - 0.000644p_b(t_d - t_w) \text{ kPa} \tag{2.3a}$$

2. *Relative humidity*—ratio of vapor pressures of air at given conditions and at saturation, with temperature constant (note that relative humidity and degree of saturation are not numerically equal):

$$\phi = \frac{p_v}{p_s} \times 100\% \tag{2.4}$$

3. *Specific humidity*—weight of water vapor contained per unit weight of *dry* air:

$$W = 0.622 \frac{p_v}{p_b - p_v} \text{ lb/lb (kg/kg) dry air} \times 7000 = \text{grains/lb dry air} \tag{2.5}$$

* When two versions of an equation appear, the second (denoted by the letter "a") is for use with SI units.

4. *Degree of saturation*—ratio of weights of water vapor in air at given conditions and at saturation, with temperature constant (usually specific humidities are employed):

$$\mu = \frac{W}{W_s} \times 100\% \qquad (2.6)$$

5. *Specific volume*—volume per unit weight of *dry* air (not equal mathematically to the reciprocal of the specific weight of mixture):

$$v = \frac{RT_d}{p_a} \text{ ft}^3/\text{lb (m}^3/\text{kg) dry air} \qquad (2.7)$$

where R is the gas constant for air and T_d is absolute dry-bulb temperature (units of p_a are in psi or Pa).

6. *Specific weight*—specific weight of moist air or mixture:

$$w = \frac{1}{v}(W + 1) \text{ lb/ft}^3 \text{ (kg/m}^3) \qquad (2.8)$$

Or if steps 3 to 5 omitted, calculate by

$$w = \frac{1.325}{T_d}(p_b - 0.378p_v') \text{ lb/ft}^3 \qquad (2.9)$$

$$w = \frac{1}{0.287\, T_d}(p_b - 0.378\, p_v') \text{ kg/m}^3 \qquad (2.9a)$$

7. *Enthalpy*—total heat content of air, the sum of enthalpies of dry air and water vapor, per unit weight of *dry* air:

$$h = h_a + h_v = c_p t_d + W(h_{fg} + h_f)$$
$$= 0.24t_d + W(1060 + 0.45t_d) \text{ Btu/lb dry air} \qquad (2.10)$$

$$h = 1.005t_d + W(2.5016 + 0.001884t_d) \text{ kJ/kg} \qquad (2.10a)$$

where c_p is specific heat of air at constant pressure, h_a is enthalpy of dry air (the sensible-heat component), h_v is enthalpy of water vapor (the latent-heat component), h_f is heat of the liquid, and h_{fg} is heat of vaporization ≈ 1060 Btu/lb at normal temperatures.

Notice that the term "dry" air has been used as a reference base in defining several of the psychrometric properties of air (h, W, and v). This imaginary standard (1 lb or kg of dry air) is employed because it is the only property that remains constant when air undergoes thermodynamic changes during air conditioning processes (i.e., when a change of state occurs and water vapor condenses from or evaporates into the air). It is of considerable convenience and simplifies calculations in temperature–humidity control.

An example will demonstrate the procedure when calculating psychrometric properties from the tables.

Example 2.1   Given t_d = 70°F (21.1°C), t_w = 50°F (10.0°C), p_b = 29.921 in. Hg (101.04 kPa), find all remaining psychrometric properties of air at the state point using Appendix Table A.2.

Solution: At t_w, read p_s' = 0.3624 in. Hg (1.224 kPa). At t_d, read p_s = 0.7392 in. Hg (2.496 kPa). Using Eq. 2.3,

$$p_v = 0.3624 - \frac{(29.921 - 0.3624)(70 - 50)}{2800 - (1.3)(50)} = 0.1463 \text{ in. Hg (0.494 kPa)}$$

At p_v, read dew-point temperature from table, t_{dp} = 27.5°F or −2.5°C ($p_v = p_s$ at t_{dp}).
Using Eq. 2.4,

$$\phi = \frac{0.1463}{0.7392} \times 100 = 19.8\%$$

Using Eq. 2.5,

$$W = \frac{(0.622)(0.1463)}{(29.921 - 0.1463)} = 0.00306 \text{ lb/lb (kg/kg)}$$

At t_d, read W_s = 0.01577 lb/lb (kg/kg). Using Eq. 2.6,

$$\mu = \frac{0.00306}{0.01582} \times 100 = 19.3\%$$

Using Eq. 2.1,

$$p_a = 29.921 - 0.1463 = 29.775 \text{ in. Hg (100.54 kPa)}$$

Using Eq. 2.7,

$$v = \frac{(53.3)(460 + 70)}{(29.775)(0.491)(144)} = 13.42 \text{ ft}^3/\text{lb (0.8378 m}^3/\text{kg)}$$

Using Eq. 2.8,

$$w = \frac{1}{13.42} (0.00306 + 1) = 0.0747 \text{ lb/ft}^3 \ (1.1965 \text{ kg/m}^3)$$

or from Eq. 2.9,

$$w = \frac{1.325}{530} (29.921 - 0.378 \times 0.1463) = 0.0747 \text{ lb/ft}^3 \ (1.1965 \text{ kg/m}^3)$$

Using Eq. 2.10,

$$h = (0.24)(70) + (0.00306)(1060 + 0.45 \times 70) = 20.14 \text{ Btu/lb} \ (46.85 \text{ kJ/kg})$$

Psychrometric Chart

Determination of the various psychrometric properties of air by table and formula, although precise, is tedious even when facilitated by computers or programmable calculators. In air conditioning problems where less precision can be tolerated or only occasional determinations are made, it is customary to employ a *psychrometric chart,* which is a graphic plot relating all the psychrometric properties, similar to a Mollier diagram for steam calculations. Not only can the properties be read directly without computation, but—more important—the psychrometric process that the air is undergoing can be represented on the chart also. This is invaluable in solving air conditioning problems, and hereafter the psychrometric chart will be employed mainly for such work in this book. It should be emphasized, however, that mines employing air conditioning are well advised to program all psychrometric computations on a calculator or computer and forego use of the chart, except as a check and to plot processes.

The makeup of the psychrometric chart is such that the state point of the air can be located at a given barometric pressure if any two properties, usually temperatures, are known. The construction and coordinates of the chart are explained by Fig. 2.1. Note that the h and t_w lines do not quite coincide; a correction h_{corr} must be applied to the enthalpy as read from the graph. (If the true h lines were plotted, they would slope less steeply than the t_w lines, which indicates that the enthalpy of air at saturation is greatest. This is due to the sensible heat of the additional moisture present.) A psychrometric chart plotted at standard barometric pressure and normal temperatures is provided for problem solving in Fig. 2.2. Its use can best be demonstrated by solving the previous example graphically. (*Note:* Since specific weight is not plotted on the psychrometric chart, use Fig. 2.3 to solve Eq. 2.9 graphically.)

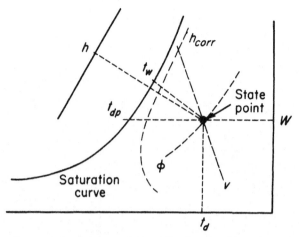

FIGURE 2.1   Construction of a psychrometric chart and location of the state point.

Example 2.2   Same as Example 2.1, but solve by chart.

Solution: Using Fig. 2.2,

$$t_{dp} = 27.7°F \ (-2.3°C)$$

$$\phi = 20.0\%$$

$$W = 0.0031 \ lb/lb \ (kg/kg)$$

$$h_{corr} = -0.08 \ Btu/lb \ (-0.19 \ kJ/kg)$$

$$h = 20.31 - 0.08 = 20.23 \ Btu/lb \ (47.05 \ kJ/kg)$$

$$v = 13.42 \ ft^3/lb \ (0.8378 \ m^3/kg)$$

$$w = 0.0748 \ lb/ft^3 \ (1.1981 \ kg/m^3)$$

by air specific-weight chart (Fig. 2.3)

$$W_s = 0.0158 \ lb/lb \ (kg/kg)$$

$$\mu = \frac{0.0031}{0.0158} \times 100 = 19.6\%$$

The answers in the two examples compare well.

Conventional psychometric charts have one major drawback, aside from

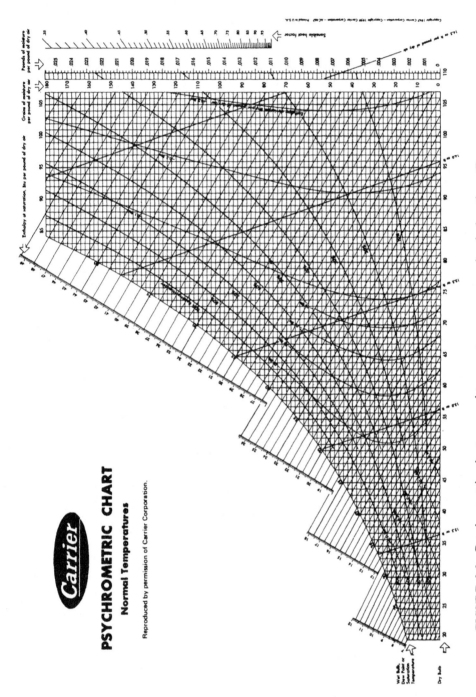

FIGURE 2.2 Psychrometric chart, normal-temperature range, sea-level elevation. (By permission from Carrier Corp. Copyright 1959, Carrier Corp., Syracuse, NY.)

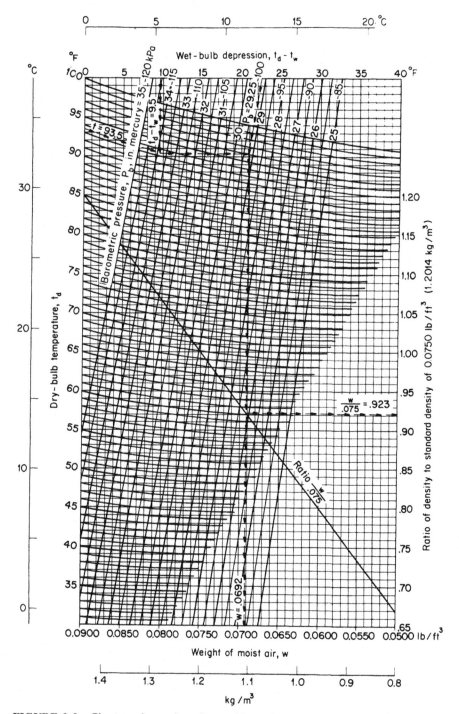

FIGURE 2.3 Chart to determine air specific weight and ratio to standard specific weight ($w = 0.0750$ lb/ft³ or 1.2014 kg/m³). (After McElroy, 1935.)

graphic limitations of accuracy: They are designed only for a single barometric pressure or elevation and a limited temperature range. Laborious corrections must be applied when using them for conditions differing by more than about 1 in. Hg (3.4 kPa) or 1000 ft (300 m) from sea level. Properties most affected are h, W, and v. For extensive work in psychometry, a series of charts covering the range of anticipated elevations or pressures and temperatures is necessary. Three are provided in the Appendix, as Figs. A.4–A.6, two for low barometric pressures (i.e., elevations above sea level) and one for high temperatures, conditions commonly encountered in mining. Others are available in the literature or can be purchased (McElroy, 1947; Barenbrug, 1965; Whillier, 1971). If charts for the needed elevations cannot be obtained, however, then the advantage of using the psychometric chart for problems is lost, and it is preferable to employ the psychometric tables and a programmable calculator than to compute corrections to the sea-level chart.

2.3   GAS LAWS: BEHAVIOR OF AIR

The following gas laws, although strictly correct only for the hypothetical *ideal* gas, are sufficiently accurate for *normal* air to employ in all routine air conditioning calculations. They are basic equations, drawn from elementary chemistry and physics.

Boyle's Law

The volume or specific volume of a gas v is inversely proportional to the absolute pressure p at constant temperature:

$$p_1 v_1 = p_2 v_2 \tag{2.11}$$

In the case of normal air, p is the absolute partial pressure of either dry air or water vapor.

Charles' Law

At constant pressure, the volume or specific volume of a gas is directly proportional to the absolute temperature T:

$$\frac{v_1}{v_2} = \frac{T_1}{T_2} \tag{2.12}$$

Alternatively stated, at constant volume, the absolute pressure of a gas is directly proportional to the absolute temperature:

$$\frac{p_1}{p_2} = \frac{T_1}{T_2} \tag{2.13}$$

General (Combined) Gas Law

The volume or specific volume of a gas varies directly as the absolute temperature and inversely as the absolute pressure:

$$\frac{p_1 v_1}{T_1} = \frac{p_2 v_2}{T_2} \tag{2.14}$$

This law can also be written in the format of Eq. 2.7.

Dalton's Law

The total pressure p exerted by a gaseous mixture is equal to the sum of the partial pressures of the individual gases. For normal air, the barometric (total) pressure p_b is expressed mathematically by Eq. 2.1.

Graham's Law

The rate of diffusion of a gas into air is inversely proportional to the square root of the ratio of the specific weights of the gas w_g and air w, or to the square root of the specific gravity of the gas s_g:

$$\text{Diffusion rate} \propto \sqrt{\frac{w}{w_g}} \propto \sqrt{\frac{1}{s_g}} \tag{2.15}$$

In other words, a gas lighter than air will diffuse faster than one heavier than air; and the smaller its specific gravity, the more rapid the diffusion. Diffusion is aided by turbulence and temperature.

In the mine atmosphere, and particularly one that is stagnant or quiescent, stratification of gaseous impurities may occur. A heavy gas (e.g., carbon dioxide) will stratify against the bottom and a light one (e.g., hydrogen) near the roof. Convection currents and the ventilation airstream, however, tend to promote the diffusion of any foreign gas that is introduced into the mine atmosphere. Once gases have diffused, they will not restratify. Whether a gaseous impurity will be found stratified or diffused is a function of the relative rates of inflow and diffusion, air movement and temperature, geometry of the mine opening, and layering (Section 3.5).

Effect of Altitude

The relationship between the specific weights of air at any elevation w_2 and at sea level w_1, at constant temperature, may be found from the following expression (Mancha, 1946):

$$\frac{w_2}{w_1} = e^{-Z/RT} \tag{2.16}$$

where Z is elevation above sea level in ft (m). Where a change in temperature occurs (temperature varies inversely with elevation, up to 35,000 ft or 10,700 m), the following equation may be used to determine specific weight at a desired elevation (Madison, 1949):

$$\frac{w_2}{w_1} = \left(\frac{288 - 0.00198Z}{288}\right)^{4.256} \tag{2.17}$$

Appendix Table A.1 lists air specific weights, barometric pressures, and temperatures for various elevations, based on Eqs. 2.16 and 2.17.

2.4 PRESSURE/HEAD RELATIONSHIP

Mine ventilation commonly employs both pressure and head measurements, pressures to express barometric readings and heads to measure airflow potentials. *Pressure* (basic units: psi or Pa) is defined as the force exerted by air per unit area, whereas *head* (basic units: in. or mm) is the height of a column of water or mercury equivalent to the pressure exerted by the air. While related, the two terms are not synonymous and should not be used interchangeably.

There are inconsistencies in practice, however, that the reader may find confusing. For example, in barometric pressure measurements, the equivalent height of a column of mercury (in in. or mm) is usually used as the unit. Similarly, in expressing heads, inches of water is employed as the English unit but pascals as the SI unit. *By custom, these exceptions are followed in mine ventilation terminology and are adopted in this text.*

The need for pressure/head conversions arises frequently in air conditioning work. Without making reference to a table of conversion factors, the equivalency can always be computed by the relation

$$p = w_1H_1 = w_2H_2 \tag{2.18}$$

where p is pressure, w is specific weight, and H is head.

Example 2.3 Calculate the pressure, in psf, equivalent to a head of 1 in. (25.4 mm) water.

Solution: Using Eq. 2.18,

$$p = wH = (62.4 \text{ lb/ft}^3)\frac{(1 \text{ in.})}{(12 \text{ in./ft})} = 5.2 \text{ psf } (248.9 \text{ Pa})$$

Example 2.4   Oil of specific gravity 0.85 is used in a vertical manometer. If the deflection is 14 in. (355.6 mm) when measuring air pressure in a duct, what is the equivalent head in in. (mm) water (specific gravity = 1)?

Solution: Use specific gravity s rather than specific weight in Eq. 2.18:

$$s_1 H_1 = s_2 H_2$$

$$(1)H_1 = (0.85)(14)$$

$$H_1 = 11.9 \text{ in. (302.3 mm) water}$$

For routine conversion of pressures and heads, the following are useful:

1 in. water = 5.2 psf (248.84 Pa or 25.4 mm water)
1 psi = 2.036 in. mercury = 27.7 in. water (6.8948 kPa or 703.6 mm water)
1 in. mercury = 0.491 psi = 13.6 in. water (3.3768 kPa or 345.4 mm water)

PROBLEMS

2.1   Calculate all the psychometric properties of air at 88°F (31.1°C) dry-bulb temperature, 85°F (29.4°C) wet-bulb temperature, and 28.0 in. Hg (94.55 kPa) barometric pressure. Use psychometric tables.

2.2   Repeat for 75°F (23.9°C) dry-bulb temperature, 65°F (18.3°C) dew point, and 29.92 in. Hg (101.03 kPa) barometric pressure.

2.3   Solve Problem 2.2 using the psychometric chart.

2.4   Solve Problem 2.1 by the psychometric chart, using the one for 2500 ft (762 m) elevation.

2.5   What is the height (in ft or m) of a column of dry air equivalent to a head of 1 in. (25.4 mm) water? Assume standard conditions.

2.6   Calculate the pressure in psf (kPa) equivalent to a head of 1 in. (25.4 mm) mercury. Solve in two ways to check your answer.

2.7   What would be the reading (actual displacement of fluid) of a vertical manometer containing oil of specific gravity 0.8 when connected to a duct maintained at 3.0 in. (76.2 mm) water?

2.8   Calculate the specific volume of dry air at standard conditions.

2.9   Determine the specific weight of air at an elevation of 2500 ft (762 m), assuming that the air at sea level has a specific weight of 0.0750 lb/ft^3

(1.2014 kg/m^3). Assume that the air at elevation and the air at sea level are at the same temperature, 70°F (21.1°C). Check your answer by Appendix Table A.1.

2.10 Solve Problem 2.9 again, assuming that the temperature varies inversely with the elevation. Check your answer by Appendix Table A.1.

PART II

MINE AIR-QUALITY CONTROL

CHAPTER 3

Mine Gases

The mine ventilation/air conditioning engineer must be concerned not only with the quantity of air the mine ventilation system can deliver but also with the chemical composition of the air. That part of total air conditioning dealing with the purity of air is termed *quality control.*

While designing or working with a mine ventilation system, the control of air quality is often one of the more important problems. Unlike an industrial environment where impurity sources are localized and the ventilation system can be and is designed to isolate the contaminant source, all the underground mine workings contain the potential for release of air contaminants such as strata gas, dust, blasting gases, and diesel exhaust.

The same passageways in which the air contaminants are generated or released must be used to transport air for underground workers to breathe. In addition, the variety and quantity of the impurities generated underground add to the complexity of the situation. Before the problem of maintaining the quality of the air can be solved, it is essential for the air conditioning engineer to be familiar with the properties of the impurities that may be encountered.

3.1 CONTAMINANTS

Broadly defined, a *contaminant,* as used in ventilation and air conditioning, is any undesirable substance not normally present in air or present in an excessive amount. Contaminants or impurities may be either *nonparticulate* (gases and vapors) or *particulate* (liquids and solids). *Liquid particulate contaminants* include mists and fogs, and *solid contaminants* include dust, fumes, smoke, and organisms (bacteria, pollen, etc.).

The most common types of air contaminants found underground are gases and dusts. These two classes of contaminants represent the major problems in quality control and therefore will be covered in detail. Familiarity with their origin, characteristics, and control is necessary to resolve contaminant problems.

3.2   THRESHOLD LIMIT VALUES

Threshold limit values (TLVs), as recommended by the American Conference of Governmental Industrial Hygienists (ACGIH), refer to airborne concentrations of substances and conditions to which it is believed that nearly all workers may be repeatedly exposed day after day without adverse health effects. Because of wide variation in individual susceptibility, however, a small percentage of workers may experience discomfort from some substances at concentrations at or below the threshold limit; a smaller percentage may be affected more seriously by aggravation of a pre-existing condition or by development of an occupational illness.

TLVs are based on available information from industrial experience, from experimental human and animal studies, and, when possible, from a combination of the three. The basis on which the values are established may differ from substance to substance; protection against impairment of health may be a guiding factor of some, whereas reasonable freedom from irritation, narcosis, nuisance, or other forms of stress may form the basis of others. Health impairments considered include those that shorten life expectancy, compromise physiological function, impair the capability for resisting other toxic substances or disease processes, or adversely affect productive function or development processes.

The amount and nature of the information available for establishing a TLV varies from substance to substance; consequently, the precision of the estimated TLV is also subject to variation, and the latest TLV *Documentation* from ACGIH (Anon., 1996) should be consulted in order to assess the extent of the data available for a given substance.

As presented by ACGIH, these limits are intended for use in the practice of industrial hygiene as guidelines or recommendations in the control of potential health hazards and for no other use. However, where used directly or by inference in federal and/or state laws, they themselves become law. Mine Safety and Health Administration's (MSHA) existing standards incorporate exposure limits from the 1972 ACGIH (Anon., 1972d) documents as these were "current" at the time of setting the standards.

When utilizing TLVs, three values are actually involved; they are discussed below and summarized in Table 3.1, together with important properties, effects, and sources of mine gases.

1. *Threshold Limit Value–Time Weighted Average* (TLV-TWA). As the name implies, TLV-TWA is the time-weighted average concentration for a normal 8-h workday or 40-h work week to which nearly all workers may be repeatedly exposed, day after day, without adverse effect. In industry, it is the TLV-TWA that is commonly referred to as "the TLV."

2. *Threshold Limit Value–Short-Term Exposure Limit* (*TLV-STEL*). According to the ACGIH, this is the maximum concentration for expo-

sures up to 15 min without suffering from (a) irritation, (b) chronic or irreversible tissue change, or (c) narcosis of sufficient degree to increase accident proneness, impair self-rescue, or materially reduce work efficiency. This is provided that (a) no more than four exposures per day are permitted, with at least 60 min between exposures, and (b) the daily TLV-TWA is not exceeded.

3. *Threshold Limit Value–Ceiling* (TLV-C). This is the concentration that should not be exceeded even instantaneously. Substances that are fast-acting are best controlled by a ceiling limit.

3.3 MINE GASES

Underground mine air is a mixture of several gases. The air found underground seldom contains the exact concentrations of gases listed in Section 2.1 because, as it circulates through the mine, it loses some of its oxygen and gains other gases from various sources such as the strata, blasting, and internal combustion engines. A listing of the properties of gases found in mines is provided in Table 3.1.

Oxygen

Of all the gases that are discussed here, oxygen is by far the most important. The human respiratory system requires oxygen in varying amounts to maintain life. The quantity of oxygen required is a function of physical activity. That is, the more active the individual, the higher the respiratory rate and the larger the volume of oxygen consumed. This is shown in Table 3.2. As indicated, the oxygen inhaled (21% by volume of the air inhaled) is much greater than the oxygen consumed. When one breathes normal air, the air that is exhaled contains approximately 16% O_2, 79% N_2, and 5% CO_2. Federal regulations require that all active workings in coal mines be ventilated by a current of air containing not less than 19.5% oxygen and not more than 0.5% carbon dioxide. Using the information presented in Table 3.2, the air quantity necessary to maintain the quality of the air can be calculated. The last column in Table 3.2, the *respiratory quotient,* gives the ratio of carbon dioxide expelled to the oxygen consumed. In other words, a miner working at a moderate rate (respiratory quotient = 0.9) would consume 0.07 cfm (3.3 $\times$ 10^{-5} m^3/s) of oxygen and expel 0.063 cfm (2.97 $\times$ 10^{-5} m^3/s) carbon dioxide.

When determining the amount of air that must be supplied to satisfy only the respiratory needs of the underground worker, it is necessary to consider both the oxygen required and the carbon dioxide produced, since carbon dioxide is a contaminant.

TABLE 3.1 Properties of Mine Gases

Name	Symbol	Specific Gravity (Air = 1)	Specific Weight lb/ft³ (kg/m³)	Other Physical Properties	Harmful Effects	Primary Sources	TLV-TWA, %	TLV-STEL, %	TLV-C, %	Explosive Range, %
Oxygen	O_2	1.1056	0.083 (1.33)	Odorless, colorless, tasteless	Nontoxic	Normal air	—	—	—	—
Nitrogen	N_2	0.9673	0.073 (1.17)	Odorless, colorless, tasteless	Nontoxic, simple asphyxiant	Normal air, strata	—	—	—	—
Carbon dioxide	CO_2	1.5291	0.115 (1.84)	Odorless, colorless, slight acid taste	Asphyxiant, increased respiration	Breathing, strata, fire, IC engines, blasting	0.5	3.0	—	—
Methane	CH_4	0.5545	0.042 (0.67)	Odorless, colorless, tasteless	Asphyxiant, explosive	Strata, blasting, IC engines	—	—	—	5–15
Carbon monoxide	CO	0.9672	0.073 (1.17)	Odorless, colorless, tasteless	Toxic, explosive	Fires and explosions, IC engines, oxidation	0.0025	0.04	—	12.5–74

32

Hydrogen sulfide	H_2S	1.1912	0.89 (1.43)	Rotten egg odor, colorless, acid taste	Toxic, explosive	Strata, strata water, blasting	0.001	0.0015	—	4–44
Sulfur dioxide	SO_2	2.2636	0.170 (2.72)	Irritating, colorless, acid taste	Toxic	Burning sulfide ore, IC engines	0.0002	0.0005	—	—
Oxides of nitrogen	NO, NO_2, N_2O	1.5895	0.119 (1.91)	Irritating odor, red-brown color, bitter taste	Toxic	Blasting, IC engines	0.0003	—	0.0005	—
Hydrogen	H_2	0.0695	0.005 (0.08)	Odorless, colorless, tasteless	Explosive	Water on a hot fire, batteries	—	—	—	4–74
Radon	Rn	7.665	0.575 (9.21)	Odorless, colorless, tasteless	Radioactive	Strata	1 WL	—	—	—

Sources: Forbes and Grove (1954), Rock and Walker (1970), and Anon. (1996).

TABLE 3.2   Respiratory Requirements

Type of Activity	Respiratory Rate, breaths/min	Air Inhaled/ Respiration, in.^{3} (10^3 mm^3)	Air Inhaled, in.^{3}/min (10^{-4} m^3/s)	O$_2$ Consumed cfm (10^{-5} m^3/s)	Respiratory Quotient
At rest	12–18	23–43	300–800	0.01	0.75
		(377–705)	(0.82–2.18)	(0.47)	
Moderate	30	90–120	2800–3600	0.07	0.9
		(1476–1968)	(7.64–9.83)	(3.3)	
Very vigorous	40	150	6000	0.10	1.0
		(2460)	(16.4)	(4.7)	

Source: Forbes and Grove (1954).

Oxygen Depletion

Unlike the other gases that will be considered, oxygen under normal conditions is not a contaminant. It is the only gas that must be maintained at as high a concentration as possible. The process in which the oxygen content of the air declines, for any of several reasons, is termed *oxygen depletion,* and the environment is said to be *oxygen-deficient.*

The processes by which the oxygen content of the air is reduced as it travels through a mine are many and are usually associated with the presence of one or more other gases. These processes include dilution by other gases, high-temperature oxidation such as internal-combustion engines and open flames, and low-temperature oxidation of wood and minerals. Usually associated with oxidation processes is the liberation of other gases such as carbon dioxide and carbon monoxide. The physiological effects of an oxygen-deficient environment vary from individual to individual and with the length of exposure. However, the following effects have been observed (Forbes and Grove, 1954):

% Oxygen in Air	Effect
17	Faster, deep breathing
15	Dizziness, buzzing in ears, rapid heartbeat
13	Possible loss of consciousness with prolonged exposure
9	Fainting, unconsciousness
7	Life endangerment
6	Convulsive movements, death

Carbon Dioxide

Carbon dioxide is a colorless, odorless, noncombustible gas that may have an acid taste when present in high concentrations. It is heavier than air and is therefore usually found in low places near the floor. Although a constituent of normal mine air (0.03%), carbon dioxide is most often found in abandoned and unventilated areas of a mine. Therefore, extreme care must be exercised when mining into areas that are inaccessible for inspection, and remote gas-sampling techniques should always be used to test the environment prior to cutting through. Sources of carbon dioxide underground include the rock strata, oxidation, fire and explosions, blasting, and the human respiratory process. Increased concentrations of carbon dioxide result in increasing lung ventilation, and individuals exposed to 0.5% carbon dioxide in otherwise normal air will breathe a little deeper and a little faster than they would if breathing normal air. When 3% carbon dioxide is present, lung ventilation is doubled, whereas a 5% concentration will result in a 300% increase in the respiratory rate. A concentration of 10% can be tolerated for only a few minutes even though such a mixture normally contains approximately 18.9% oxygen. Death occurs rapidly at 18% CO_2. Quite often, a mixture of carbon dioxide and air is referred to in mining as "blackdamp." An example will demonstrate how minimum breathing requirements may be calculated on the basis of (1) oxygen depletion and (2) carbon dioxide contamination. Flow-rate equations establish the quantity balance in each case.

Example 3.1 Assuming vigorous activity, an oxygen content of 21%, and a carbon dioxide content of 0.03% in the intake air, find the quantity of air Q in cfm that must be supplied per individual if the downstream air current is to be maintained at acceptable levels (i.e., 19.5% O_2 and 0.5% CO_2).

Solution: Required $O_2 = 0.1$ cfm (4.7×10^{-5} m³/s) (Table 3.1). Respiratory quotient $= 1.0$ (Table 3.2)

$$(O_2 \text{ in intake}) - (O_2 \text{ consumed}) = (O_2 \text{ downstream})$$

$$0.21 \, Q - 0.1 = 0.195 \, Q \tag{3.1}$$

$$Q = 6.7 \text{ cfm } (3.2 \times 10^{-3} \text{ m}^3/\text{s})$$

$$(CO_2 \text{ in intake}) + (CO_2 \text{ produced}) = (CO_2 \text{ downstream})$$

$$0.0003 \, Q + 1.0 \, (0.1) = 0.005 \, Q \tag{3.2}$$

$$Q = 21.3 \text{ cfm } (0.01 \text{ m}^3/\text{s})$$

Carbon dioxide thus governs, and a minimum quantity of 21.3 cfm (0.01 m³/s) per individual must be supplied.

As indicated by the preceding example, the quantity of air required to meet respiratory requirements is quite small compared to the quantity of air normally circulated through a mine. In this example, however, the only source of oxygen depletion is the respiratory process, which is seldom if ever the case.

Methane

The most common contaminant gas found in coal mines is methane. Although frequently associated with coal and other carbonaceous rocks, methane is also found in some noncoal mines, most notably in trona mines and in some potash, limestone, oil shale, and salt mines. Small amounts of methane have also been detected in some copper, tungsten, iron, gypsum, marble, and gold and silver mines (Thimons et al., 1979). Methane is colorless, odorless, tasteless, nontoxic, highly flammable, and lighter than air. This last attribute results in methane accumulations forming along rooflines and in high areas of mines.

During the formation of a coalbed (coalification), methane is produced along with carbon dioxide, higher hydrocarbons, and other inert gases. Increasing pressure and temperature during coalification tends to proceed toward early total elimination of oxygen with concurrent removal of some hydrogen and carbon, and then total removal of hydrogen with concurrent removal of some carbon (Hargrave, 1973). While carbon dioxide is a significant constituent of seam gas in brown coal (Ettinger and Sulla, 1964), the main constituent of higher-rank coal seam gas is methane. The volume of these byproduct gases increases with the rank of coal and is the highest for anthracite, where, for every ton of coal formed, nearly 20,000 ft^3 (565 m^3) of carbon dioxide and 27,000 ft^3 (765 m^3) of methane is produced. A large fraction of these gases is lost during burial of the decayed plant material. The amount of gas retained per ton of coal is known as the *seam gas content.*

One popular index of seam gas content in mines has been *specific methane emission,* expressed as the volume of methane emitted per ton of coal produced. A good correlation between specific methane emission and seam gas content has been established. However, the factors governing the total flow of methane in a specific mine are too diverse for any simple generalization. Several direct and indirect methods to measure seam gas content have been developed (Kissell et al., 1973; McCulloch and Diamond, 1976). In the direct method, the linear relationship existing between the amount of gas described and the square root of time of desorption is used. Here a freshly cored sample of the coal seam is placed in an airtight container and desorbed for several weeks until the desorption rate is insignificantly small. The residual gas in the coal sample is determined by crushing it to minus 200 mesh. The amount of gas lost in the coring operation is obtained graphically by estimating the time between penetrating the coals seam and placing the cored sample in container. The sum of the lost gas, desorbed gas, and residual gas is the gas content for the weight of the sample. Seam gas content can also be determined indirectly. Most of the gases contained in coal are adsorbed on the coal matrix. Langmuir (1918) has derived an equation for the idealized monolayer adsorption of methane on coal matrix as follows:

$$Y = \frac{Y_c Bp}{1 + Bp} \tag{3.3}$$

where Y is volume of gas adsorbed on the coal matrix in ft^3 (m^3), Y_c is the volume required to cover and saturate the surface completely in ft^3 (m^3), B is a characteristic constant of the coal seam in atm^{-1} (reciprocal atmospheres), and p is reservoir pressure in atm. Gas contents can be determined if the constants Y_c and B and the reservoir pressure are known.

The pressure of the gas in the seam shows a linear relationship with the depth, with a gradient of 0.333 psi/ft or 7.53 kPa/m (Thakur and Davis, 1977). It is also correlated with the rank of coal (Muche, 1975). The maximum pressure observed in an anthracite seam was 711 psi (4.9 MPa), whereas in a steam–coal seam, it was only 232 psi (1.6 MPa). This relationship appears to be valid for U.S. coals as well (Fig. 3.1).

Substituting the depth of coal seam for reservoir pressure, Eq. 3.3 is modified to

$$Y = \frac{Y_c bd}{1 + bd} \tag{3.4}$$

where d is depth of seam in ft (m) and b is a modified seam characteristic.

The gas contents, composition, and calorific values for five U.S. coal seams are shown in Table 3.3. The amount of gas in a coalbed depends on temperature, pressure, degree of fracturing, and permeability of the coal and the surrounding strata. It can exist as free gas in cracks and fissures in the coalbed or be adsorbed on the surface of the coal itself (Kim, 1973). The gas is released when there is a decrease in pressure caused particularly by nearby mining activity in the same seam or adjacent formation.

The rate at which methane is liberated is highly variable among coalbeds and may reach 2 to 4 cfm ($9-19 \times 10^{-4}$ m^3/s) for each ft^2 (0.093 m^2) of fresh coal face exposed, gradually diminishing and even stopping if adjacent mining bypasses a particular face. Wide variability is also noted in the instantaneous rate of methane emission at a face, the steady rate ranging from 40 to 400 cfm (0.02–0.2 m^3/s) at highly gassy mines in West Virginia. It is estimated that approximately 700 billion ft^3 (19.8 billion m^3) of methane is contained in the Mary Lee coalbed in Alabama (Diamond et al., 1976). The gas potential of this coalbed—as with other coalbeds—is related to the depth of overburden; the potential at present mining depths is approximately 400 ft^3/ton (12.5 $\times$ 10^3 mm^3/g). The actual gas liberation rate, however, is usually 6–9 times greater than this potential, because methane is liberated not only by the coal being removed but by the coal that is left behind as ribs and pillars, as well as by the adjacent strata (Kissell et al., 1973).

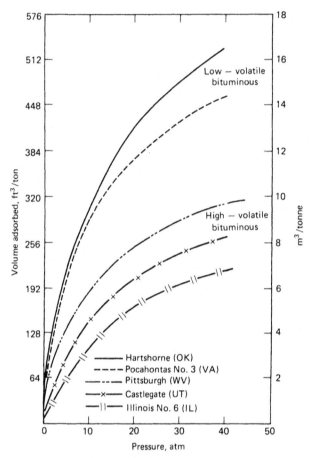

FIGURE 3.1   Variation of seam gas content with the rank of coal. (After Kissell et al., 1973.)

Prior to 1969, some coal mines were thought to be methane-free or to liberate methane in such small amounts as to be referred to as *nongassy*. With passage of the Federal Coal Mine Health and Safety Act of 1969, however, the "nongassy" classification was eliminated, and all coal mines are now considered to be *gassy*. Noncoal mines, however, retain a classification scheme consisting of several categories into one of which a mine must be placed. The criteria for classifying a noncoal mine are given in the Code of Federal Regulations, 30 *CFR* 57, subpart T (Anon., 1995a). Although the explosive range of methane is 5–15% with a minimum oxygen requirement of approximately 12%, the fact that only 0.25% needs to be detected in a noncoal mine to place the mine in category I.A is an indication of the degree of respect that this gas commands. Figure 3.2 shows the explosibility relation-

TABLE 3.3 Seam Gas Content and Composition

| Seam | Gas Content (Direct Method), ft³/ton (m³/t) | Composition, % | | | | Caloric Value, Btu/ft³ (MJ/m³) |
		CH₄	HC	CO₂ + N₂	Others: O₂, He, etc.	
Pittsburgh (WV)	220 (6.9)	90.8	0.3	8.8	0.1	922 (34.3)
Pocahontas (VA)	450 (14.0)	96.9	1.4	1.7	—	1003 (37.4)
Mary Lee (AL)	430 (13.4)	96.1	0.0	3.5	0.4	970 (36.1)
Blue Creek (AL)	513 (16.0)	96.2	0.0	3.8	—	971 (36.2)
Beckley (WV)	262 (8.2)	99.2	0.0	0.6	0.2	1002 (37.4)
Natural gas (WV)	—	94.4	4.9	0.4	0.3	1040 (38.7)

Source: Thakur and Dahl (1982).

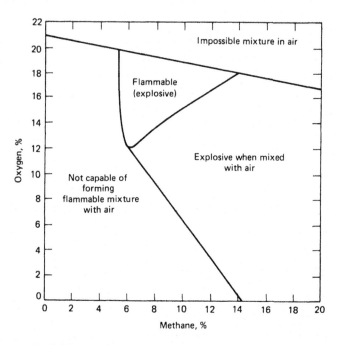

FIGURE 3.2 Explosibility (Coward) diagram for methane. (After Coward and Jones, 1952.)

ship between methane and oxygen in air. A mixture of methane in air is often referred to in the mining industry as "firedamp."

Methane Emission Models   The rate of gas emission from coal seams is generally acknowledged to be influenced by Fick's and Darcy's laws (Cervik, 1969; Kissell and Bielecki, 1972). For mathematical analysis, the coal seam is assumed to be composed of small spheres, at the surface of which methane diffusion takes place. Fick's law describes the diffusion of methane from these hypothetical spheres along the concentration gradient. For a spherical geometry, Fick's law can be written

$$\frac{\partial C}{\partial \tau} = D\left[\frac{\partial^2 C}{\partial r^2} + \frac{2\partial C}{r\partial r}\right] \tag{3.5}$$

where r is the radial coordinate in ft, τ is time in seconds, C is concentration of methane in coal in ft^3/ft^3 (m^3/m^3) of coal, and D is the coefficient of diffusion or diffusivity in ft^2/s (m^2/s). Solutions of Eq. 3.5 for various initial and boundary conditions are generally available (Crank, 1975). The size of the hypothetical sphere where diffusion takes place is important, since it determines the degree of fragmentation of the coal seam. Combining the radius of this sphere a with the coefficient of diffusion D, a new diffusion parameter D/a^2 is developed that determines both the rate at which a coal seam will diffuse methane and the fraction of seam gas content that can be drained in a given time.

Darcy's law, on the other hand, describes gas transport through the fracture system in the coal where the driving force is the pressure gradient. For a linear case in a homogeneous medium, the transport equation for laminar flow of methane for low pressures can be written as

$$\frac{\partial^2 p^2}{\partial x^2} = \frac{\mu\phi}{kp}\frac{\partial p^2}{\partial \tau} \tag{3.6}$$

where p is gas pressure in atmospheres, x is distance into the coalbed from the working face in ft (m), μ is absolute viscosity of methane in lb mass/ft·s (Pa·s), k is permeability of the coal in millidarcies (m^2), and ϕ is pseudoporosity, defined as the ratio of the volume of gas adsorbed on the coal surface per atmosphere per volume of coal.

It is clear from Eqs. 3.5 and 3.6 that the net rate of emission under idealized conditions is controlled by the diffusivity, reservoir pressure, permeability, and seam gas content of the coal seam. Depending on the magnitude of these parameters, either Fick's law or Darcy's law will control the flow of methane. Naturally, the slower process of the two will determine the net rate of methane emission. An example will illustrate this point better. A typical value of D/a^2 for the Pocahontas No. 3 seam is 1×10^{-6} s^{-1}, which is nearly 100 times larger than that for the Pittsburgh seam. This means that in a given time, Pocahontas No. 3 coal can emit a much larger fraction of its original gas content than Pittsburgh coal. However, typical permeability for the Pittsburgh seam is on the order of 10 millidarcies (9.9×10^{-15} m^2), which is at least 10 times higher than that of the Pocahontas seam. This explains why the flow from horizontal holes in the Pittsburgh seam is 140–280 ft^3/day·ft (13–26 m^3/day·m) compared to only 32–75 ft^3/day·ft (3–7 m^3/day·m) for holes in the Pocahontas seam, even though the seam gas content and reservoir pressure are much higher for the latter seam. It appears that the rate of emission for the Pittsburgh seam is controlled by the diffusivity of coal, but the permeability of coal governs the flow in the Pocahontas seam.

Carbon Monoxide

Carbon monoxide is a colorless, odorless, tasteless, toxic, and flammable gas produced by the incomplete combustion of carbonaceous material. Carbon monoxide is both poisonous at very low concentrations and explosive over a wide range (12.5–74% in air). It is formed underground by mine fires and explosions, blasting, frictional heating prior to open burning, low-temperature oxidation, and internal-combustion engines.

Although explosive, the property that makes carbon monoxide one of the gases most feared by the underground miner is its extreme toxicity. Carbon monoxide acts as a type of asphyxiant by displacing the oxygen normally carried by the hemoglobin of the blood. The affinity of the blood for carbon monoxide is approximately 3000 times that for oxygen (Forbes and Grove, 1954). Therefore, if the air breathed into the lungs contains only a small amount of carbon monoxide, the hemoglobin will absorb it in preference to the oxygen present. The compound formed when oxygen joins with hemoglobin is called *oxyhemoglobin,* and the compound formed by carbon monoxide and hemoglobin is known as *carboxyhemoglobin* (COHb). When discussing the poisoning or toxic effects of carbon monoxide, the percentage COHb or blood saturation level is usually used. At levels as low as 5% COHb, the first effects of carbon monoxide poisoning appear (Anon., 1972a). The COHb blood level is dependent on the carbon monoxide concentration, the length of exposure, and the level of activity of the individual exposed. At blood saturation levels of 70% COHb and above, the amount of oxygen being carried by the blood is insufficient to sustain life. Table 3.4 lists the symptoms caused by various levels of blood saturation.

The relationship between the blood saturation and duration of exposure to various concentrations of carbon monoxide is shown in Fig. 3.3. Figure 3.4 shows the toxicity of carbon monoxide as a function of the concentration and exposure time.

The currently recognized TLV-TWA for an 8-h exposure to carbon monoxide is 50 ppm (Anon., 1972d). The lowest published lethal concentration (the lowest concentration in air that has been reported to have caused death

TABLE 3.4   Effects of Carbon Monoxide Poisoning

Blood Saturation % COHb	Symptoms
5–10	First noticeable effect, loss of some cognitive function
10–20	Tightness across forehead, possible headache
20–30	Headache, throbbing in temples
30–40	Severe headache, weakness, dizziness, dimness of vision, nausea and vomiting, and collapse
40–60	Increased likelihood of collapse and unconsciousness, coma with intermittent convulsions
60–70	Coma, possible death
70–80	Respiratory failure, death

Sources: Forbes and Grove (1954) and Anon. (1972a).

after exposure of less than 24 h) is 4000 ppm (Christensen and Luginbyhl, 1975). Although colorless, carbon monoxide is sometimes referred to as "whitedamp."

Hydrogen Sulfide

Hydrogen sulfide, often called "stinkdamp" because of its odor, which resembles that of rotten eggs, is a colorless, toxic, and explosive gas formed

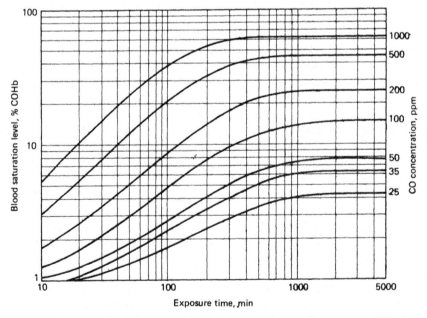

FIGURE 3.3   Carbon monoxide blood saturation. (After Anon., 1972a.)

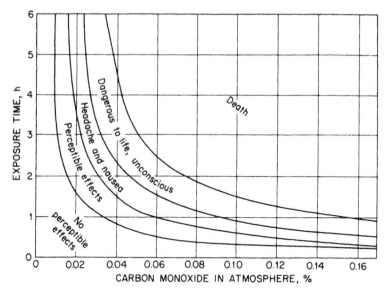

FIGURE 3.4 Toxicity of carbon monoxide as a function of concentration and time. (After Bryom, 1957. By permission of *Engineering and Mining Journal*, Chicago, IL.)

by the decomposition of sulfur compounds. Low concentrations may be found in air from heated gobs or may be released from water seeping in from the strata. Large concentrations occur in natural gas and oil fields and in some sulfur and gypsum mines. Hydrogen sulfide is quite soluble in water and may be carried into active mine workings by groundwater. It is slightly heavier than air and is explosive in air in the range of 4–44%. Hydrogen sulfide is extremely toxic, and the TLV-TWA for an 8-h exposure has been set at 10 ppm, with a TLV-STEL of 15 ppm (Anon., 1972d). Although hydrogen sulfide has a distinctive odor, the sense of smell cannot be relied upon as a means of detection, because after one or two inhalations, the olfactory nerves become paralyzed and the odor can no longer be detected. The lowest published lethal concentration is 600 ppm. The physiological effects of exposure to hydrogen sulfide are presented in Table 3.5.

Sulfur Dioxide

Sulfur dioxide is a colorless, nonflammable, toxic gas formed whenever sulfur or sulfur compounds are burned. Underground, it may be formed during the blasting of certain sulfur ores, during fires involving sulfur compounds such as iron pyrite, and from internal-combustion engines. It is significantly heavier than air, and in very low concentrations is irritating to the eyes, nose, and throat. The present TLV-TWA is 5 ppm for an 8-h exposure with

TABLE 3.5   Physiological Effects of Hydrogen Sulfide

Concentration	Symptom
0.025 ppm	Threshold of odor
0.005–0.010%	Slight symptoms such as eye and respiratory–tract irritation after 1 h
0.010%	Loss of odor after 15-min exposure
0.02–0.07%	Increased eye irritation, headache, dizziness, nausea, dryness, and pain in nose, throat, and chest
0.07–0.10%	Unconsciousness, cessation of respiration and death
0.10%	Death in a few minutes

Sources: Forbes and Grove (1954) and Sunshine (1969).

a TLV-STEL of 5 ppm (Anon., 1972d). Some of the physiological effects of sulfur dioxide are presented in Table 3.6.

Oxides of Nitrogen

Nitrogen, occurring as it does in normal air, is physiologically inert; however, under certain conditions it will form several oxides, some of which are extremely toxic. The most common are nitric oxide and nitrogen dioxide. Oxides of nitrogen are formed underground during blasting and from the operation of internal-combustion engines. Nitric oxide is rapidly oxidized to nitrogen dioxide in the presence of moisture and air and is therefore seldom found in significant amounts underground. Nitrogen dioxide is not only the more common of the two but also the more toxic. The accepted TLV-TWA for both an 8-h exposure and short-term exposure is 5 ppm (Anon., 1972d). The toxic oxides of nitrogen react with moisture to form nitrous and nitric acid. In this manner, relatively small quantities of these gases may cause

TABLE 3.6   Physiological Effects of Sulfur Dioxide

Concentration, ppm	Effect
0.3–1	Detectable by taste (acidic)
3–5	Detectable by odor (sulfur)
20	Irritation of eyes, nose, throat
50	Pronounced irritation of eyes, throat, and lungs, possible to breathe for several minutes
400–500	Immediately dangerous to life

Source: Miller and Dalzell (1982).

TABLE 3.7 **Effects of Exposure to the Toxic Oxides of Nitrogen**

Concentration, ppm	Effect
3	Current TLV-TWA
60	Least amount causing immediate throat irritation
100	Least amount causing coughing
100–500	Dangerous even for short exposure
200–700	Rapidly fatal

Sources: Forbes and Grove (1954) and Christensen and Luginbyhl (1975).

death by combining with the moisture in the lungs and corroding the respiratory passages. Death from exposure to the oxides of nitrogen may be very quick if the exposure level is high or may occur several days later as a result of pulmonary edema (water in the lungs) or even weeks later as a result of infectious pneumonia. Some of the physiological effects at various concentrations are listed in Table 3.7.

Hydrogen

Hydrogen is colorless, odorless, tasteless, nontoxic, and the lightest of all gases found underground. The sources of hydrogen underground (all somewhat uncommon) are the charging of batteries, the action of water or steam on hot materials, and the action of acid on metals. Hydrogen is extremely explosive, having an explosive range of approximately 4–74% in air. Although methane requires at least 12% oxygen for ignition, hydrogen can explode when the oxygen content of the air is as low as 5%. This is illustrated in Fig. 3.5.

Radon

Radon is a gaseous, chemically inert, radioactive product of the disintegration of radium. Found primarily in uranium mines, although present in trace amounts in other types of mine including coal mines (Rock et al., 1975), radon diffuses from the rock strata into the mine environment, where the decay process continues. The disintegration process for uranium-238 to become lead-206 is shown in Table 3.8, including the type of radiation given off by each decay process and the half-life of each element in the series. The half-life of a radioactive substance is the time required for a given amount of that substance to lose one-half of its radioactivity. As indicated (Table 3.8), the half-life of uranium-238 is approximately 4.5 billion years; radium, 1622 years; and radon, 3.8 days.

Once radon is released into the mine environment, the decay process continues with the formation of radium A, which decays to radium B, which

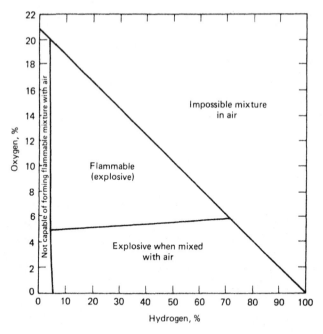

FIGURE 3.5 Explosibility diagram for hydrogen. (After Coward and Jones, 1952.)

TABLE 3.8 Uranium Disintegration Process

Common Name or Symbol	Isotope	Type of Radiation	Half-life
Uranium	$^{238}_{92}$Uranium	Alpha	4.49×10^9 yr
UX$_1$	$^{234}_{90}$Thorium	Beta	24.1 days
UX$_2$	$^{234}_{91}$Protactinium	Beta	1.17 min
Uranium-234	$^{234}_{92}$Uranium	Alpha	2.48×10^5 yr
Ionium	$^{230}_{90}$Thorium	Alpha	8×10^4 yr
Radium	$^{226}_{88}$Radium	Alpha	1622 yr
Radon	$^{222}_{86}$Radon	Alpha	3.825 days
Radium A	$^{218}_{84}$Polonium	Alpha	3.05 min
Radium B	$^{214}_{82}$Lead	Beta, gamma	26.8 min
Radium C	$^{214}_{83}$Bismuth	Beta, gamma	19.7 min
Radium C'	$^{214}_{84}$Polonium	Alpha	2.73×10^{-6} min
Radium D	$^{210}_{82}$Lead	Beta, gamma	22 yr
Radium E	$^{210}_{83}$Bismuth	Beta	5.02 days
Radium F	$^{210}_{84}$Polonium	Alpha	138.3 days
Radium G	$^{206}_{82}$Lead	—	Stable

Source: Miller and Dalzell (1982).

produces radium C, and so forth. The products formed by the decay of radon are referred to as *radon daughters*. The radon-daughter products are atoms of solid matter having relatively short half-lives. During the decay process, either alpha or beta particles are emitted. These emissions may also be accompanied by gamma-ray activity. It is the short-lived alpha particles and potential alpha emitters such as radon and its daughters that are of prime concern to the ventilation engineer. Because it is a gas and has a relatively long half-life, inhaled radon is exhaled before large amounts of alpha particles are emitted. The daughter products, however, because they are particles of solid matter, attach themselves to the dust that is present in the environment and when inhaled, tend to be deposited and concentrated in the respiratory system. It has been estimated that when both radon and radon daughters are inhaled, only about 5% of the alpha radiation received is contributed by the radon (Holaday et al., 1957).

During radioactive decay, the individual members in the series are decaying and being formed at the same time. At some point in time, equilibrium is reached, and the quantity of each member in the series remains constant. At this time, each member of the series is being generated at the same rate at which it is decaying. The time required for the radon daughters through radium C' to reach equilibrium from a given quantity of radon is approximately 3 h. In about 40 min, the alpha energy reaches approximately 50% of maximum (Rock and Walker, 1970). For this reason, the ventilation engineer must insure that unventilated or poorly ventilated areas do not exist.

Exposure to excessive concentrations of radon and radon daughters has been linked with a high incidence of lung cancer. The maximum exposure limit for radon daughters has been set at 1.0 working level (WL), with a yearly cumulative exposure of 4 working-level months (WLMs). A *working level* is defined as that concentration of short-lived radon-daughter products in a liter of air that will yield 1.3×10^5 million electron volts (MeV) of alpha energy in decaying through radium C' (Holaday et al., 1957). The unit designated WLM is a cumulative measure of exposure that is calculated by multiplying the average working level of exposure during a given time period by the time of exposure and dividing by 173 (the number of working-level hours per working-level month). An example will aid in understanding this concept.

Example 3.2   Given the following exposures during a shift:

$$
\begin{array}{ll}
5\ \text{h} & 0.5\ \text{WL} \\
2\ \text{h} & 0.2\ \text{WL} \\
1\ \text{h} & 0.9\ \text{WL}
\end{array}
$$

find the working-level months of exposure.

Solution:

$$\text{Average WL} = \frac{(5 \text{ h})(0.5 \text{ WL}) + (2 \text{ h})(0.2 \text{ WL}) + (1 \text{ h})(0.9 \text{ WL})}{8 \text{ h}}$$

$$= 0.48 \text{ WL}$$

$$\text{WLM} = \frac{(0.48 \text{ WL})(8 \text{ h})}{173 \dfrac{\text{WLH}}{\text{WLM}}}$$

$$= 0.022 \text{ WLM}$$

3.4 GAS DETECTION AND MONITORING

The state of the art of gas detection and gas monitoring has advanced rapidly during the last few years. Using today's portable equipment, it is possible to get a direct readout of contaminants at levels well below the ambient levels found in most mines.

The types of instrumentation available to the mining industry for measuring gas concentrations fall into four basic classes: handheld detectors, machine-mounted monitors, area monitors, and personal dosimeters. Detection methods are discussed first.

Detection Methods

The principles on which the detection methods are based include catalytic oxidation, electrochemical, optical (nondispersing infrared and interferometer type), electrical conductivity (semiconductors), and chemical adsorbents.

Catalytic-oxidation detectors are used to measure the concentration of combustible gases, most notably methane and carbon monoxide, by measuring either the heat generated during the oxidation process or the change in resistance in an electrical circuit (Wheatstone bridge). In the Wheatstone bridge principle, one leg of the bridge is used to burn the gas, thus heating that leg and causing an imbalance in the bridge resistance, which is proportional to the concentration of the combustible gases present.

Electrochemical sensors have found application in determining the concentration of oxygen, carbon monoxide, hydrogen sulfide, and oxides of nitrogen. In these sensors, the gas being measured reacts with a special electrode in an electrolyte. This reaction generates an electrical current that is proportional to the concentration of the gas present.

Optical detectors are basically of two types. The *nondispersive infrared detector* is based on the principle that different gases absorb light at specific and distinct wavelengths. By passing light through a gas mixture and measuring the amount of absorption, the concentration of the gas is determined.

The second type, the *interferometer*, is based on the difference in the index of refraction between two gases. Basically, a beam of light is split, with one part passing through a chamber filled with air and the other passing through a chamber filled with an unknown gas mixture. The difference in the velocity of the two beams is proportional to the concentration of the gas of interest (usually methane).

The newest method of gas detection, *electrical conductivity*, uses special types of elements (semiconductors) that change resistance in the presence of certain gases.

The last method of gas detection, *stain tubes*, uses the reactive properties of gases and chemicals to cause color changes in the chemical. These color changes are proportional to the gas concentration, measured as either the length or the intensity of stain.

Handheld Detectors

The most commonly used type of instrument in the mining industry, handheld detectors, are small and lightweight. They are used to spotcheck the air quality at various locations underground. Those that are labeled permissible have been tested by and met the requirements of MSHA for electrical safety. In general, the permissibility label signifies only that the instrument is safe for use in a methane-air mixture. Further, detectors that are tested in a methane–air mixture should not be used to make checks for gases that are more explosive than methane. For example, a methane detector should not be used to check for hydrogen in a battery-charging station.

Safety Lamps   The flame safety lamp is the oldest type of gas detector. For many years it was the only means available to check for methane. It is also useful for checking for oxygen deficiency. A flame safety lamp will not burn in methane-free air having an oxygen content of less than approximately 16%. In environments containing methane, a flame safety lamp will continue to burn at lower oxygen concentrations; however, at oxygen contents of less than 13%, the lamp will be extinguished regardless of the methane content of the air. It should be recalled that at an oxygen content of approximately 13%, there may be a loss of consciousness if the exposure is prolonged.

Flame safety lamps should be clean, in good condition, and properly assembled, because numerous accidents and explosions have reportedly been caused by defective lamps. Flame safety lamps should never be opened underground and should be relit only in intake air. One common type of flame safety lamp is shown in Fig. 3.6. Today the flame safety lamp is seldom used except for assessing oxygen deficiency.

Methane Detectors   Handheld methane detectors or methanometers use two basic methods of detection: catalytic oxidation and optical interferometry. Catalytic-oxidation detectors operate on the Wheatstone bridge princi-

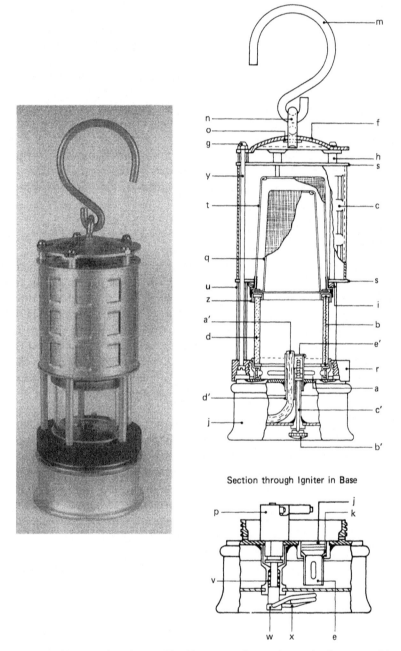

FIGURE 3.6 Flame safety lamp, Koehler type. Legend: *a*, air ring assembly; *b*, asbestos washer; *c*, bonnet; *d*, burner glass; *e*, cotton spreader; *f*, dome top; *g*, dome top nut; *h*, dome top support; *i*, expansion spring; *j*, filler plug; *k*, filler washer; *l*, fount assembly; *m*, handle; *n*, eyelet; *o*, eyelet washer; *p*, igniter assembly; *q*, inner

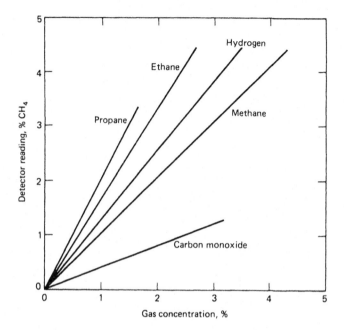

FIGURE 3.7 Response of catalytic-oxidation methane detector to gases other than methane. (After Ferber and Wieser, 1972.)

ple. In addition to detecting methane, this type of detector is sensitive to higher hydrocarbons such as ethane and propane, hydrogen, and other flammable gases. If these gases are present, they cause erroneous high readings (Fig. 3.7). In addition to being affected by other flammable gases, the catalytic-oxidation detector is also sensitive to oxygen deficiency. Tests have shown that this type of detector should not be used when the oxygen content of the air falls below approximately 10%.

Handheld optical detectors used for detecting methane are of the interferometer type. These detectors compare the velocity of light through pure air with the speed through the air being tested. The two beams of light are combined, producing interference fringes. The position of these fringes is indicative of the methane concentration. As with the catalytic-oxidation methane detector, the interferometer also is sensitive to gases other than

---

gauze, brass; *r*, lock ring assembly; *s*, middle ring; *t*, outer gauze, brass; *u*, spacer; *v*, spring; *w*, swivel screw; *x*, thumb spring; *y*, tie rod; *z*, washer retainer; *a'*, round wick; *b'*, wick adjuster button; *c'*, wick adjuster screw; *d'*, round wick holder; *e'*, round wick stop. (By permission from National Mine Service, Inc., Indiana, PA.)

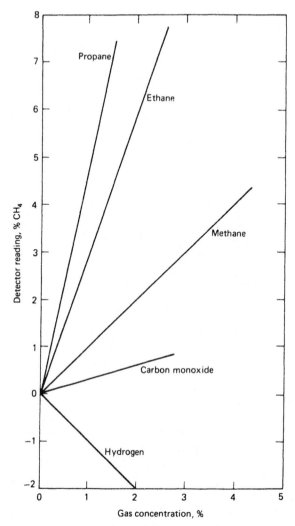

FIGURE 3.8   Response of interferometer methane detector to gases other than methane. (After Ferber and Wieser, 1972.)

methane (Fig. 3.8). For example, propane and ethane cause high readings, whereas carbon monoxide causes a reading that is lower than the actual concentration of methane (Watson et al., 1966). When exposed to a hydrogen–air mixture, this type of methane detector indicates a negative gas concentration. The nature of this response to hydrogen is such that a mixture containing 1.0% hydrogen and 1.0% methane in air results in an interferometer reading of approximately zero. This type of detector is also sensitive to carbon dioxide, water vapor, and oxygen deficiency. The indices of refrac-

TABLE 3.9 Performance Standards for Indicating Methane Detectors

Methane Mixture, %	Minimum Indication, %	Maximum Indication, %
0.25	0.10	0.40
0.50	0.35	0.65
1.00	0.80	1.20
2.00	1.80	2.20
3.00	2.70	3.30
4.00	3.70	4.30

Source: Anon. (1995a).

tion of carbon dioxide and water vapor are comparable to that of methane. One percent carbon dioxide gives a methane reading of approximately 1%. Therefore, carbon dioxide and water vapor must be scrubbed from the sample. Decreased oxygen content causes the interferometer-type detector to read high. Each percent decrease in oxygen results in approximately a 2% methane indication.

Methane detectors are the only instruments used for making gas determinations that are approved for both electrical safety and performance. Performance standards for indicating methane detectors are given in Table 3.9.

Oxygen Detectors   As noted previously, the most commonly used oxygen detector is the flame safety lamp. In addition to the flame safety lamp, which indicates only an oxygen-deficient environment, the actual oxygen concentration can be measured using liquid absorption devices, stain tubes, paramagnetic analyzers, and electrochemical or fuel-cell detectors. Instruments are now available that combine methane detectors and oxygen detectors in one package.

Detectors for Carbon Monoxide and Other Toxic Gases   All toxic gases found underground are detectable using colorimetric indicator tubes (stain tubes). These devices usually consist of a mechanical pump, a limiting orifice that regulates the airflow rate, and a glass tube that contains a chemical reagent that is sensitive to the gas of interest. Two basic types of stain tubes are used in the mining industry. The first utilizes the length of the stain generated by the reaction of the gas of interest with the chemical reagent in the tube as a measure of the gas concentration. The other type compares the intensity of the color created by the reaction with a standard color chart for the tube. One difficulty with stain tubes is the cross-sensitivity of some of the tubes. That is, some tubes are sensitive to more than one gas. For this reason, the person making the measurement should know characteristics of the tube being used.

TABLE 3.10   Gas Detection Methods

Gas	Symbol	Detection Methods
Methane	CH_4	Flame safety lamp
		Catalytic oxidation
		Thermal conductivity
		Optical (IR and interferometer)
Oxygen	O_2	Liquid absorption
		Stain tubes
		Paramagnetic analyzers
		Electrochemical sensors
Carbon dioxide	CO_2	Liquid absorption
		Stain tubes
		Optical interferometer
Carbon monoxide	CO	Electrochemical sensors
		Catalytic oxidation
		Hopcolyte optical IR
		Metal oxide semiconductor
		Stain tubes
Oxides of nitrogen	NO_x	Electrochemical sensors
		Stain tubes
Hydrogen sulfide	H_2S	Electrochemical sensors
		Metal oxide semiconductors
		Stain tubes
Sulfur dioxide	SO_2	Electrochemical sensors
		Stain tubes
Hydrogen	H_2	Stain tubes
Radon	Rn	Radiation detectors

Source: Miller and Dalzell (1982).

The available methods of detection for each of the gases discussed are summarized in Table 3.10.

Monitoring

The term *gas monitoring* implies continuous or cyclic measurement of a gaseous contaminant as opposed to *gas detection*, which implies intermittent checks for contaminants. Two types of monitoring systems are used underground: machine-mounted monitors and area monitors. Monitors may be used for protection from a safety hazard such as fires or explosions or for protection from a health hazard such as a toxic gas.

Machine-mounted methane monitors are required in coal mines on all-electric face-cutting equipment, continuous miners, longwall face equipment, and loading machines. The function of these monitors is to detect methane liberated during the mining process and to protect the miner by giving warning at a methane concentration of 1.0% and by deenergizing the equipment on which it is installed when the methane concentration reaches 2.0%. Recently developed monitors are equipped with dual sensing elements located remotely to reduce the likelihood of having an undetected methane accumulation. In operation, only the higher of the two concentrations is displayed, and a high concentration at either location will activate the warning and/or deenergizing functions of the monitors.

Monitors are also being used on diesel-powered equipment. The function of these monitors is to measure continuously a contaminant concentration within the breathing zone of the equipment operator. The gases of interest for this type of system are carbon monoxide, carbon dioxide, oxygen, and the oxides of nitrogen.

In area monitoring, contaminant concentrations, airflow quantities, temperature, and relative humidity are measured at fixed locations. In addition to the monitors, these systems incorporate telemetry equipment to transmit the signals to other locations where data analysis and output and alarm monitoring take place. In one system, low-level carbon monoxide monitors are used for fire detection in belt-haulage entries (Miller et al., 1980). Continuous monitoring of methane concentrations along underground gas-transmission lines is needed as an integral part of in-mine methane drainage systems. In some European mines, carbon monoxide is being monitored as a means of detecting the onset of spontaneous heating. In the United States, carbon monoxide is being monitored in belt entries where it is necessary to use belt air to ventilate the working face (Miller, 1978; Whitt, 1980).

Laboratory Analysis and Sampling

In some instances, a more exact analysis of the mine environment is needed. On these occasions, grab-sampling techniques are used to collect samples underground in suitable containers that are then returned to the laboratory for analysis. Such techniques are also used to aspire samples through boreholes or tubes from areas that are inaccessible to direct sampling. When collecting samples, care should be taken to insure that the sample is representative of the environment and is not contaminated prior to analysis (Berger and Schrenk, 1948). Once collected, the sample should be analyzed as quickly as practical. The primary method of analysis used today is gas chromatography.

3.5 CONTROL OF GASES UNDERGROUND

Once a contaminant gas is identified, its source located, and its release rate determined, the ventilation engineer must plan the ventilation system to control the gas within its maximum allowable level. The control techniques available range from simple dilution with the main ventilation airstream to complex drainage systems designed to remove the gas prior to mining. The control technique selected depends on the source of the gas and nature of its occurrence (continuous or intermittent liberation, stationary or mobile source). The following are the techniques used to control gases in mines, listed in the preferred order of their application to a given situation (Section 1.2):

1. Prevention
   a. Proper procedure in blasting
   b. Adjustment and maintenance of internal-combustion (IC) engines
   c. Avoidance of open flames, and so forth
2. Removal
   a. Drainage in advance of mining
   b. Drainage by bleeder entries
   c. Local-exhaust ventilation
   d. Water infusion in advance of mining
3. Absorption
   a. Chemical reaction in IC engine conditioner
   b. Solution by air–water spray in blasting
4. Isolation
   a. Sealing off abandoned workings of fire areas
   b. Restricted blasting or off-shift blasting
5. Dilution
   a. Local dilution by auxiliary ventilation
   b. Dilution by main ventilation airstream
   c. Local dilution by diffusers and water sprays

Dilution is the only control technique that is universally applicable. Other control measures are suited to particular conditions, such as the source and occurrence of gas. A combination of techniques is often most cost-effective.

Control of Strata Gases

The control of gases emanating from the ore body or the surrounding strata is the most common and serious gas problem encountered underground. Of particular concern is the potential for accumulation of strata gases such as methane and carbon dioxide, which can lead to oxygen-deficient atmospheres.

A dangerous effect called *layering* has been observed in coal mines with methane (Leach and Thompson, 1968). If the ventilation airstream generates insufficient turbulence, any methane present remains stratified and forms a persistent layer moving along the roof. Worse, if the mine opening is inclined and the airflow is "downhill" (descentional), a methane layer can back "uphill" against the airstream. Ample air velocity and turbulence are the primary safeguards against layering.

To forewarn of the likely occurrence of methane layering, Leach and Thompson have developed an empirical indicator, *layering number* N_l, which can be calculated from this relationship:

$$N_l = \frac{V_u}{41}\left(\frac{b}{Q_g}\right)^{1/3} \tag{3.7}$$

$$N_l = \frac{V_u}{1.8}\left(\frac{b}{Q_g}\right)^{1/3} \tag{3.7a}$$

where V_u is air velocity in the upper half of the airway in fpm (m/s), Q_g is methane inflow in cfm (m³/s), and b is airway width in ft (m). Tests have shown that layering can be controlled if the air velocity is sufficient to maintain N_l at values ≥ 5 in a horizontal airway, 5 in an airway with uphill flow (8 if steep), and 3 in an airway with downhill flow (5 if steep). Irrespective of airway slope, a safe approximate value of N_l to use on normal grades is thus 5. In typical coal mine openings with moderate gas inflows, high air velocities (≥ 200 fpm) (≥ 1 m/s) may be required to counteract the layering effect.

Example 3.3 Given the following data, $V_u = 30$ fpm (0.152 m/s), $Q_g = 15$ cfm (0.00708 m³/s), $b = 13$ ft (3.96 m), determine the layering number and whether the air velocity is adequate to control layering in a horizontal airway.

Solution: Using Eq. 3.7,

$$N_l = \frac{30}{41}\left(\frac{13}{15}\right)^{1/3} = 0.70$$

which is substantially less than the value (~ 5) needed to control layering under the given conditions.

The perception of the potential hazardous conditions arising from the strata gases must be developed through training. The actual hazard must be controlled through proper ventilation and other means. In addition to dilution, several other control measures are used to control strata gases. The most important is to *design the mining and ventilation systems* with these gases in mind. Through proper planning, a sufficient air quantity can be provided, gas can be removed close to the point of emission through the use of bleeder entries, and areas that have been totally mined out can be isolated from the remainder of the mine through sealing. System design is covered in detail in Chapters 13 and 14.

Draining in advance of mining can be practiced in coal mines to reduce the inflow of methane into the active workings. It was first used in the Ruhr district of Germany in 1943 and is now widely used in Europe and in the United States.

Auxiliary ventilation is necessary for the control of methane at the working face in coal mines and in some noncoal mines. It is also used to provide ventilation of active working faces in uranium mines. Auxiliary ventilation systems may utilize either line brattice or fans and tubing installed in either blowing or exhausting modes. Occasionally, combination systems (push–pull) are used in an attempt to take advantage of the good methane-dilution capability of the blower system and the good dust-collection capability of the exhaust system. In ultragassy coal mines, diffusers and specially designed water sprays or venturi devices are sometimes used to assist the auxiliary ventilation system. Quality control aspects of auxiliary ventilation are discussed in Chapter 11.

Control of Blasting Gases

The following control measures are applicable to blasting gases.

Prevention or *reduction* in the amount of gases liberated from blasting is possible through the proper selection of explosives and proper blasting techniques. In coal mines and in gassy noncoal mines, only explosives designated as permissible may be used. Proper stemming is essential to reduce gases.

Removal of blasting gases through local-exhaust systems or auxiliary ventilation is used quite often and is considered good practice.

Absorption of some ingredients of the gas formed in blasting is achieved by an air–water spray. The spray is mounted in the drift or raise to produce a curtain of fine mist across the opening, some distance behind the face, and is turned on by the miners prior to blasting. It is fairly effective for water-soluble gases, such as sulfur dioxide, hydrogen sulfide, and nitrogen dioxide, but is ineffective for carbon monoxide.

Blasting off-shift, or at restricted times, and *localizing* the blast are often practiced as control measures. Localizing or isolating the effects of blasting can be effective control measures where workings are isolated or sectionalized in the ventilation system.

Control of Internal-Combustion Engine Exhaust

The use of gasoline engines in most underground mines is prohibited by law. Diesel engines, however, are permitted. They are widely employed in noncoal mines and are used to a lesser extent in coal mines. At the current time, extensive regulations with regard to the use of diesel-powered equipment in coal mines are under consideration (Anon., 1989a, 1992a). The regulations address the storage and handling of diesel fuel underground and the control of the exhaust gases and sparks from the exhaust into an explosive or flammable atmosphere of gas or dust. As part of the approval process, the maximum allowable fuel:air ratio is established, the ventilation requirements are determined and the maintenance schedules are specified. The air quantity requirement is based on the most undesirable and hazardous condition of operation and is designed to insure that the following gas concentrations in the dilute exhaust mixture are not exceeded: carbon dioxide, 0.25%; carbon monoxide, 0.005%; and oxides of nitrogen, 0.00125%. Additionally, the oxygen content of the diluted mixture must not be less than 20%. A standard for diesel-exhaust particulates in the mine atmosphere is under consideration (Anon., 1992a).

Diesel equipment used in noncoal mines, other than those in gassy noncoal mines, is not required to be approved; however, sufficient dilution air must be provided to maintain the exhaust gases at acceptable levels.

Absorption of certain exhaust components can be achieved with an appropriate conditioning device on the engine discharge. Scrubbers containing a water or reagent bath or granule filter are used to remove many of the gases or reduce their concentration. For effective absorption, intimate contact between the gases and the reagents in the scrubber is essential. Catalytic converters are used, where appropriate, for reducing carbon monoxide and nitric oxide in the exhaust.

An engine in good condition will produce less smoke and toxic gas. Therefore, one of the best control measures is a comprehensive *maintenance* program.

Control of Fires and Explosions

The most effective control measure against fires and explosions is *prevention*. Once ignition has occurred, *isolation* of the zone of conflagration is the most important action to be taken to contain the blaze and gaseous products generated and to shut off the oxygen supply. For control of fires and explosions, see Chapter 15.

Control of Battery Gas

The hydrogen liberated in charging conventional storage batteries underground can be controlled by *isolation* of the charging station and providing adequate ventilation, including a *separate split of air* to ensure dilution of the discharge.

3.6 DETERMINING DILUTION REQUIREMENTS

Dilution by general ventilation is the most useful method of gas control practiced in mining. In dilution ventilation, contaminated air is diluted with uncontaminated air to keep the concentration of the contaminants below the maximum permissible. It contrasts with a local exhaust system in which an airflow of sufficient velocity is created to remove the contaminated air from around its source and discharge it directly into places where it can cause no harm to human beings. A versatile means for control of any mine gas, regardless of occurrence, dilution by general ventilation has no equal.

Dilution ventilation has four limiting factors: (1) the quantity of contaminant generated must not be excessive as otherwise the volume of air necessary for dilution will be impractical, (2) either the workers must be far enough away from the source of contaminant or else the contaminant must be in sufficiently low concentration so that the workers will not experience an exposure above acceptable concentrations, (3) the toxicity of the contaminant must be low, and (4) the evolution or generation of the contaminant must be reasonably uniform. Obviously, where toxicities and/or the generation of contaminants are very high, methods of control other than dilution ventilation must be used.

Dilution Ventilation Calculation

The amount of diluting air required for dilution ventilation is a function of three factors: (1) the maximum allowable concentration (MAC) for the contaminant, (2) the rate at which the contaminant is flowing into the mine, and (3) the concentration of the contaminant in the incoming air. While TLVs may be the maximum allowable concentrations from a legal point of view, mine management may set values for maximum allowable concentrations for certain contaminants below the corresponding TLVs to provide additional margins of safety. *Permissible exposure limit* (PEL) is yet another term that is commonly used in place of TLV and has legal connotation.

Steady-State Dilution   The quantity of dilution air required for a steady-state situation is calculated as follows:

MAC $=$ maximum allowable concentration, fraction

B $=$ concentration of the gas in the incoming air, fraction

Q_g $=$ inflow rate of the gas into the mine atmosphere, cfm (m³/s)

Q $=$ inflow rate of the incoming air, cfm (m³/s)

To be in compliance, the following volumetric relationship should hold:

$$Q \geqslant Q_g \frac{(1 - \text{MAC})}{(\text{MAC} - B)} \tag{3.8}$$

Where the concentration of the contaminant gas in the incoming air is very small as compared to the MAC (i.e., very nearly zero), then

$$Q \geq \frac{Q_g}{(MAC - B)} - Q_g \qquad (3.9)$$

The quantity of air Q (cfm or m^3/s) flowing in an airway is calculated from the average velocity V (fpm or m/s) and the cross-sectional area of the airway A (ft^2 or m^2) at the point of measurement:

$$Q = VA \qquad (3.10)$$

Example 3.4 A strata gas flows into a workplace at a rate of 90 cfm (0.04247 m^3/s); its concentration in the intake air is 0.25%. Assuming that the TLV for the gas is 1%, compute the quantity of intake air required to dilute the gas.

Solution: Use Eq. 3.8:

$$Q = \frac{90(1 - 0.01)}{(0.01 - 0.0025)} = 11,800 \text{ cfm } (5.606 \text{ m}^3/\text{s})$$

Example 3.5 Slow oxidation of a sulfide ore in a stope liberates 0.015 cfm (7.079×10^{-6} m^3/s) of a gaseous contaminant. Given a TLV of 5 ppm, approximate the quantity of fresh air required for dilution.

Solution: Use Eq. 3.9:

$$Q = \frac{(0.015)}{(0.000005)} = 3000 \text{ cfm } (1.415 \text{ m}^3/\text{s})$$

When two or more pollutants are involved, the dilution ventilation requirement for each pollutant must be calculated. If there are no synergistic effects, the minimum air required will be the maximum of the quantities calculated. Where synergism is known or suspected, the following summation can be used to ascertain the condition of the atmosphere (Anon., 1996):

$$\sum_{i=1}^{N} \frac{C_i}{TLV_i} \qquad (3.11)$$

where C_i = concentration of the gaseous specie i
  TLV_i = threshold limit value for the specie i
  N = number of gases that have synergistic effects

When the summation is less than 1.0, the atmosphere can be regarded as safe.

Time-Dependent Growth or Decay   Where the concentration buildup occurs over a period of time, then it becomes important to include the time
factor. Examples of such buildup in the mining process can occur during
blasting, during seepage of gas into ventilated areas, and/or removal of gas
after an initial outburst. Let

Y = volume of the workings, ft^3 (m^3)

Q = incoming air, cfm (m^3/s)

Q_g = gas inflow, cfm (m^3/s)

B = concentration of the contaminant in the incoming air, fraction

x_0 = concentration in the working at time τ_0, fraction

x_τ = concentration at time τ, fraction

The increase in the volume of gas in the workings in the time interval τ to
$\tau + d\tau$ is

$$Y dx = (QB)d\tau + Q_g d\tau - (Q + Q_g) x d\tau \qquad (3.12)$$

Rearranging and integrating, it can be seen that

$$\tau = \frac{Y}{Q + Q_g} \ln\left[\frac{QB + Q_g - (Q + Q_g) x_0}{QB + Q_g - (Q + Q_g) x_\tau}\right] \qquad (3.13)$$

A variety of situations can be investigated with Eq. 3.13. For example, in
the case of blasting, fumes are generated instantly with the blast, and there
is no further emission of blasting fumes ($Q_g = 0$). Also the incoming air does
not contain any concentration of blasting fumes (i.e., $B = 0$).

Example 3.6   Consider a breast stope 200 ft wide, 3 ft high, with 200 ft
between the levels (volume $Y = 120{,}000$ ft^3). The quantity of air circulating
through the stope is 3000 cfm. The intake air has no concentration of blast
fumes. The concentration immediately after a blast is about 2000 ppm. Calculate the time required for the concentration to reach a value of 50 ppm.

Solution: Noting that $x_0 = 2000$ ppm, $x_\tau = 50$ ppm, $B = 0$ ppm, $Y = 120{,}000$
ft^3, $Q_g = 0$ cfm, and $Q = 3000$ cfm, we can solve using Eq. 3.13.

$$\tau = \frac{120{,}000}{0 + 3000} \ln\left[\frac{3000(0) + 0 - (3000 + 0)\, 0.002}{3000(0) + 0 - (3000 + 0)\, 0.00005}\right]$$

$$\tau = 147.6 \text{ min}$$

Example 3.7 An inflow of 2.0 cfm of a strata gas occurs whenever cutting and loading operations are conducted in a stope. The inflow ceases with the stoppage of cutting and loading. The maximum allowable concentration (MAC) of this strata gas is 500 ppm. At the current time, the cutting and loading operations are frequently interrupted due to gassing out (exceeding the MAC).

Maximum cutting and loading cycle time, 35 min
Concentration of the gas in the intake air to the stope, 50 ppm
Stope volume, 100,000 ft^3
Initial concentration of the gas in the stope, 75 ppm
Intake-air quantity to the stope, 2000 cfm

Calculate the length of time from the start of work after which work must be stopped due to the gas concentration exceeding the MAC.

Solution: Noting that $x_0 = 75$ ppm, $x_\tau = 500$ ppm, $Q = 2000$ cfm, $Q_g = 2.0$ cfm, $B = 50$ ppm, and $Y = 100,000$ ft^3, we can calculate the value of τ as follows (Eq. 3.3):

$$\tau = \frac{100,000}{2000 + 2} \ln\left[\frac{(2000 \times 0.00005 + 2) - (2000 + 2)(0.000075)}{(2000 \times 0.00005 + 2) - (2000 + 2)(0.0005)}\right]$$

$$\tau = 28.6 \text{ min}$$

Example 3.8 In Example 3.7, what are the options to ensure that the cycle can be completed under 28.6 min?

Solution: The options are to either singly or in combination increase the quantity of air flowing (Q), increase the volume of the stope (Y), decrease the initial concentration in the stope (x_0), decrease the concentration in the intake air (B), and so forth. Increasing the quantity of air flowing is probably the more readily applicable option.

Under steady-state conditions, note that $(dx/d\tau) = 0$, and that Eq. 3.12 reduces to

$$(QB + Q_g) - (Q + Q_g) x = 0 \tag{3.14}$$

The quantity of air Q required to reach the steady-state concentration x as $\tau \to \infty$ is

$$Q = \frac{Q_g (1 - x)}{(x - B)} \tag{3.15}$$

These dilution equations are valid for the solution of problems involving any contaminant, particulate (dust) as well as nonparticulate (gas). Weight flow rate G can be substituted for Q_g if the emission rate of contaminant in weight per unit time is known (mg/min). Likewise, the quantity Q to dilute a dust condition can be determined when the concentration and MAC are both expressed on a weight or number basis per unit volume of air (mg/m^3). Equation 3.9 is normally preferred to Eq. 3.8 for particulate calculations (since the MAC for a dust on a percentage basis is extremely small, it can be neglected in the numerator of Eq. 3.9). The similarity between Eqs. 3.8 and 3.15 must be noted; Eq. 3.15 is the more general form.

3.7 METHANE DRAINAGE

A systematic study of methane flow and the initiation of control programs are important for safe, uninterrupted operations underground, particularly where large volumes of methane are anticipated. Control techniques considered for use in coal mines can be divided into three categories: (1) dilution ventilation, (2) blocking or diverting gas flow in the coalbed by means of seals, and (3) removing relatively pure or diluted methane through boreholes. As already explained, dilution ventilation is most widely used to reduce the concentration of methane to a safe level in underground mines. Larger mines in the Pittsburgh coalbed circulate between 5 to 20 tons of air per ton of coal extracted. In most cases, this is adequate to keep the methane concentration at the mine exhaust well below the maximum level of 1%. However, in newly developing mines or in mines that are driving development headings into virgin coal, dilution ventilation alone may not be enough, and development may have to be curtailed or stopped. Furthermore, since the methane emission rate depends on both the rate and depth of coal production, it would be expected that even more air will be needed as mining rates and depths increase unless some of the methane is removed before mining proceeds.

The techniques considered to date in attempts to prevent the entry of methane into the ventilating airstream are summarized in Table 3.11 for four regions in a mine: the working space, the gob, the coal seam, and the adjacent strata. The latter three regions are also considered because they can and do influence the atmosphere in the working space. Furthermore, as these regions represent the seat of the problem, they have been the focus of all the initial methane control procedures.

Several studies have been conducted by the U.S. Bureau of Mines and others on improved packers for use underground, the composition of coalbed gas, the movement of methane in coal, the determination of the methane content of a coalbed for cores, degasification through vertical and horizontal boreholes, the use of isolated panels, the effects of oil and gas wells, methane control by water infusion, gob degasification through vertical boreholes, the effects of bleeder systems, methane emission from operating mines, and the

TABLE 3.11 Methane Control Techniques Considered for Underground Use

| | Methane Control Technique | |
Region	Methods of Diluting with Air	Removal of CH_4
Working space	Ventilate; isolate and ventilate	Chemical, physical, biological processes
Gob	Ventilate; use bleeders	Surface boreholes
Coal	—	Vertical surface boreholes; horizontal and sloped underground boreholes; multipurpose borehole
Adjacent strata	—	Boreholes

ventilation of deep mines (Duel and Kim, 1986). These studies have shown quite conclusively that many U.S. coalbeds can be successfully degasified before mining commences (Fig. 3.9).

Methane drainage is the process of removing the gas contained in the coal seam and surrounding strata through pipelines. The drained gas is mostly methane. Methane drainage has become an important part of both methane control and overall ventilation planning.

The most important advantage of methane drainage is reduced gas concentrations in the workplace. This lessens the chances of ignition and explosions considerably. Almost 50% of the potential gas volumes emitted at the face have been recovered by methane-drainage boreholes in solid coal (Thakur and Davis, 1977). Even higher percentages of recovery have been claimed for gob areas (Kimmins, 1971). Second, methane drainage also reduces ventilation air requirements for the mine. In highly gassy mines without methane drainage, mining activities may have to be slowed down or even completely stopped for a while to let the gas bleed from the freshly exposed areas. This loss of production can certainly be prevented by a well-planned methane drainage scheme. The third advantage of methane drainage is that the recovered gas, which usually has a high calorific value, can be commercially used by industry. Gas recovered from solid coal has a calorific value above 900 Btu/ft^3 (33.4 MJ/m^3), and gob gas has a variable clorific value of 300–650 Btu/ft^3 (11.1–24.2 MJ/m^3). The amount of methane emitted by all underground coal mines in 1995 is approximated from past estimates (Hogan, 1993) at about 220 billion ft^3 (Bcf) or 6.3 billion cubic meters (Bcm). The amount of methane recovered and used by coal mines is estimated by the Environmental Protection Agency (EPA) (Kruger, 1994) at about 30 Bcf (0.92 Bcm). If an all-out effort is made to drain methane ahead of mining, over 25% of the emitted gases can be beneficially recovered (Thakur and Umphrey, 1979). Precise information for the reservoir properties of the coal seam and the surrounding strata as well as careful planning are necessary for optimal methane recovery.

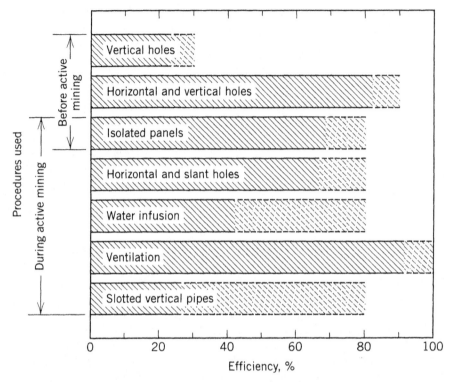

FIGURE 3.9   Efficiencies of various methane control techniques in eastern U.S. coal mines (After Zabetakis, 1973).

Techniques of Methane Drainage

There are five methods of recovering methane from underground coal mines. These are the vertical degas system, vertical gob degas system, horizontal borehole degas system, cross-measure borehole system, and packed-cavity system. In the horizontal borehole and the cross-measure borehole systems, the captured methane has to be transported by a pipeline through the mine to a borehole to the surface.

Vertical Degas System   The vertical degas well system consists of vertical holes drilled from the surface to the virgin areas of the coal seam and equipped with the necessary pump, pipe, instrumentation, and control system to extract the methane gas (Fig. 3.10). To increase the production of gas, the coal seam may be hydraulically fractured. In general, no serious damaging effects due to hydrofracturing on the roof or floor of the coal seam, which could affect subsequent coal extraction, have been reported.

Vertical Gob Degas System   The vertical gob degas well system consists of vertical holes that communicate with the gob (Fig. 3.11) and extract the

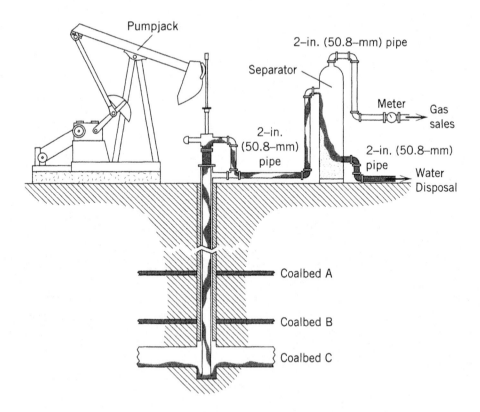

FIGURE 3.10 Vertical degas well.

gas from the gob area. Successful methane drainage from gob areas requires precise information on the reservoir properties of the various strata constituting the gas-emission space and their dependence on time and stress concentration. The boundary of gas-emission space established by various authors varies as illustrated in Fig. 3.12. The height of this space above the working seam varies from 300 to 600 ft (90–180 m), whereas the depth below the seam varies from 80 to 300 ft (24–90 m). On a horizontal plane, gas-emission-space dimensions can be expected to exceed those of the longwall panel by at least 200 ft (60 m).

The gas-emission space for a 500 × 5000-ft (150 × 1500-m) longwall panel would be an ellipsoid, with a total volume of 0.72–1.72 Bcf (20–49 × 10⁶ m³). The percentage of initial gas content contained in this space that is lost to the mine workings can be estimated from Fig. 3.12. Some variation in these estimates can be expected because of local geology, mining methods, percentage extraction, and the nature and thickness of gas-bearing strata in the gas-emission space.

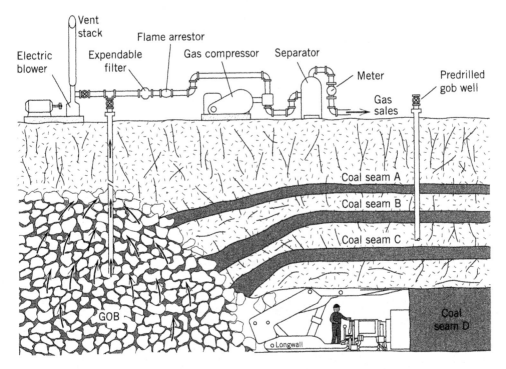

FIGURE 3.11   Vertical gob gas well.

Coal extraction by pillaring or a longwall causes cracking of the overlying strata and results in improved methane production in boreholes. Vacuum pumps are often added to further improve the flow, and in some cases, to prevent the reversal of flow in boreholes. Two or three gob degas holes are generally sufficient for a panel 5000 ft (1500 m) long. The first hole is within 500 ft (150 m) of the starting line for the panel, but other holes could be 2000 ft (600 m) apart. Flow tests on these degas holes show a definite and rapid communication with the mine workings. The production varies from a few kcfd (m³/min) to over 1000 kcfd (20 m³/min), depending on the characteristics of the gas-emission space. The boreholes over pillar panels and those over longwalls make up two distinct families; the latter are more permeable and in direct communication with mine air. Consequently, these holes also produce less pure (generally less than 50%) methane. In pillar-extracted gobs, the production of gob degas holes seems to depend on the percentage of extraction. Methane purity generally exceeds 50%. Methane capture ratios vary from 30 to 50%.

Horizontal Borehole Degas System   The third system is the horizontal borehole degas system, where long horizontal holes are drilled into the coal

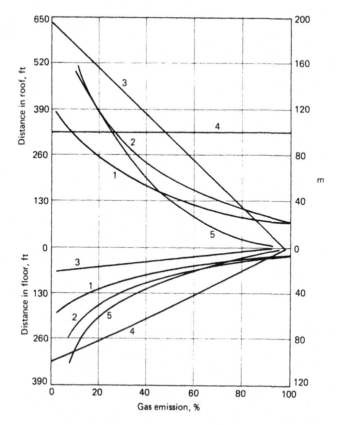

FIGURE 3.12 Influence of mine workings on gas emission and gas-emission space: (1) Stuffken, (2) Winter, (3) Lidin, (4) Gunther, (5) Patteiski. (After Winter, 1975.)

seam from the mine workings and the gas is extracted through a system consisting of a compressor, monitoring and control equipment, pipelines in the mine, and in a vertical hole (Fig. 3.13). A horizontal drilling system can be divided into three subsystems: the *drill rig,* the *drillbit-guidance system,* and *borehole-surveying instruments* (Thakur and Poundstone, 1980). The drill rig provides the thrust and torque necessary to drill holes 3–5 in. (76–127 mm) in diameter up to 2000 ft (600 m) deep and contains the mud circulation and gas and cuttings separation systems. The drillbit-guidance system guides the bit horizontally and vertically as desired. Borehole-surveying instruments measure the pitch, roll, and azimuth of the borehole assembly. Some instruments also indicate the thickness of coal between the borehole and the roof or floor of the coal seam. Thus instruments become powerful tools for locating the presence of faults, clay veins, sand channels, and the thickness of coal in advance of mining. In recent years, many other potential uses of horizontal boreholes have come to light, such as in situ gasification, longwall

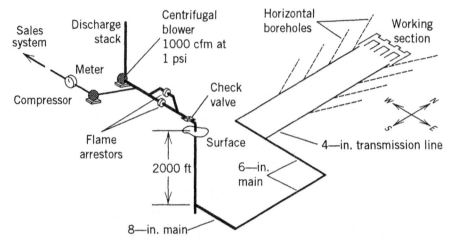

FIGURE 3.13 Horizontal borehole drainage project. (Conversion factors: 1 in. = 25.4 mm, 1 ft = 0.3048 m, 1 cfm = 4.7195×10^{-10} m³/s, 1 psi = 6.8948 kPa.)

blasting, improved auger mining, and oil and gas production from shallow deposits. The production of methane from a horizontal degas hole drilled ahead of a development section in the Pittsburgh seam is shown in Fig. 3.14. The rates of rib and face emissions and the total emission rates for the section are also shown.

Productivity of horizontal holes, typically expressed as thousands of cubic feet per day (kcfd) per 100 ft (m³/min·m) of the horizontal boreholes, varies

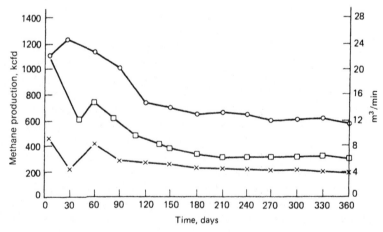

FIGURE 3.14 Methane production history for development section in the Pittsburgh seam ($\square$ = face and rib emission, X = borehole production, $\bigcirc$ = total methane production). (After Thakur and Umphrey, 1979.)

TABLE 3.12   Methane-Drainage Characteristics of U.S. Coal Seams

Seam	Depth ft (m)	Reservoir Pressure psi (MPa)	Initial Productivity kcfd/100 ft(m³/day·m)	Average Productivity kcfd/100 ft(m³/day·m)
Pittsburgh (WV)	700 (213)	275 (1.90)	25 (21.6)	15 (13.0)
Beckley (WV)	900 (274)	300 (2.07)[a]	50 (43.2)	15 (13.0)[a]
Pocahontas (VA)	1700 (518)	650 (4.48)	17 (14.7)	10 (8.6)
Mary Lee (AL)	1500 (457)	500 (3.45)[a]	20 (17.3)	10 (8.6)[a]
Lower Sunnyside (UT)	2000 (610)	670 (4.62)[a]	32 (27.6)	12 (10.4)[a]

Source: Thakur and Dahl (1982).

[a] Estimated values.

from seam to seam. Productivity data and other methane-drainage characteristics for some gassier U.S. coal seams are listed in Table 3.12. The success of methane drainage from solid coal is highly dependent on the permeability of coal. The gas in the coal seam is usually saturated with water. A typical 1000-ft (300-m) horizontal hole in the Pittsburgh seam produces 1–3 gpm (0.06–0.19 L/s) of water. As the water phase is depleted, the relative permeability to gas improves and is reflected in increased production of gas. The permeability of a coal seam appears to decrease with increasing depth to the seam, that is, with increasing ground stress. It is generally believed that productivity of horizontal holes will decline as the depth increases beyond 2000 ft (600 m).

Underground Methane Transport   Realization of the benefits of the horizontal borehole degas system—safety, productivity, and perhaps energy conservation through utilization of the gas as a fuel—requires that the gas be conducted safely and dependably out of the mine to the pipeline of the gas company. In the United States, the underground methane-transport schemes must be approved by MSHA. The details of a typical underground methane-transport installation are shown in Fig. 3.15.

The pipeline connected to a horizontal borehole actually extends into the borehole in the form of a steel casing, which is grouted into the rib at the onset of the drilling operation. Several pipeline branches with their attendant valving and other hardware are manifolded to form one main line that connects to the venthole.

In designing a pipeline system for transporting methane, detailed consideration should be given to many potential safety problems. Roof and rib falls, floor heaving, electrochemical corrosion, and the likelihood of leakage due to improper assembly are factors to be dealt with in the design process. In addition, necessary features include a means of removing water and solids from the pipeline; valving, including a check valve to prevent taking in air

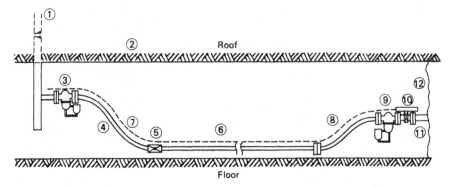

(1) Vertical borehole
(2) Mine roof
(3) Separator and trap, 4 in. (102 mm)
(4) Stainless steel flex line, 4 in. (102 mm)
(5) Butterfly valve, 4 in. (102 mm)
(6) Aldyl "A" pipe, 4 in. (102 mm)

(7) Nitrogen line, ¼ in. (6.4 mm)
(8) Stainless steel flex line, 3 in. (76 mm)
(9) Separator and trap, 3 in. (76 mm)
(10) Automatic shutoff valve, 2 in. (51 mm)
(11) Stand pipe in rib
(12) Mine rib

FIGURE 3.15   Underground methane-pipe installation.

from the surface; lighting and flame protections at the surface; and some allowance for relative motion between system segments.

Each horizontal borehole is equipped with a spring-closed shutoff valve, which is held open by pressurized air or nitrogen acting on a pneumatic actuator. The air or nitrogen is supplied through a fragile, plastic-tubing manifold running parallel to the methane pipeline. Severing of the tubing by a roof fall or other occurrence causes the valve to close, the logic being that the pipeline may have been damaged by the same occurrence.

Steel pipe was used in early installations. However, steel pipe is heavy, is difficult to handle underground, and is subject to external and internal corrosion. A very desirable alternative to steel pipe is a high-molecular-weight polyethylene pipe; it has proved to be very suitable for underground applications just as it has for commercial natural-gas distribution systems on surface.

Cross-Measure Borehole System   This system of methane drainage is popular in deep European mines. Small-diameter [2–2½-in. (51–64-mm)] boreholes are drilled from the tailgate of longwall faces to intercept the overlying strata at an angle of 30–40° from the vertical, parallel in plan to the line of the face, and also inclined over the gob. Holes drilled downward into the floor strata also sometimes provide appreciable flows of gas. The depth of these holes is generally 120–150 ft (36–45 m), with a spacing of 60–90 ft (18–27 m). To minimize the entry of air into the drainage system, a 3-in. (76-mm)-diameter standpipe is inserted into the mouth of the hole and grouted in with cement. All the boreholes are connected to a main drainage range, typically 6–8 in. (152–203 mm) in diameter. Exhausters are used to maintain a suction of 2–4 in. (51–102 mm) water. This overcomes the resistance of the pipeline to gas flow and improves production. The productivity of these cross-measure holes varies considerably depending on properties of the gas-emission space. Over 150 cfm (4.2 m³/min) are frequently measured, and occasionally flows of more than twice this have been recorded. The reduction in gas flow in the ventilation current as a result of methane drainage is normally 50–60%, although figures in excess of this are sometimes encountered.

Packed-Cavity System   In this system of methane drainage, gas is drawn from corridors left and supported in the gob as the face advances. Drainage pipes are inserted into the corridors from a mine airway and are then connected to methane-drainage mains. The success of this technique depends on an effective seal against the ingress of air from ventilation currents and good communication with broken strata overlying the coal seam. It is ideal for mines where gobs are packed with solid material to prevent subsidence. The total quantity of gas drained and hence the methane capture ratios are generally less than the cross-measure-borehole method or the vertical gob degas holes. The packed-cavity system is not recommended for coal seams susceptible to spontaneous combustion.

The key features of the three principal methane recovery methods used in U.S. coal mines are summarized in Table 3.13. In practice, for successful

TABLE 3.13   Summary of Methods for Recovering Methane from Underground Mines

Method	Description	Methane Quality	Recovery Efficiency[a]	Current Use in U.S. Coal Mines
Vertical wells	Drilled from surface to coal seam several years in advance of mining	Recovers nearly pure methane	≤70%	Used by at least 3 U.S. mining companies in about 10 mines
Gob wells	Drilled from surface to a few feet above coal seam just prior to mining	Recovers methane that is sometimes contaminated with mine air	≤50%	Used by over 30 mines
Horizontal boreholes	Drilled from inside the mine to degasify the coal seam	Recovers nearly pure methane	≤20%	Used by over 10 mines
Cross-measure boreholes	Drilled from inside the mine to degasify surrounding rock strata	Recovers methane that is sometimes contaminated with mine air	≤20%	Not widely used in the United States

Source: Kruger (1994).

[a] Percentage of methane recovered that would otherwise be emitted.

longwalling in very deep and very gassy mines, all three methods are employed: vertical degas holes several years ahead of mining, horizontal holes from the mine development sections, and gob wells drilled just prior to mining.

PROBLEMS

3.1 Gas adsorption curves in Fig. 3.1 are described by Eq. 3.3, but a simpler representation is $Y = Y_c p^n$. Plot each curve on log–log paper, and determine Y_c and n for each coal seam. Does the rank of coal have any influence on Y_c and n?

3.2 Crank's solution for Eq. 3.5, when the surface concentration is assumed to vary as $\phi(\tau) = k\tau$, is given as

$$\frac{3M_\tau}{4\pi a^3 k} = \left(\tau - \frac{a^2}{15D}\right) + \frac{6a^2}{\pi^4 D} \sum_{n=1}^{\infty} \frac{1}{n^4} \exp\left(\frac{-n^2\pi^2 D\tau}{a^2}\right) \qquad (3.16)$$

where M_τ is volume of gas desorbed in time τ, k is a constant, and τ is time in days. The left-hand-side (LHS) term is the fraction of original gas content desorbed. Determine the time taken to desorb 80% of the original gas content for lumps of Pocahontas and Pittsburgh seam coal. Assume D/a^2 for Pocahontas as 10^{-6} s^{-1} and for Pittsburgh coal as 10^{-8} s^{-1}.

3.3 A development section is stopped because of high methane emissions.
(a) Plot the pressure profile against distance into the solid coal at time intervals of 1, 15, 30, and 180 days, assuming an approximate solution for Eq. 3.6 as

$$\frac{p^2(x, \tau) - p_{seam}^2}{p_{atm}^2 - p_{seam}^2} = \text{erfc}\,\frac{1}{2(\tau_d)^{1/2}} \qquad (3.17)$$

where $p(x, \tau)$ is pressure at a distance x from the face at time τ, and the term erfc is the complementary error function. Assume $p_{seam} = 275$ psia (lb/in.^{2} absolute) (1.90 MPa) and $p_{atm} = 15$ psia (103 kPa); then

$$\tau_d = \frac{k}{\phi}\frac{\tau\bar{p}}{\mu x^2} \qquad (3.18)$$

Assume $k/\phi = 0.5$; $\bar{p} = (1/2)(p_{seam} + p_{atm})$.
(b) Plot the same pressures profiles, assume $k/\phi = 0.2, 0.7$, and 0.9. What influence do permeability and gas content have on the gas-pressure gradient?

3.4 Nitrogen is liberated at the rate of 8.5 cfm (0.004012 m³/s) as a strata gas in a workplace. What quantity of intake air must be delivered to the face to maintain the nitrogen content at a concentration of not more than 80%?

3.5 How many hours, theoretically, could a miner remain alive after a fall of ground that caved the heading where that miner worked to within 30 ft (9.144 m) of the face? The heading is 5 × 8 ft (1.524 × 2.438 m). Assume a respiratory quotient of 0.9, and investigate from the standpoints of (a) oxygen depletion and (b) suffocation from carbon dioxide. Which governs?

3.6 (a) Calculate the quantity of ventilating air needed to dilute the toxic fumes of a 100-bhp (74.57-kW) diesel engine. Sufficient air should be provided to ensure that no constituent concentration exceeds its TLV-TWA. An analysis of the exhaust reveals the following production of gases per engine bhp (kW):

Nitrogen oxides	0.0015 cfm (0.19493 × 10⁻⁶ m³/s)
Carbon monoxide	0.0006 cfm (0.3797 × 10⁻⁶ m³/s)
Carbon dioxide	0.2670 cfm (168.983 × 10⁻⁶ m³/s)

(b) Assuming that this engine were left running in a dead-end entry with 5000 cfm (2.36 m³/s) of intake air provided by an auxiliary ventilation system, calculate the time required for each of the gases to exceed its TLV-TWA. Assume the same exhaust gases as part (a) and uncontaminated intake air. The entry dimensions are 10 × 25 × 500 ft (1.83 × 5.49 × 76.20 m).

3.7 Blasting in a raise liberates 200 ft³ (5.664 m³) of toxic fumes and smoke. The raise is 4 × 6 ft (1.219 × 1.829 m) in cross section and 40 ft (12.192 m) above the level. If auxiliary ventilation supplies 800 cfm (0.3776 m³/s) of fresh air to the face, how long will it take to dilute the fumes to a safe concentration of 100 ppm so the miners can return to work?

3.8 An inflow of 1.2 cfm of strata gas occurs whenever cutting and loading operations are conducted in a stope. The following data are provided for the stope:

Stope volume	100,000 ft³
Intake-air volume	2000 cfm
Concentration of the strata gas in intake air	50 ppm
Concentration of the strata gas in stope prior to start of work	75 ppm

(a) The cycle times are normally distributed with a mean of 34 min and a standard deviation of 3.5 min. Calculate the probability of being gassed out during a particular cycle given that the maximum allowable concentration for the strata gas is 400 ppm.

(b) A remotely operated cutting and loading machine is being considered. Assume that no miners will be in the stope return where the concentration may be higher than the MAC, calculate the steady-state concentration in the stope. Calculate the time at which 95% of this concentration will be reached.

3.9 The instantaneous liberation rate of methane in a room being advanced by a continuous miner varies from 50 cfm (0.0236 m³/s) to 250 cfm (0.1180 m³/s). Calculate the quantity of fresh intake air containing no methane that must be delivered to the workplace to maintain continuously the methane concentration at the face at less than 1%.

3.10 A section in a coal mine is ventilated with 15,000 cfm (7.079 m³/s). The methane concentrations measured in the intake airway and at the working face are 0.14 and 0.98%, respectively. Assuming no fluctuation in the methane content of the intake air, and neglecting air leakage in the section, calculate the intake-air quantity that is required to maintain the methane concentration at 0.65% at the working face.

3.11 A rock drill liberates dust at the rate of 1 lb/min (0.00756 kg/s). Given that the explosive concentration for that particular dust is 4 grains/ft³ (9153 mg/m³), calculate the required quantity of ventilation to dilute the dust to that level. The dust concentration in the intake ventilating air is 0.5 grains/ft³ (1144 mg/m³).

CHAPTER 4

Dusts and Other Mine Aerosols

4.1 AEROSOL TYPES AND DEFINITIONS

An *aerosol* may be defined as any mass of solid or liquid particles suspended in a gas. These colloidal mixtures can take a number of forms; the most common are the following:

Dust. Dust consists of solid particulate matter suspended in a gas. Dusts constitute the most common aerosol problem in the mineral industries and are usually formed by fragmentation processes such as drilling, crushing, and grinding but can also result from resuspension due to equipment operation or air movement. Dusts particles vary from 1 to 100 μm (micrometer) in diameter, but the size range is normally 1–20 μm. This is a result of the fact that particles below 1 μm are not formed in abundance, and particles above 20 μm are usually relatively quick to settle.

Fumes. Fumes are the solid products that result from the physicochemical processes of combustion, sublimation, or distillation (Reist, 1993, p. 3). Particles that result from such processes are usually less than 1 μm in diameter. They are an important category of aerosol in mining because diesel engines generate fume material known as *diesel particulate matter* (DPM).

Smoke. This aerosol is formed by incomplete combustion and ordinarily consisting of particles 0.01–1.0 μm in size. Smoke particles are usually visible and are distinguished from fumes by the fact that they do not result from condensation processes.

Fog. Fog is an aerosol of liquid particles in a gas formed by the condensation of liquid or the dispersal of small liquid droplets. Essentially the same as a mist, fog is ordinarily made up of liquid particles of a few micrometers to 100 μm in diameter. Fogs normally occur in underground mines as a result of condensation of moisture due to temperature changes in the mine air.

Smog. A combination of smoke and fog, smog often contains photochemical reaction products combined with water vapor (Reist, 1993, p. 4). Most often associated with populated areas, smog normally consists of submicrometer (<1 μm) particles.

Haze. Similar to a smog, a haze consists primarily of submicron solid particles associated with water vapor.

The generally accepted size ranges for these aerosols, their sampling, and their control are outlined in Fig. 4.1 (Lapple, 1961). This chart has been adopted by the American Society of Heating, Refrigerating & Air Conditioning Engineers in their handbook of fundamentals (Anon., 1993, p. 11.3). Because of this usage, the aerosol size ranges in the figure tend to be the standard adopted in most engineering applications. (It should be noted that aerosol particle parameters are customarily stated only in SI units because the SI system has become a worldwide standard for dust sizing and particle concentration calculations.)

In mines and mineral processing operations, dusts are the most important aerosol that require engineering controls. The aerosol category next most troublesome to the mine ventilation engineer is fumes. With diesel equipment being used in a large proportion of mines, control of both combustion gases and fumes is also assuming more importance. At times, other aerosols may become of concern in the mine atmosphere, and the mining engineer should be aware of their existence and behavior. Additional information on aerosols can be found in texts on aerosol science by Hinds (1982), Vincent (1989), and Reist (1993).

4.2 DYNAMIC BEHAVIOR OF AEROSOLS

To practicing mine ventilation engineers, the prevention and control of aerosols in the mine atmosphere is their primary concern with particulates. Behavior of the aerosol must be understood in order to enhance the control processes or provide other means of dealing with the mine aerosol. In the following, some behavioral characteristics of aerosols are described.

Brownian Motion

When aerosol particles are less than 0.1 μm in diameter and suspended in a quiescent fluid, the process of *Brownian motion* controls their behavior. The random movements of the particles are caused by the movements of gas molecules that nudge the particles first this way, then another way, resulting in a random series of movements. The resulting mixing of the particles and gases is known as *diffusion* and can be used to collect the particles through agglomeration or by impaction on solid surfaces. In addition, diffusion can be used to help understand the behavior of contaminants in a moving airstream.

The behavior of dust or other particles due to Brownian motion is reasonably well understood. A detailed presentation is provided by Hinds (1982, pp. 133–150). A number of the theoretical equations developed there may

CHARACTERISTICS OF PARTICLES AND PARTICLE DISPERSOIDS

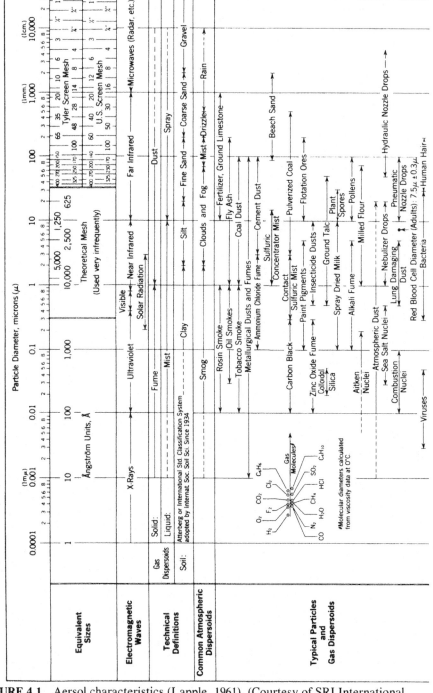

FIGURE 4.1 Aersol characteristics (Lapple, 1961). (Courtesy of SRI International, Menlo Park, CA; used with permission.)

Methods for Particle Size Analysis

- Sieving
- Electroformed Sieves
- Microscope
- Impingers
- Ultramicroscope
- Electron Microscope
- Centrifuge
- Elutriation
- Sedimentation
- Ultracentrifuge
- Turbidimetry++
- Permeability++
- X-Ray Diffraction+
- Adsorption+
- Light Scattering++
- Nuclei Counter
- Visible to Eye
- Machine Tools (Micrometers, Calipers, etc.)
- Scanners
- Electrical Conductivity

+ Furnishes average particle diameter but no size distribution.
++ Size distribution may be obtained by special calibration.

Types of Gas Cleaning Equipment

- Settling Chambers
- Centrifugal Separators
- Liquid Scubbers
- Cloth Collectors
- Packed Beds
- Common Air Filters
- Impingement Separators
- Mechanical Separators
- High Efficiency Air Filters
- Thermal Precipitation (used only for sampling)
- Electrical Precipitators
- Ultrasonics (very limited industrial application)

	Reynolds Number
Terminal Gravitational Settling* [for spheres, sp. gr. 2.0] In Air at 25°C 1 atm	Settling Velocity, cm/sec.
In Water at 25°C	Reynolds Number
	Settling Velocity, cm/sec.

Particle Diffusion Coefficient,* cm²/sec.	In Air at 25°C 1 atm
	In Water at 25°C

*Stokes-Cunningham factor included in values given for air but not included for water.

Particle Diameter, microns (μ)

0.0001   0.001 (1mμ)   0.01   0.1   1   10   100   1,000 (1mm.)   10,000 (1cm.)

PREPARED BY C E LAPPLE

Reprinted from Stanford Research Institute Journal. Third Quarter. Single copies 8-1/2 by 11 inches free or $15 per hundred. 20 by 26 inch wall chart $15 each

SRI International   333 Ravenswood Ave. • Menlo Park, California 94025 • (415) 326-6200

FIGURE 4.1 *(Continued)*

80

be useful to the particulate specialist. For example, the author provides equations for the depositional velocity of fine-sized particles and the rate of deposition of particles on a surface located near a aerosol cloud subject to diffusion alone.

Stokes' Law

Stokes' law is used to determine the settling velocity of dust particles greater than 1 μm in diameter that are falling in a quiescent fluid with Reynolds numbers less than 1 (Hinds, 1982, p. 40). The settling velocity is determined by equating the drag force on a falling spherical particle and the force of gravity. This provides an estimate of the settling velocity with less than 10% error (for the conditions described above) and is adequate for most dust settling calculations. If the particles are not spherical, then Stokes' diameter can be used instead. *Stokes' diameter* is the diameter of a hypothetical spherical particle that has the same density and settling velocity as a given nonspherical particle.

The most useful aerosol equation derived from Stokes' law is the formula for the terminal settling velocity V_t in m/s of particles in air (Hinds, 1982, p. 44):

$$V_t = \frac{\rho_p D_p^2 g}{18 \; \mu} \tag{4.1}$$

where ρ_p is density of the particle in kg/m^3, D_p is diameter of the particle in m, g is the gravity constant $= 9.807$ m/s^2, and μ is viscosity of air $= 1.81 \times 10^{-5}$ Pa·s. This equation applies if the particle diameter D_p is greater than 1 μm and the particle Reynolds number is less than or equal to 1.0. The calculation of the terminal settling velocity is accompanied by calculation of the particle Reynolds number N_{Re} as follows:

$$N_{Re} = \frac{\rho_p V_t D_p}{\mu} \tag{4.2}$$

where V_t is the relative velocity of the particle in the fluid. (*Note:* The variable N_{Re} is the particle Reynolds number, which is defined differently than the Reynolds number for fluid flow in Section 5.4.) This is performed to insure that the conditions meet the assumptions necessary for use of Stokes' law. For Reynolds numbers exceeding 1, a different equation should be used (Hinds, 1982, p. 51).

Example 4.1   A blast is performed in a stope with no ventilating air. If the stope is 6 m high, how long will it take a 3-μm particle to settle from the roof to the floor if the particle has a specific gravity of 2.10?

Solution: For the 3-μm particle, the terminal velocity is calculated using Eq. 4.1 as follows:

$$D_p = 3 \ \mu m = 3 \times 10^{-6} \ m$$

$$\rho_p = 2.10 \ g/cm^3 = 2100 \ kg/m^3$$

$$V_t = \frac{(2100) \ (3 \times 10^{-6})^2 \ (9.807)}{(18) \ (1.81 \times 10^{-5})} = 0.00057 \ m/s$$

To determine the time of fall τ, the distance L is divided by the velocity:

$$\tau = \frac{L}{V_t} = \frac{6 \ m}{0.00057 \ m/s} = 10,500 \ s \ \text{or} \ 175 \ min$$

The Reynolds number is now calculated using Eq. 4.2:

$$N_{Re} = \frac{(2100)(0.00057)(3 \times 10^{-6})}{1.81 \times 10^{-5}} = 0.198 < 1.0$$

This number tells us that the conditions are acceptable for application of the equation, and we can feel confident that the calculations are reasonably accurate.

When the particle diameter falls below the 1-μm mark, Eq. 4.1 loses accuracy as a result of slippage that develops at the surface of the particle. This causes the particle to fall faster than the equation predicts. Under this set of conditions, it is necessary to apply the *Cunningham correction factor* C_c to the velocity calculation, which can be calculated (Hinds, 1982, p. 45) as follows:

$$C_c = \frac{1 + 2.52\lambda}{D_p} \qquad (4.3)$$

In this equation, λ is mean free path of the molecules of the gas expressed in meters. The mean free path is the average distance that a molecule of a gas travels before colliding with another molecule. The mean free path for air is 0.066 μm (6.6×10^{-8} m) at 20°C and atmospheric pressure. This allows the use of Stokes' law for particles as small as 0.1 μm in diameter. The correction factor is then used in the expression

$$V_t = \frac{\rho_p D_p^2 g \ C_c}{18 \ \mu} \qquad (4.4)$$

This procedure is applied below.

Example 4.2   If a 0.5-μm diesel particle is released into quiescent air at a height of 2 m above the floor of a mine opening, how long will it take for the particle to fall to the floor of the opening if it has a specific gravity of 0.6? Assume that the mean free path for air is 6.6×10^{-8} m.

Solution: The problem above is approached by first using Eq. 4.3 to determine the value of the Cunningham correction factor as follows:

$$C_c = 1 + \frac{2.52\,(6.6 \times 10^{-8})}{0.5 \times 10^{-6}} = 1.333$$

The value of the correction factor allows the use of Eq. 4.4 to calculate the settling velocity:

$$V_t = \frac{(600)\,(0.5 \times 10^{-6})\,(9.807)\,(1.333)}{(18)\,(1.81 \times 10^{-5})} = 6 \times 10^{-6}\text{ m/s}$$

The time to settle is then computed to be

$$\tau = \frac{2}{6 \times 10^{-6}} = 332{,}000 \text{ s} = 92.3 \text{ h}$$

This large value indicates why DPM is best kept from entering the atmosphere as it will not readily settle out of the air.

Behavior of Aerosols in a Moving Airstream

Most of the aerosols that are encountered in the underground mine atmosphere are found in a working area that is swept by a flow of ventilating air. In these cases, the dust concentrations downwind of a dust source are normally a function of several factors: rate of convection, diffusion, tendency of the dust to agglomerate, and other depositional factors. As a result, the dust concentration can be difficult to predict in a given situation. However, a number of empirical studies have been performed that can be used to predict the behavior in general terms (Bhaskar and Ramani, 1986, 1988). These studies should be consulted for an idea of dust behavior under the relevant conditions.

4.3   CLASSIFICATION OF MINERAL DUSTS AND OTHER RELEVANT AEROSOLS

The composition of dusts in mines is determined by the mineral content of the ore (and waste) mined and thus is quite variable. In addition, diesel engines, spray equipment, and other mechanical equipment may emit fumes or other aerosols into mine air. Only diesel fumes are common to today's mines, but other aerosols may be present at times because of the use of chemicals for mine support, stopping construction, or other applications. In each category of the following classification, the aerosols are listed in order of decreasing harm; some appear in several categories:

1. Fibrogenic dusts (capable of producing fibrosis or scarring of the lung surfaces)
   a. Silica (quartz, cristobalite, tridymite, chert)
   b. Silicates (asbestos, talc, mica, sillimanite)
   c. Metal fumes (nearly all)
   d. Beryllium ore
   e. Tin ore
   f. Iron ores (some)
   g. Carborundum
   h. Coal (bituminous, anthracite)
2. Carcinogenic aerosols
   a. Asbestos
   b. Radon daughters (attached to any dust)
   c. Arsenic
   d. Diesel particulate matter (a suspected carcinogen)
   e. Silica (a suspected carcinogen)
3. Toxic aerosols (poisonous to body organs, tissue, etc.)
   a. Dusts of ores of beryllium, arsenic, lead, uranium, radium, thorium, chromium, vanadium, mercury, cadmium, antimony, selenium, manganese, tungsten, nickel, silver (principally the oxides and carbonates)
   b. Mists and fumes of organic and other body-sensitizing chemicals
4. Radioactive dusts
   a. Ores of uranium, radium, and thorium (injurious because of alpha and beta radiation).
   b. Dusts with radon daughters attached (source of alpha radiation)
5. Explosive dusts (combustible when airborne)
   a. Metallic dusts (magnesium, aluminum, zinc, tin, iron)
   b. Coal (bituminous, lignite)
   c. Sulfide ores
   d. Organic dusts
6. Nuisance dusts (little adverse effect on humans)
   a. Gypsum
   b. Kaolin
   c. Limestone

As noted, nuisance dusts are generally of little consequence to human beings. However, they cannot be considered totally benign. When sufficient quantities of these dusts are present in the atmosphere, the burden on the particle-clearing mechanism of the lungs is increased, which has a detrimental effect on the humans subject to such a condition. *It is therefore imperative that the engineer reduce dust concentrations to their lowest achievable levels, even if the dusts are not considered to be fibrogenic or otherwise harmful.*

4.4 PHYSIOLOGICAL EFFECTS OF MINERAL DUSTS

Human Respiratory System

The harmful effects of mine dusts, particularly those with fibrogenic potential, are best understood in the light of knowledge of the components and functioning of the human respiratory system, shown in simplified form in Fig. 4.2. Air is introduced into the respiratory tract through either the nose or the mouth. Along with that air, aerosols (such as dust, pollen, bacteria, fumes, etc.) may be brought into the body. As the particulate matter passes through the nasal passages, hair and mucus help to filter out the larger particles. The air then flows into the *nasopharynx* region, where it is warmed before it moves into the deeper parts of the respiratory system. This air, and any air breathed through the mouth, is then passed through the *trachea* (windpipe), the *bronchi* (the two short branches off the trachea), and the *bronchioles* (branches off the bronchi), and into the *alveoli* (the terminal lung sacks where oxygen is transmitted into the bloodstream). Along the trachea, bronchi, and bronchioles, additional particles of medium size are impacted on the mucous layer lining the openings. These particles are normally swept upward by the *cilia* (hairlike cells that line the openings) and deposited in the throat. This material is then either coughed up or passes through the digestive system of the body.

The dust particles that reach the alveoli are of smaller size and can be deposited on the lung surfaces through settling, impaction, Brownian motion, or other processes. The body's defense mechanism for this dust consists of phagocytes (wandering scavenger cells) called *alveolar macrophages* that act to engulf the invading particles and either isolate them in the lung or transport them to the lymph nodes for disposal. How successful this natural protection is depends on the dust concentration, the composition of the dust, a person's health, and other variables. The human body has evolved into an efficient organism designed to eliminate particles, but only for levels that are found in a normal or natural environment. A dusty mine atmosphere may overload the respiratory system, resulting in an inadequate capability to handle all the dust that is deposited in the lungs. It is then that lung diseases normally develop.

The terminology applied to the dust that invades the body during breathing is also of importance here. This has been defined by Vincent (1989, p. 2). *Inspirable particles* are those particles that are capable of being inhaled into the nose or mouth during breathing. These particles are smaller than 100 μm

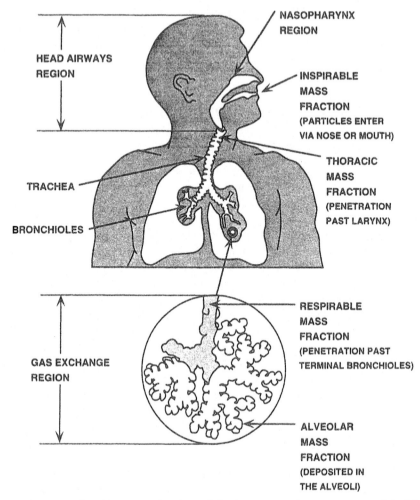

FIGURE 4.2   Graphic representation of the human respiratory system. (Modified after Lioy, 1995, from *Air Sampling Instruments,* 8th Ed., ACGIH, Cincinnati, OH. Adapted with permission.)

in diameter. Once in the respiratory system, the particles that reach the thoracic region (the area past the mouth and nasal passages) are known as the *thoracic aerosol*. These particles will normally be less than 25 μm in size. In the trachea and bronchi, many of the particles will be impacted or otherwise be deposited on the mucous membranes. These are known as the *tracheobronchial particles* and cover the size range 5–25 μm. The portion of the particle load that enters the alveolar region of the lungs is called *respirable dust*. The size of respirable dust is often defined as being less than 5 μm. Some of the dust that penetrates into the alveolar region of the lungs is exhaled rather than deposited in the lungs. The particles that are deposited in the alveoli are known as the *alveolar aerosol*. While any particles that are deposited can cause problems, the alveolar aerosol is the primary cause of lung diseases.

The percentage of particles that is respirable is a function of the size of the particles. Several organizations [such as the British Medical Research Council (BMRC) and the American Conference of Governmental Industrial Hygienists (ACGIH)] have published curves that are meant to approximate the percentage of particles that are respirable. These are shown in Fig. 4.3, where the rate of penetration is provided as a function of the aerodynamic diameter of the particles. The *aerodynamic diameter* is defined as the diameter of a particle with a density of 1 g/cm^3 that has the same aerodynamic properties as a given particle of arbitrary shape. The curves indicate that 50% of the particles of 5 μm in aerodynamic diameter and nearly 100% of 1-μm particles will penetrate the lungs.

After entering the alveolar region of the lungs, particles are not always deposited. The deposition curve (Fig. 4.4) has been experimentally determined (Stahlhofen et al., 1980) to provide an approximation of the percentage of particles at each size that are deposited in the alveolar region. The deposition rates shown apply to two specific breathing patterns, but similar curves occur for other breathing patterns that were experimentally tested. As can

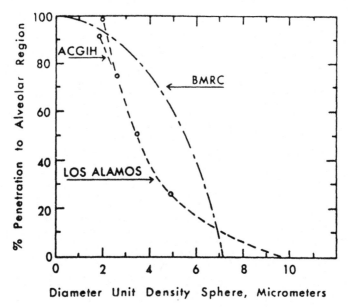

FIGURE 4.3 Characteristic curves for penetration of aerosols into the alveolar region of the lungs. [Copyright (1970), American Industrial Hygiene Association, reprinted with permission).]

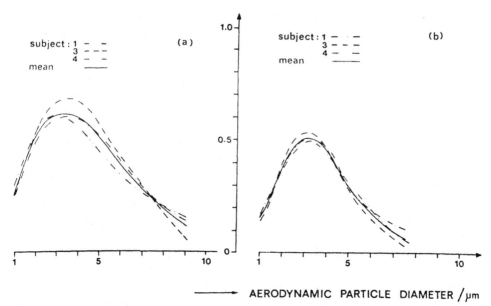

FIGURE 4.4   Deposition curve for aerosol particles in the alveolar region of the human lung for two breathing patterns: (*a*) breathing rate = 250 cm^3/s; (*b*) breathing rate = 750 cm^3/s. (For other variables of the breathing patterns, see original article.) (After Stahlhofen et al., 1980, copyright American Industrial Hygiene Association, used with permission.)

be seen, the maximum deposition rate occurs at the 3-μm size and drops off on either side of that value. The net result is that most of the particles deposited in the lungs are between 1 and 5 μm in aerodynamic diameter.

Historical Background of Health Effects

The history of health effects from aerosols, particularly dusts, in mining and related industries is a most revealing and thought-provoking record. Diseases caused by mine dust were first reported by Pliny the Younger in his writings on natural history in the first century CE. The effects of mineral dusts were also described by Georgius Agricola (1556, p. 214) in the first scientific treatise on mining. For several centuries thereafter, medical ailments due to exposure to dust in the trades, particularly mining and foundry jobs, were recognized but were not well understood.

It was not until 1896 that knowledge of lung diseases was advanced significantly. At that time, the x-ray machine was invented by Roentgen, providing more capability to analyze the specifics of lung diseases. While x-rays provided more medical knowledge of industrial lung diseases, impetus for social action did not come until later. The galvanizing event in the United States came during the 1930s after a tunnel was driven through a sandstone mountain near Gauley Bridge, WV. Dust conditions in this tunnel were deplorable due to lack of concern for the workers' health. Dust controls were seldom used in the tunnel, and several hundred workers died in the ensuing years as a result of acute silicosis (Seaton, 1975a, p. 82). As a consequence, Congress held hearings and states began to adopt laws to protect workers in industries where silica was mined or used.

The next milestone event in the fight against diseases from dust occurred when it was discovered that asbestos fibers were responsible for cancers of the lungs. While cancer due to asbestos was suspected for some time, the public alarm was sounded in the 1970s when more convincing data became available (Webster, 1970; McDonald, 1973). Today's stringent rules on asbestos use are the result of this public awareness of the hazards of asbestos.

One additional historical event that is of importance in the concern for aerosols in the workplace came about in the 1980s. In was during this period that several researchers began to see a link between DPM and cancers in laboratory animals (Heinrich et al., 1986; Brightwell et al., 1986; Ishinishi et al., 1986; Iwai et al., 1986; Mauderly et al., 1987). These findings have led to the call by the National Institute for Occupational Safety and Health (NIOSH) for control of DPM in the mine atmosphere (Anon., 1988).

Respiratory Ailments and Diseases

The physical ailments that result from aerosols in the mine environment range from simple irritation of the respiratory system due to overloading the lungs with dust to death from lack of respiratory function or cancer. Readers who are interested in a more complete discussion should consult the references by Morgan and Seaton (1975) and NIOSH (Merchant, 1986). The limited discussion here centers on the most common diseases and those that require knowledge on the part of practicing mine ventilation engineers.

The most common form of lung dysfunction found among miners is *pneumoconiosis,* a condition characterized by the formation of fibrous tissue in the lungs arising from dust particles deposited in the lungs. Pneumoconiosis is a general term; specific names are generally attached to the resultant maladies, depending on the specific causal material. Some of the specific terminology for the more prevalent mineral-related diseases and the dusts believed to be causal are as follows (Merchant, 1986):

Silicosis (dusts of quartz, tridymite, and cristobalite)

Silicate pneumoconiosis (dusts of silicate minerals such as kaolin, talc, tremolite, actinolite, and anthophyllite)

Asbestosis (asbestiform dusts such as dusts of amosite, chrysotile, and crocidolite)

Coal workers' pneumoconiosis (coal dust)

Beryllium disease (dusts of beryllium compounds including ores)

Siderosis (dusts of iron including ores)

Of these diseases, silicosis and coal workers' pneumoconiosis are of most importance in mines.

Silicosis is the disease that results when dust particles containing silica are trapped in sufficient quantities in the alveoli of the lungs. The disease is characterized by fibrotic tissues being formed around quartz particles or particles containing quartz, particularly in the areas of the lungs around the central bronchioles. These fibrotic lesions generally consist of collagen fibers in the central part of the lesions with reticulin fibers peripherally (Seaton, 1975a, p. 100). A more severe form of the disease, termed *acute silicosis*, is recognized as a separate disease characterized by proteinosis of the lungs and is normally associated with exposure to extremely high levels of silica over short periods of time.

Silicosis has been widely recognized in the mining, foundry, stone-working, and construction trades where quartz in various forms has been processed. The occurrence of the disease as a result of mining activity has decreased significantly since public awareness of the affliction came about in the 1930s. This is due to regulations controlling quartz in the mine environment, a change from dry to wet drilling in underground mining operations, and a reduction in the amount of silica-related minerals mined using underground methods.

While knowledge of the control of silicosis is readily available, inadequate levels of protection and lack of awareness of workers concerning the hazards of silica still allow cases of silicosis to develop. One of the areas where this is occurring is in the oil equipment and construction industries where sandblasting is used (Weisenfeld et al., 1993; Short et al., 1993). In addition, a significant number of cases of silicosis have been appearing among the workers on surface mine drilling equipment where dry drilling is the practice (Linch and Cocalis, 1994). This appears to result from the lack of proper engineering control of silica-laden dusts, a deplorable condition considering the severity of the disease and the current state of dust control technology.

Coal workers' pneumoconiosis (CWP) is often called "black lung" and is quite prevalent in miners who worked for long periods in the coal industry before the institution of dust-control regulations that restrict the concentration of coal dust. Like silicosis, CWP is a lung disease resulting from the formation of lesions in the lung around dust particles that overload the lungs. In the case of coal particles, the lesions formed consist mainly of reticulin fiber rather than collagen fiber, thus producing less scarring of the lung surfaces (Wright, 1978). This is the primary reason why CWP is not as deadly as silicosis. However, CWP is more prevalent and is associated with higher costs to the public, making it a significant health and economic problem that the mining industry must address.

Because CWP is widely recognized in the various coal mining countries throughout the world, and because of the significance of the health and economic effects of CWP, the International Labor Organization (ILO) has established and maintained a disease classification scheme based on the size of the opacities occurring on a miner's lung x-ray. This classification recognizes two main categories of CWP: *simple pneumoconiosis* (with discrete opacities less than 10 mm in size) and *progressive massive pneumoconiosis* (also called *complicated pneumoconiosis*) with larger opacities.

The effects of simple pneumoconiosis on the health of miners are rarely severe, but progressive massive pneumoconiosis is more debilitating. This form of the disease can result in reduced pulmonary capacity and can be a significant problem to a miner with other cardiopulmonary problems. In addition, CWP can resemble silicosis if the coal dust involved is high in silica. This worsens the health effects of the disease, increasing the possibility that the disease can be life-shortening. Consequently, reduced dust concentrations are often legislated in coal mines possessing higher percentages of silica in their dusts.

In addition to the pneumoconioses, some mineral dusts and other mine aerosols can cause cancer. This was revealed when significant numbers of asbestos miners and workers were reported to be suffering from lung cancer as a result of their exposure to the fibrous silicates often referred to as *asbestos*. Other aerosols were also found to be a cancer hazard, although the degree of hazard is less. To date, the following mineral-related aerosols are thought to be carcinogenic:

Asbestos (specifically the minerals chrysotile, crocidolite, asbestiform actinolite, asbestiform tremolite, amosite, and anthophyllite)

Radon (when attached to respirable dust particles)

Diesel particulate matter

Silica

The case of asbestos is outlined in detail in the publication by Dement et al. (1986). This mineral has been intensively studied, and it appears clear that occurrences of lung cancer are associated with asbestosis (the pneumoconiosis caused by asbestos) to such an extent that all asbestos workers are candidates for lung cancer unless they can be protected from breathing asbestos dusts. This explains the heavy emphasis on asbestos control in mining, milling, and the construction industry and the move toward removal of asbestos from public buildings.

Radon gas (Section 3.3) is a product of the decay of radioactive materials that can occur in the mine environment. The further decay of the radon gas results in the generation of alpha particles, a low-energy deterioration product. Alpha particles cannot penetrate deeply into body tissue. However, radon can become resident in the lungs because of their attachment to dust particles. This puts them in close contact with sensitive lung tissue over a long period of time, constituting a hazard to the human body. Radon gas is generated in uranium mines and, to a lesser degree, in other mines that have small amounts of uranium in the mineral being mined or in mines that release radon from the infiltrating groundwater. In addition, it can occur in homes and buildings from the decay of trace amounts of uranium in the surrounding soils and rocks. The cancer hazard from radon gas has been statistically demonstrated to result in higher incidences of lung cancers; hence the need for control over radon in the mine atmosphere is important.

The cancer hazard from DPM in the atmosphere has only recently been recognized. The official warning was sounded by NIOSH (Anon., 1988) as a result of laboratory testing of rats and other small animals. More recent indications are that the health hazard from DPM is relatively modest (Mauderly, 1995). However, the carcinogenic potential is thought to be statistically significant for those whose exposure is heavy and chronic. Regulations limiting the exposure to DPM in mine atmospheres are pending.

A carcinogenic risk associated with breathing of silica dust has also been suggested during the last decade. The first indication of this hazard was published by the International Association on Research on Cancer (Anon., 1987), which reported suspicions that a relationship existed. Further research of the epidemiologic type was then performed that seems to be supportive of such a relationship. The final word is not yet available on this topic. However, the fact that silicosis is such a threat in itself makes the avoidance of silica in the workplace a must.

Other aerosol health effects have been suggested that may be of concern to the mine ventilation engineer. Because so many concerns exist, and the potential for causing disease through the breathing of mineral dusts is so great, the mine ventilation engineer has more than sufficient reason to ensure that the working environment is free of aerosol contaminants.

4.5 FACTORS THAT DETERMINE DUST HARMFULNESS

Various aerosols have been discussed with little explanation of how they vary in harmfulness. In this section, variables that are significant in determining the health effects on humans are discussed in order of decreasing importance.

Composition

The chemical or mineralogical makeup of a given dust or aerosol is probably the most significant variable in the determination of whether an aerosol is harmful to human beings. It is clear that some mineral dusts are relatively benign while others are quite harmful. However, it is not clear whether the differences are due to the chemical makeup of the minerals or their physical form. For example, there are several schools of thought as to why asbestos minerals are so harmful. Three important mechanisms have been proposed (Seaton, 1975b, p. 144): the first is mechanical (long, sharp particles cause severe fibrosis), another is chemical (gradual solution of the asbestos causes certain chemicals to affect the lung, causing fibrosis), and the third relates to autoimmune hypotheses (the fibrous mineral produces abnormal globulins).

The chemical makeup of DPM is another important example of how the composition plays a part in the harmfulness of an aerosol. The particles emitted from a diesel engine are primarily submicron particles of unburned carbon. However, it is possible to find a significant percentage by mass of polynuclear aromatic hydrocarbons (PAHs) on the particle surfaces. These chemicals are believed to be the source of the cancer hazard to humans.

Concentration

The importance of concentration of dusts and other aerosols in the workplace air was demonstrated in a landmark study on CWP by Jacobsen et al. (1970) using the workforce at 25 British coal mines as a medical cohort. An analysis of dust exposures was performed along with periodic x-rays to monitor the health of 30,000 bituminous coal miners. After 15 years of work, a statistical pattern began to emerge from the data as shown in the dose–response curves in Fig. 4.5. Both simple and complicated pneumoconiosis showed a strong relationship with the average mass concentration of the coal mine dusts to which the workers were exposed.

While investigation of the incidence of CWP is not as easy in U.S. coal mines (because miners cannot be required to participate), recent research results published by NIOSH (Attfield et al., 1984a, 1984b) provide results similar to those from the British study. These studies were complicated by more variables and were not based on as large a population of miners. However, they add evidence that similar disease incidences are found in both U.S. and British coal mines.

Other minerals are known to demonstrate a similar average concentration–incidence relationship. While the average concentration allowed for various minerals is different, reducing concentrations of any given dust will ordinarily also reduce the incidence of lung disease. Thus every attempt should be made to provide reduced dust levels in the working areas of a mine.

Particle Size

As a general rule, the size of dust particles plays a part in harmfulness to the human body because it controls the location in the respiratory tract where the dust particles will reside. Particles less than 5 μm are most likely to penetrate into the lungs and become trapped in the alveoli. However, these finer particles are also the most difficult to filter or otherwise remove from

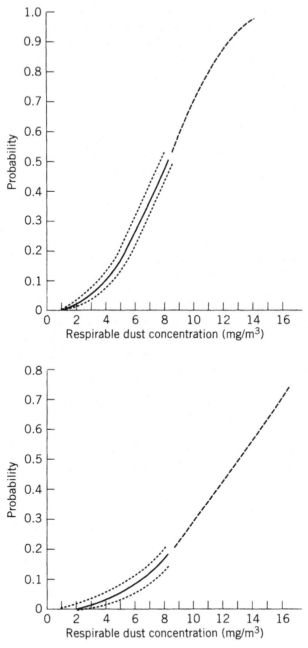

FIGURE 4.5 Probability of contracting simple pneumoconiosis (*a*) category 1/0 or greater and (*b*) category 2/1 or greater after 35 yr of exposure. (After Jacobsen et al., 1970, used with permission.)

the environment. Any attempt to filter the air of aerosol contaminants will be fruitless, therefore, if the procedure does not filter out the <5-μm material.

Exposure Time

Most of the diseases associated with breathing mineral dusts manifest themselves over a relatively long period of time, normally 10 years or more of occupational exposure. Even asbestosis, one of the more dreaded forms of mineral-induced diseases, is thought to take about 10 years to develop (Kane, 1993, p. 354). One exception to this general rule is silicosis, which can develop after only a few years of exposure (Linch and Cocalis, 1994).

An additional factor to consider in the assessment of mineral-related lung disease is the tendency of some of these diseases to cease progressing if exposure to the dust is terminated. This appears to be the tendency in CWP (Lapp, 1981) where the progression in the degree of lung opacities is slowed when the worker is removed from the dusty atmosphere. This removal strategy is used in fighting CWP in the United States, where coal miners can opt for low-exposure jobs if they are known to have incurred certain levels of pneumoconiosis.

Individual Susceptibility

The susceptibility of an individual when subjected to years of occupational dust exposure is certainly a variable that will affect whether that person will suffer from lung disease. This is evident from the statistics that result from various epidemiologic studies. Individual workers will not know ahead of time whether they are predisposed to develop lung disease or will be part of the group that escapes serious medical effects. It is therefore important for the company to protect its workers and—equally so—for the workers to protect themselves.

One important modifier that impacts on individual susceptibility is the smoking habits of the worker. It is clear that this will affect the probability of serious medical consequences of exposure to aerosols. For example, it is known that there is a synergistic effect between cigarette smoking and exposure to asbestos minerals (Dement et al., 1986), and that the resulting incidence of cancer will be greater than would be found by adding the incidences due to each cause.

4.6 EXPLOSIVE DUSTS

Many organic and metallic dusts are explosive if suspended in air at a high concentration. The dispersed particles of a dust in air have a very large surface area that potentially can result in a powerful chemical reaction as the particles rapidly oxidize as a result of some ignition stimulus. Dust ignition can occur because of a point source (e.g., an open flame, an electric arc, a discharge of explosives, or an ignition of methane) or autoignition (self-ignition) from the dust being heated above some critical temperature. Point sources are most common in mining and must be controlled to ensure against dust explosions in coal mines.

Explosions in air–methane mixtures propagate through the mixture at about the speed of sound [1117 fps (340 m/s)]. These are termed *detonating explosions*. The explosions that occur in air/dust mixtures normally advance much slower with the flame speed being about 30–35 fps (10 m/s). This is termed a *deflagrating explosion* and has the tendency to be self-feeding as the shock wave from the reaction moves away from the explosion source faster than the flame, thus permitting the shock wave to raise dust from mine surfaces to feed the flames (Hinds, 1982, p. 355).

Dust explosions have been widely investigated because of their importance in a number of industries. The conditions that allow an explosion to occur and corresponding methods for preventing them have been outlined in detail by Field (1982), Cross and Farrer (1982), Nagy and Verakis (1983), and Cashdollar and Hertzberg (1987). These references are invaluable to those interested in a thorough knowledge of dust explosions. Some aspects of this science for those in mine ventilation are outlined in the paragraphs that follow.

Coal Dust Explosions

Fortunately, explosions in coal mines, in processing facilities, and in coal silos are being reduced over time because of stricter regulations and more knowledge concerning the conditions that cause such events. The U.S. Bureau of Mines (USBM) has performed extensive coal dust experiments to establish the range of explosibility of dusts (Nagy and Verakis, 1983; Cashdollar and Hertzberg, 1987). The subject of explosibility is complex and is affected by many variables; the more important of these is discussed in the following.

Composition   Chemical variations in coal affect the temperature at which coal autoignites. For example, Conti and Hertzberg (1987) state that autoignition temperatures for some lignites with volatile-matter content of 30–43% range from 797 to 1112°F (425–600°C) while anthracite coals with volatile-matter content of 4–8% have autoignition temperatures of 1247–1346°F (675–780°C). Bituminous and subbituminous coals have autoignition temperatures between those of lignite and anthracite. The net result is that the more volatile matter a coal possesses, the lower its ignition temperature, and the more readily it will explode.

Particle Size   The particle size of coal in a dust cloud will clearly affect the explosibility. Particles larger than 100 μm will not readily explode at any concentration. Smaller particles (<20 μm) will explode at much lower concentrations than particles 20–100 μm in diameter (see Hertzberg and Cashdollar, 1987, p. 23). Unfortunately, much of the coal dust occurring in a coal mine is less than 20 μm, making diligent efforts at controlling dusts in mines of great importance.

Concentration The lower explosive limit of airborne coal dust by weight is approximately 60 g/m³ or 60,000 mg/m³. This concentration is almost never found under normal coal mining conditions. However, such concentrations can occur if a methane ignition occurs underground. Under these conditions, the pressure wave from the methane ignition can raise settled dust into the air and then ignite the dust. Thus control of the ignitibility of settled dust and control of methane is as important in the prevention of dust explosions in coal mines as is control of dust itself. The preceding comments, however, do not apply to dust conditions in coal silos. Dust concentrations in silos or other storage facilities may exceed explosibility limits, and extra care must be taken therein to insure against explosions set off by open flames, welding arcs, or other potential ignition sources.

Flammable Gas The presence of methane alters the explosibility of coal in mine air. Figure 4.6 provides a flammability limit on highly volatile coal–methane mixtures and indicates the danger of having methane in the mine atmosphere. At a level of 5%, for example, the methane itself can ignite, and any coal dust in the air will simply provide more fuel for the reaction. Under these conditions, methane control emerges again as an important issue in the prevention of explosions in underground coal mines.

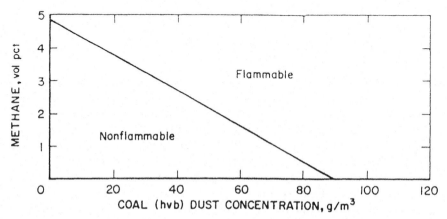

FIGURE 4.6 Lean flammability limit for mixtures of methane and coal dust. (After Hertzberg and Cashdollar, 1987. Copyright ASTM. Reprinted with permission.)

Moisture Effects   The effects of moisture in the air and moisture in the coal play some part in the explosibility of coal dust. However, the moisture level in the coal itself is more important as increased moisture will reduce the explosibility and the pressure generated by an explosion. Moisture in the air has little effect on explosibility.

Presence of Incombustible Material   The presence of incombustible dust mixed with coal dust in air greatly affects the reactions that takes place during an ignition. Any inert dust can be used, but the standard practice is to use <200-mesh limestone dust because it also virtually lacks any pulmonary risk. In the mine, this material is commonly called *rock dust*. Current regulations in the United States require that 65% by weight of the dust on the roof, ribs, and floor in coal mine sections be rock dust while 80% of the dust in returns be rock dust. This is to prevent ignition if the dust is raised by a methane ignition or some other event in the mine.

The capability of rock dust to reduce the explosibility of a dust mixture is related to the heat-absorbing capability of the dust. The ignition energy of a coal–rock dust mix is much greater than that for pure coal dust, and the pressure and energy release rates are much lower for coal–rock dust mixtures. The explosibility data in Fig. 4.7 show this effect for varying amounts of rock dust in the suspended dust.

Other Dusts   Other combustible mineral-related dusts are similar to coal in terms of their behavior but differ in their lower explosive limit, ignition temperature, and maximum pressure generated (Table 4.1). Those that are of concern in underground mining operations are oil shale, sulfides, and gilsonite. Generally, explosions of noncoal dusts are relatively rare. Sulfide ore dusts are known to explode during blasting operations, although little technical information is available (Hall et al., 1989a).

4.7   THRESHOLD LIMIT VALUES

Threshold limit values (TLVs) are guidelines that refer to the airborne concentrations of various substances to which it is believed that nearly all workers may be repeatedly exposed day after day and suffer no adverse health effects (Anon., 1996). ACGIH provides a comprehensive list of these substances (both chemical and aerosol) that serves industrial hygienists and other health and safety personnel with targets to achieve in their environmental work. Definitions of TLV values are provided in Section 3.2.

TLVs that are related to aerosols in the mine environment are listed in Table 4.2. Keep in mind that the limits specified by regulations may be different. Recent history indicates that the TLVs change frequently. Ventilation and industrial hygiene personnel should check current values often.

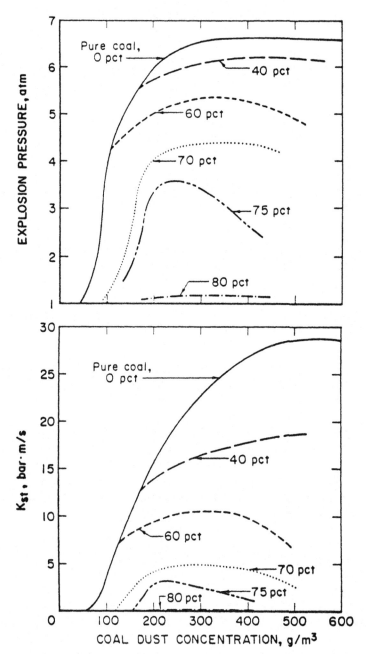

FIGURE 4.7 Explosibility data for mixtures of coal dust and limestone dust for varying amounts of rock dust. (After Hertzberg and Cashdollar, 1987. Copyright ASTM. Reprinted with permission.)

TABLE 4.1   **Explosibilities of Mineral-Related Dusts**

Dust	Lower Explosive Limit 10^{-3} oz/ft^3 (10^3 mg/m^3)	Ignition Temperature °F (°C)	Maximum Pressure psi (kPa)	Source
Aluminum	25 (25)	1193 (645)	89 (614)	Hartmann (1948a, 1948b)
Iron	250 (250)	797 (425)	36 (248)	" " "
Magnesium	20 (20)	986 (530)	72 (496)	" " "
Titanium	45 (45)	896 (480)	52 (359)	" " "
Zinc	480 (480)	1112 (600)	36 (248)	" " "
Coal	60 (60)	1130 (610)	46 (317)	" " "
Sulfide ore	—	1022 (550)	—	Conti and Hertzberg (1987)
Oil Shale	120 (120)	887 (475)	—	" " "
Sulfur	20 (20)	374 (190)	78 (538)	Dorsett and Nagy (1968)
Gilsonite	20 (20)	1076 (580)	78 (538)	Nagy et al. (1965)

4.8 AEROSOL MEASUREMENT

Dust Monitors

Methods of measuring and sampling aerosols are quite numerous but can be classified into several distinct categories. The first of these is measurement with instruments of the direct-reading type in which a sample is not collected. These instruments ordinarily are termed monitors and are used to spotcheck the dust concentrations in workplaces and to periodically record concentrations over some period of time. Several types of instruments are available to determine aerosol concentrations on this basis (Glenn and Craft, 1986). However, only the light-scattering-type photometer is commonly used in monitoring underground mine aerosol concentrations.

Two versions of light-scattering instruments have been used fairly extensively in mine aerosol monitoring. The first is the SIMSLIN II (Safety in Mines Scattered Light Instrument), which was developed in Great Britain for use in mines, and the real-time aerosol monitor, Model RAM-1, developed in the United States. The successor to the RAM-1 (the RAM-1-2G), manufactured by MIE Inc. (Billerica, MA), uses similar technology. This portable instrument operates by drawing 2.0 L/min of the surrounding atmosphere through an inlet, where it is processed through a Dorr–Oliver 10-mm nylon cyclone to take out the nonrespirable aerosol. The remaining aerosol is then analyzed for concentration by measuring the intensity of near-forward-scattered electromagnetic radiation in the near-infrared range. The instrument must be calibrated for each aerosol type and is still subject to a certain amount of error. However, it is available in an intrinsically safe form and

TABLE 4.2   ACGIH Threshold Limit Values for Some Mine-Related Aerosols

Substance	TLV-TWA	Comments
Arsenic, elemental	0.01 mg/m^3	Confirmed carcinogen
Asbestos, amosite	(0.5 fiber/cm^3)^a	Confirmed carcinogen, applies to fibers >5 μ long
Asbestos, chrysotile	(2.0 fibers/cm^3)^a	Confirmed carcinogen, applies to fibers >5 μ long
Asbestos, crocedolite	0.2 fiber/cm^3	Confirmed carcinogen, applies to fibers >5 μ long
Asbestos, other forms	(2.0 fibers/cm^3)^a	Confirmed carcinogen, applies to fibers >5 μ long
Calcium carbonate	10 mg/m^3	If no asbestos and <1% silica present
Coal dust, respirable	2 mg/m^3	For coal dust with <5% silica; otherwise silica limit applies
Graphite (natural)	2 mg/m^3	All forms except fibers; respirable fraction only
Kaolin, respirable	2 mg/m^3	
Magnesite	10 mg/m^3	For total dust with no asbestos and <1% silica
Mica, respirable	3 mg/m^3	
Nusiance dusts	10 mg/m^3	For total dust with no asbestos and <1% silica
Oil mist, mineral	5 mg/m^3	Excludes vapor; TLV-STEL is 10 mg/m^3
Perlite	10 mg/m^3	For total dust with no asbestos and 1% silica
Portland cement	10 mg/m^3	For total dust with no asbestos and 1% silica
Silica, crystalline		
Cristobalite	0.05 mg/m^3	Respirable fraction
Quartz	0.1 mg/m^3	Respirable fraction
Tridymite	0.05 mg/m^3	Respirable fraction
Soapstone, respirable dust	3 mg/m^3	With no asbestos and <1% silica
Soapstone, total dust	6 mg/m^3	With no asbestos and <1% silica
Talc	2 mg/m^3	With no asbestos; otherwise use asbestos values
Welding fumes	5 mg/m^3	Composition of welding fumes varies; may be subject to other TLVs

Source: Anon. (1996).

^{a} Proposed for change to 0.2 fiber/cm^3.

has been used for reading concentrations of both coal mine dusts and diesel particulate matter.

The RAM-1-2G instrument is portable and self-contained and can be used on both a spot basis and for periodic readings over a fixed time period. Readings can be obtained directly from the instrument, but processing of data is often performed using a data logger available from Metrosonics, Inc. (Henrietta, NY). The data logger allows the user to study to entire pattern of periodic concentration readings and to process the data on a computer. Because of the instrument features and the data processing capabilities, the RAM-1 has been widely used for both research and dust control applications.

Aerosol Samplers

Aerosol samplers are devices that gather a sample of the aerosol from the atmosphere in order to read indirectly the concentration or characteristics of the aerosol. This can be performed using instantaneous samplers or samplers that cumulate (integrate) the sample over time. Most of the samplers used for mine environmental purposes gather an integrated sample. Samplers can be used in two principal ways. The first of these is *personal sampling,* in which an aerosol sample is taken on or near a worker to determine the aerosol concentration or dose that the worker is experiencing during the workday. The second is *area sampling* in which a workplace or equipment location is sampled to establish the aerosol concentrations that exist at that point or to measure the effectiveness of controls.

Many different types of samplers are available in the marketplace. The major types are listed in Table 4.3 with a description of the collection principle and some applications. Except for thermal and electrostatic precipitators, which do not find routine use for measurement of aerosols in mining, all the samplers listed have been applied in monitoring and controlling aerosol levels in mines.

The most familiar sampling package for mine use combines an elutriator or cyclone and a filter assembly to gather an integrated (cumulative) dust sample over a work shift. The cyclone or elutriator scalps off the larger-than-respirable dust, and the filter is used to collect the respirable dust for determination of the shift-long, time-weighted average (TWA) dust concentration. An example is shown in Fig. 4.8, which illustrates a Mine Safety Appliances (Pittsburgh, PA) gravimetric dust sampling system. This type of system is used to measure the TWA dust concentration in all U.S. coal mines. The pump can be worn on the miner's belt, and the stainless-steel sampler assembly (containing a 10-mm nylon cyclone and a filter cassette) can be attached to the miner's outer garment.

In using the gravimetric sampler, the pump operates at a flow rate of 2.0 L/min so that the 10-mm nylon cyclone approximates the ACGIH curve for respirable dust (see Fig. 4.3). The pump has a feedback-control circuit to ensure that the flow remains constant as the filter resistance increases with

TABLE 4.3   Aerosol Sampling Instruments

Type	Collection Principle	Examples with Mine Application	Additional Information
Elutriators	Gravitational separation of larger particles from smaller particles	Hexhlet horizontal elutriator	Hering (1989, p. 381)
		MRE gravimetric dust sampler	" " "
Cyclones	Centrifugal separation of larger particles from smaller particles	Dorr–Oliver 10-mm cyclone	Hering (1989, pp. 379–380)
Impactors	Impaction of flowing particles on a solid surface	Anderson/Marple Model 298 multistage impactor	Hering (1989, pp. 373–374)
		MOUDI (microorifice uniform deposit impactor)	Hering (1989, p. 372)
Impingers	Impingement of particles into a liquid	Greenburg–Smith impinger	Hering (1989, p. 382)
		Midget impinger	" " "
Filters	Combination of inertial impaction, interception, diffusion, electrostatic, attraction, and gravitational forces	Fiber filters	Lippman (1989, pp. 305–336)
		Cellulose	" " "
		Glass	" " "
		Paper	" " "
		Membrane	" " "
		Millipore	" " "
		Nuclepore	" " "
		Silver	" " "
		Polyvinyl chloride	" " "
Electrostatic precipitators	Electrical charging of particles with collection on electrode of opposite polarity	Seldom used in mines	Swift and Lippman (1989, pp. 387–403)
Thermal	Particle movement toward a cooler surface in a thermal gradient	Seldom used in mines	" " "

Source: Modified after Glenn and Craft (1986) and First (1989).

dust load. The filter cassette is enclosed in a sealed plastic housing containing a 37-mm polyvinylchloride (PVC) filter that is preweighed so that the dust collected on the filter can be determined by reweighing the filter cassette after use. The PVC filter is used because it is less hygroscopic than other filters and because it produces no significant ash when combusted. This allows the filter and the coal dust to be ashed in a low-temperature asher

FIGURE 4.8   Mine Safety Appliances (MSA) gravimetric sampler and pump. (Photo courtesy of MSA, Pittsburgh, PA.)

and analyzed for quartz content using the P-7 infrared analysis method developed by MSHA (Anon., 1984).

   With the dust collecting system shown in Fig. 4.8, the filter gathers only respirable dust and is therefore called a *selective* sampling system. If the cyclone from the filter assembly in Fig. 4.8 were removed from the system, and all the particles were drawn through and collected on the filter, then the sampling system would be *nonselective* and the particles collected would be called a *total dust sample*.

   In addition to cyclones, elutriators, and filters, which are used rather frequently in mining operations, the category of samplers known as *impac-*

tors has grown in use, particularly where size-selective samplers are needed. This allows a size distribution of the dust to be produced for analysis. This is ordinarily performed using a multiple-stage impactor where the jets at each stage are made successively smaller so that the jet velocities are successively higher. This means that ever-smaller dust particles are collected as the air goes through the impactor, and a size distribution can then be determined gravimetrically. The usefulness of impactors is further discussed in Section 4.9.

Impingers were once used with regularity in the mining industry. This has changed in recent decades because the method requires laborious counting of dust particles in a microscopic cell. This is both expensive and unnecessary with the emphasis on mass concentration of dusts in current regulations. As a result, impingers are not as popular as they once were.

Measures of Aerosol Exposure

The engineer or scientist who attempts to measure dust exposures should also understand the relationship between exposure, dose, and TWA values. *Exposure E* is simply the level or concentration of an aerosol that a person is subjected to at a given moment in time. *Dose* is the total or cumulative exposure to which a person is subjected over time. If the exposure can be expressed as a continuous function of time $E(\tau)$, then the dose or cumulative exposure to which a person is exposed is defined as

$$\text{Dose} = \int_0^T E(\tau)d\tau \tag{4.5}$$

This relationship can also be expressed as a discrete cumulative function if the exposure $E_i(i = 1, 2, \ldots, n)$ and the corresponding time periods τ_i are known. For this case, the dose can be determined as

$$\text{Dose} = \sum_{i=1}^n E_i\tau_i \tag{4.6}$$

If the dose actually deposited in the alveolar regions of the lungs is needed, then an approximation to the deposited dose is calculated as

$$\text{Deposited dose} = IP_pP_d \int_0^T E(\tau)d\tau \tag{4.7}$$

where I is inhalation (breathing) rate in m^3/h, P_p is the probability that the measured dust will penetrate into the alveolar regions of the lungs, P_d is the probability that the dust penetrating the alveolar regions of the lungs will be deposited, $E(\tau)$ is instantaneous exposure rate in mg/m^3, and time τ is measured in hours. The deposited dose will then take the units of mg. If the concentrations for discrete time periods are known, then

$$\text{Deposited dose} = IP_pP_d \sum_{i=1}^{n} E_i\tau_i \qquad (4.8)$$

If the TWA exposure is to be calculated, it can always be expressed as a function of the dose as follows:

$$E_{TWA} = \text{dose/total exposure time} \qquad (4.9)$$

This can be illustrated by means of examples.

Example 4.3 A coal miner working on a longwall is studied using an aerosol monitor for an entire shift and is found to have the following exposures:

Time (h)	Activity	Respirable Dust Exposure (mg/m^3)
0.5	Travel to longwall	0.2
1.0	Prepare to mine	0.5
3.0	Advance shields	3.5
0.5	Lunch	0.4
2.5	Run shearer	1.2
0.5	Travel to portal	0.3

Assume that the aerosol monitor is equipped with a 10-mm nylon cyclone and is calibrated for coal. What is the dose of respirable dust that the miner receives, and what is the TWA exposure to the respirable dust? Does this level of dust exposure meet the current regulations? If the miner were wearing an MSA gravimetric dust sampler, would the sampler provide the same value as the calculation?

Solution: The dose can be calculated using Eq. 4.6.

Time (h)	Exposure (mg/m^3)	Product $(mg{\cdot}h/m^3)$
0.5	0.2	0.10
1.0	0.5	0.50
3.0	3.5	10.50
0.5	0.4	0.20
2.5	1.2	3.00
0.5	0.3	0.15
8.0		14.45

The dose is read directly as

$$\text{Dose} = 14.45 \text{ mg·h/m}^3$$

The TWA exposure then becomes

$$E_{TWA} = \frac{14.45 \text{ mg·h/m}^3}{8.0 \text{ h}} = 1.81 \text{ mg/m}^3$$

This level of exposure would meet the current dust regulations if the quartz content were 5% or less. Theoretically, the MSA gravimetric dust sampler should provide the same exposure level since both instruments use an identical cyclone to record the respirable dust levels. Some minor differences may be experienced since the aerosol monitor measures periodically and the MSA system samples continuously. In addition, the calibration of the aerosol monitor may result in some imprecision. The two values, one based on a series of measurements and one based on continuous sampling, should be close.

Example 4.4 A miner is known to have been subjected to an average respirable dust concentration of 1.2 mg/m^3 for a given working year. What is the miner's deposited dose of respirable dust for the year if that miner worked for 8 h per day for 240 days and did not wear a respirator? Assume that the average inhalation rate is 500 cm^3/s, the probability of dust penetration into the alveolar regions is 0.6, and the probability of deposition on entry into the alveolar regions is 0.3.

Solution: The deposited dose can be calculated using Eq. 4.7. The total time of exposure is

$$\tau = 240 \text{ days} \times 8 \text{ h/day} = 1920 \text{ h}$$

$$I = 500 \text{ cm}^3/\text{s} \times 3600 \text{ s/h} \times (1 \text{ m}/100 \text{ cm})^3 = 1.8 \text{ m}^3/\text{h}$$

The deposited dose using Eq. 4.7 is then

$$\text{Deposited dose} = (1.8)(0.6)(0.3) \int_0^{1920} 1.2 \, d\tau$$

$$= (0.324)(1.2)(1920) = 746 \text{ mg}$$

The value calculated is less than 1 g. This does not seem like much, but 746 mg of dust contains approximately 500 million particles.

4.9 SAMPLING OF DIESEL PARTICULATE MATTER

One aerosol that is assuming greater importance in the mining industry is diesel particulate matter. Pending legislation will probably limit the amount of DPM in the mine working environment. This is of concern for several reasons. First, this adds another material in the workplace that must be controlled. Second, the measurement of DPM is complicated because of the difficulty of differentiating it from coal and other mineral matter. This section contains some information that will help the mine ventilation engineer understand the assessment of DPM in the mine environment.

Description of DPM

Diesel particulate consists of fumes or soot resulting from incomplete combustion. These fumes are normally rounded particles less than 1 μm in diameter. The particles sometimes agglomerate and form clusters that resemble bunches of grapes under microscopic analysis. The primary constituent of DPM is carbon, but other chemical compounds are often absorbed on the particles. These compounds often contain PAHs that are suspected to be mutagenic or carcinogenic.

In the mine atmosphere, or for that matter anywhere diesels operate, the DPM is considered to be a health risk because of the aerosol's small size, allowing most particles to penetrate deep into the lungs. While legislation of diesel emissions for underground mines is pending, it is not yet clear what methods for measuring the DPM will be adopted.

Measurement of DPM

In the last decade, a great deal has been learned about mass concentrations of DPM in coal mines where diesel engines are used and about the size distribution of diesel particles (Rubow et al., 1990; Cantrell and Rubow, 1991). In these mines, 10-stage impactors of the MOUDI type have been used to determine details of the aerosol size distribution of mines where only coal and related mineral dusts occur and mines where both coal and diesel aerosols exist. Figure 4.9 shows the results of similar experiments performed under laboratory conditions.

The three curves indicate the difficulty of collecting samples of DPM in the mine atmosphere to determine the amount of DPM present. The size distribution curve for DPM shows two modes with the submicrometer mode being dominant. The second mode, an agglomeration mode, generally constitutes less than 10% of the DPM mass. On the other hand, coal is mostly material greater than 1 μm in size. Unfortunately, when the coal and DPM aerosols are collected in the same atmosphere, the differentiation of DPM and coal by a size-selection process is difficult due to the overlap of the DPM and coal size distributions. However, a size separation at 0.8 μm provides a reasonable estimate of the coal and DPM portions of the total or respirable

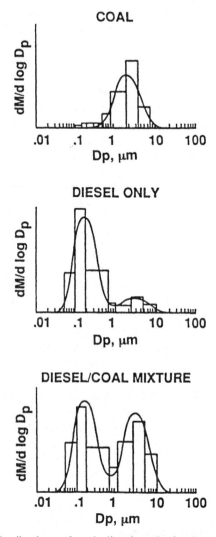

FIGURE 4.9 Size distributions of coal, diesel, and mixed aerosols measured under laboratory conditions. (Cantrell and Rubow, 1991; used with permission.)

aerosol. This method is utilized because no simple or inexpensive analytical techniques are currently available.

DPM Control Technology

While permanent regulations placing limits on DPM have not yet been instituted, the development of technology to control the amount of DPM in the mine atmosphere has been pursued for a decade. Both U.S. and Canadian firms have been actively engaged in testing and improving methods for control of diesel emissions. Two methods of controlling DPM have been rather successful. The first of these is the use of a ceramic filter on the diesel exhaust of underground equipment (Bickel et al., 1992). A simple concept, the ceramic element is constructed in a honeycomb fashion with many thin, porous walls that are capable of trapping DPM particles (Fig. 4.10). This ceramic element is placed in the exhaust as close to the engine as possible so that the DPM will eventually combust and regenerate the filter. To enhance the capability for doing this, the filter is normally catalyzed to reduce the temperature at which the regeneration process is initiated. Filtering efficiencies of greater than 90% have been achieved; however, the efficiency is significantly lower under some engine operating conditions. While simple in concept, ceramic filters are expensive, difficult to maintain, and currently suitable only for noncoal mines.

The second technology in current use has been successful for coal mine applications. This is the use of a low-temperature disposable filter in the diesel exhaust stream (Ambs and Hillman, 1992). These filters consist of a pleated filter media similar to paper, which is encased in a metal frame for proper sealing into the exhaust system. The filters have been used on diesel haulers, coupled with a waterbath scrubber system to keep the exhaust temperatures at a safe level. These filters can achieve efficiencies of 90% when properly used, greatly reducing the DPM in the mine air. The main disadvantage is their relatively short life of a shift or so. Additional types of filters that can be cleaned and reused have also been investigated. Only the disposable filter has been widely used, however.

Because of the governmental movement to clean up the exhaust of highway diesels, additional DPM control technology may be forthcoming. Of particular interest are electronic engine controls and catalytic emission-con-

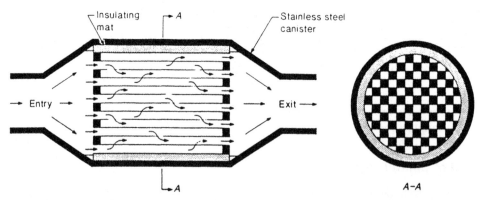

FIGURE 4.10   Schematic of a ceramic diesel particulate filter (Ambs and Hillman, 1992).

trol systems. This technology may be quite helpful in the future to provide control over the amount of DPM that occurs in the mine environment.

4.10 SOURCES OF DUSTS IN MINES

To control dust levels in mines, the sources of dust must be known. In many cases, the major sources will be known based on a study of dusty locations. However, it is important to evaluate all sources of dust as each may contribute significantly to the total dust level. In some cases, one source or another may be easier to control or possess a strategic reason (e.g., silica content) for control.

Underground Coal Mining

Specific sources of dust vary with the type of mining systems utilized. Levels of dust generation are best known in coal mining as the most stringent dust standards are applied here, and extensive research has been performed to meet those standards. In the tabular lists that follow, potential dust sources in underground coal mines are outlined. If an operation produces or creates a dust through some disintegration process, it is termed a *primary* dust source. If it agitates or redisperses dust already created, it is termed a *secondary* source. Because the existence of quartz dust in coal mines is often due primarily to one or more sources, the contribution to quartz generation is also listed. For purposes of differentiating between different dust contributions, a plus ($+$) signifies a major source, a minus ($-$) signifies a minor source, and a zero (0) signifies a negligible source.

Longwall mining:

Operation/Equipment	Primary Source	Secondary Source	Quartz Source
Shearer/plow	+	+	+
Stageloader–crusher	+	–	–
Roof supports	–	0	–
Belt conveyor	0	–	0
Outby equipment	0	–	0

Continuous mining:

Operation/Equipment	Primary Source	Secondary Source	Quartz Source
Continuous miner	+	–	+
Roof bolter	+	+	+
Shuttle cars	0	+	–
Feeder–breaker	–	–	–
Belt conveyor	0	–	0
Outby equipment	0	–	0

With strict regulation of dust in coal mines currently being emphasized, it may be necessary to control each of these sources to achieve overall success in dust control efforts.

Underground Noncoal Mining

In underground metal and nonmetallic mines, the sources of dust are not as easy to identify because of the variation in mining practices. Regulations are not as difficult to meet in most cases, but quartz content of the ore may be more of a concern in some segments of the industry. Statistical analysis (Watts and Parker, 1987) shows that 16.6% of the respirable dust samples taken during 1980–1984 in underground mines exceeded the TLV for respirable quartz. This is indicative of the need for enhanced dust control in underground mines in general and in mines with ores containing quartz in particular. The tendency to exceed the TLV occurred in all segments of the industry but was highest in the nonmetallics industry.

The following list provides some idea of where dust is generated in underground noncoal mines:

Operation/Equipment	Primary Source	Secondary Source
Blasting	+	+
Cutting, continuous mining	+	−
Drilling, roof bolting	+	0
Mucking, loading	−	+
Slushing	−	+
Drawing chutes	−	+
Loading skips	−	+
Blowing holes	0	+
Crushing	+	−
Belt transfer, discharge	0	−
Haulage	0	−
Timbering	0	−
Barring down	0	−

Quartz content in dust occurs fairly often in noncoal mines. Dealing with a quartz dust problem requires that the quartz-rich sources be identified so that they can be controlled.

Surface Mining

In surface mines, dust problems differ from those underground, particularly in terms of the sources of dust and in the exposure of workers to the dust. Sources of surface mine dust often exist out of the breathing atmosphere of the workers. This condition often brings on a sense of complacency and a lack of diligence in pursuing dust control. Some of the more important sources of dusts in surface mines are listed below.

Operation/ Equipment	Primary Source	Secondary Source
Drilling	+	−
Blasting	+	−
Draglines	−	+
Loaders, shovels	−	+
Haulage roads	0	+
Crushers	+	−
Belt conveyors	0	−
Waste, leach dumps	0	−
Tailings ponds	0	−

As indicated, surface mines can have many sources of quartz dusts, but many of the problems associated with surface mines occur in drilling operations, both in coal and noncoal mining. For this reason, significant attention must be paid to dust control in drilling operations.

Other Dust Sources Related to Mining

While the primary discussion here is oriented toward mining operations and their dust problems, the mine ventilation engineer may also be called on to provide atmospheric control in milling and bagging operations. A report that summarized dust exposures for 1980–1984 in various minerals operations (Watts and Parker, 1987) pinpointed milling and bagging occupations as having the highest percentage of dust samples above the TLV. The data indicate a strong need for dust control in processing operations as well as in mining operations.

4.11 AEROSOL CONTROL TECHNOLOGY

The control of aerosols, particularly dust, in the mine atmosphere is one of the most important objectives of the ventilation engineer because of its relationship with the health of workers. There are many control technologies that apply in mining (Section 1.2). Some of the strategies and methods for controlling dust in mineral operations appear in Table 4.4. The strategies are listed from the most desirable (prevention) to the least desirable (dilution). Some of the common methods for accomplishing control strategies are listed with an assessment of their cost and effectiveness. It should be noted, how-

TABLE 4.4   Basic Dust Control Strategies and Methods

Strategy	Method	Expense	Efficiency
Prevention	Water/steam infusion	High	Moderate
	Foam infusion	High	Moderate
	Wet drilling	Low	High
Removal	Dust collectors—wet	Moderate	High
	Dust collectors—dry	Moderate	High
Suppression	Water sprays	Low	Moderate
	Wet cutting	Low	Moderate
	Foam	Moderate	Moderate
	Cutting variable optimization	Low	Moderate
	Deliquescent chemicals	Moderate	Moderate
	Rock dusting	Moderate	Moderate
	Waterjet-assisted cutting	High	Moderate
Isolation	Enclosed cabs	Moderate	Moderate–high
	Enclosed dust generation	Moderate	Moderate
	Exhaust ventilation	Low	Moderate
	Blasting off-shift	Low	Moderate
	Control of airflow		
	Separate air split	Low	Moderate
	Sprayfans	Low	Moderate
	Air curtains	Moderate	Moderate
	Control of personnel location		
	Remote control	Low	Moderate
	Unidirectional cutting	High	Moderate
Dilution	Main ventilation stream	Moderate	Moderate
	Local ventilation dilution	Low	Moderate

ever, that other methods may exist and that the cost and effectiveness of the methods are a function of the application and can vary significantly.

The cardinal rule of dust control is to prevent dust from occurring. This can be accomplished by altering the mined material or the mining process to keep dust from forming, or to suppress dust immediately to prevent it from becoming airborne. Once the dust becomes airborne, other techniques further down the list in Table 4.4 must be utilized. However, they should be deferred until necessary as they are often less effective and more expensive. It should also be noted that several control procedures can be used in combination for better dust control.

Various control methods that are effective in mine dust control are presented in the following paragraphs. Most of the discussion is oriented toward controls that are useful in mining and mineral processing operations.

Water-Based Control Measures

The use of water is one of the most widespread and effective methods of controlling dust in mining operations. Many variations exist in the way that water is used to control dust. The most fundamental of these utilizes water or steam for infusion of the mineral deposit prior to the mining process so that the mineral is wetted and less likely to become airborne. This has been applied primarily in relatively permeable coal seams and has a rather long history of use in European mines. Cervik (1977) reports that water infusion in European coal mines has been 50–95% effective in reducing dust. In the United States, water infusion has been applied periodically on a limited basis to both room and pillar mines and to operations using longwall. The success in these mines has been toward the lower end of the scale, with typical efficiencies around 50% (Cervik, 1977; McClelland et al., 1987).

The infusion process is normally accomplished through boreholes in the coal seam drilled ahead of the mining operation. Figure 4.11 shows one plan for infusing the coal ahead of a room-and-pillar development operation. This process can be disruptive in such an operation because infusion must be repeated numerous times as the section advances, potentially interrupting the mining process. The infusion process in a longwall is less problematic, with drilling and infusion conducted in the gateroads. Water infusion is normally expensive, and this is perhaps the most serious drawback to the method.

Other methods for using water to control dust levels in mineral operations

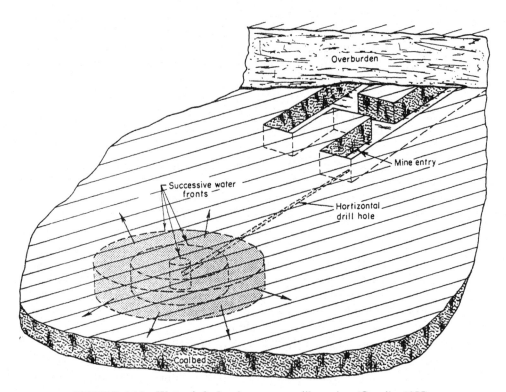

FIGURE 4.11 Water infusion in a room–pillar mine (Cervik, 1977).

are more universal, modestly priced, and applicable to a variety of sources. The first of these is the use of water in drilling, cutting, and continuous mining operations. Wet drilling became the norm in many mines after the problem of silicosis was publicized in the 1930s. The method can achieve efficiencies of nearly 100% in many situations. In underground mines where soluble ores are extracted, the widespread use of water may not be as feasible because excessive moisture in the mineral product must be avoided. In this situation, water is applied under controlled conditions. For example, Page (1982) found that using foam through the drill rod reduced total dust on a gypsum face by 95%, while water mist through the drill rod reduced total dust by 91%. Use of a controlled amount of water on the cutter bar of a cutting machine in similar mines was shown to reduce total dust by about 60 to 70%.

Once dust gets in the air, use of water becomes less efficient as the particulate matter becomes more difficult to wet. A variety of sprays are available as an aid in dust control. The most common types are shown in Fig. 4.12 (Jankowski, 1995). A general description of each type follows:

1. *Solid Stream* (Fig. 4.12a). This spray nozzle provides a solid stream of water with a spray angle of 0°. It is used as a shearer drum spray on longwalls, including duty as a pick–point spray.

2. *Flat Spray* (Fig. 4.12b). This spray provides a narrow, approximately rectangular, spray pattern with spray angles of up to 110°. It provides large droplets and is used to wet coal before belt discharge points and to clean coal particles from the side of the belt in contact with the drive and idlers.

3. *Full Cone* (Fig. 4.12c). The full-cone nozzle produces a uniform distribution of large droplets over a circular area, finding use in wetting minerals on conveyor belts and for cleaning the nonconveying side of conveyor belts.

4. *Hollow Cone* (Fig. 4.12d). A hollow-cone spray is characterized by a conical spray pattern with a hollow center and with most of the droplets concentrated around the perimeter. It is utilized at transfer points, on shearers, and in crushers to knock down dust that has become airborne.

5. *Atomizing Spray* (Fig. 4.12e). Atomizing sprays are also useful for reduction of airborne dusts, using higher pressures to produce very small water droplets. The resulting mist is less likely to create a flowline pattern that will move dust particles rather than collide with them, achieving greater efficiency in dust knockdown. In some mining applications, atomizing sprays are more effective if they are generated by two-phase (air–water) spray systems.

6. *Venturi Spray* (Fig. 4.12f). The venturi nozzle is a conventional nozzle mounted inside a venturi shroud. The spray entrains air in the shroud and results in the movement of air with the water droplets, thus provid-

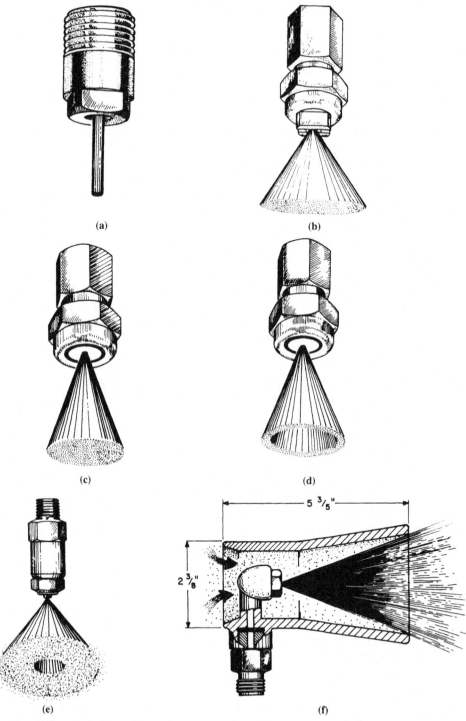

FIGURE 4.12 Principal types of water sprays used in mine dust control (Jankowski, 1995).

ing the capability of aiding in methane removal at a face as well as keeping dust down.

The efficiencies of a number of different spray nozzle types are shown in Fig. 4.13 (Shirey et al., 1985). While the graph gives an indication of the comparative efficiency of the different types of sprays, the overall efficiency of sprays is more important. This has been reported at 20–60% for typical coal mining situations, with an average efficiency of about 30% (Courtney and Cheng, 1977). Further, while spray technology is constantly improving, the use of sprays alone may not be sufficient to adequately address all dust problems in a mining system. As a result, they are often used as a supplement to other methods.

The use of high-pressure water jets for cutting coals and other earth materials is a technology that is becoming more useful and practical. For many mining operations, the cost of breaking coal with high-pressure jets is greater than simply using carbide cutting tools. However, the ability to suppress dust using water jets is quite good, with 85% reduction reported on a longwall shearer (Taylor et al., 1986). As a result, high-pressure jets have an important industrial hygiene advantage over competitive cutting technology. This may result in the use of more water jet cutting in the future, particularly in coal mining.

As a means of increasing the effectiveness of water, foams and surfactant chemicals are sometimes used. Foam generators can be used to contain or envelop dust on continuous miners, shearers, drills, cutting machines, and crushers. While the use of foam is effective, the cost of using foam is often cited as the reason it is not more widely used (Bhaskar and Gong, 1992). Surfactants (wetting agents) also fall in the category of supplemental means for improving dust control using water. The effectiveness of surfactants varies, and there is disagreement as to whether the agents act mostly to prevent dust from becoming airborne or improve the collection of the dust after it becomes airborne (Shirey et al., 1985). It is agreed, however, that they can be of help in the dust-control process.

Dust Collectors

Dust collectors have been increasingly used in industry, including mining and mineral processing, over the last several decades as both health concerns about dust and social consciousness have risen. Table 4.5 lists various types of dust collectors that may find application in mining operations. It provides some of the characteristics of dust collectors under average or design conditions.

As far as application is concerned, wet scrubbers and filters have been rather widely applied in underground mines, and cyclones are common in surface mines. Wet scrubbers have often found use in the processing of minerals as well as in underground mining (Divers and Janosik, 1978; Or-

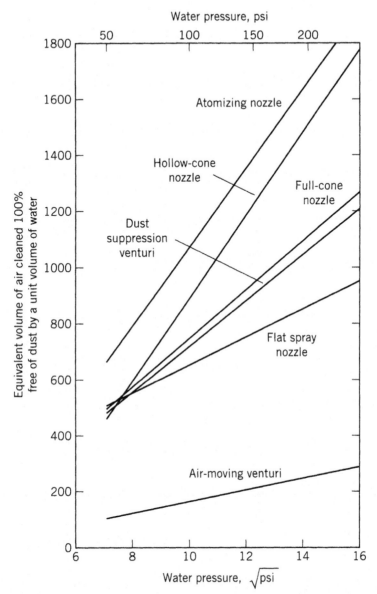

FIGURE 4.13 Airborne dust capture efficiency for six types of spray nozzles (Shirey et al., 1985).

TABLE 4.5 Characteristics of Some Dust Collectors

Principle and Type	Concentration Handled	Typical Collection Efficiency, Wt%				Head Loss (in. water)	Power Required	Remarks
		1 μm	1-5 μm	5-10 μm				
Settling: settling chamber	Medium	10	10	10		0.2-0.6	Low	Large space required; collects coarse dust only; precleaner
Inertial								
High-throughput cyclone	High	10	10	40-80		0.6-3.0	Low	Cheap; good efficiency at 40-80 μm; precleaner
High-efficiency cyclone	Medium	10	10-40	80-95		2.0-6.0	Medium	Good efficiency at 5-10 μm
Wet scrubbers								
Wet cyclone	Medium	10	20-80	90-95		2.0-5.0	Medium	Uses centrifugal forces
Wetted fan	Medium	20-60	60-95	95-98		1.0-5.0	Medium	Relatively small
Wetted brush	Medium	50-70	60-95	95-98		3.0-5.0	Medium	Relatively small

Flooded fibrous bed	Medium	50–85	90–98	98–99.9+	4.0–6.0	Medium	High efficiency, modest cost
Venturi	Medium	80–90	99	99	30–40	High	Used mostly for fumes
Water-powered scrubber	Medium	...........74–84% overall............				Low	Low cost and efficiency; for spot application
Electrostatic precipitator	Low	80–99	80–99	99	0.2–0.8	Low to medium environment	Used in hot gases, not for use in a methane
Filters:							
Cloth or paper cartridge	Medium	99	99.9+	99.9+	2.0–6.0	Medium	Used with roof bolters
HEPA[a]	Low	99.9+	99.9+	99.9+	1.0–2.0	Low	Highest efficiency, used to particulate-filter radioactive dusts

Source: Adapted from Goodfellow (1985), Divers and Janosik (1978), and Organischak et al. (1983).

[a] High-efficiency particulate aerosol.

ganischak et al., 1983; Divers and Cecala, 1990). For underground application, the most prolific success story has been the use of flooded-fibrous-bed scrubbers, often referred to as *flooded-bed* scrubbers (Campbell, 1988). These scrubbers are now widely used on continuous miners and are highly successful in reducing the amount of dust that is released at the working face in coal mining operations. Efficiencies for removal of respirable dust typically run from 90 to 95%, providing that the screens are washed relatively often (Colinet et al., 1990).

The general layout of a flooded-bed scrubber is shown in Fig. 4.14. Water sprays are located near the inlet and keep a steady spray of water on the surface of the scrubber screen. The screen is made of multiple layers of stainless steel or synthetic fiber mesh in a stainless-steel frame. The impaction of the dust and water droplets on the screen surface wets the dust particles, which are then collected in the sump. Additional water droplets and dust particles are eliminated from the airflow by the demister, which collects the liquid and solid aerosol by impaction on the demister surfaces. The specific layout of the scrubber on a continuous miner is shown in Fig. 4.15. The scrubber is normally designed to gather dusty air from both sides of the cutting head, discharging the clean air out one side of the miner where it can be swept by airflow behind a brattice. Some continuous miners allow for discharge of the clean air from either side of the continuous miner.

The second dust collector type that has made a major difference in reducing dust levels in underground mines is the exhaust filter system used for roof drilling in underground coal mines. In these systems, roof drills are

SCRUBBER CROSS-SECTION

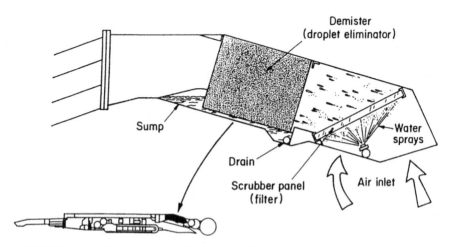

FIGURE 4.14 Cross section of a flooded-fibrous-bed scrubber (Anon., 1983).

MACHINE MOUNTED DUST COLLECTOR

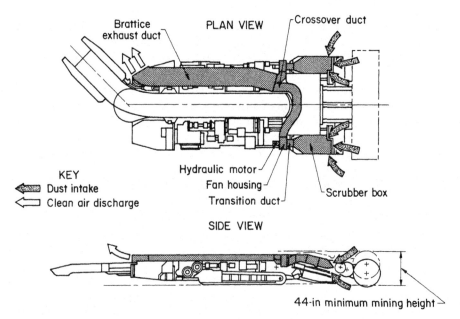

FIGURE 4.15 Flooded-fibrous-bed scrubber mounted on a continuous miner (Anon., 1983).

provided with suction through the drill rods, which draws the drill cuttings through the rod and into a dust collection system, consisting of a precleaner and a high-efficiency cloth or paper cartridge filter, and then through the fan. Most of the dust is gathered in the dust box and is removed once or twice a shift by the operator. The collection of this dust from the roof bolter is important since it is usually of high quartz content. The suction-type drill collection system works rather efficiently, but only if it is well maintained and the operator empties the dust box with care.

In surface mines, much airborne dust is generated in the drilling operation, which is often performed dry. Dust collection on these drills is often carried out by an exhaust system that consists of a shroud around the blasthole collar, a fan to exhaust the dusty air away from the collar, and a cyclone-type dust collector to gather much of the dust. As mentioned earlier in this chapter, drilling operations are often the source of problems because of poorly maintained shrouds, inappropriate location of the dust collector exhaust, and lack of interest by operators in reducing ambient dust levels. Because of the threat of silicosis in many drilling operations, this will likely remain an area of concern in the future.

Local and General Controls

In many situations, the use of water and dust collectors for control of dust in the mine environment is either impossible or impractical. Under these conditions, local controls of a number of types are often both useful and successful. Some of these local and general controls are outlined here.

While the use of water sprays and dust collectors has become common on coal mine faces, the optimization of cutting parameters is also an important consideration in the reduction of dust on both continuous miner and longwall faces. This includes optimizing the rotational speed of cutting drums, depth of cut on an individual cutting bit, spacing of bits on the drum, direction of rotation of a shearer drum, shape and other construction features of cutting bits, and arrangement of water sprays on the cutting drums. In addition, the web depth and the internal design of the drums in longwall shearing play a significant part in determining the amount of dust that is generated on a longwall face. When considering the longwall mining method, both the equipment manufacturer and the user must strive jointly to achieve the lowest possible dust generation rate.

Other local controls that merit attention are air curtains and cabs. Air curtains are streams of air directed along designed paths that form a wall or barrier to help restrict the flow of contaminants into a given space, such as the space around an operator's seat. Air curtains have been used in a number of applications of air and contaminant control outside of mining. They have not been widely used in mining but may find increased use in the future. Enclosed cabs are a second method of attempting to isolate personnel from the contaminants that may exist in the adjacent environment. Cabs have been used on surface equipment to keep operators out of a dust cloud, particularly on large drills. To be effective, these cabs should provide air conditioning and a flow of fresh filtered air under positive pressure.

Another control technique is the use of exhaust ventilation systems to remove dusts and other contaminants from the breathing atmosphere of personnel. In these methods, the logic is to remove the contaminants and isolate personnel from them. Both face and auxiliary ventilation systems that exhaust the dust fall in this category. Additional logic on isolation of dust is discussed in Chapter 11.

One additional control procedure of a general nature is the use of deliquescent and binding chemicals to prevent dust from becoming airborne. These chemical treatments are normally used on haul roads, tailings ponds, railcars, stockpiles, and other areas where equipment movements or wind conditions may result in the raising of dust from surface locations (Olson and Veith, 1987). Deliquescent chemicals, such as calcium chloride and magnesium chloride, are used because they absorb moisture from the atmosphere and keep the surfaces from becoming dry and dusty. Binding chemicals such as petroleum derivatives or resin compounds exhibit the tendency to bind individual particles to others, thus reducing the likelihood of particles becoming airborne. These chemical treatments are used both underground and on the surface, but surface use predominates because of the number of possible applications.

Dilution by Ventilation

When other methods fail to achieve the dust concentrations that are desired in the working areas of a mine, it may be necessary to increase the ventilation in order to dilute the dust to TLV levels. This can often be achieved by using or improving auxiliary ventilation or by increasing the main ventilation quantity. When changing the ventilation quantity, both the minimum and maximum ventilation velocities should be considered. It is necessary to maintain 13–40 fpm (0.07–0.20 m/s) to provide turbulent flow in a typical mine opening (Section 5.4). However, federal regulations require that at least 60 fpm (0.3 m/s) be maintained in the working areas of coal mines to insure that the flow will adequately sweep methane and dust from the working faces. In some mine openings, such as longwalls, the effective face velocity is thought to be 400–600 fpm (2.0–3.0 m/s), with increases in dust concentration occurring above 600 fpm (3.0 m/s) because of the tendency of high velocities to raise settled dust (Jankowski et al., 1993). This places an upper limit on the air velocity that can normally be utilized for dust control purposes. However, higher velocities are used on longwalls when methane dilution governs.

Quantity requirements for dilution of gaseous airborne contaminants were discussed in Chapter 3. For use with dusts and other aerosols, the dilution equation may be modified as follows, expressed in SI units (Mateer, 1981):

$$\frac{(G + BQ) - xQ}{(G + BQ) - x_0Q} = e^{-(Q/Y)\tau} \tag{4.10}$$

where G is dust generation rate in mass per unit time (mg/s); Q is quantity of ventilation air in m^3/s; Y is volume of the space (m^3); τ is time in seconds; and B, x, and x_0 are airborne dust concentrations (mg/m^3). The dust concentration B represents the concentration in the intake air, x in the mixture at any time τ (may be set to the TLV value), and x_0 in the mixture at time $\tau = 0$.

Equation 4.10 may be simplified as time approaches infinity, resulting in this relationship for quantity of diluting air:

$$Q = \frac{G}{TLV - B} \tag{4.11}$$

This provides the steady-state equation for dust dilution. It is used whenever the dust generation rate is constant or where an average value is known.

(Equations 4.10 and 4.11 can also be used with English units if both sides of the equation contain consistent units.)

Example 4.4   A longwall face has been upgraded for better production and is experiencing dust concentrations above the regulatory limit of 2.0 mg/m^3. The water sprays, shearer cutting parameters, and feeder–breaker dust-control systems have been improved, but respirable dust is still being generated at the rate of 2.5 g/min (average) along the face, and the intake air contains a dust concentration of 0.4 mg/m^3. What quantity of air will be required to reduce the dust downwind of the shearer to 2.0 mg/m^3? If the cross-sectional area of the longwall face averages 8 m^2, what face velocity will this quantity require? Is this velocity above the recommended maximum air velocity?

Solution: Using Eq. 4.11 and expressing concentrations in units of mg/m^3, the following quantity can be determined:

$$Q = \frac{2500}{2.0 - 0.4} = 1560 \text{ m}^3/\text{min} = 26.0 \text{ m}^3/\text{s} \text{ (44,070 cfm)}$$

On the basis of this quantity and the cross-sectional area of the longwall face, the average face velocity will be

$$V = \frac{26.0}{8} = 3.25 \text{ m/s (637 fpm)}$$

This velocity is higher than the 3.0-m/s (600-fpm) maximum suggested above and may itself be the source of airborne dust.

Multiple Controls: Longwall Case Study

Longwall mining provides an interesting case study for dust control; it is an area where an intense effort has been made by mine operators and dust researchers to provide better control over the dust generated. However, while attempts were being made to control dust, the productivity of the longwall mining method was increasing to many times its level of 20 years previous. Thus improvements in dust control were not enough to provide control over the dust on longwall sections. Very significant progress was made, however, by emphasizing dust reduction for every controllable dust source.

Today's longwall operation is likely to use the following technology in its efforts to control dust (Jankowski et al., 1993; Haney, 1995):

1. Ventilation air velocity of 400–450 fpm (2.0–2.3 m/s) to keep dust concentrations down.

2. Unidirectional cutting (in many cases) to reduce the portion of the shift when shield operators must be downwind of the shearer.

3. Gob air curtains in the headgate to keep the ventilating airstream moving across the longwall and out of the gob.

4. Cutout curtains in the headgate to reduce the dust that occurs when the shearer cuts into the headgate entry.

5. Belt entry air coursed across the face to improve the ventilation quantity.

6. Enclosed crusher–stageloader with 80-psi (0.55-Pa) sprays or a small dust collector.

7. One spray nozzle per bit on the drum with sprays operated at 80 psi (0.55 Pa).

8. Air-moving jetsprays on the shearer body to keep dust away from the operator. This is known as a shearer–clearer system and operates at 125 psi (0.86 Pa).

9. Shields are sprayed automatically during movement or manually to suppress dust generation during shield movement.

10. Remote control is used on the shearer and/or shields to keep the shearer operator and jacksetters out of the worst dust.

Measures used to suppress dust on longwalls vary considerably. However, one thing is certain. To be successful in keeping dust down on a modern longwall, a multiple-control approach is necessary. Only through such an approach can this difficult dust problem be kept at bay.

4.12 PERSONAL PROTECTION DEVICES

While it is generally considered the responsibility of the employer to keep dust levels under the appropriate TLV, there are times when this may be impossible. When this occurs, workers should do whatever is necessary to protect themselves against dusts or other aerosols in the breathing environment. If the only method of further reducing the person's exposure to the dust is to wear a respirator or other personal protection, then it is prudent to do so.

Personal protection from dust and other aerosols can be achieved through the use of respirators, powered positive-pressure respirators, and air helmets. A respirator is a device to protect the respiratory tract from any irritating or toxic gases, fumes, smokes, or dusts in the environment. Respirators come in many forms, depending on the types of aerosol or gas from which the human body must be protected. For protection from dusts, the simplest type of respirator is a filter mask that fits tightly over the mouth and nose to filter the air taken in by breathing. If a disposable filter mask is used, only about 60% of the dust is filtered out because leakages around the filter allow the escape of significant quantities of dust past the mask (Divers and Cecala, 1990).

For improved efficiency, disposable masks are replaced by respirators with a replaceable filter. These masks are designed to fit better around the nose and mouth to more effectively seal out dusty air. One study (Cole, 1984) indicated that such respirators provided 80–92% efficiency in reducing dust exposures. However, filters of this type are not always so efficient. It is important that the wearer be dedicated to proper fit and use of the respirator; otherwise, the respirator efficiency goes down significantly.

If very dusty conditions exist, such as in dusty processing plants or bagging operations, a powered positive-pressure respirator may be used. A typical respirator of this type is shown in Fig. 4.16. The unit consists of a pump and filter located on the person's belt, which is connected to a face mask. The pump supplies filtered air under positive pressure to minimize the amount of dusty air that leaks past the face mask. Using this type of respirator, efficiencies of more than 90% can be achieved (Divers and Cecala, 1990).

The final personal protective device to be discussed here is the air helmet

FIGURE 4.16   Positive-pressure respirator (Divers and Cecala, 1990).

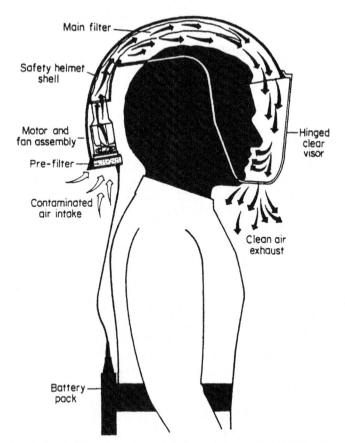

Main filter

Safety helmet
shell

Motor and
fan assembly

Pre-filter

Contaminated
air intake

Hinged
clear
visor

Clean air
exhaust

Battery
pack

FIGURE 4.17   Diagram of an air helmet (Divers and Cecala, 1990).

(Fig. 4.17). This unit consists of a helmet and a battery pack worn on the belt. The helmet is somewhat cumbersome but provides both cranial and respiratory protection. The battery powers a fan in the helmet that provides filtered air to the plastic face enclosure. If the face enclosure fits the contour of the face fairly closely, dust exposure reductions greater than 90% are possible (Divers and Cecala, 1990). Air helmets are used for many dusty industrial jobs and serve well for personal protection in bagging operations, dusty processing plants, and longwalls. Helmets have several disadvantages such as cost, bulkiness, and reduced efficiency in high-velocity airstreams. However, they can significantly reduce the dose of dust to which a worker is subjected. The cost may be quite small when compared with the cost of not providing protection. Conditioned as well as filtered air can be provided by microclimate cooling systems (Section 17.5). These systems are not commonly used in dust control but are effective at reducing wearer exposure when used.

4.13   MEDICAL AND LEGAL MEANS OF DUST CONTROL

The Federal Coal Mines Health and Safety Act of 1969 serves as an example of both medical and legal means of reducing the health effects due to dust in the coal mine atmosphere. The regulations inherent in that act required that respirable dust levels be reduced in all coal mines to 3.0 mg/m³ by June 1970 and to 2.0 mg/m³ after December 1972. Because 1968/69 levels of dust in continuous miner and cutting machine faces of coal mines averaged about 6.5 mg/m³ (Jacobson, 1972), these regulations challenged the ability of U.S. coal mining companies to improve their dust-control practices.

After initial reluctance to accept the new standards, mining companies began to work with the Mine Safety and Health Administration and the USBM to improve their dust-control practices, and rapid progress was made in reducing dust levels in coal mines. Today, while production is much higher in coal mines, dust levels are well below the regulated concentration limits on average, and only longwall sections have difficulty in meeting dust standards. On longwalls, productivity improvements have been an overwhelming challenge to dust-control technology. The mismatch may continue, although longwall automation is expected to greatly reduce this problem.

The 1969 Act also provided for a medical means to control dust. It mandated lung x-rays for new miners and provided for the right of miners to opt for less dusty jobs if x-rays indicate that they are suffering from CWP. The objective of this medical measure was to arrest the progression of the disease through reduction in the dose to which the affected miner was subjected. Because billions of dollars have already been paid in black lung benefits to miners who suffer from black lung, the results of the legislation should be favorable to society as well as the miner.

PROBLEMS

4.1   Assume that a blast in a sublevel stoping operation puts dust particles in the air. If no ventilation air is moving through the stope, how long will it take 5-μm particles with a specific gravity of 2.60 to settle out of the stope air if it is 200 ft (61 m) high? How long would it take for 1-μm particles?

4.2   Compare the settling time of the following particles in a quiescent mine atmosphere. Assume that the particles are released at a height of 2 m.
  (a)   A rock particle with a diameter of 3 μm and a specific gravity of 2.40.
  (b)   A coal particle with a diameter of 3 μm and a specific gravity of 1.40.
  (c)   A diesel particle 0.3 μm in diameter with a specific gravity of 0.50. Assume that the mean free path for air is 0.066 μm.

4.3 A combined dust survey and time study is made in a mining stope. The following data are obtained:

Start Time	End Time	Operation	Respirable Dust Concentration (mg/m^3)
8:00	8:20	Barring down	1.0
8:20	10:45	Scraping	2.1
10:45	11:30	Timbering	0.4
11:30	12:00	Lunch	0.2
12:00	2:20	Drilling	2.9
2:20	3:00	Loading holes	0.3
3:00	3:15	Blasting	3.5

Determine the TWA concentration in the stope over the shift and the total yearly respirable dust dose that a person working 220 days per year would be exposed to in the stope.

4.4 A gravimetric dust sampler is used to monitor the shift exposure of a miner operating a continuous miner in a coal seam. Sampling time is 8 h, and the total weight of the dust is 1.7 mg. If the sampler operates at a rate of 2.0 L/min, what is the average shift exposure? Does this exposure exceed the TLV for respirable coal dust?

4.5 A gravimetric dust sampler, operated at a sampling rate of 2.0 L/min, is operated for 8 h in a coal mine section. If a total mass of 1.30 mg of dust is gathered, and 120 μg of that total is found to be quartz, does the sample meet the TLVs for both coal mine dust and quartz?

4.6 A miner is working in a mine atmosphere that contains an average respirable coal dust concentration of 1.2 mg/m^3. If the miner works 8 h per day for an average of 240 days per year for 40 yr, what is the total dose over the working life of the miner?

4.7 Assume that a person is working 8 h a day in an atmosphere containing 0.5 mg/m^3 of respirable dust for 220 days/yr. In a year's time, how much dust is respired into the body during the person's working hours? If the person breathes at an average rate of 600 cm^3/s, what mass of dust reaches the alveolar regions of the lungs, and what mass is deposited in the lungs? Assume that 65% of the dust reaches the alveolar region and that 35% is deposited.

4.8 A loading operation has been sampled and found to have a downwind concentration of 1.8 mg/m^3 of respirable dust. Because of the quartz content of the dust, the mine environmental engineer wishes to reduce the concentration to 0.5 mg/m^3 using an increased ventilation quantity. If the loading location has an airflow of 8000 cfm (3.78 m^3/s) of air containing 0.2 mg/m^3, how much air must be provided to meet the new dust concentration objective? Assume that the dust concentration of the intake air does not change.

4.9 A longwall face is experiencing a high average dust concentration of about 2.8 mg/m^3 downwind of the shearer. If the intake air consists of 50,000 cfm (23.6 m^3/s) of air containing 0.25 mg/m^3 of dust, how much air must be coursed through the face to reduce the dust concentration downwind of the shearer to 2.0 mg/m^3? If the cross-sectional area of the longwall face is 77.5 ft^2 (7.2 m^2), what is the average velocity of the air through the face area after the ventilation is increased? Is this an acceptable velocity?

PART III

MINE VENTILATION

CHAPTER 5

Airflow through Mine Openings and Ducts

As defined in Chapter 1, the function of controlling air movement, as well as its direction and magnitude, is termed *quantity control*. The principal process of total air conditioning concerned with control of air circulation is *ventilation*. The forerunner of modern mine air conditioning was mine ventilation, and this familiar term is still used loosely to refer to many underground atmospheric-control functions.

Basically, the purpose of ventilation is to supply enough air for human comfort and product needs. In mines, where comfort air conditioning is uppermost in importance, ventilation requirements for the sustenance of human life are modest—about 20 cfm (0.01 m³/s) per person (Section 3.3). However, the requirements soon multiply, because ventilation serves many functions. In the control of both the chemical and physical quality of the air, clean fresh air must be supplied and contaminants (gas, dust, heat, and moisture) removed by the ventilation system. Considering the entire mine, ventilation requirements far outstrip the minimal 20 cfm (0.01 m³/s) per person, usually exceeding 200 cfm (0.1 m³/s), and on occasion, 2000 cfm (1 m³/s) per person. Circulation of 10–20 tons of air per ton of mineral mined is not unusual today under adverse conditions.

Without question, ventilation is the most useful process of total mine air conditioning. Because of the extent of the demand and the distance that air must travel from the surface to the working face, ventilation may also become both complicated and costly. Mine passages constitute the openings through which the air must travel, and the path may be long and tortuous.

An understanding of the theory of airflow requires a knowledge of fluid mechanics. Mine ventilation, as stated in Chapter 1, applies the principles of fluid dynamics to the flow of air in mine openings and ducts. Even though air is a gas, and hence a compressible fluid, in nearly all mine ventilation work, air may be treated as incompressible. This is an important simplification for calculations.

5.1 ENERGY CHANGES IN FLUID FLOW

General Energy Equation

Mine ventilation is normally an example of a *steady-state* process, that is, one in which none of the variables of flow changes with time. Transitions

135

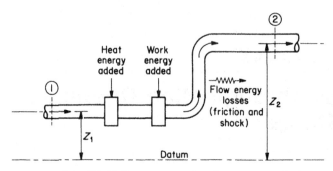

FIGURE 5.1   Fluid-flow system showing energy relations.

and losses in energy are involved in such a process, and it is important to understand their nature and to be able to express them mathematically. Energy changes are basic to the calculation of the mine quantity and head, one of the ultimate objectives of mine ventilation engineering. An expression relating the energy variables may be developed as follows [see a fluid mechanics textbook for the complete derivation (e.g., Munson et al., 1994)].

The total energy at any section in a moving fluid consists of the sum of the internal, static, velocity, potential, and heat energies at that section. Assume a real fluid moving in a conduit, and consider the energy changes that occur between any two sections in the system (Fig. 5.1). The heat change is generally negligible compared to the other terms, except in deep mines or ones naturally ventilated, and the addition of mechanical energy (e.g., by a fan) is usually considered separately. Omitting these terms for now, the total energy at section 1 equals the total energy at section 2, plus the flow energy losses occurring between 1 and 2, or

$$\text{Total energy}_1 = (\text{total energy})_2 + (\text{flow energy losses})_{1-2} \quad (5.1)$$

Substituting expressions for the various energy terms and disregarding the minor change in internal energy, the following general energy equation for fluid flow results:

$$\frac{p_1}{w} + \frac{V_1^2}{2g} + Z_1 = \frac{p_2}{w} + \frac{V_2^2}{2g} + Z_2 + H_l \quad (5.2)$$

where p is pressure, w is specific weight, V is velocity, p/w is static energy, $V^2/2g$ is velocity energy, Z is potential energy, and H_l is flow energy loss.

Equation 5.2 is recognized as the familiar Bernoulli equation, applicable to all fluid flow processes. In this form, it applies only to an incompressible fluid, which air is assumed to be in nearly all mine ventilation because of minor changes in air specific weight (Section 5.7).

Each term in the equation is actually a *specific energy*, in units of ft·lb/lb, or ft (m). Since ft (m) is a measure of fluid head, these terms are referred to simply as *heads*. In dealing with air, it is customary to employ in. (mm) of water rather than ft (m) of air as the unit of head, for two reasons: (1) the minuteness of the measurements and (2) the still-prevalent use of a manometer (or "water" gage) to measure heads.

Accepting the equivalency of specific energy and head, the general energy equation as written in Eq. 5.1 can also be expressed

$$H_{t_1} = H_{t_2} + H_l \tag{5.3}$$

where H_t is total head; and Eq. 5.2 can be expressed

$$H_{s_1} + H_{v_1} + H_{z_1} = H_{s_2} + H_{v_2} + H_{z_2} + H_l \tag{5.4}$$

where H_s is static head, H_v is velocity head, and H_z is elevation or potential head. All heads have the unit of in. (mm) water.*

These versions of the Bernoulli equation (Eqs. 5.3 and 5.4) are both basic and general—and the most appropriate to apply in mine ventilation. In relating the static, velocity, potential, and total heads plus the losses in flow, the energy equation permits the writing of an expression encompassing all flow variables between any two points in the ventilation system. These points may be selected at the beginning and end of the system (the entrance and discharge of the mine for the air circuit), enabling calculation of the characteristics for the entire system (the mine heads). More generalized circumstances are considered in Section 5.8.

Modified Energy Equation

The potential energy term H_z in Eq. 5.4 can complicate calculations in mine ventilation systems because of sizable differences in elevation that frequently occur. This difficulty normally can be avoided and the elevation change compensated for if a modified procedure in solving ventilation problems is adopted.

To analyze the effect of a change in elevation, Eq. 2.18 is used to calculate the elevation change equivalent to 1.0 in. (25.4 mm) water. The resulting elevation difference, at standard air specific weight, is

$$H_1 = \frac{w_2 H_2}{w_1} = \frac{(62.4 \text{ lb/ft}^3) (1.0 \text{ in.})}{0.0750 \text{ lb/ft}^3} = 832 \text{ in.} = 69.3 \text{ ft } (21.1 \text{ m}) \text{ air}$$

* Heads in the SI convention can also be expressed directly in units of pascals (Pa), a measure of pressure, rather than in head of a column of fluid, such as millimeters of water (1 mm water = 0.797 Pa). See Section 2.4.

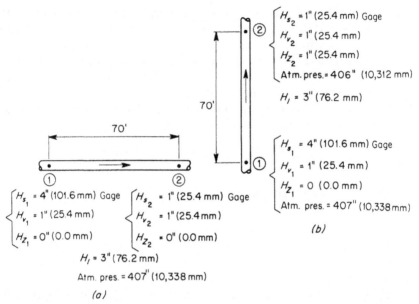

FIGURE 5.2 Arrangement of a duct (a) horizontally and (b) vertically to demonstrate the effect of elevation.

Thus, at sea level, a 69.3 ft (21.1 m) increase in elevation will *increase* the potential head term H_z by 1.0 in. water (25.4 mm), and the static head term H_s will *decrease* by 1.0 in. water (25.4 mm) in compensation. In many calculations, a conversion factor of 70 ft of air for 1.0 in. (0.83 m of air for 1 mm) of water (10 Pa per mm) is considered sufficiently accurate.

It is important to recognize that the static head term H_s is expressed on an absolute-pressure basis when the potential head term H_z is used in Eq. 5.4. The static head is thus provided with respect to a moving datum, that is, expressed as the difference above or below the atmospheric pressure. The term H_s can have both positive or negative values in this case.

To demonstrate these points (Berry, 1963), assume a straight mine opening or duct that is placed first horizontally and then vertically (Fig. 5.2). Let the head loss H_l between points 1 and 2 equal 3 in. (76 mm). Assume sea-level location of point 1 (1 atm = 14.7 psi = 407 in. or 10.3 m water at sea level) and values for the various heads as indicated. Then the general energy equation can be written in the form of Eq. 5.4 to equate total heads and head loss for the system, substituting numerical values.

For the duct in the horizontal position (Fig. 5.2a) and employing absolute pressures (heads), the equation expressed in in. (mm) water is

$$(4 + 407) + 1 + 0 = (1 + 407) + 1 + 0 + 3$$

$$412 = 412 \text{ (or } 10,465 \text{ mm} = 10,465 \text{ mm)}$$

Or using gage pressures (heads), the equation becomes

$$4 + 1 + 0 = 1 + 1 + 0 + 3$$

$$5 = 5 \text{ (or 127 mm} = 127 \text{ mm)}$$

On either basis, the equation checks.

Now place the duct in a vertical position (Fig. 5.2*b*), and repeat the procedure. Writing the energy equation with absolute pressures, the equation balances:

$$(4 + 407) + 1 + 0 = (1 + 406) + 1 + 1 + 3$$

$$412 = 412 \text{ (or 10,465 mm} = 10,465 \text{ mm)}$$

but using gage pressures, it does not:

$$4 + 1 + 0 \neq 1 + 1 + 1 + 3$$

$$5 \neq 6 \text{ (or 127 mm} \neq 152 \text{ mm)}$$

The latter is incorrect because it neglects the change in datum that occurs with a change in elevation. The barometer (static head) changes to compensate for altitude (potential head).

This difficulty could be avoided by always basing calculations on absolute pressures. However, this is a nuisance and not the customary practice in ventilation computations. A convention will be adopted that permits the use of gage pressures at no sacrifice in accuracy for most mine ventilation circuits: *Omit the H_z term from all calculations, and employ a gage-pressure basis.*

Using this technique, check the preceding example again. When horizontal, the energy equation for the duct in in. (mm) water is

$$4 + 1 = 1 + 1 + 3$$

$$5 = 5 \text{ (or 127 mm} = 127 \text{ mm)}$$

When the duct is vertical, the same result is obtained:

$$4 + 1 = 1 + 1 + 3$$

$$5 = 5 \text{ (or 127 mm} = 127 \text{ mm)}$$

This time, on the basis of gage pressures, the calculations are correct, because compensating changes in both elevation and the datum for static pressure are omitted.

Accordingly, revising Eq. 5.4, a modified and simplified general energy equation for mine ventilation may be written

$$H_{t_1} = H_{t_2} + H_l$$

$$H_{s_1} + H_{v_1} = H_{s_2} + H_{v_2} + H_l \tag{5.5}$$

This version is correct for ducts in any position, so long as all static-head measurements and calculations are made on a gage-pressure basis in reference to the atmospheric pressure existing at the point of measurement. Measurement procedures (Chapter 6) for both heads and the difference in heads have been developed in conformance with this requirement.

An exception occurs in the measurement and calculation of natural ventilation heads, in which elevation differences cannot be omitted (see Section 5.8 and Chapter 8).

5.2   HEAD LOSSES AND MINE HEADS

Head Losses in Fluid Flow

In mine ventilation, as in hydraulics and other fields applying fluid mechanics principles, there is more interest in pressure differences (differences in head) than in pressures. Flow occurs because a pressure difference is created between two points in the system. Energy that is supplied to a steady-state process by either natural or mechanical means and creates this pressure difference is consumed in overcoming flow losses, represented by H_l. As with the general energy equation, the unit of the various pressure losses is actually fluid head, measured in in. (mm) water.

The head loss in fluid flow is made up of two components, friction loss H_f and shock loss H_x:

$$H_l = H_f + H_x \tag{5.6}$$

all normally expressed in in. (mm) water. *Friction losses* represent head losses in flow through ducts of constant area, whereas *shock losses* are losses resulting from changes in direction of flow or area of duct. Shock losses also occur at the inlet or discharge of a system, at splits or junctions of two or more currents of air, and at obstructions in airways.

Head losses must be overcome by the energy supplied to the fluid system. This energy is available in the form of static head, and the losses, both friction and shock, cause corresponding decreases to occur in the static head of the fluid. However, when shock losses occur because of change of duct area, some velocity head is converted into static head (if the area increases) or static head into velocity head (if the area decreases). *Thus static and velocity heads are mutually convertible, although the conversion is accompanied by shock loss.*

This conversion of heads may cause an apparent increase of static head, in spite of shock loss. *Hence it is required to charge all changes in head against the total head, rather than the static head, to reflect head losses due to either friction or shock.* These changes, according to Eq. 5.3, are always decreases.

Overall or Mine Heads

In many situations, it is necessary to sum all the various flow-energy losses to determine the amount of head that must be supplied to overcome the losses and produce a desired flow. In a mine ventilation system with a single fan or other pressure source and a single discharge, the cumulative energy consumption is termed the *mine head*. The mine head is in reality a difference in head, determined in accordance with Bernoulli's principle, that must be supplied to move the desired quantity of air. In mines with multiple energy sources (fans) or discharges, the definitions supplied in the following paragraphs cannot be applied in the same fashion (see Chapter 7). However, much can be learned concerning the nature of quantity control by defining first the mine heads and analyzing the energy source for a simple mine ventilation circuit.

Mine static head (mine H_s) represents the energy consumed in the ventilation system to overcome all flow head losses. It includes all the decreases in total head that occur between the entrance and discharge of the system and may be expressed simply

$$\text{Mine } H_s = \Sigma H_l = \Sigma(H_f + H_x) \tag{5.7}$$

This applies to a series circuit or to the equivalent series circuit in a network.

Mine velocity head (mine H_v) is taken as the velocity head at the discharge of the system. Throughout the system, the velocity head changes with each change in duct area or number and is a function only of air specific weight and velocity of airflow ($H_v = V^2/2g$; see Eq. 5.2). It is not a cumulative head loss. Strictly speaking, however, the velocity head for the system or mine H_v is a loss, because the kinetic energy of the air is discharged to the atmosphere and wasted. Hence it must also be considered a loss to the system in determining overall energy loss.

Mine total head (mine H_t) is the sum of all energy losses in the ventilation system. Numerically, it is the total of the mine static and velocity heads:

$$\text{Mine } H_t = \text{mine } H_s + \text{mine } H_v \tag{5.8}$$

The significance of these overall heads for the entire system will soon become apparent. Methods of calculating head losses and the mine heads are discussed in Sections 5.3 and 5.5. Sometimes it is necessary to determine the head requirements of only a portion of the entire mine ventilation system. These also may be found by Eqs. 5.7 and 5.8, substituting the word *airway* or *duct* for *mine*.

5.3 HEAD GRADIENTS

In representing the various head components of the general energy equation graphically, pressure or *head gradients* are obtained. Figure 5.3 shows the gradients for a basic system involving any fluid in any conduit. This is simply a pictorial way of expressing the Bernoulli energy equation and depicts the head relations of Eqs. 5.3 and 5.4. Three distinct gradients—elevation, static plus elevation (including atmospheric pressure), and total—appear in Fig. 5.3; the static and velocity heads are shown as the difference between the gradients and not as separate gradients. Note that the cumulative head relations at any point in the system can be read from the graph.

Usually in mine ventilation, only two head gradients—static and total—need to be plotted, in accordance with Eq. 5.5, the modified version of the general energy equation. The elevation effect is omitted, and the datum employed parallels the barometric pressure line for the system, which rises and falls inversely with elevation. Heads at any point are gage values, relative to the atmospheric pressure at that point.

Example 5.1 Plot head gradients for the duct of Fig. 5.2 in an inclined position. Assume the same head relations and difference in elevation. Plot on the basis of (a) absolute pressures and (b) gage pressures.

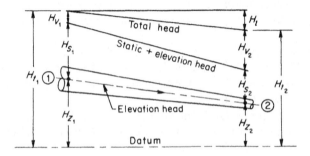

FIGURE 5.3 Head gradients for a basic fluid-flow system.

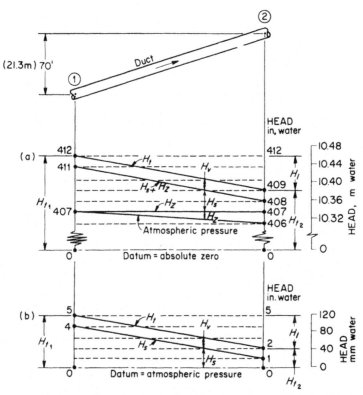

FIGURE 5.4 Head gradients for the duct of Fig. 5.2 arranged in an inclined position, on basis of (*a*) absolute pressure and (*b*) gage pressure (see Example 5.1).

Solution: See Fig. 5.4. Note that Bernoulli's equation is satisfied by both graphs. Normally, the gradients on a gage basis, (b), suffice.

Because head changes in mine ventilation systems are varied and frequent, it is helpful to employ gradients in picturing these changes. A simple sketch of the head gradients can generally reduce even a complicated system to an understandable level. Determination of the mine heads is also considerably expedited by the use of gradients.

Head gradients for the three basic types of ventilation systems (classified on the basis of the location of the energy source, usually a fan), assuming a series circuit, are illustrated and discussed individually.

Blower Systems

A *blower system* (Fig. 5.5) is one in which the energy source (= fan) is located at the inlet and raises the head of the mine or duct air above atmospheric.

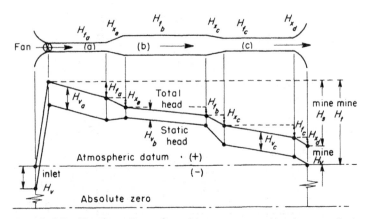

FIGURE 5.5 Head gradients for a blower system in mine ventilation.

An assumed mine ventilation system together with the corresponding head gradients is illustrated; head losses and velocity heads for each section or point in the airway are indicated. For velocity head H_v and friction losses H_f, the subscript for the appropriate duct section (a, b, or c) appears; and for shock losses H_x, the subscripts denote inlet (i), expansion (e), contraction (c), or discharge (d).

Note that the inlet velocity head drops below the datum. This occurs because a suction condition must exist here in order for air to flow into the system. In the same manner, the discharge velocity head lies above the datum, because the air is in motion when it leaves the system and is discharged with that amount of kinetic energy remaining. Since this is lost to the system, the discharge velocity head should be reduced as much as possible before the air is released to the atmosphere. This can be accomplished by converting some of the velocity head to static head in a diffuser duct or evasé discharge (Section 10.3).

By convention, gage readings of heads (static or total) are expressed as positive if above the atmospheric datum line and negative if below it. It would not be necessary to employ plus and minus signs if absolute pressures were used (reference to absolute zero as the datum), but since gage pressures are customary, this convention of signs is necessary. Special consideration is required in exhaust systems. Note that in a blower system, all heads are normally positive (with respect to atmospheric), except at the entrance, or an expansion near the discharge. With other types of systems, negative total and static heads will be encountered, but the velocity head will always remain positive.

It is well to employ the following rules in plotting a head gradient for any type of ventilation system:

1. The total head is always zero at the inlet to the system but positive and equal to the velocity head at the discharge.

2. The static head is always negative and equal to the velocity head at the inlet to the system but zero at the discharge.
3. The total head at any point is plotted first, and the static head is then plotted as the total head less the velocity head.

With a blower system, plotting is facilitated by starting at the discharge and working toward the intake. This permits the H_s line to begin at zero (the datum line). When all individual head losses have been determined previously, this can be done readily.

Note that at any point in the system,

$$H_t = H_s + H_v$$

in accordance with Eq. 5.5. The mine heads can easily be found by Eqs. 5.7 and 5.8 and are indicated in Fig. 5.5. The individual head losses are either measured or calculated. They are plotted along the total head line, not the static line. There is no contradiction here, however, because head losses should be reflected in change in total head, which is constantly dropping in the direction of airflow (except at the fan), whereas the static gradient fluctuates with conversions in velocity head. Inlet loss does not appear in a blower system, because any shock here must be absorbed by the fan. No allowance or correction need be made for elevation, so long as gage pressures are employed.

Example 5.2 For the blower ventilation system shown in Fig. 5.6, plot and label the head gradients. The values of heads and losses are as follows:

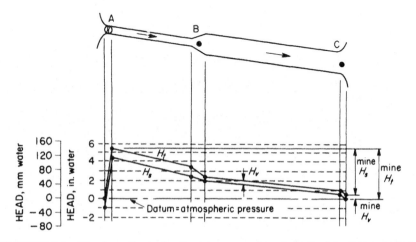

FIGURE 5.6 Head gradients for the blower system described in Example 5.2.

For duct AB: Friction loss = 2.0 in. (50.8 mm) water
                Velocity head = 1.0 in. (25.4 mm)
For duct BC: Friction loss = 1.5 in. (38.1 mm)
                Velocity head = 0.5 in. (12.7 mm)
For inlet A: Shock loss = 1.0 in. (25.4 mm), without fan
For expansion B: Shock loss = 1.0 in. (25.4 mm)
For discharge C: Shock loss = 0.5 in. (12.7 mm)

Assume the velocity head at discharge equals that in duct BC. Also calculate and show the mine heads.

Solution: See Fig. 5.6. Start plotting gradients from discharge at C, working backward to fan at A. Shock loss due to inlet can be disregarded. Values of mine heads calculated by Eqs. 5.7 and 5.8 are

$$\text{Mine } H_s = H_{f_{AB}} + H_{f_{BC}} + H_{x_A} + H_{x_B} + H_{x_C}$$

$$= 2.0 + 1.5 + 0 + 1.0 + 0.5$$

$$= 5.0 \text{ in. } (127.0 \text{ mm) water}$$

$$\text{Mine } H_v = H_{v_{BC}} = 0.5 \text{ in. } (12.7 \text{ mm)}$$

$$\text{Mine } H_t = 5.0 + 0.5 = 5.5 \text{ in. } (139.7 \text{ mm)}$$

Exhaust Systems (Fig. 5.7)

If the energy source in the mine ventilation system of Fig. 5.5 is relocated at the discharge without changing the direction of airflow, an exhaust system

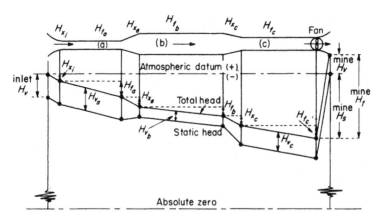

FIGURE 5.7 Head gradients for an exhaust system in mine ventilation.

results. The gradients are plotted as before, except the intake serves as the starting point, and they present the same general appearance.

Since all losses are determined on a gage basis and are negative, almost the entire gradient falls below the atmospheric datum line. This, of course, is because an energy source, when located in the exhaust position, receives air below atmospheric and discharges it at atmospheric pressure. Note, however, that the H_s and H_t lines begin and end at the same points as in the blower system.

In keeping with the sign convention with reference to atmospheric pressure, the gradients represent *negative* static and total heads. At any point in the system, however, the algebraic equation (5.5)

$$H_t = H_s + H_v$$

still holds. The static head is negative since the fan creates a suction in the system, resulting in pressures below atmospheric. The velocity head is positive, resulting in the total-head line being above the static head (but still negative in sign, except at discharge). This is the normal or expected position, which does not change because H_s is negative.

To avoid the awkward convention of signs in overall head calculations and to retain the use of gage pressures, *mine heads are always considered positive, regardless of the type of system or position of the datum.* It is quite apparent that with the same quantity of air flowing in a blower and in an exhaust system, the head losses and mine heads must be equal. This may be verified by comparing the gradients and mine heads of Figs. 5.5 and 5.7. Note that the mine H_s (the sum of all head losses) is plotted in both cases on the total-head gradient, running from the highest to the lowest points on that gradient.

Actually, the mine H_s and H_t values for a system operated either as blower or exhaust will not be exactly the same, because a shock loss at the inlet (H_{x_i}) occurs only in an exhaust system and at the discharge (H_{x_d}) only in a blower system. These two losses are seldom equal in magnitude, and therefore the total head losses in the two systems are not precisely equal. The difference is slight, and for all practical purposes, the mine heads are generally taken as equal.

Booster Systems

Now let the same mine ventilation system be arranged as a booster system, with the energy source located at some point between the inlet and the discharge (Fig. 5.8). In a booster system, the fan normally receives air below atmospheric pressure and discharges it above atmospheric pressure.

The mine static head is made up of two components: H_{s_i} on the intake side of the energy source and H_{s_d} on the discharge side. The sum of the two is the mine H_s, which, for a constant airflow, is essentially equal to that for

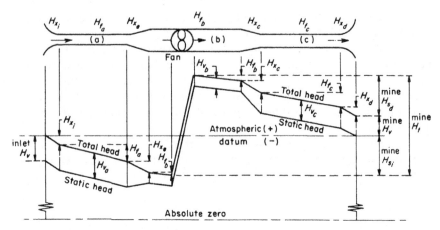

FIGURE 5.8   Head gradients for a booster system in mine ventilation.

the blower and exhaust systems discussed previously. Again the mine velocity head is the H_v at discharge, and the mine total head is the sum of the mine H_s and the mine H_v.

As with the other systems, the mine heads are plotted along the same gradients and taken as positive. Plotting proceeds from both ends of the system toward the fan. It should be noted that the booster system combines the characteristics of blower and the exhaust systems. It is a hybrid system, partially blower and partially exhaust. Notice that both inlet and discharge losses occur.

5.4   STATE OF AIRFLOW IN MINE OPENINGS

In fluid mechanics, two distinct states of fluid flow with a transitional zone between are recognized: laminar, intermediate, and turbulent (Munson et al., 1994). Accordingly, in fluid flow calculations, it is necessary to identify the state of flow that prevails, because the fluid exhibits different characteristics and head losses in each state.

The dimensionless criterion used in establishing boundaries for each state is called the *Reynolds number* N_{Re}. Laminar flow exists up to $N_{Re} = 2000$ and turbulent flow above $N_{Re} = 4000$. These boundaries are only approximate, and the region between is known as the *intermediate range*. The Reynolds number in fluid flow is a function of the fluid properties and can be determined as follows:

$$N_{Re} = \frac{\rho DV}{\mu} = \frac{DV}{\nu} \tag{5.9}$$

where ρ is fluid mass density ($= w/g$) in lb·s^2/ft^4 (kg/m^3), ν is kinematic viscosity in ft^2/s (m^2/s), μ is absolute viscosity ($= \rho\nu$) in lb·s/ft^2 (Pa·s), D is diameter of conduit in ft (m), and V is velocity in fps (m/s). For air, $\nu = 1.6 \times 10^{-4}$ ft^2/s (14.8×10^{-6} m^2/s) at normal temperatures, and Eq. 5.9 reduces to

$$N_{Re} = 6250DV \qquad (5.10)$$

$$N_{Re} = 67,280DV \qquad (5.10a)$$

The fluid velocity corresponding to $N_{Re} = 4000$, the lower boundary of turbulent flow for a conduit of given size, is called the *critical velocity* V_c. If the fluid velocity exceeds V_c, then the state of flow is always turbulent. The critical velocity can be found easily from the last relation above, solving for V_c in fpm (m/s) and setting $N_{Re} = 4000$:

$$V_c = \frac{60N_{Re}}{6250D} = \frac{(60)\,(4000)}{6250D} = \frac{38.4}{D}$$

or approximately

$$V_c \simeq \frac{40}{D} \qquad (5.11)$$

$$V_c \simeq \frac{0.06}{D} \qquad (5.11a)$$

In mine openings, it is important that turbulent flow always prevails. As explained in the section on quality control, this ensures satisfactory dispersion and removal of contaminants produced in the workplaces. Although the critical velocity to ensure turbulent flow varies with the size of the opening or duct, it is obvious from Eq. 5.11 and from a brief consideration of duct and airway dimensions that *turbulent flow will nearly always prevail in mine openings*. Ventilation pipe or tubing less than 1 ft (0.3 m) in diameter is seldom used; thus velocities over 40 fpm (0.2 m/s) will always produce turbulent flow in vent tubing. Mine openings rarely have an equivalent diameter smaller than 3 ft (0.9 m), and therefore velocities exceeding 13 fpm (0.07 m/s) will cause turbulent flow in mine headings, raises, and other openings. Exceptions where laminar flow may be encountered are in leakage through doors and stoppings in airways and in exhaust through caved or filled areas.

Effect of State of Flow on Velocity Distribution

One way that the state of flow affects the dynamic characteristics of a fluid is in the velocity distribution over the cross section of the conduit. Different velocity distributions in a circular conduit for the same average velocity of airflow over varying Reynolds numbers are shown in Fig. 5.9.

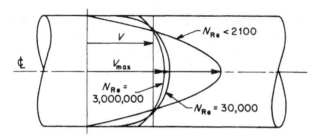

FIGURE 5.9 Velocity distributions in circular conduits, average velocity constant. (After Rouse, 1937. Reprinted from *Trans. Amer. Soc. Civil Engl.*, Vol. 102, p. 163 with permission of the ASCE.)

The maximum velocity V_{max} occurs at the center of the conduit, but it varies in magnitude with the Reynolds number. The usual aim in ventilation is the measurement of average velocity V, not maximum velocity; and therefore centerline measurements of velocity alone are not sufficient (Chapter 6). The variation of V with V_{max} is determinable as a function of the Reynolds number, however, as shown in Fig. 5.10. This graph enables one to estimate the average velocity when only one measurement along the centerline has been made. Discretion should be exercised in its use in mine ventilation, however, since mine openings are rarely circular and the many irregularities of the walls tend to produce a nonsymmetrical flow pattern. Because the Reynolds number in mine ventilation generally exceeds 10,000, it is customary to assume for approximate work that

$$V \simeq 0.8\ V_{max} \tag{5.12}$$

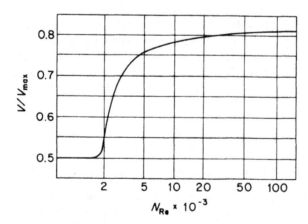

FIGURE 5.10 Relation of velocity ratio V/V_{max} and Reynolds number. (After Vennard, 1940. Reprinted by permission of John Wiley & Sons, Inc.)

5.5 CALCULATIONS OF HEAD LOSSES

Velocity Head

Although not a conventional head loss within the ventilation circuit, the velocity head nevertheless represents kinetic energy that has to be supplied to maintain flow and is lost to the system at discharge (Section 5.2). There is frequently the need to determine, by calculation or measurement, the velocity head at various points in the system. For example, the velocity head at discharge must be known in order to determine the mine heads. On other occasions, measurement of velocity head permits the velocity of airflow to be calculated.

An equation for velocity head involving conventional units is desirable. Starting with the basic relation appearing as Eq. 5.2,

$$H_v = \frac{V^2}{2g}$$

with V in fps (m/s) and H_v in ft (m) of fluid, and applying Eq. 2.18, the following may be obtained:

$$H_v = \frac{wV^2}{(5.2)(64.4)(60)^2} = w\left(\frac{V}{1098}\right)^2 \tag{5.13}$$

with V in fpm and H_v in in. water. For standard air at sea level ($w = 0.0750$ lb/ft^3), this becomes

$$H_v = \left(\frac{V}{4009}\right)^2 \tag{5.14}$$

Using SI units, Eq. 5.13 takes the form

$$H_v = \frac{\rho V^2}{2} = \frac{wV^2}{2g} \tag{5.13a}$$

with H_v in Pa, V in m/s, and ρ and w/g in kg/m^3. Equation 5.14 states that a velocity of about 4000 fpm (20.3 m/s) is equivalent to a velocity head of 1 in. water (249 Pa). These equations may also be used to calculate the velocity when the head is known. To facilitate conversions of velocity and velocity head, the nomograph in Appendix Fig. A.1 may be employed for estimation purposes.

Friction Loss

The friction losses in airflow through mine openings constitute 70–90% of the sum of the head losses in a mine ventilation system. They are therefore of greater practical importance than shock losses and deserving of more care and precision in calculation.

Atkinson Equation for Friction Loss As a loss in static pressure that occurs in flow as a result of the drag or resistance of the walls of the opening or duct and the internal friction of the fluid itself, the friction loss in a mine airway is a function of the velocity of flow, the interior-surface characteristics of the conduit, and the dimensions of the conduit. (In addition, it would be expected from fluid mechanics that the friction loss would be dependent on the state of flow of the fluid. It has already been demonstrated, however, that in mine ventilation systems, airflow can nearly always be considered turbulent.)

The fluid mechanics (Darcy–Weisbach) equation for calculating friction loss in any circular conduit is (Munson et al., 1994)

$$H_l = f \frac{L}{D} \frac{V^2}{2g} \tag{5.15}$$

where H_l is head loss in ft (m) of fluid, L is length in ft (m), D is diameter in ft (m), V is velocity in fps (m/s), and f is coefficient of friction. A more versatile equation, one applicable to any shape of conduit, may be obtained by expressing the head loss in terms of the hydraulic radius R_h, the ratio of the area A to the perimeter O of the duct. For a circular conduit,

$$R_h = \frac{A}{O} = \frac{(\pi/4)D^2}{\pi D} = \frac{D}{4} \tag{5.16}$$

Substituting Eq. 5.16 in Eq. 5.15, there results

$$H_l = f \frac{L}{4R_h} \frac{V^2}{2g} \tag{5.17}$$

From this version of the Darcy–Weisbach equation for general fluid mechanics, the Atkinson equation for friction loss in mine ventilation can be derived as follows (Weeks, 1926):

$$H_f = \frac{f}{5.2} \frac{L}{4R_h} \frac{0.0750V^2}{2g\,(60)^2} = \frac{K}{5.2} \frac{L}{R_h} V^2 = \frac{KOLV^2}{5.2A} \tag{5.18}$$

$$H_f = \frac{KOLV^2}{A} \tag{5.18a}$$

where H_f is friction loss in in. water (Pa), V is velocity in fpm (m/s), and K is an empirical friction factor in lb·min^2/ft^4 (kg/m^3). The equation is frequently seen in the following form, although it is seldom so used directly:

$$H_f = \frac{KSV^2}{5.2A} \qquad (5.19)$$

$$H_f = \frac{KSV^2}{A} \qquad (5.19a)$$

where S is rubbing surface area in ft^2 (m^2) $= O \cdot L$.

Often the air velocity is not known, but the quantity Q is given. Since $V = Q/A$ (Eq. 3.10), and rather than compute V separately, a convenient form of the friction loss equation to employ is

$$H_f = \frac{KOLQ^2}{5.2A^3} \qquad (5.20)$$

$$H_f = \frac{KOLQ^2}{A^3} \qquad (5.20a)$$

Equation 5.20 is the most useful form of the Atkinson formula and most used in this text.

It should be noted that K takes units of lb·min^2/ft^4 (kg/m^3) in Eqs. 5.18–5.20 as a result of the units assigned to the other variables. It has often been American practice to ignore the units of K, because little in the way of physical meaning can be attached to the units. However, it is important to specify the proper units, particularly when it is desirable to convert K values from one system of units to another. To emphasize the proper units and to overcome the improper practice of expressing K without units, the units lb·min^2/ft^4 (kg/m^3) are attached to K throughout this book. Engineers interested in converting K and other parameters to other unit conventions (e.g., SI units) should consult Appendix B or the references by Hunt (1960), McPherson (1971), and Rahim et al. (1976). It should also be pointed out that K is not a constant but varies directly with air specific weight; values of K are commonly expressed in tables at standard air specific weight.

The friction factor K in mine ventilation corresponds to the coefficient of friction in general fluid flow. Mathematically, from Eq. 5.18, the two are related approximately (for standard air specific weight):

$$K \simeq (800)(10)^{-10}f \qquad (5.21)$$

$$K \simeq 0.148f \qquad (5.21a)$$

Actually, in turbulent flow, f is not a constant for a given conduit but varies with the Reynolds number. In mine ventilation, K is assumed constant for a given airway, regardless of the Reynolds number. This is only an approximation, and on occasion the error can be sizable (Falkie, 1958). However, for the usual range of Reynolds number (50,000 to 2 million) encountered in mine workings, the error is probably not excessive in view of the variability of the other factors involved. Hence it is ignored in all but precision measurements or research-type investigations.

The Atkinson equation, by assuming friction factor constant, states that friction loss is a function of the square of the velocity (see Eq. 5.18). From fluid mechanics, H_f is actually known to vary with V raised to a power between 1.75 and 2, because f is a function of N_{Re} and therefore of V. When flow is laminar, such as prevails with leakage, the relation approaches linearity. In mine ventilation, where flow is otherwise nearly always turbulent, any departure from the square-law relation is generally disregarded. The proportionality of H to V^2 (and Q^2) has useful applications (Chapter 7).

Determination of Airway Friction Factor

The only accurate way to determine the friction factor for a given airway is to compute it (by Eq. 5.18) from the pressure drop measured underground. For estimation or projection purposes, friction factors may have to be selected from experience. The most widely used are the values of friction factor listed in Table 5.1, on the basis of exhaustive tests in a classic study by the U.S. Bureau of Mines (McElroy, 1935). Although these factors were designed principally for metal mine openings, they have had some use and acceptance for airways in coal mines as well.

In recent years, other investigators have conducted empirical tests using models or actual mine openings to extend the range of McElroy's friction factors to other conditions, especially in coal mines (Pursall, 1960; Skochinsky and Komarov, 1969; Rahim et al., 1976). Perhaps more important than these others, because it was conducted in coal mines in the United States, was the study of Kharkar et al. (1974). Results are displayed in Table 5.2 in a format similar to McElroy's. *Smooth lined* corresponds to openings driven by boring-type continuous miners and tunneling machines and *unlined* to openings produced by conventional methods. Openings not specified *timbered* are assumed to be rock-bolted. Kharkar's values for conditions comparable to McElroy's tend to be lower by 5–20%. Wala (1991) has also measured friction factors in coal mines. His values are in general agreement with those of Kharkar.

Two recent publications provide friction factors for mine situations that are not covered in Tables 5.1 and 5.2. McPherson (1987) outlines friction factors for longwall faces under different conditions. His measurements show friction factors from 200×10^{-10} to 350×10^{-10} lb·min²/ft⁴ [0.0370–0.0750 kg (m³)]. An additional publication by McPherson (1985) provides friction

TABLE 5.1 Friction Factor K for Noncoal Mine Airways and Openings

Values of $K \times 10^{10}$ [a]

Type of Airway	Irregularities of Surfaces, Areas, and Alignment	Straight			Slightly			Sinuous or Curved — Moderately			High Degree		
		Clean (Basic Values)	Slightly Obstructed	Moderately Obstructed	Clean	Slightly Obstructed	Moderately Obstructed	Clean	Slightly Obstructed	Moderately Obstructed	Clean	Slightly Obstructed	Moderately Obstructed
Smooth lined	Minimum	10	15	25	20	25	35	25	30	40	35	40	50
	Average	15	20	30	25	30	40	30	35	45	40	45	55
	Maximum	20	25	35	30	35	45	35	40	50	45	50	60
Sedimentary rock	Minimum	30	35	45	40	45	55	45	50	60	55	60	70
	Average	55	60	70	65	70	80	70	75	85	80	85	95
	Maximum	70	75	85	80	85	95	85	95	100	95	100	110
Timbered (5-ft centers)	Minimum	80	85	95	90	95	105	95	100	110	105	110	120
	Average	95	100	110	105	110	120	110	115	125	120	125	135
	Maximum	105	110	120	115	120	130	120	125	135	130	135	145
Igneous rock	Minimum	90	95	105	100	105	115	105	110	120	115	120	130
	Average	145	150	160	155	160	165	160	165	175	170	175	195
	Maximum	195	200	210	205	210	220	210	215	225	220	225	235

Source: McElroy (1935).

[a] To provide correct values of K, the numerical values obtained from the table are multiplied by 10^{-10} and units of $\text{lb} \cdot \text{min}^2/\text{ft}^4$ attached. K is based on standard air specific weight ($w = 0.0750 \text{ lb/ft}^3$). Recommended values are in italics. To convert K to SI units (kg/m^3), multiply table values by 1.855×10^6.

TABLE 5.2   Friction Factor K for Coal Mine Airways and Openings

	Value of $K \times 10^{10}$ [a]					
	Straight			Curved		
Type of Airway	Clean	Slightly Obstructed	Moderately Obstructed	Clean	Slightly Obstructed	Moderately Obstructed
Smooth lined	25	28	34	31	39	43
Unlined (rock-bolted)	43	49	61	62	68	74
Timbered	67	75	82	85	87	90

Source: Kharkar et al. (1974).

[a] To provide correct values of K, the numerical values obtained from the table are multiplied by 10^{-10} and units of lb·min^2/ft^4 attached. K is based on standard air specific weight ($w = 0.0750$ lb/ft^3). To convert K to SI units (kg/m^3), multiply table values by 1.855×10^6.

factors for shafts with different wall constructions, structural obstructions, and conveyances. These references may be of use when these types of openings are encountered in mine ventilation circuits.

Observe the following precautions in selecting a value of friction factor from the tables (5.1 and 5.2) for use in calculations:

1. To provide correct values of K, multiply the numerical values obtained from the table by 10^{-10}, and attach units of lb·min^2/ft^4. (For K in SI units of kg/m^3, multiply table values by 1.855×10^6.)

2. Employ a value of K determined or checked experimentally, if at all possible. This should be based on the results of actual tests conducted underground in the mine opening or ventilation duct involved. Careful measurement of H_f over a known length of airway of constant cross section will allow a reliable experimental value of K to be calculated.

3. Tables 5.1 and 5.2 list values of K based on standard air specific weight. Since K is proportional to w, correct K for actual w by the formula

$$\text{Corrected } K = (\text{table } K)\left(\frac{w}{0.0750}\right) \qquad (5.22)$$

$$\text{Corrected } K = (\text{table } K)\left(\frac{w}{1.201}\right) \qquad (5.22a)$$

before using in Eqs. 5.18 to 5.20.

4. Select K carefully for the conditions (rock type, straightness, cleanliness, irregularity, etc.) prevalent in the airway. When in doubt, use average values. Italicized values commonly occur and are safe to use

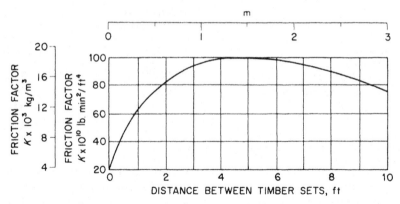

FIGURE 5.11 Effect of spacing of timber sets on friction factor K. (After McElroy, 1935.)

in calculations. For noncoal mines, select K from Table 5.1; for coal mines, use Table 5.2.

5. If the airway is timbered and the sets are spaced on other than 5-ft. (1.5-m) centers, modify K according to Fig. 5.11. If roof bolting is used in place of timbering, assume an unlined airway.

Example 5.3 Select the friction factor for a highly sinuous, slightly obstructed drift in igneous rock.

Solution: From Table 5.1, read $K = 120–225 \times 10^{-10}$ lb·min²/ft⁴ for minimum to maximum irregularities. If no other information is available, and assuming standard air specific weight, use average value $K = 175 \times 10^{-10}$ lb·min²/ft⁴ (0.0325 kg/m³).

Example 5.4 Select the friction factor for a straight, clean, unlined entry in coal, given $w = 0.0650$ lb/ft³ (1.041 kg/m³).

Solution: From Table 5.2, read $K = 43 \times 10^{-10}$ lb·min²/ft⁴ (0.0080 kg/m³) Corrected $K = (43 \times 10^{-10})(0.0650/0.0750) = 37 \times 10^{-10}$ lb·min²/ft⁴ (0.0068 kg/m³)

Determination of Friction Factor for Vent Pipe

Friction factors to use with different types of ventilation pipe or tubing varying with the material and its condition. The following are satisfactory for routine calculations (based on $w = 0.0750$ lb/ft³):

Pipe or Tubing	Friction Factor, $K \times 10^{10}$ lb·min^2/ft^4 (kg/m^3)	
	Good, New	Average, Used
Steel, wood, fiberglass (rigid)	15 (0.0028)	20 (0.0037)
Jute, canvas, plastic (flexible)	20 (0.0037)	25 (0.0046)
Spiral-type canvas	22.5 (0.0042)	27.5 (0.0051)

Estimation of Friction Loss by Graph

The approximate determination of friction losses in mine ventilation is simplified by the use of graphs. The most versatile to use for mine openings appears in Appendix Fig. A.2. Knowledge of the velocity, friction factor, and hydraulic radius permits the friction loss per 100 ft (30 m) of airway to be read directly. Friction factor should be corrected for air specific weight by Eq. 5.22 beforehand. Charts that require less manipulation have also been devised for various airway shapes and friction factors (Lee and Ember, 1946; Anon., 1955).

For ventilation pipe or tubing, use Appendix Fig. A.3. Values of friction loss are read per 100 ft (30 m) of length and standard air specific weight (corrected H_f = graph $H_f \times w/0.0750$, or $w/1.201$ in SI units). This graph is based on circular sheetmetal duct in good condition and is used without correction for new, steel vent pipe. Correct as follows for other tubing or condition:

Pipe or Tubing	Correction Factor for Pipe Condition	
	Good, New	Average, Used
Steel, wood, fiberglass (rigid)	1.00	1.33
Jute, canvas, plastic (flexible)	1.33	1.67
Spiral-type canvas	1.50	1.83

Actually, values of friction loss read from the graph are often more accurate than those calculated by formula (Eq. 5.18 or 5.20). The graph was constructed to compensate for the drop in friction factor with increasing Reynolds numbers, and values are correct within a few percent (Anon., 1993).

Calculation of Friction Loss by Formula

Friction loss in a mine duct or airway can be calculated by Eq. 5.18 or 5.20. A separate calculation is necessary for each airway of different characteristics (K) or cross-sectional dimensions (A, O) and for each different airflow (V or Q). For different airflows in a given airway, solve $KOL/5.2A^3$ separately and multiply by Q^2. For different airway lengths or friction factors, multiply H_f by the ratio of lengths or friction factors, respectively.

Example 5.5 Determine the friction loss in a mine airway having the following characteristics: unlined airway in coal, curved, moderately obstructed. Cross section is 4 × 12 ft. (1.22 × 3.66m), L = 3000 ft (914 m), Q = 48,000 cfm (22.6 m³/s), w = 0.0750 lb/ft³ (1.201 kg/m³). Solve the example using (a) Eq. 5.20, (b) using the nomograph, and (c) using Eq. 5.20a. Use the answer in part (c) to check the other friction loss values.

Solution:
(a) Using Table 5.2, select K = 74 × 10^{-10} lb·min²/ft⁴ (0.0137 kg/m³)

$$H_f = \frac{(74 \times 10^{-10})(32)(3000)(48,000)^2}{(5.2)(48)^3} = 2.85 \text{ in. water (709 Pa)}$$

(b) Using the nomograph (Fig. A.2 in Appendix A):

$$R_h = \frac{48}{32} = 1.5 \text{ ft,} \qquad V = \frac{48,000}{48} = 1000 \text{ fpm}$$

Read H_f = 0.094 in./100 ft.

$$H_f = (0.094)\left(\frac{3000}{100}\right) = 2.82 \text{ in. water (702 Pa)}$$

(c) For use of Eq. 5.20a, K must be converted to SI units:

$$K = 74 \times 10^{-10} \times 1.855 \times 10^6 = 0.0137 \text{ kg/m}^3$$

$$H_f = \frac{(0.0137)(9.76)(914)(22.6)^2}{(4.465)^3} = 701 \text{ Pa (2.82 in. water)}$$

Check:

2.85 in. water (709 Pa) = 2.82 in. water (702 Pa)
                        = 2.82 in. water (701 Pa)

Check is acceptable.

Example 5.6 Determine the friction loss in mine vent tubing, by formula and nomograph, under the following conditions: plastic tubing, average condition, D = 48 in. (1.219 m), L = 3000 ft (914 m), Q = 48,000 cfm (22.6 m³/s), w = 0.0750 lb/ft³ (1.201 kg/m³).

Solution:
1. Using Eq. 5.20:

Select $K = 25 \times 10^{-10}$ lb·min²/ft⁴ (0.0046 kg/m³)

$$A = 12.566 \text{ ft}^2, \ O = 12.566 \text{ ft}$$

$$H_f = \frac{(25 \times 10^{-10})(12.566)(3000)(48,000)^2}{(5.2)(12.566)^3} = 21.04 \text{ in. water (5.236 kPa)}$$

2. Using Appendix Fig. A.3:

$$\text{Head } H_f = 0.30 \text{ in. water/100 ft}$$

$$H_f = (0.30)(1.67)\left(\frac{3000}{100}\right) = 15.03 \text{ in. water (3.740 kPa)}$$

Noting the poor agreement of the two methods in Example 5.6, it is preferable to use the answer obtained in (2), because Appendix Fig. A.3 considers variation of K with N_{Re}. In addition, because of the high head loss, it would be advisable to consider the compressibility effect (see Section 5.7).

As a calculation aid in using Eq. 5.20 with large quantities, Q may be expressed in units of 100,000 cfm. Thus Q^2 is written as an interger $\times 10^{-10}$, which cancels the 10^{-10} associated with K in the equation (Ramani, 1992a).

Shock Loss

Shock losses occur in mine ventilation in addition to friction losses and are caused by changes in the direction of airflow or the shape or size of the duct. Obstructions cause shock loss by reducing the duct area. Although generally constituting only 10–30% of the total head loss in mine ventilation systems, shock losses should always be considered in exact calculations, in major airways, or in short lengths of duct with many bends or area changes.

Shock losses do not lend themselves to precise calculation because of the great range of variability in occurrence and because of a lack of understanding about their actual nature. Basically, for a given source of shock, the head loss varies as the square of the velocity or directly as the velocity head. Calculation of individual shock losses can be carried out in several ways.

Calculation of Shock Loss Directly

The amount of shock loss H_x in in. water (Pa) can be calculated from the velocity head H_v (McElroy, 1935):

$$H_x = XH_v \tag{5.23}$$

where X is shock loss factor (dimensionless). This factor, roughly analogous to the friction factor K in friction loss, is a constant only for a given, constant set of conditions. In other words, every bend, area change, and obstruction has its own shock-loss factor, depending on dimensions and characteristics.

Individual calculation of shock losses is rarely warranted in mine ventilation because of their relatively small magnitude, the inaccuracy of X values, and the inordinate amount of time involved. Where necessary, for exact field or laboratory work, shock-loss factors may be determined by empirical formulas and graphs developed by McElroy (1935) and Hartman (1960). A summary listing is contained in the second edition of this book (Hartman et al., 1982, App. A).

Calculation of Shock Loss by Increase in Friction Factor

Either a calculated or estimated increment in the friction factor K may be applied to allow for shock losses in each airway in a mine ventilation system. Although an inexact procedure, it greatly simplifies calculations and can yield acceptable results when used properly, especially in a large system where the sources of shock are numerous and repetitive. Note that Table 5.1 includes several allowances for shock loss: sinuosity (curvature), obstructions, and timber sets. Judgment in selecting the appropriate value of K for calculations is critical, if shock losses of these types are to be evaluated accurately. McElroy (1935) also discusses a method of calculating an increment in K for various types of shock losses.

Calculation of Shock Loss by Equivalent Length Method

The recommended method of determining shock loss considered most useful in mine ventilation calculations is that expressing each significant loss in terms of the equivalent length of straight airway (McElroy, 1935). In other words, an increment in the length of airway is determined, similar to the increment in K discussed above.

A shock-loss expression for the equivalent length of straight airway can be found by equating the formulas for friction loss and shock loss (Eqs. 5.18 and 5.23):

$$H_x = H_f$$

$$XH_v = \frac{KLV^2}{5.2R_h}$$

$$X\frac{wV^2}{(1098)^2} = \frac{KLV^2}{5.2R_h}$$

TABLE 5.3   Equivalent Lengths for Various Sources of Shock Loss

Source	ft	(m)	Source	ft	(m)
Bend, acute, round	3	(1)	Contraction, gradual	1	(1)
Bend, acute, sharp	150	(45)	Contraction, abrupt	10	(3)
Bend, right, round	1	(1)	Expansion, gradual	1	(1)
Bend, right, sharp	70	(20)	Expansion, abrupt	20	(6)
Bend, obtuse, round	1	(1)	Splitting, straight branch	30	(10)
Bend, obtuse, sharp	15	(5)	Splitting, deflected branch (90°)	200	(60)
Doorway	70	(20)	Junction, straight branch	60	(20)
Overcast	65	(20)	Junction, deflected branch (90°)	30	(10)
Inlet	20	(6)	Mine car or skip (20% of airway area)	100	(30)
Discharge	65	(20)	Mine car or skip (40% of airway area)	500	(150)

Simplifying and solving for L but using the special symbol L_e to represent equivalent length:

$$L_e = \frac{5.2\ wR_hX}{K(1098)^2} = \frac{3235\ R_hX}{10^{10}\ K}\ \text{ft} \tag{5.24}$$

$$L_e = \frac{wR_hX}{2gK}\ \text{m} \tag{5.24a}$$

Calculation of, first, the shock loss factor X and then the corresponding equivalent length L_e for a given shock loss by formula is time-consuming. *Sufficient accuracy for routine calculations is obtained by selecting the appropriate value of L_e for a shock loss from* Table 5.3, then using it (in Eq. 5.25—see following) when computing overall head losses in airways. The table lists approximate average values of L_e calculated by formula and based on an airway with values of $K = 100 \times 10^{-10}$ lb·min²/ft⁴ (0.0186 kg/m³) and $R_h = 2$ ft (0.61 m) and on standard air specific weight.

Certain precautions are necessary in calculating shock losses by this method:

1. Values of L_e from Table 5.3 need *not* be corrected for K, R_h, or other conditions in most problems (the accuracy of the data and the shock-loss formulas generally do not warrant it).

2. With a change in area (splitting not involved) or an inlet, include the shock loss in the airway section *following* the change. This also applies to a bend in conjunction with area change. Discharge is an exception; include it in the section *preceding* the change.

3. At splits and junctions in airways, use only the portion of the total flow involved in a change of direction or area. Values from Table 5.3 assume an even division of flow and allow for bend and area change. Include loss at split or junction *within* the pressure drop for the particular branch.

4. Judgment must be exercised in making proper allowance for unusual sources of shock loss (e.g., obstructions).

5. Exclude shock loss due to inlet or discharge if a fan is located there.

Combined Head Losses and Mine Heads

The equivalent length method of handling shock loss permits a single calculation (from Eqs. 5.6 and 5.20) of the overall head loss for a given airway. Rewriting Eq. 5.20 to include equivalent length due to shock loss:

$$H_l = H_f + H_x = \frac{KO(L + L_e)Q^2}{5.2A^3} \qquad (5.25)$$

$$H_l = H_f + H_x = \frac{KO(L + L_e)Q^2}{A^3} \qquad (5.25a)$$

The mine heads are then determined by cumulating the airway head losses, according to Eqs. 5.7 and 5.8. *This procedure is recommended for all routine ventilation calculations.*

Example 5.7 Calculate the combined friction-shock loss for each airway shown in Fig. 5.12, and determine the head losses in the system and the

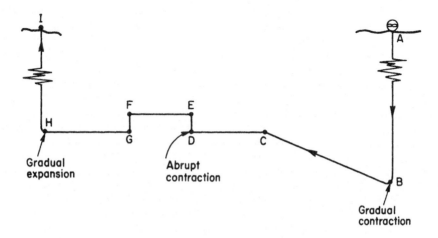

FIGURE 5.12 Mine ventilation system in Example 5.7.

mine heads. Given the following characteristics (with dimensions as tabulated below):

$$K = 125 \times 10^{-10} \text{ lb·min}^2/\text{ft}(0.0232 \text{ kg/m}^3)$$

$$Q = 20,000 \text{ cfm } (9.44 \text{ m}^3/\text{s})$$

$$w = 0.0750 \text{ lb/ft}^3(1.201 \text{ kg/m}^3)$$

Solution: Review Example 5.2. Select equivalent lengths from Table 5.3 and compute head loss by Eq. 5.25. A single calculation suffices for airways having the same dimensions. Sample calculation, AB + HI:

$$H_l = \frac{(125 \times 10^{-10})(60)(1677)(20,000)^2}{(5.2)(200)^3} = 0.0121 \text{ in. water}$$

Find velocity head in airway HI by Eq. 5.14:

$$V = 20,000/200 = 100 \text{ fpm}; \ H_v = (100/4009)^2 = 0.0006 \text{ in. water}$$

Airway	Size, ft	L, ft	L_e, ft[a]	$L + L_e$, ft	A, ft^2	O, ft	H_L, in. water (Pa)
AB	10 × 20	810	(None)	—	—	—	—[b]
BC	8 × 8	800	(b) (c) 3 + 1 = 4				
CD	8 × 8	350	(b) 15	1169	64	32	0.1372 (34.1)
DE	5 × 7	100	(b) (c) 70 + 10 = 80				
EF	5 × 7	250	(b) 70				
FG	5 × 7	100	(b) 70				
GH	5 × 7	400	(b) 70	1140	35	24	0.6136 (152.7)
HI	10 × 20	800	(b) (c) (d) 1 + 1 + 65 = 67	1677	200	60	0.0121 (3.0)
			mine $H_s = \Sigma H_L =$			0.7629 in. water (189.8 Pa)	
			mine H_v =			0.0006 (0.1)	
			mine H_t =			0.7635 (190.0)	

[a] Shock losses (L_e): (b) bend, (c) contraction, (d) discharge, (e) expansion.
[b] Computed with airway HI.

The answer can be rounded off to 0.764 in. water (190 Pa).

5.6 AIR POWER

Since power is the time rate of doing work, the power required to overcome the energy losses in an airstream, called the *air power* P_a, is, in basic units

$$P_a = pQ = 5.2HQ \text{ ft·lb/min}$$

and in more customary units

$$P_a = \frac{5.2HQ}{33,000} = \frac{HQ}{6346} \text{ hp} \tag{5.26}$$

$$P_a = \frac{HQ}{1000} \text{ kW} \tag{5.26a}$$

with H expressed in in. water (Pa), p in psf (Pa), and Q in cfm (m³/s).

If the total air power is desired, then the head H should be the mine total head H_t. If the mine static head H_s is used in Eq. 5.26, then only the static air power is obtained. The application of the power concept in mine ventilation becomes more evident in Chapters 7, 10, and 12, but for the present, it is worth noting that the power P_a varies directly as H but as the cube of Q (because $H \propto Q^2$).

Example 5.8 Calculate the air power in hp (kW) required for the ventilation system of Example 5.7.

Solution: Use Eq. 5.26:

$$P_a = \frac{(0.7635)(20,000)}{6346} = 2.406 \text{ hp } (1.794 \text{ kW})$$

5.7 COMPRESSIBILITY EFFECTS

Although compressibility has been disregarded to this point, air begins to demonstrate compressibility effects at relatively low pressures. The response is evident mainly in volume (quantity) and temperature (heat content), in accordance with Boyle's and Charles' laws, Eqs. 2.11 and 2.13. It also occurs in pressure (= head loss).

Accepting 5% as a mathematically significant effect, a barometric pressure difference of

$$p = (29.92)(0.05) = 1.5 \text{ in. (38 mm) Hg}$$

or a head loss of

$$H = (1.5)(13.6) = 20.3 \text{ in. (517 mm) water (or 5.06 kPa)}$$

or an elevation difference of

$$Z = (20.3)\left(\frac{5.2}{0.0750}\right) = 1411 \text{ ft (430 m)}$$

above or below atmospheric datum will produce measurable differences in air quantity, specific weight, and head loss. Thermodynamic effects on the heat content of air are more complex and involve autocompression as well (Section 5.8 and Chapter 16).

To demonstrate the compressibility effect numerically, refer again to the mine ventilation tubing problem of Example 5.6. Based on the system head loss (16.03 in. water) and standard atmospheric conditions (29.92 in. Hg = 407 in. water), the error due to neglect of compressibility is

$$\text{Error} = 1 - \frac{407}{407 + 15.03} = 3.6\%$$

That head (pressure) effect produces a change in quantity of

$$Q = (0.036)(48{,}000) = 1730 \text{ cfm } (0.82 \text{ m}^3/\text{s})$$

representing a flow contraction in a blower system and an expansion in an exhaust system:

Discharge of blower, $Q = 48{,}000 - 1730 = 46{,}300$ cfm (21.9 m³/s)

Intake of exhaust, $Q = 48{,}000 + 1730 = 49{,}700$ cfm (23.5 m³/s)

Only in the exhaust system, however, would a change of fan duty occur; the fan would need to be larger in size to accommodate the expanded flow at intake [49,700 cfm (23.5 m³/s)]. In a blower system, the fan takes in the same quantity [47,700 cfm (22.7 m³/s)], regardless of the compressibility effect, and there is no correction.

In high-pressure (>20 in. or 5.0 kPa) ventilation systems, it is advisable to consider compressibility effects by employing the head-loss formula for compressed airflow instead of the Atkinson equation, developed for air as an incompressible fluid (Weeks, 1926):

$$p_1^2 - p_2^2 = \frac{K_c Q^2 L}{D^5} \tag{5.27}$$

where p is absolute pressure in psia (Pa absolute) at two points in the system (1 and 2), Q is quantity of free air in cfm (m^3/s), D is diameter in in. (mm), and K_c is compression friction factor in $lb^2{\cdot}in.{\cdot}min^2/ft^7$ ($kg^2/s^2{\cdot}m^4$). Here, because of their magnitude, fluid flow potentials are expressed in units of pressure rather than head. Application of Eq. 5.27 is deferred until the topic of auxiliary ventilation is introduced in Chapter 11.

In summary, then, *there are two circumstances in mine ventilation where compressibility effects have to be considered and corrections applied to quantity and head:* (1) long ventilation-pipe installations in mines and tunnels where the pressure drop exceeds 20 in. (510 mm or 5.0 kPa), and (2) deep shafts or raises where the difference in elevation exceeds 1400 ft (427 m).

Seldom does the entire ventilation system for a mine require consideration of compressibility effects. Generally, the mine static head is under 10 in. (2.5 kPa) and seldom reaches 20 in. (5.0 kPa). When the mine is large, very deep, and thermodynamic and natural ventilation effects large, then analysis of the ventilation system using an indicator (p–v) diagram may be warranted, as in natural ventilation.

5.8 THERMODYNAMIC APPROACH TO MINE VENTILATION

As discussed in Sections 5.1 and 5.7, it is common to analyze airflow through mines as an incompressible flow problem on the assumption that changes in the air specific weight in mine ventilation are minor. However, this assumption is not reasonable where there is significant exchange of heat and moisture between the air and the mine surroundings, particularly in deep and hot mines. Here, as it flows through the mine, air undergoes wide variations in heat content and temperature and in pressure and density. Because the airflow processes in these cases are the same as the processes in a heat engine, a thermodynamic analysis of the mine ventilation system is possible. As mines or subsurface excavations for other purposes go deeper—and especially when natural ventilation is involved—the thermodynamic approach is not only desirable but may become essential. Hinsley (1950/51) developed the thermodynamic approach to mine ventilation, comparing the ventilation system to a heat engine. An extensive treatment of the thermodynamic approach to mine ventilation analysis can be found in McPherson (1993). Other useful references include Williams (1960a) and Hall (1967, 1981). Only a brief introduction is presented here.

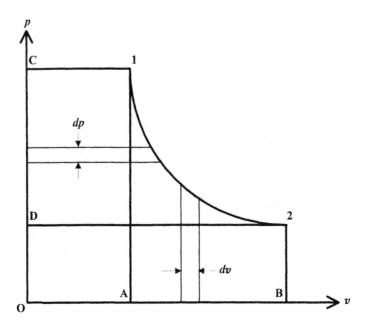

FIGURE 5.13   Indicator diagram for the system in Fig. 5.1.

Consider 1 lbm (kg) of air, whatever its volume, flowing between points 1 and 2 as shown in Fig. 5.1. Assume that no external mechanical work is done on the air (i.e., work energy added = 0), and no heat transfer takes place through the walls (i.e., external heat energy added or heat lost to the outside = 0). As the air flows from point 1 to point 2, there is a drop in the absolute pressure from p_1 to p_2 and, in turn, an increase in specific volume from v_1 to v_2. The indicator diagram (p–v diagram) for the process is shown in Fig. 5.13, the line from 1 to 2 representing the flow path.

Now analyze the work done by and on the 1 lbm (kg) of air flowing from point 1 to point 2. In entering the system and overcoming pressure p_1, the work done = p_1v_1 ft·lb/lbm (J/kg). In expanding from v_1 to v_2 and overcoming changes in elevation, changes in velocity, and frictional resistance of flow, the work done = $\int_{v_1}^{v_2} p\ dv$. In discharging from the system against pressure p_2, the work done on the air = p_2v_2. Therefore, the net work done by the air $p_1v_1 + \int_{v_1}^{v_2} p\ dv - p_2v_2$. From the indicator diagram, it may be seen that $\int_{p_1}^{p_2} v\ dp = p_1v_1 + \int_{v_1}^{v_2} p\ dv - p_2v_2$. The term $\int_{p_1}^{p_2} v\ dp$ is known as the flow work, and $\int_{v_1}^{v_2} p\ dv$ is called the non–flow work.

The Bernoulli equation for incompressible fluid flow (Eq. 5.2) follows, incorporating the external mechanical energy input between points 1 and 2. In this form, it is known as the mechanical energy equation for incompressible flow:

$$\frac{p_1}{w} + \frac{V_1^2}{2g} + Z_1 + \dot{W}_{12} = \frac{p_2}{w} + \frac{V_2^2}{2g} + Z_2 + H_{l_{12}} \qquad (5.28)$$

where $\dot{W}_{12}$ is mechanical energy in ft·lb/lb (J/kg) of air added between points 1 and 2, and $H_{l_{12}}$ is friction and shock energy (head) losses expended in ft·lb/lb (J/kg) of air between 1 and 2. In Eq. 5.28, any addition of heat energy between 1 and 2 does not appear explicitly nor is it required. The added heat energy will, however, be reflected in the mechanical energy values at 2, preserving the energy balance. Equation 5.28 can be rewritten as

$$\left(\frac{V_1^2}{2g} - \frac{V_2^2}{2g}\right) + (Z_1 - Z_2) + \dot{W}_{12} = v(p_2 - p_1) + H_{l_{12}} \qquad (5.29)$$

where the specific volume $v = (1/w)$ lb/ft³ (kg/m³), assuming dry air. In compressible flow, the change in pressure from p_1 to p_2 takes place in infinitely small steps, changing the specific volume accordingly. Therefore, Eq. 5.29 for compressible flow can be written

$$\left(\frac{V_1^2}{2g} - \frac{V_2^2}{2g}\right) + (Z_1 - Z_2) + \dot{W}_{12} = \int_{p_1}^{p_2} v \, dp + H_{l_{12}} \qquad (5.30)$$

The mechanical energy relation, Eq. 5.28, can incorporate the heat energy added between points 1 and 2 and the internal energy of the air. Noting that friction and shock energy losses result in changes in the internal and mechanical energy terms, ensuring a balance between the mechanical energy and the internal energy in the system, the energy relation equivalent to Eq. 5.28 is

$$v_1 p_1 + I_1 + \frac{V_1^2}{2g} + Z_1 + \dot{W}_{12} + \dot{q}_{12} = v_2 p_2 + I_2 + \frac{V_2^2}{2g} + Z_2$$

$$(5.31)$$

where I is internal energy and $\dot{q}_{12}$ is amount of heat energy added per unit weight of air between points 1 and 2. The traditional unit for internal and heat energies is the Btu (J). With appropriate conversion factors (1 Btu = 778 ft·lb), I and $\dot{q}$ can be expressed in specific energy (head) units (ft·lb/lb or J/kg).

Since the term $(vp + I)$ is the enthalpy h of the air, a state property (Section 2.2), Eq. 5.31 can be restated as

$$\left(\frac{V_1^2}{2g} - \frac{V_2^2}{2g}\right) + (Z_1 - Z_2) + \dot{W}_{12} = (h_2 - h_1) - \dot{q}_{12} \qquad (5.32)$$

where h_1 and h_2 are enthalpies at points 1 and 2, respectively. From Eqs. 5.30 and 5.32, it is evident that

$$\left(\frac{V_1^2}{2g} - \frac{V_2^2}{2g}\right) + (Z_1 - Z_2) + \dot{W}_{12} = \int_{p_1}^{p_2} v \, dp + H_{l_{12}} = (h_2 - h_1) - \dot{q}_{12}$$

$$(5.33)$$

Equation 5.33 explicitly states the balance that must exist among the mechanical, thermal, and internal energies of the mine ventilation system.

Using Eq. 5.30, an expression for the head losses can be written:

$$H_{l_{12}} = \left(\frac{V_1^2}{2g} - \frac{V_2^2}{2g}\right) + (Z_1 - Z_2) - \int_{p_1}^{p_2} v \, dp + \dot{W}_{12} \qquad (5.34)$$

The integral $\int_1^2 v \, dp$ can be evaluated in terms of measurable parameters from the relationships between p and v (the polytropic equation $pv^n = C$ and the general gas law $pv = RT$):

$$n = \frac{1}{1 - \dfrac{\ln(T_2/T_1)}{\ln(p_2/p_1)}} \qquad (5.35)$$

$$\int_{p_1}^{p_2} v \, dp = R(T_2 - T_1)\left(\frac{\ln(p_2/p_1)}{\ln(T_2/T_1)}\right) \qquad (5.36)$$

$$\int_{p_1}^{p_2} v \, dp = \left(\frac{n}{n-1}\right) R(T_2 - T_1) \qquad (5.37)$$

Thus, knowing the absolute pressures p, absolute temperatures T, air velocities V, and elevations Z at points 1 and 2, both the index of the process n and the friction and shock energy (head) losses H_l can be calculated for the flow between points 1 and 2, using the following relationship:

$$H_{l_{12}} = \left(\frac{V_1^2}{2g} - \frac{V_2^2}{2g}\right) + (Z_1 - Z_2) - \left(\frac{n}{n-1}\right) R(T_2 - T_1) \qquad (5.38)$$

If the mass flow in the airway and the airway dimensions are known, both the resistance of the airway and the coefficient of friction can also be calculated. The following example illustrates the calculations involved.

Example 5.9   The following measurements have been made at stations 1 and 2 along an inclined airway. The quantity of air measured at station 1 is 90,000 cfm (42.48 m³/s).

	Station 1	Station 2
Air velocity V, fps (m/s)	6.5 (1.98)	11.5 (3.51)
Absolute pressure p, psi (kPa)	13.5 (93.08)	13.9 (93.84)
Temperature t, °F (°C)	83 (28.33)	86 (30.00)
Elevation Z, ft (m)	500 (152.40)	−400 (121.93)
	above datum	below datum

Assume dry air conditions and that the airflow follows a polytropic law (pv^n = constant).

(a) Draw the indicator diagram and verify the relationship between $\int_1^2 v\, dp$ and $\int_1^2 p\, dv$.

(b) Calculate the friction and shock energy losses and the head loss between stations 1 and 2.

Solution: Using Eq. 5.35, the value of the process index is first calculated:

$$n = \cfrac{1}{1 - \cfrac{\ln(546/543)}{\ln(13.9/13.5)}} = 1.232$$

The specific weights and volumes of the air at 1 and 2 are then calculated:

$$w_1 = \frac{p_1}{RT_1} = \frac{(13.5)(144)}{(53.35)(460 + 83)} = 0.0671 \text{ lb/ft}^3 \ (1.0748 \text{ kg/m}^3)$$

$$v_1 = \frac{1}{w_1} = \frac{1}{0.0671} = 14.9031 \text{ ft}^3/\text{lb} \ (0.9304 \text{ m}^3/\text{kg})$$

$$w_2 = \frac{p_2}{RT_2} = \frac{(13.9)(144)}{(53.35)(460 + 86)} = 0.0687 \text{ lb/ft}^3 \ (1.1005 \text{ kg/m}^3)$$

$$v_2 = \frac{1}{w_2} = \frac{1}{0.0687} = 14.556 \text{ ft}^3/\text{lb} \ (0.9087 \text{ kg/m}^3)$$

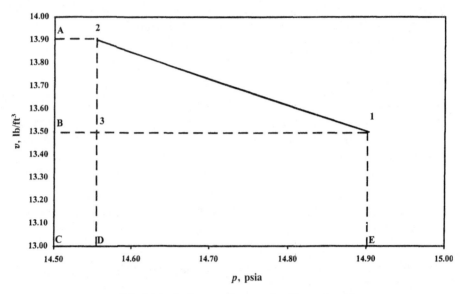

FIGURE 5.14 Indicator diagram for Example 5.9.

(a) The indicator diagram for the process is shown in Fig. 5.14. Assuming the process line 12 is a straight line, the following can be calculated:

$$p_1 v_1 = (144)(13.50)(14.90) = 28{,}965.60 \text{ ft·lb/lb } (86{,}578.18 \text{ J/kg})$$

$$p_2 v_2 = (144)(13.90)(14.56) = 29{,}143.30 \text{ ft·lb/lb } (87{,}109.32 \text{ J/kg})$$

$$\text{area } 1231 = \tfrac{1}{2}(14.90 - 14.56)(13.90 - 13.50)(144) = 9.792 \text{ ft·lb/lb } (29.29 \text{ J/kg})$$

$$\int_1^2 v \, dp = \text{area } 12AB1 = \text{area } 1231 + (144)(13.90 - 13.50)(14.56)$$

$$= 9.792 + 838.656 = 848.448 \text{ ft·lb/lb } (2536.00 \text{ J/kg})$$

$$\int_1^2 p \, dv = \text{area } 12DE1 = \text{area } 1231 + (144)(13.50)(14.90 - 14.56)$$

$$= 9.792 + 660.960 = 670.752 \text{ ft·lb/lb } (2004.88 \text{ J/kg})$$

From the indicator diagram

$$\int_1^2 v \, dp = p_2 v_2 - p_1 v_1 + \int_1^2 p \, dv = 29{,}143.30 - 28{,}965.60 + 670.752$$

$$= 848.552 \text{ ft·lb/lb } (2536.32 \text{ J/kg})$$

(b) The friction and shock energy losses per lb of air between stations 1 and 2 using Eq. 5.38 is

$$H_{l_{12}} = -1.39 + 900 - 849.92$$

$$= 48.69 \text{ ft·lb/lb} \ (145.53 \text{ J/kg})$$

Note that the last term in Eq. 5.38 is $\int_1^2 v \, dp = 849.92$ ft·lb/lb (2540.41 J/kg).

The head loss at standard specific weight ($w = 0.0750$ lb/ft^3 or 1.20 kg/m^3)

$$H_{l_{12}} = (48.69)(0.0750)$$

$$= 3.65 \text{ psf}$$

$$= 0.70 \text{ in. water} \ (174.36 \text{ Pa})$$

Head losses at w_1, w_2, or $\overline{w} = (w_1 + w_2)/2$ can be similarly calculated. Also other useful information can be obtained from the survey data:

1. Mass flow rate of air $G = (90,000)(0.0671) = 6039$ lb/min $= 100.65$ lb/h (45.65 kg/h).

2. Quantity of flow Q at station 2 $= (6039 \text{ lb/min}/0.0687 \text{ lb/ft}^3) = 87,904$ cfm (41.47 m^3/s).

3. Enthalpies of the air at 1 and 2 are, respectively, $c_p t_1$ and $c_p t_2$ (Eq. 2.10). Therefore, the change in enthalpy $= h_1 - h_2 = 0.24 \ (t_1 - t_2)$ $= 0.24 \ (83-86) = -0.72$ Btu/lb (-1674.72 J/kg), indicating an increase in the enthalpy from 1 to 2. Noting that 1 Btu $= 778$ ft·lb, the change in enthalpy in energy units $= 187.0159 \ (t_1 - t_2)$ in ft·lbf/lbm $= 187.015$ $(83-86) = -561.05$ ft·lbf/lbm (-1676.98 J/kg).

4. From Eq. 5.33, the heat energy input between stations 1 and 2 into the air is given by $\dot{q}_{12} = (h_2 - h_1) - \int_1^2 v \, dp - H_{l_{12}} = 561.05 - 849.92$ $- 48.69 = -337.56$ ft·lb/lbm (-1008.97 J/kg). Also note from Eq. 5.33 that heat input can be expressed as

$$\dot{q}_{12} = (h_2 - h_1) - \left(\frac{V_1^2}{2g} - \frac{V_2^2}{2g} \right) - (Z_1 - Z_2)$$

$$= 561.05 + 1.39 - [500 - (-400)]$$

$$= -337.56 \text{ ft·lb/lb} \ (-1008.97 \text{ J/kg})$$

The negative sign indicates that heat is being transferred from the air to the strata.

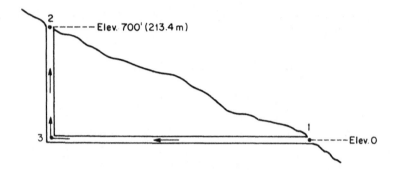

FIGURE 5.15   Mine ventilation system in Problems 5.1 and 5.2.

5. Rate of heat transfer $= (-337.56)(90,000)(0.0671)$ ft·lb/min

$$= -2,038,525 \text{ ft·lb/min}$$

$$= -61.77 \text{ hp } (-46.07 \text{ kW})$$

Essentially, with data on absolute pressures and temperatures at selected points in the mine and mass flow at a point in the system, and following the procedure of Examples 5.9 and 5.7, the mine ventilation system can be analyzed fully. Application of the thermodynamic approach to mine ventilation is further demonstrated under the topic of natural ventilation in Chapter 8.

PROBLEMS

5.1   Write and balance the Bernoulli general energy equation in terms of specific energy between points 1 and 2 of the mine ventilation system shown in Fig. 5.15, on the basis of (a) absolute pressure and (b) gage pressure, given the following:

$$
\begin{array}{ll}
H_{s_1} = 5.0 \text{ in. (127 mm) water} & \text{Drift } H_l = 3.5 \ (88.9) \\
H_{v_1} = 1.0 \ (25.4) & \text{Bend } H_l = 0.5 \ (12.7) \\
H_{s_2} = 0 & \text{Shaft } H_l = 1.5 \ (38.1) \\
H_{v_2} = 0.5 \ (12.7) &
\end{array}
$$

Disregard shock losses at inlet and discharge.

5.2   Plot and label the head gradients for (a) and (b) in Problem 5.1 showing the fan in each of three possible locations (blower, exhaust, and booster). Calculate the mine heads for each arrangement. Assume the booster is located at point 3.

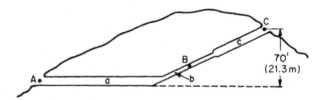

FIGURE 5.16 Mine ventilation system in Problems 5.3 and 5.4.

5.3 Plot and label the head gradients for the system shown in Fig. 5.16 with the energy source located (a) as a blower (at A), (b) as an exhauster (at C), and (c) as a booster (at B). The following heads and head losses occur:

Friction	
Duct a	0.5 in. (12.7 mm) water
Duct b	2.0 (50.8)
Duct c	1.0 (25.4)
Velocity	
Duct a	0.5 (12.7)
Duct b	1.5 (38.1)
Duct c	1.0 (25.4)
Shock	
Inlet	0.5 (12.7)
Contraction	1.0 (25.4)
Bend	0.5 (12.7)
Expansion	0.5 (12.7)
Discharge	1.0 (25.4)

5.4 Label and calculate the mine static, velocity, and total heads in each system of Problem 5.3.

5.5 (a) Calculate the mine static, velocity, and total heads for a ventilation duct system as follows: Air quantity is 3000 cfm (1.416 m^3/s), air specific weight is 0.0734 lb/ft^3 (1.176 kg/m^3), and the duct is steel, average condition, diameter 18 in. (457 mm), and length 750 ft (229 m). Disregard shock loss and leakage.
(b) Check the mine static head by a graphic solution.

5.6 Using Eq. 5.24 as a point of departure, develop an equation that can be used to correct the values of equivalent length in Table 5.3 for nonstandard values of K and w and other values of hydraulic radius.

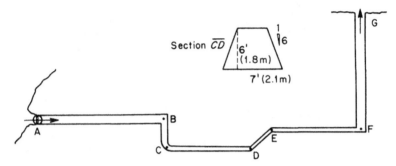

FIGURE 5.17   Mine ventilation system in Problems 5.7 and 5.9.

5.7 (a) Calculate the air quantity and mine static, velocity, and total head requirements for the mine ventilation system shown in Fig. 5.17, assuming that the air specific weight is 0.0650 lb/ft³ (1.041 kg/m³) and the velocity in drift EF is 1975 fpm (10.0 m/s). Disregard shock losses. All workings are in igneous rock; all cross sections are rectangular except CD. Dimensions and condition of the openings are as follows:

Airway	Size, ft (m)	Length, ft (m)	Condition
AB	8 × 12 (2.44 × 3.66)	650 (198.1)	Bare, average
BC	8 × 8 (2.44 × 2.44)	100 (30.5)	Timbered, average
CD	6 × 7 (1.83 × 2.13)	480 (146.3)	Timbered, bad
DE	4 × 6 (1.22 × 1.83)	90 (27.4)	Timbered, average
EF	4 × 6 (1.22 × 1.83)	485 (147.8)	Timbered, average
FG	10 × 14 (3.05 × 4.27)	560 (170.7)	Concreted, average

    (b) Check the mine static head by a graphic solution.

5.8 Assume that an overcast exists in a coal mine entry that possesses a K of 50×10^{-10} lb·min²/ft⁴ (0.00928 kg/m³). If the entry is 7 × 12 ft (2.13 × 3.66 m), what is the equivalent length for the shock loss due to the overcast using the value provided in Table 5.3 (adjust the equivalent length to reflect the change in the K value and hydraulic radius)?

5.9 Solve Problem 5.7 considering all shock losses. Compare answers and compute the error that occurs in neglecting shock losses.

5.10 Determine the power cost in $/yr that is due to the existence of the overcast in Problem 5.8. Assume that the unit power cost is $0.04/ kWh, operation is continuous, and 25,000 cfm (11.80 m³/s) passes through the overcast.

5.11 Calculate the air power for Problems 5.7 and 5.9.

5.12 Plot and label the head gradients for Problem 5.9.

5.13 Engineers in nearly all other countries use SI units in their calculations. If friction loss calculations are performed using Eq. 5.20a, what are the correct SI units of K?

5.14 Assume that an airway has a K value of 100×10^{-10} lb·min^2/ft^4. Determine (a) the multiplier that must be used to convert the K value into the units used in Problem 5.13 and (b) the numerical value of K for the airway above when expressed in SI units.

CHAPTER 6

Ventilation Measurements and Surveys

6.1 INTRODUCTION

To ensure effective and efficient mine ventilation, reliable data must be obtained on various aspects of the ventilation system. Such data form the basis for evaluation and planning and for improvements or changes that may be required for quantity and quality control. The direct measurement of air properties and/or calculation of some air properties from other measurements are the only means for obtaining reliable data.

The most important properties for quantity control are air velocity and air pressure or head. When these properties are known at various points in the mine, the overall quantity-control problem can be resolved by utilizing analytical and/or numerical methods of fluid flow. Other properties of air such as temperature and specific weight also must be known for quantity control. They are, however, more important in air conditioning and temperature–humidity control.

The status of the mine ventilation system can be assessed with regular spotchecks of the direction and quantity of airflow. Critical sections with unusual ventilation problems may require a continuous air-quality and air-quantity monitoring system to ensure a safe mine environment. These measurements are no substitutes for comprehensive ventilation surveys, which must be performed periodically to (1) obtain knowledge of the extent and adequacy of the existing ventilation system in meeting specific needs, standards, and regulations; (2) provide information for use in emergencies or disasters, such as fires, explosions, major cave-ins, or floodings; (3) plan for the improvement of present environmental conditions or the efficiency of the present ventilation system; and (4) make provisions for mine expansion or modifications, new fan installations, changes in airways or circuits, and new air shafts. Four major areas are included under the general heading of ventilation surveys: air quantity, pressure or head, temperature, and air quality.

In temperature surveys, measurements may include virgin-rock temperatures and the air dry- and wet-bulb temperatures. Because the specific weight of air is affected by the amount of water vapor in it, when pressure surveys are conducted, it is necessary to know the dry- and wet-bulb temperatures. Also, in airways where there is an appreciable change in air velocities, allowance must be made for changes in velocity heads. Therefore, it is common practice to conduct air quantity, pressure or head, and temperature surveys concurrently.

6.2 TEMPERATURE MEASUREMENT

The major purpose of temperature measurements in mines is to assess the specific weight, humidity, and cooling power of the air. Additionally, in situ rock temperature data are necessary to determine the heat input from the surrounding rock into the mine atmosphere.

Air Temperature Measurements

The temperature of the mine air can be readily measured by conventional *mercury-type thermometers* to obtain what is commonly referred to as the *dry-bulb temperature* t_d. In hot, humid environments, the *wet-bulb temperature* t_w is measured by a *wet-bulb thermometer*. In principle, the dry-bulb temperature is a measure of the sensible heat of the atmosphere, whereas the dry-bulb temperature minus the wet-bulb temperature is a measure of the evaporative rate of the air.

Dry- and wet-bulb temperatures are measured simultaneously by a sling psychrometer or whirling hygrometer. Stationary hygrometers, some of the recording type, are occasionally used at main-fan installations, but for routine work, the sling psychrometer is widely used (Fig. 6.1). It consists of two identical thermometers; one measures the dry-bulb temperature, and the other, with a silk or cotton sleeve around its bulb, measures the wet-bulb temperature. Both thermometers are mounted side by side in a rigid frame; they are generally 6–10 in. (152.4–254.0 mm) in length, graduated from 0 to 120°F (-18 to 49°C). A water reservoir, filled with distilled or tap water, keeps the wet-bulb sleeve wet at all times during measurement. The frame is attached by a swivel connection to a handle for whirling.

With saturated air, the wet- and dry-bulb temperatures are the same, and evaporation is inhibited. With unsaturated air, water evaporates from the surface of the wet bulb at a rate that is inversely proportional to the vapor pressure of the moisture in the ambient air; for example, if the air contains little moisture, its vapor pressure is low and evaporation takes place rapidly. Evaporation cools the bulb, and the temperature drops. Equilibrium is achieved when the amount of heat transferred to the sleeve by convection equals the amount of latent heat needed to sustain the evaporation. At this stage, the correct wet-bulb temperature is reached.

The recommended measurement procedure is as follows. The operator, facing upstream and holding the handle of the instrument at arm's length in

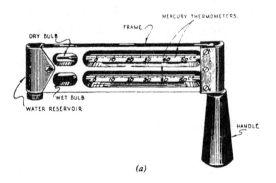

(a)

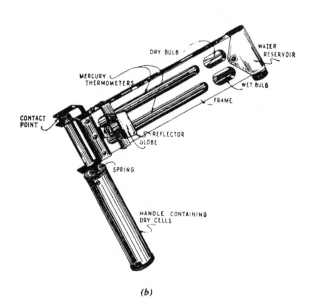

(b)

FIGURE 6.1 Sling psychrometers for measuring dry- and wet-bulb temperatures: (*a*) conventional psychrometer; (*b*) psychrometer with illuminated scales. (After Anon., 1972b. By permission from Chamber of Mines of South Africa.)

front of his/her body, whirls the psychrometer at 2–3 rev/s to attain a linear bulb velocity of at least 250 fpm or 1.27 m/s (Parczewski and Hinsley, 1956/ 57). Whirling also minimizes radiation effects on the wet-bulb thermometer from surrounding objects, such as the operator himself, machinery, the walls of the opening, and so forth. At the recommended whirling speed, the effect of radiation on the wet-bulb thermometer reading is also fully eliminated. Whirling is not necessary when the existing air velocity substantially exceeds the required linear bulb velocity. After whirling for at least a minute, the two temperatures are read rapidly, the wet-bulb first, since it tends to rise when whirling stops. Several observations should be taken until consistent temperatures and a minimum wet-bulb reading are obtained. For maximum precision, readings should be estimated to the nearest fraction of a degree depending on the thermometer scales.

Rock Temperature Measurement

Knowledge of the rock temperature is important for temperature–humidity air conditioning (Chapter 17). To find the geothermal gradient in the rock mass and the rate of heat flow from the wall rock to the airways, it is necessary to determine the temperature distribution in the rock adjacent to the opening, including the *virgin-rock temperature* (temperature unaffected by the mine opening and mining operations). Analytical and empirical methods may yield approximate values, but for an accurate temperature profile, in situ temperature measurement is necessary. Rock temperatures are usually measured by using any of the following instruments: mercury-in-glass thermometer, resistance thermometer, or thermocouple.

Although a conventional thermometer is simple to use, some type of *maximum-reading thermometer* with metallic protective case is preferred. In the constrictive-type thermometer, a constriction in the glass column retains the mercury at the point of its highest rise; whereas in the float-type thermometer, the float remains at the maximum reading for a sufficient time to allow retrieval and reading of the thermometer before subsiding.

Resistance thermometers, made of a good metallic conductor such as copper, nickel, platinum, or silver, increase in electrical resistance when heated. On the other hand, the *thermistor,* a variation of the resistance thermometer, decreases in resistance as the temperature increases. The thermistor, made from a small bead of semiconducting material such as oxide of nickel, manganese, or cobalt, has a high gage factor, the ratio of resistance change with temperature. The ratio is, however, nonlinear, and calibration is required. *Thermocouples* are more frequently used in underground temperature measurement. On the basis of the thermoelectric-effect principle, junction points of wire pairs at different temperature generate voltages that are proportional to the temperature difference. When a thermocouple is connected to a servobalance recorder, an analog or continuous record of temperature can be obtained, making it highly suitable for temperature logging of boreholes as well as indicating when the virgin-rock temperature is reached.

Rock temperatures are measured from boreholes that are drilled either from the surface or from underground openings (Vost, 1976). Measurements may be scheduled either during breaks in drilling or after the entire hole is completed. In either case, it is a good practice to allow sufficient time for the rock temperature in the borehole to stabilize and eliminate the heating and cooling effects of drilling and drilling fluids. Measurements during breaks in drilling are usually carried out at the bottom of the borehole, where the time for temperature equilibrium to be reestablished is usually short, since the heating and cooling effects tend to nullify each other (Mullins and Hinsley, 1958). Waiting periods of 1–8 h yield sufficient accuracy in temperature measurements at the bottom of the borehole.

The temperature profile may also be determined after completion of the borehole. Equilibrium temperature is usually reached within 8 h after drilling stops. Other investigators use the time required to drill the borehole as the time required to reach equilibrium temperature. With a completed borehole, one or more thermal sensors are set at predetermined intervals and allowed to stabilize for at least one hour to yield the rock temperatures at those depths. A plot of temperature versus depth yields the temperature distribution and geothermal gradient in the borehole. When the temperature becomes asymptotic, the virgin-rock temperature is reached. A formula has been derived for determining the virgin-rock temperature from a number of measurements (Morris, 1953a):

$$t_r = \frac{t_{z/2}^2 - t_z t_0}{2t_{z/2} - (t_z + t_0)} \tag{6.1}$$

where t_r is virgin-rock temperature, t_0 and t_z are temperatures measured at the rock surface and at depth z in ft (m), respectively, and $t_{z/2}$ is the temperature at depth $z/2$ in °F (°C).

The science of thermometry and temperature measurements is discussed in great detail by Van der Walt and Hemp (1989).

6.3. AIR SPECIFIC-WEIGHT DETERMINATIONS

The specific weight w (density in SI units) of normal air is required in ventilation and air conditioning calculations, precise ventilation surveys, and selection of mine ventilation equipment. It is not usually measured directly but instead is determined from other air properties. From the general gas laws, the following formula can be derived for air specific weight (Section 2.2):

$$w = \frac{1.325}{T_d} (p_b - 0.378 p_v') \tag{6.2}$$

$$w = \frac{1}{0.287 \, T_d} (p_b - 0.378 p_v') \tag{6.2a}$$

where w is air specific weight in lb/ft^3 (kg/m^3), T_d is absolute dry-bulb temperature in °R (K), p_b is atmospheric pressure in in. mercury (kPa), and p_v' is the vapor pressure at dew point in in. Hg (kPa). Other formulas can also be derived relating air specific weight with other psychometric properties of air. For convenience, graphic solutions and charts to determine w have been devised, and a suitable one for mine ventilation work is shown in Fig. 2.3 (McElroy, 1935). The chart relates t_d, t_w, and p_b of the air with its specific weight.

Humidity Calculations

Knowing the dry- and wet-bulb temperatures and the barometric pressure, one can determine specific or relative humidities using psychometric charts or tables (Section 2.2). Additionally, Table 6.1 can be used to estimate relative humidities with negligible error, for altitudes up to 3000 ft above sea level. The table also provides the value for 0.378 p_v' in Eq. 6.2.

6.4 VELOCITY MEASUREMENT

Whether as a part of a ventilation survey or of routine determinations mandated by the mining regulations, air velocity is the property most frequently measured in mine airways and workplaces. Several instruments are currently available for determining air velocity, and additional ones are being conceived and developed, based on modern and advanced principles of electronic mensuration (Hardcastle et al., 1993). These instruments cannot measure air velocity directly, but instead measure the physical effects of air or gas in motion, from which air velocity is inferred. Among the physical effects commonly measured are (1) the mechanical effect of air on moving objects, (2) the pressure effect produced by moving air, and (3) the cooling effect of air in motion. Some are accurate only in certain air-velocity ranges, but a few are suitable for use over a broad range of air velocities. For convenience in classifying instruments, the following arbitrary velocity ranges are adopted: (1) low air velocity (less than 100 fpm or 0.508 m/s), (2) intermediate air velocity (100–750 fpm, or 0.508–3.81 m/s), and (3) high air velocity (over 750 fpm or 3.81 m/s). The characteristics of velocity-measuring instruments that are suitable for use underground are summarized in Table 6.2. Except for the hot-wire anemometer, all of these instruments have the common features of simplicity, compactness, and safety for use in any environment. However, approval for their use in mining must be obtained.

Smoke Tube The *smoke tube* method, based on the mechanical effect of air in motion, is commonly used for determining the presence of moving air, the direction of flow, and the approximate velocity in low-velocity flows. The device consists of a glass or plastic tube containing a smoke-generating reagent, such as pumice stone saturated with anhydrous tin or titanium chloride. One manufacturer currently uses ethylene diamine plus acetic acid.

TABLE 6.1 Relative Humidities for 30.0 in. Hg Barometric Pressure

Air temperature, t_d °F	$0.378 \times p'_v$	— Depression of wet-bulb thermometer ($t_d - t_w$), °F																			
		0.5	1.0	1.5	2.0	2.5	3.0	3.5	4.0	4.5	5.0	5.5	6.0	6.5	7.0	7.5	8.0	8.5	9.0	9.5	10.0	10.5
20	0.039	92	85	77	70	62	55	48	40	33	26	19	12	5								
21	.041	92	85	78	71	63	56	49	42	35	28	21	15	8	1							
22	.043	93	86	78	71	65	58	51	44	37	31	24	17	11	4							
23	.045	93	86	79	72	66	59	52	46	39	33	26	20	14	7	1						
24	.047	93	87	80	73	67	60	54	47	41	35	29	22	16	10	4						
25	.049	94	87	81	74	68	62	55	49	43	37	31	25	19	13	7	1					
26	.051	94	87	81	75	69	63	57	51	45	39	33	27	21	16	10	4					
27	.054	94	88	82	76	70	64	58	52	47	41	35	29	24	18	13	7	2				
28	.057	94	88	82	76	71	65	59	54	48	43	37	32	26	21	15	10	5				
29	.059	94	88	83	77	72	66	60	55	50	44	39	34	28	23	18	13	8	3			
30	.062	94	89	83	78	73	67	62	56	51	46	41	36	31	26	21	16	11	6	1		
31	.065	94	89	84	78	73	68	63	58	52	47	42	37	33	28	23	18	13	8	4		
32	.068	95	89	84	79	74	69	64	59	54	49	44	39	35	30	25	20	16	11	7	2	
33	.071	95	90	85	80	75	70	65	60	56	51	46	41	37	32	27	23	18	14	9	5	0
34	.074	95	90	86	81	76	71	66	62	57	52	48	43	38	34	29	25	21	16	12	8	3
35	.077	95	91	86	81	77	72	67	63	58	54	49	45	40	36	32	27	23	19	14	10	6
36	.080	95	91	86	82	77	73	68	64	60	55	51	46	42	38	34	29	25	21	17	13	9
37	.083	95	91	87	83	78	74	69	65	61	57	53	48	44	40	36	31	27	23	19	15	11
38	.086	96	91	87	83	79	75	70	66	62	58	54	50	46	42	37	33	29	25	21	17	14
39	.090	96	92	87	83	79	75	71	67	63	59	55	51	47	43	39	35	31	27	24	20	16
40	.093	96	92	87	83	79	75	71	68	64	60	56	52	48	45	41	37	33	29	26	22	18
41	.097	96	92	88	84	80	76	72	69	65	61	57	54	50	46	42	39	35	31	28	24	20
42	.101	96	92	88	85	81	77	73	69	65	62	58	55	51	47	44	40	36	33	30	26	23
43	.105	96	92	88	85	81	77	73	70	66	63	59	55	52	48	45	42	38	35	31	28	25
44	.109	96	93	89	85	81	78	74	71	67	63	60	56	53	49	46	43	39	36	33	30	26
45	.113	96	93	89	86	82	78	74	71	67	64	61	57	54	51	47	44	41	38	34	31	28
46	.117	96	93	89	86	82	79	75	72	68	65	61	58	55	52	48	45	42	39	35	32	29
47	.122	96	93	89	86	82	79	75	72	69	66	62	59	56	53	49	46	43	40	37	34	31
48	.126	96	93	90	86	83	79	76	73	69	66	63	60	57	54	50	47	44	41	38	35	32
49	.131	96	93	90	86	83	80	76	73	70	67	64	61	57	54	51	48	45	42	39	36	34

		59	61	62	64	66	68	70	72	74	75	77	79	81	83	85	87	89	91	94	96	98
50	.136	35	38	41	43	46	49	52	55	58	61	64	67	71	74	77	80	83	87	90	93	96
51	.141	36	39	42	45	47	50	53	56	59	62	65	68	71	75	78	81	84	87	90	94	97
52	.146	37	40	43	46	49	51	54	57	60	63	66	68	72	75	78	81	84	87	90	94	97
53	.152	39	41	44	47	50	52	55	58	61	63	66	69	72	75	78	81	84	87	90	94	97
54	.158	40	42	45	48	50	53	56	59	61	64	67	70	73	76	79	82	85	88	91	94	97
55	.163	41	43	46	49	51	54	57	59	62	65	68	70	73	76	79	82	85	88	91	94	97
56	.169	42	44	47	50	52	55	57	60	63	65	68	71	73	76	79	82	85	88	91	94	97
57	.176	43	45	48	50	53	55	58	61	63	66	69	71	74	77	80	82	85	88	91	94	97
58	.182	44	46	49	51	54	56	58	61	64	66	69	72	74	77	80	83	85	89	91	95	97
59	.189	45	47	49	52	55	57	59	62	65	67	70	72	75	78	80	83	86	89	92	95	97
60	.196	46	48	50	53	55	58	59	63	65	68	70	73	75	78	81	83	86	89	92	95	97
61	.203	47	49	51	54	56	58	60	63	65	68	71	73	76	78	81	84	86	89	92	95	97
62	.210	47	50	52	54	57	59	61	64	66	69	71	74	76	79	81	84	87	90	92	95	97
63	.217	48	50	53	55	57	60	61	64	67	69	71	74	77	79	82	84	87	90	92	95	97
64	.225	49	51	53	56	58	60	62	65	67	70	72	74	77	79	82	84	87	90	92	95	97
65	.233	50	52	54	56	59	61	63	66	68	70	72	75	77	80	82	85	87	90	92	95	97
66	.241	51	53	55	57	59	61	63	66	68	71	73	75	78	80	82	85	88	90	92	95	97
67	.250	51	53	56	58	60	62	64	66	69	71	73	76	78	80	83	85	88	91	92	95	97
68	.259	52	54	56	58	60	62	64	67	69	71	74	76	78	80	83	85	88	91	92	95	97
69	.263	53	55	57	59	61	63	65	67	70	72	74	76	79	81	83	86	89	91	93	95	97
70	.277	53	55	57	59	61	63	65	68	70	72	74	77	79	81	84	86	89	91	93	95	98
71	.286	54	56	58	60	62	64	66	68	70	72	75	77	79	81	84	86	89	91	93	95	98
72	.296	55	57	59	61	63	64	66	69	71	73	75	77	79	82	84	86	89	91	93	95	98
73	.306	55	57	59	61	63	65	67	69	71	73	75	78	80	82	84	87	89	91	93	95	98
74	.317	56	58	60	61	63	65	67	69	71	74	76	78	80	82	84	87	89	91	93	95	98
75	.328	56	58	60	62	64	66	68	70	72	74	76	78	80	82	84	86	89	91	93	96	98
76	.339	57	59	61	62	64	66	68	70	72	74	76	79	80	82	84	87	89	91	93	96	98
77	.350	57	59	61	63	65	67	69	71	72	74	77	79	81	83	85	87	89	91	93	96	98
78	.362	58	60	62	63	65	67	69	71	73	75	77	79	81	83	85	87	89	91	93	96	98
79	.374	58	60	62	64	66	68	69	71	73	75	77	79	81	83	85	87	89	91	93	96	98
80	**.387**	**59**	**61**	**62**	**64**	**66**	**68**	**70**	**72**	**74**	**75**	**77**	**79**	**81**	**83**	**85**	**87**	**89**	**91**	**94**	**96**	**98**

Source: McElroy and Kingery (1957)

TABLE 6.2 Instruments for Measuring Velocity in Mines

Instrument	Velocity Range, fpm	Sensitivity, fpm	Accuracy	Features
Smoke tube	20–120 (low)	5–10	70–90%	Indirect, approximate
Vane anemometer	150–2000 (intermediate to high)	10–25	80–90%	Needs calibration, needs maintenance
	2000–10,000 (very high)	50–100		
Velometer	30–3000 (low to high) multirange	5–10 25–50	3% of upper-scale reading	Rapid, direct reading, delicate, needs maintenance
Thermoanemometer Thermometer	10–500 (low to intermediate)	2–10	80–95%	Slow, delicate, requires power (6 V), safe
Hot-wire	10–300 100–3000 (low to high) multirange	1–2 10–20	90–95%	Rapid, direct reading, delicate, requires power, needs maintenance
Kata thermometer	100–1500 (intermediate to high)	10–25	70–90%	Indirect, slow, delicate
Pitot tube	750–10,000 (high)	10–25	90–98%	Slow, indirect, accurate

Conversion factor: 1 fpm = 0.00508 m/s.

One end of the tube is attached to a rubber aspirator bulb, and the other end is open to the atmosphere. In operation, the bulb is squeezed a number of times, causing the tube to emit a cloud of smoke into the airway. It is good practice to hold the tube normal to the airway axis when generating smoke in order to prevent inducing flow or inaccurate measurements. The presence and direction of moving air is indicated by the motion of the smoke, and the air velocity can be determined by timing the travel of the smoke over a premeasured distance. Since measurement usually occurs near the centerline of the airway, where air velocity is highest, a correction factor must be applied to determine the average air velocity (Section 5.4). If the Reynolds number is known, the correction factor can be calculated, but for approximate work, a value of 0.8 is usually adopted. When the device is not in use, it is advisable to either cap the tube or disconnect the tube and plug both ends.

Vane Anemometer The *vane anemometer* is the standard instrument for general mine ventilation work. More frequently simply termed the *anemome-*

ter, it is a small windmill-like instrument consisting of a number of radial blades or impellers, gearing mechanism, clutch system, and pointer (Fig. 6.2*a*). In operation, it can be handheld but is preferably mounted on an extension rod or wand and is held with the blades oriented normal to the direction of air motion. The gearing mechanism translates the rotation speed of the blades directly into linear airflow in ft (m), which is read on a marked dial. The start and the termination of a measurement is accomplished by engaging or disengaging the clutch. Disengaging stops the recording and locks the pointer at the current value of the total linear airflow during the time of measurement. Most anemometers are provided with a switch for setting the readings to zero before the start of measurement. Digital anemometers provide direct readouts of the linear feet of air travel (Fig. 6.2*b*).

Anemometers are used mainly in the intermediate- and high-velocity ranges, although specially designed anemometers are available for low or very high air-velocity ranges (2000–10,000 fpm, or 10.16–50.8 m/s). Vane anemometers can register a velocity within plus or minus 10% of the true velocity. However, anemometers are usually accurate only within a narrow range of air velocities, so that routine mine ventilation measurements at a wide range of air velocities must use a calibration chart to correct observed velocities (Boshkov and Wane, 1955). These charts are usually developed and provided by the manufacturers with the instrument. For more accurate and reliable measurements, self-calibration is recommended following a well-established simple procedure (Haney, 1980).

The vane anemometer has been the primary instrument for airflow measurement in mines since the early 1900s. In order to obtain accurate velocity measurements, it is necessary to know the proper measurement techniques and understand the source and nature of errors inherent in vane anemometry. The errors are broadly classified as (1) human errors, (2) errors due to the design of the instrument, and (3) errors due to variation in airflow.

For correct measurements, the anemometer should be held with the blades oriented normal to the direction of air motion. In practice, especially in continuous traversing, it is difficult to maintain its perpendicular orientation to the airstream, and a certain yaw is inevitable. Several investigators, however, have found that the accuracy of velocity readings is not significantly affected when the yaw is less than 20° (Swirles and Hinsley, 1953/54; Boshkov and Wane, 1955).

Swirles and Hinsley (1953/54) have compared anemometer readings when the body position was downstream, upstream, or to the side of the anemometer. The tests were carried out in openings with cross-sectional areas greater than 60 ft^2 (5.57 m^2), so that the reduced area due to the presence of the operator did not have a significant effect on the mean velocity. Their study showed that if the operator stays more than 4 ft (1.22 m) away from the anemometer, the error in velocity readings will be less than 2%. For ease of measurement, however, the downstream location is recommended.

These investigators also compared velocities obtained with an anemome-

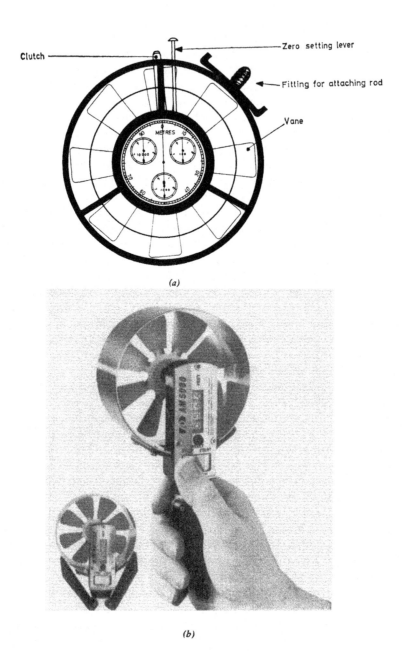

(a)

(b)

FIGURE 6.2 Vane anemometers for measuring air velocity. (*a*) Dial type. (After Anon., 1972b. By permission from Chamber of Mines of South Africa.) (*b*) Digital type. [Airflow Developments, Ltd. (Canada), Richmond Hill, Ontario.]

ter held by an operator in the plane of measurement with velocities obtained without an operator present. The tests, carried out in openings more than 70 ft² (6.50 m²) in area, showed a 15% increase in velocity with the handheld anemometer. Tests in a smaller area of 30 ft² (2.79 m²) showed that handheld anemometer readings were unacceptably high and unpredictable. By using an extension rod, errors due to handling the anemometer and relative body position of the operator are largely eliminated. The extension rod also facilitates measurement in large openings.

In traversing with the vane anemometer, the movement of the instrument itself affects the velocity measurement. The resulting error can be significant in low-velocity cases. To reduce the error from this source, a uniform, relatively slow traversing speed of less than 40 fpm (0.203 m/s) is recommended (Swirles and Hinsley, 1953/54).

Errors in instrument design occur mainly as a result of the inertia of the gears. At the start of measurement, a certain time is required for the vanes to accelerate to the rotational speed that is proportional to the air velocity. By allowing sufficient time for measurement, not less than one minute per reading, this problem is substantially eliminated. Inertial effects are also involved in measurements of pulsating flow. Anemometers are insensitive to sudden decreases, and to a lesser degree increases, in velocity that accompanies pulsating flow, and the indicated readings are usually high. Investigators have found, however, that if the amplitude of pulsation is less than 14% of the mean observed value, the errors are negligible; but when the amplitude exceeds this figure, the errors rise rapidly and must be corrected (Drummond, 1974). If practicable, measurements in areas of pulsating flow should be avoided.

Errors due to variations in airflow are due to specific weight differences and nonuniform velocity distributions. Difference in air specific weight when the anemometer is calibrated and when it is used underground have been found to affect the accuracy of anemometer readings, since air specific weight relates the mass flow with quantity (Swirles and Hinsley, 1953/54; Ower, 1966). When the specific weight difference exceeds 5%, especially for low air velocities, the following correction should be applied:

$$R_a' = R_a + C_a \sqrt{\frac{w_c}{w_m}} \tag{6.3}$$

where R_a' is the corrected anemometer reading, R_a is the indicated anemometer reading, C_a is the correction obtained from the anemometer calibration curve, and w_c and w_m are air specific weights at the time of calibration and at the time of measurement, respectively.

The velocity distribution varies not only across the mine opening but also across the smaller cross-sectional area of the anemometer. This variation is more pronounced in smaller mine openings. Thus during measurements, the blades of the anemometer are acted on by different forces at different sectors of the anemometer. Analysis of the problem showed that accurate results can be obtained only when the ratio of diameters of the airway and the anemometer is greater than 6 (Ower, 1966). Experiments comparing measurements in a mine gallery with an orifice-plate standard showed that turbulence, itself unpredictable, is the major source of error in anemometry (Teale, 1958).

Even with all these problems and error sources, the continuous traverse method with anemometers is still the most suitable technique for rapid determinations of air velocity. (Mbuyikamba et al., 1992). However, all air measurements with anemometer should be checked by repeat measurements until a satisfactory comparison of velocities is obtained. Check measurements within 2% can be obtained in coal mines (Kingery, 1960).

Velometer   The *velometer,* a direct-reading velocity meter, is used widely in industrial surveys of ducts. The device, based on the mechanical effect of moving air, has many advantages for use in underground work. It consists of a jewel-mounted and spring-held vane located in a rectangular passage with a pointer and dial, all housed in a molded case. In operation, air enters a port on one side of the instrument, deflecting the vane. The amount of deflection, directly proportional to the air velocity, is indicated by the pointer on a marked dial. The velometer is a versatile instrument, since it may be supplied with interchangeable jet probes and multiple scales, permitting accurate measurements of velocities over a broad range as well as in large airways, pipes, and grills. Its principal shortcoming is its inability to measure an integrated, average velocity for an airway. For use in dusty mine atmospheres, a special filter is attached to the intake port of the instrument. Because of its directional sensitivity, the correct reading is the maximum reading obtained when it is rotated gently from side to side.

Pitot Tube   The pitot-static tube, commonly called the *pitot tube,* is a widely used device for measurements in the high-velocity range underground and in laboratory testing installations. The device, based on the pressure effect of moving air, consists of two concentric tubes bent in an L shape. The inner tube is open at both ends, whereas the outer tube is closed at one end but perforated with small openings a short distance back from the facing tube. Each tube is connected to a manometer in three different ways for different head measurements (Fig. 6.3). In operation, the instrument is pointed directly into the airflow, and the pressure effect of moving air depresses the fluid in the manometer. The difference in elevation in the fluid is a measure of the pressure or head of the air.

For air-velocity measurements, each tube is connected to opposite legs of a manometer, as shown in Fig. 6.3a. The indicated reading is a measure of the velocity head H_v. The corresponding air velocity can then be determined from nomographs relating velocity and velocity heads (Appendix Fig.

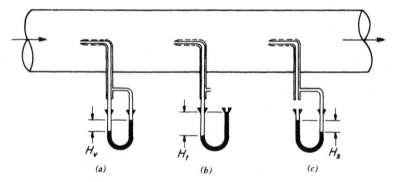

FIGURE 6.3   Pitot-tube and manometer connections to measure different heads in ventilation pipe or tubing.

A.1) or from the velocity-head relation (Eq. 5.13). To obtain average velocities from a series of pitot-tube measurements, the velocity for each reading should be determined analytically or graphically and these results averaged. Calculating the average velocity from the arithmetic average of the velocity heads would be incorrect because of the nonlinear relationship of velocity head; the error, however, is small.

For measuring total head H_t, the inner tube is connected to one leg of the manometer with the other leg open to the atmosphere, as in Fig. 6.3b. This arrangement measures the total pressure, so that the manometer reading gives the total head of the air. For measuring static head H_s, only the outer tube is connected to a manometer (Fig. 6.3c). Since the openings of the outer tube are perpendicular to the direction of airflow, only the static-head component is measured. A simpler and more accurate way of measuring static head, which is essentially constant over a given cross section of duct, is by connecting the manometer directly to a smooth opening in the wall of the conduit. This is because the static perforations in the pilot tube are affected by the facing tube and stem of the device, problems that can be minimized by proper sizing of the stem and locating the perforations correctly.

The pitot tube with its lack of moving parts is inherently accurate and reliable. Very accurate instruments with errors of less than 1% are available. Such precision, however, is not generally warranted in the variable conditions encountered underground. Since the accuracy of the pitot tube is limited in practice by the sensitivity of the manometer to which it is connected, the device must be calibrated with its manometer. The pitot tube is also directionally sensitive, so that for accurate results, its head must be aligned with the direction of airflow. Studies show that the maximum reading obtained as the head is moved from side to side is the correct reading (Ower, 1966). There are also a number of factors and precautions to be observed in special applications involving high-velocity leakage currents, pulsations in the velocity head, and large pressure head.

Rotameter   *Rotameters* are high-accuracy, direct-reading velocity instruments, particularly suitable for measurements of flow rates in vent pipe or tubing. The device consists of a tapered glass tube containing a metering float or ball and a scale. In operation, air flows vertically upward, pushing the metering float until a position is reached where the weight of the float is balanced by the force of the air flowing past it. The height of the float is a function of the flow rate.

Thermoanemometer and Hot-Wire Anemometer   There are several types of anemometers, particularly designed for low-velocity ranges, that measure air velocity by the rate at which heat is removed from a heated element placed in the airflow. The thermometer anemometer, or simply *thermoanemometer,* consists of two thermometers; one has an electrically heated element (powered by a battery) connected to the bulb, and the other is a conventional thermometer. In operation, both thermometers are placed in an airflow. The heated bulb cools, and the difference in temperature between the two thermometers can be related to the air velocity, either analytically or by a conversion chart. The device is not directionally sensitive, since the heated element is usually cylindrically shaped.

The *hot-wire anemometer* consists of a probe-mounted, hot-wire filament and a battery power source, which are both connected to a measuring circuit. In one type of hot-wire anemometer, the temperature is maintained constant by varying the current supplied to the wire. In the other type, the current supplied is constant while the temperature of the wire varies. Thermoanemometers and hot-wire anemometers are very accurate instruments for low air velocities, which would make them useful for measurements in stopes and large underground openings. Unfortunately, commercially available anemometers are not well suited for underground conditions and are seldom used except for precision work. Also, the hot-wire anemometers, with exposed heated wire, cannot be used in explosive atmospheres.

Kata Thermometer   The *kata thermometer,* used mainly for determining the cooling power of air, can also be used to measure air velocity. On the basis of the cooling effect of air in motion, the kata is a specially designed alcohol thermometer with bulbs at both ends of the stem. A 5°F (2.8°C) graduated range is marked on the stem, either 95–100°F (35.0–37.8°C) for the standard kata or 130–135°F (54.4–57.2°C) for the blue kata. In use, the lower bulb is dipped in a flask of hot water, causing the alcohol to rise into the upper bulb. Measurement is carried out by suspending the kata in moving air and observing the time required for the alcohol column to drop from the upper graduation to the lower graduation [either 100°F down to 95°F (37.8–35°C), or 135°F down to 130°F (57.2–54.4°C)]. Each kata is calibrated (in metric units) by the manufacturer with a "kata factor," which is the heat liberated by the thermometer during the 5°F (2.8°C) temperature change divided by the surface area of the lower bulb. The cooling power of the air

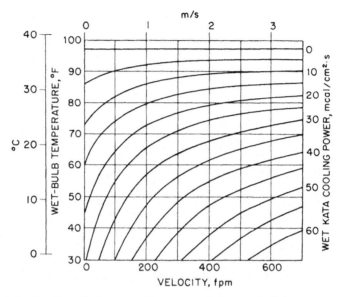

FIGURE 6.4   Graph to find wet-cooling power from wet-bulb temperature and velocity.

h_c, in Btu/ft²·h (mcal/cm²·s), is obtained by dividing the kata factor by the cooling time in seconds.

The kata bulb encased in a moistened wick, known as a *wet-kata* thermometer, is used mainly for determining cooling power of the air. Figure 6.4 shows the relationship between the wet-bulb temperature, air velocity, and wet-kata cooling power. The dry-kata thermometer is employed in determining air velocity and dry cooling power by formula or chart furnished by the manufacturer; see Fig. 6.5 (Weeks, 1926). Kata thermometers are delicate instruments but are not directionally sensitive.

Techniques of Measuring Velocity

Air velocity varies over the cross section of an opening because of the frictional resistance of the walls of the opening and variations in the state of airflow (Section 5.4). In general, the maximum velocity occurs at the geometric center of the opening and decreases gradually to a minimum toward the walls. For air-quality work involving gases and dusts, knowledge of the velocity distribution in the openings is important in designing removal or dilution control measures. For quantity control, air velocities are used mainly for calculating airflow quantities and velocity-head losses, both of which are highly dependent on the velocity distribution across the cross section. Fairly accurate results are obtainable when average velocities are used in the calcu-

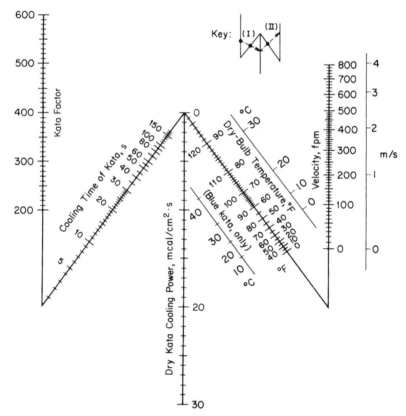

FIGURE 6.5 Graph to find dry-cooling power and velocity from dry-kata thermometer readings.

lations. The minimum or maximum air velocity in a cross section may not be useful or even be misleading for this purpose. In measuring air velocities, therefore, special efforts are necessary to obtain average values. For the smoke-tube method, which must necessarily be used in the center of the opening for air-velocity measurement, correction factors are applied to the calculated velocity values at the center in order to obtain an approximate average velocity in the cross section. With other instruments, several methods can be employed to determine the average velocity in an airstream. They may be classified as single or multiple measurements.

Single (or one-point) measurements are the fastest, simplest, but least accurate, and are made by holding or mounting the instrument at the center or maximum-velocity section of the airway. The observed velocity is then corrected by using an approximate factor, or determined more accurately if the Reynolds number is known (Fig. 5.10) (Rouse, 1937). This approximate technique is justified only for quick, rough-check measurements.

Multiple measurements are made by taking a series of readings by either the *fixed-point traversing* or *continuous-traversing* method. With readings obtained over the entire cross section of the opening, a more nearly representative and accurate average velocity for the opening is obtained. In the fixed-point traversing method, also known as *equal-area traversing,* the airway cross section is first divided into a number of imaginary equal areas with the same or similar shape as the airway. The average velocity for each equal area is then obtained by either measuring the air velocity at the center of each equal area or traversing with the instrument slowly and evenly over each equal area. The average velocity for the entire cross section is the arithmetic average of all the equal areas.

Multiple measurements by the continuous-traversing method are more rapid than fixed-point traversing and satisfactory for routine measurements. The method involves an imaginary, single, continuous, nonoverlapping track covering the entire cross section, with the lines of the imaginary track preferably uniform or equidistant from one another. With the velocity-measuring instrument turned on, the entire track is traversed slowly, evenly, and at a uniform speed. The average velocity for the entire cross section will then be the indicated reading divided by the total traverse time. It should be noted that instantaneous-velocity-measuring instruments are not suitable for continuous traversing, since the final reading corresponds to the air velocity at that point only. The instrument must be of the integrating type; thus only the vane anemometer can be used for continuous traversing. In fixed-point traversing with single measurements in each equal area, any of the instruments except the smoke tube can be used.

The accuracy of measurement by either the fixed-point or continuous-traversing method is highly dependent on the velocity gradient and the total number of measurements across the airway cross section. Obviously, the more measurements (increased number of points) or increased time of traverse, the more accurate the result will be. The use of an extension rod with the instrument to minimize interference from the body of the observer and measurement upstream or downstream from the observer is always recommended. For very accurate results such as those needed for fan testing, replicate measurements are desirable.

Examples of fixed-point and continuous-traversing methods for measuring velocity in mine airways and ducts are shown in Fig. 6.6. In Fig. 6.6*a*, continuous traversing is started near one corner of the airway section, proceeds across to the side, down by the predetermined spacing, and back to the other side. These procedures are continued in a regular fashion until the area has been covered in the allotted time, which is generally not less than one minute.

In traversing a circular cross section, horizontal and vertical centerlines are first laid out. Then concentric circles of equal area are constructed, and the intersections of alternate ones with the centerlines are located. Velocity readings are taken with the instrument placed at these intersections, as shown in Fig. 6.6*b*. Ordinarily, no reading is taken at the exact center of the

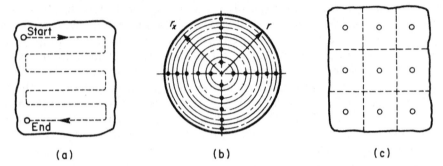

FIGURE 6.6   Methods of measuring velocity in mine airways and ducts: (*a*) continuous traversing; (*b*) fixed-point traversing in a circular opening; (*c*) fixed-point traversing in a rectangular airway.

airway. Three, four, or five alternate circles may be constructed, which results in 12, 16, or 20 readings, respectively. The number selected is governed by the accuracy required. For fan testing, at least 20 readings are taken; underground, 12 readings are usually ample.

The radii of the equal-area circles must be calculated. Since the readings are taken along the centerlines at the intersections with alternate concentric circles, it is necessary to calculate the radius distance only for the alternate circles. Mathematically, this is equivalent to saying that the radius r_x for any concentric group of readings is (Weeks, 1926)

$$r_x = r \sqrt{\frac{2x - 1}{2N}} \tag{6.4}$$

where r is radius of the airway, x is the number of that group of readings, and N is the number of alternate circles ($2N$ is the number of equal areas).

For example, if 20 readings are required, $N = 20/4 = 5$ alternate circles (10 equal areas) must be provided. Then the radius of the first circle (first group of readings) is

$$r_1 = r \sqrt{\frac{1}{10}} = 0.316r$$

the radius of the circle for the second group of readings is

$$r_2 = r \sqrt{\frac{3}{10}} = 0.548r$$

and so on:

$$r_3 = 0.707r$$

$$r_4 = 0.837r$$

$$r_5 = 0.949r$$

Radii can be calculated in a similar manner for any number of readings desired. The division of a rectangular or square cross section into equal areas similar in shape to the airway itself is shown in Fig. 6.6c. A single or multiple measurement is then conducted in each equal area, a single measurement obtained with the instrument at the center of each equal area, and multiple measurements by traversing the instrument in each equal area.

6.5 AIR QUANTITY MEASUREMENT

Calculation from Measured Velocity and Area

The quantity of air Q flowing in an airway is not usually measured directly but is calculated from the average velocity V and the cross-sectional area of the airway A at the point of measurement from Eq. 3.10:

$$Q = VA$$

with Q expressed in units of cfm (m³/s), V in fpm (m/s), and A in ft² (m²).

The accuracy of the calculated Q is dependent on the accuracy of the measured or calculated V and A. For general mine ventilation work, acceptable accuracy can be obtained in velocity measurements by utilizing the most suitable instruments and following recommended procedures for their use. The determination of areas of mine openings is mainly a practical, geometric exercise of dividing or subdividing the airway cross section into regular geometric shapes and applying applicable mensuration formulas (Weast, 1973). For highly irregular shapes with curved surfaces, protrudences, and so forth, a number of special techniques have been developed for determining the area of the openings: (1) vertical and horizontal taping, (2) vertical and horizontal offsets, (3) diagonal offsets, (4) triangulation, (5) spiked protractor, (6) full-circle protractor and profilographs, and (7) photographic methods. For routine work, vertical and horizontal taping is usually used, whereas for more accurate determination of cross-sectional areas of main airways, the photographic method is suggested (Potts, 1945/46).

Values of both air velocity and cross-sectional area are certainly more accurate when determined at specially constructed measuring stations. However, the expense of special measuring stations cannot easily be justified, unless the stations are used for additional purposes, such as ventilation doors or stoppings. In some cases, it is advisable to have special measuring stations in major ventilation openings and near large fans. If it is feasible, areas of regular and uniform shapes should be selected for measurement locations in active areas and secondary openings.

Tracer-Gas Technique

Although airflow quantities are more commonly calculated, the *tracer-gas* technique measures the quantity of airflow directly without measuring the cross-sectional area of the opening (Higgins and Shuttleworth, 1958). There are several other applications of this technique in mine ventilation (Klinowsky and Kennedy, 1991).

The tracer-gas technique is advantageous for air quantity determinations in sections where areas are irregular and the flows are not steady. The test gas is usually selected on the basis of (1) the ease with which it can be detected and analyzed, (2) its absence in the normal mine air, (3) the fact that it is not absorbed by chemical or physical means in the airway, (4) nonreactivity with other gases in mine air, and (5) nontoxicity or explosiveness. Gases considered suitable include ozone, hydrogen, carbon dioxide, and sulfur hexafluoride (SF_6).

There are two methods of using the tracer-gas technique (Thimons and Kissell, 1974). In the first method, the tracer gas is continuously metered into an airway. After thorough mixing has occurred and equilibrium has been established, air samples are taken at a point downstream. The concentration of the tracer gas in the sample is determined. The rate of airflow is calculated as

$$Q = \frac{Q_g}{C} \qquad (6.5)$$

where Q is quantity of air flowing in cfm (m³/s), Q_g is feed rate of the tracer gas in cfm (m³/s), and C is concentration of the tracer gas in ft³/ft³ (m³/m³).

In the second method, a known mass (or volume) of the gas, Q_g ft³ (m³), is injected into the airstream, and its concentration at a downstream point is sampled either continuously or as often as possible until the concentration can no longer be measured. The quantity of tracer flowing through the sampling point at any time interval (τ, $\tau + d\tau$) is given by $[QC_\tau]\,[d\tau]$, where C_τ is concentration at time τ, and Q is the airflow. Thus, over the time interval (τ_0, τ_f),

$$Q_g = \int_{\tau_0}^{\tau_f} QC_\tau d\tau \qquad (6.6)$$

where τ_0 is arrival time of first measurable concentration, and τ_f is time after which the concentration is not measurable.

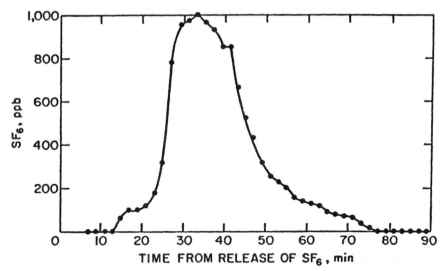

FIGURE 6.7 Concentration of SF$_6$ at 2-min intervals (Example 6.1).

If C_τ were plotted against τ, the curve would look like that shown in Fig. 6.7, where $\int_{\tau_0}^{\tau_f} C_\tau d\tau$ is the area under the curve and

$$Q = \frac{Q_g}{\int_{\tau_0}^{\tau_f} C_\tau d\tau}$$

On the other hand, the integration of $\int_{\tau_0}^{\tau_f} C_\tau d\tau$ can be performed by simply taking the average of the measured concentrations (C_{av}) and multiplying the average by the total time ($\tau_T = \tau_f - \tau_0$), during which measurable amounts of concentration are found at the sampling point. Then

$$Q_g = Q \int_{\tau_0}^{\tau_f} C_\tau d\tau = Q C_{av}[\tau_f - \tau_0] = Q C_{av} \tau_T$$

from which

$$Q = \frac{Q_g}{C_{av}\tau_T} \tag{6.7}$$

Example 6.1 Thimons and Kissell (1974) report that, when they released 0.37 ft^3 (0.011 m^3) of sulfur hexafluoride in a mine and monitored its concentration in the return, of the 45 samples taken at 2-min intervals, only 31 samples contained measurable quantities of the gas (Fig. 6.7). Average concentration in the 31 samples of the gas is calculated as 3.77×10^{-7} ft^3/ft^3

(m^3/m^3) of the air. The total sampling time was 62 min. Find the quantity of airflow.

Solution Using Eq. 6.6, the quantity of air flowing at the measurement point is calculated as

$$Q = \frac{0.371}{(3.77 \times 10^{-7})(62)} = 16,000 \text{ cfm } (7.55 \text{ m}^3/\text{s})$$

The tracer-gas technique has potential applications in measuring air quantities in problem areas such as recirculation of return into intake air, leakage from adjacent mines, "lost" intake air, and unknown transit flow times through stoped or gobbed areas.

6.6   AIR-PRESSURE MEASUREMENT

Next to air velocity, air pressure or head is the most commonly measured physical property of mine air. *Atmospheric pressure* refers to the force per unit area exerted by the weight of a column of air in the atmosphere. Atmospheric pressure varies with both time and place. *Pressure difference* is the difference in air pressure between two points, and *gage pressure* refers to the difference between the pressure at a point within a system such as a duct and atmospheric pressure at that point. Gage pressure plus the atmospheric pressure at a point is known as the *absolute pressure* at that point. Meanwhile, *head* is a unit measure of pressure in terms of a column of fluid, which in mine ventilation is customarily expressed in in. water (Pa). As explained in Section 2.4, head is the preferred form of measurement in mine ventilation.

Atmospheric-Pressure Measurement

Atmospheric (barometric) pressure p_b is measured by a *barometer* of either the *mercury, aneroid,* or *special feature* type.

Mercury Barometers   *Mercury barometers* are of two forms. The *cistern barometer* consists of a Torricellian tube and a measuring scale. Only the mercury level at the measuring column is observed, whereas the rise and fall of the mercury in the cistern is compensated for on the measuring scale. The reading obtained is proportional to the physical difference in level between the mercury in the column and mercury in the cistern. The other type of mercury barometer, called a *siphon barometer,* consists of a U-tube partly filled with mercury. One end of the tube is closed, and thus there is a vacuum above the mercury on this part of the tube. The other end is open to the atmosphere. The height of the column of mercury between the two legs is a measure of the atmospheric pressure. At sea level with the mercury at 85.4°F (14.7°C), the normal height of the mercury column is 29.92 in. (760 mm), which converts to an air pressure of 14.70 psi (101.34 kPa).

For routine air pressure measurements, mercury barometers are adequate, but for more precision, a number of corrections must be applied. The temperature correction accounts for measurements obtained at other than 32°F (0°C), the temperature at which mercury barometers are designed by convention to give the correct pressure reading. This correction considers the thermal expansion of the mercury and the materials used in the construction of the device. The gravity correction accounts for the difference in local gravity and standard gravity, the internationally agreed gravity for mercury barometers to give the correct pressure reading. The index correction, usually furnished by the manufacturer, accounts for slight variations in the manufacture of the devices.

Mercury barometers are fairly accurate when used at a fixed position. When used at different locations with different temperatures, the device should first be acclimatized for several hours. Since mercury barometers are also usually bulky and fragile, they are seldom used for mining work except for monitoring barometric pressure at a fixed point, usually on the surface, and for calibrating other types of barometers.

Aneroid Barometers   The *aneroid barometer* is widely used underground for atmospheric pressure measurements. It consists of an airtight, flexible diaphragm from which most of the air has been removed, so that the pressure inside is less than atmospheric (Fig. 6.8). The collapse and motion of the diaphragm is resisted by a spring element. When the atmospheric pressure changes, the diaphragm moves either inward or outward, depending on whether the pressure has increased or decreased. The movements of the diaphragm are magnified by a system of levers, which are connected to a pointer over a dial that is calibrated in terms of pressure.

There are a number of types of aneroid barometers available. The Askania *microbarometer,* probably the most sensitive of aneroid barometers, has an

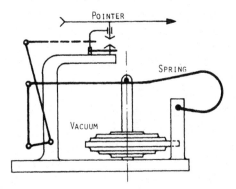

FIGURE 6.8 Aneroid barometer for measuring atmospheric pressure.

optical linkage and contains a thermometer to indicate the temperature of the core of the instrument. Its accuracy decreases as the measured pressure is reduced. Less sensitive but more widely used aneroid barometers are the *direct-reading* type, manufactured by Wallace and Tiernan, and an *indirect-reading* type, manufactured by American Paulin.

Atmospheric pressure is due mainly to elevation, with acceleration due to gravity and differences in air specific weight accounting for only about 10% of the total pressure (Allan et al., 1968). The significant influence of elevation on atmospheric pressure has led to the adoption of *altimeters,* aneroid barometers calibrated in terms of elevation for pressure measurements underground. In fact, elevation and pressure measurements are related by the following conversion formula (Allan et al., 1968):

$$Z = 62{,}583.6 \log \left(\frac{29.92}{p_b} \right) \tag{6.8}$$

$$Z = 19{,}075 \log \left(\frac{101.3}{p_b} \right) \tag{6.8a}$$

where Z is elevation above sea level in ft (m) and p_b is atmospheric pressure in in. Hg (kPa). This formula applies only to dry air at 50°F (10°C) and standard barometric pressure of 29.92 in. (760 mm) Hg. At other conditions, corrections must be made for the difference in the actual density of the air compared with the standard conditions. Cooler air is more dense, resulting in artificially lower altimeter readings, whereas warmer air is lighter, resulting in higher altimeter readings. Paulin altimeters incorporate a correction for humidity in the dial of the instrument, and a volumetric gas-expansion formula is used to correct errors due to differences in temperature (Hodgson, 1979). Wallace and Tiernan altimeters employ the specific weight ratio of standard air with the actual air to obtain the correction.

Aneroid barometers should be calibrated periodically against a mercury barometer and certainly before a major survey, since their thermal-compensation characteristic changes, and a definite zero shift or drift can continue with time. Calibration curves, one for increasing pressure and one for decreasing-pressure measurements, should also be obtained, since the indicated reading depends on whether the pressure increased or decreased to that particular value.

Special-Feature Barometers   Besides mercury and aneroid barometers, *airostats* and *differential barometers* can also determine the atmospheric pressure. The airostat, similar to a manometer, has one limb connected to a sealed chamber and a "free" limb open to the atmosphere. The sealed chamber is maintained at a constant temperature and pressure so that any reading represents the difference between the atmospheric pressure and the pressure in the sealed chamber. Factors that affect the readings of conventional barometers also affect airostats, so that similar corrections are applied. Airostat barometers have no moving parts and no hysteresis characteristics, which make them superior to aneroid barometers. Since the pressure in the sealed chamber can be varied, a wide range of pressure can readily be measured by the device.

The differential barometer consists of two connected vessels at different elevations and a measuring tube connected to the lower vessel. The vessels contain mercury, and an indicator fluid floats on the mercury in the lower vessel. The top of the upper vessel is a vacuum, and the top of the measuring limb is open to the atmosphere. Any change in the external pressure changes the mercury level and the corresponding magnified movement of the indicator fluid. The device with no zero drift is specially suitable for measuring small pressure changes. It is, however, bulky and requires corrections similar to those for mercury barometers.

Differential-Pressure Measurement

The pressure difference between two points in the mine ventilation system is more commonly measured than absolute or atmospheric pressure. These differences can be indicated as gage pressures or heads, which are measured relative to the atmospheric pressure existing at the point of measurement.

Instruments recording both gage and absolute pressures are in use for ventilation measurements. Regardless of type, they are capable only of indicating static-head differences along the static-head gradient. In other words, they record the difference in the vertical coordinate of points on the static gradient, whether the fan is blower, exhaust, or booster. Any velocity-head component or conversion particularly will introduce an unknown, erroneous effect in pressure measurements. If the velocity head changes, it should be computed or measured separately and combined algebraically or graphically with the static head to obtain the total head gradient for the airway. Head losses can then be determined correctly as changes in total head (Section 5.3).

Manometers A *manometer* measures pressure difference by the difference in elevation of two surfaces of a connected body of liquid. The liquid surface with the higher pressure is depressed and the other surface raised (Fig. 6.9).

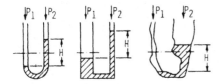

FIGURE 6.9 Differential-pressure measurements with manometers of various sizes and shapes.

Irrespective of the size and shape of the vessel, the pressure difference can be calculated from the measured vertical difference in liquid level H and the specific weight of the liquid w using the pressure-conversion formula (Eq. 2.18). In practice, the scales can be graduated directly in terms of pressure for a given liquid used with the instrument. More conveniently, pressure differences are expressed directly as heads in in. water (Pa), regardless of the liquid used.

There are several possible sources of error that must be considered in using a manometer. A major error occurs when the liquid is constrained in reaching its true position as indicated by the center of the meniscus. This problem is minimized by choosing an appropriate liquid and using as large a tubing bore as possible to allow formation of a good meniscus. The choice of the liquid should also depend on its "wetting" characteristics, visibility, and specific weight. Its wetting properties, the ease with which the liquid wets the tubing-bore surfaces, affects the true positioning of the liquid surface. The liquid specific weight must be stable and predictable over the range of temperatures likely to be encountered. For measuring small pressure differences, fluids with a small specific weight are desirable, since the displacements are larger for a given pressure change compared with a denser liquid. Using a less dense fluid reduces the range of measurement for a given size manometer, however. Visibility of the liquid surface allows ease and accuracy in taking readings, especially underground, and also allows checking for air bubbles. To maximize visibility, colored liquids are used or dyes are added to transparent liquids that do not alter their other desirable properties.

There are several types of manometers, from the simple, less accurate ones to those with more elaborate and improved features for sensitivity and accuracy. The *vertical U-tube,* the basis for all other manometers, is easy to use and fabricate from readily available materials. The use of water as a gage fluid means inferior wetting qualities and a poor meniscus; other liquids are preferred.

The *well-type* or *reservoir-type manometer* consists of a reservoir of large cross-sectional area connected to a measuring limb of smaller cross section. Higher pressure acting on the reservoir depresses it slightly, but the corresponding rise in level in the measuring limb is magnified by the ratio of their cross-sectional areas. The scale graduations, designed to take into account the slight change in level of the reservoir, are marked directly in either pressure or head. Although the measuring limb for the reservoir-type manometer is usually vertical, it can be inclined for increased sensitivity, as in Fig. 6.10. The magnification of an inclined manometer is equal to the ratio of the inclined length per in. water (mm/mm) vertical equivalent (Weeks, 1926).

Example 6.2   An oil of specific gravity $s = 0.85$ is to be used in an inclined manometer constructed with a magnification of 12. If the tubing bore D_0 is 0.15 in. (3.81 mm) and the reservoir diameter D is 1.50 in. (38.10 mm), find

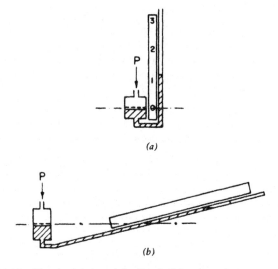

FIGURE 6.10 Vertical (a) and inclined (b) reservoir-type manometers.

the required angle of inclination θ, the error in reading only the inclined leg, and the scale correction that must be made in compensation.

Solution: Using Eq. 2.18, substituting specific gravity for specific weight (water = 1.00):

$$H_2 = H_1 \frac{s_1}{s_2} = (1)\left(\frac{1.00}{0.82}\right) = 1.177 \text{ in. oil}$$

$$\sin \theta = \frac{\text{vertical rise}}{\text{magnification}} = \frac{1.177}{12} = 0.098$$

$$\theta = 5.6°$$

For a range of 1 in. (25.4 mm) water, construct a manometer with a leg 12 in. (304.8 mm) in length, 1.177 in. (29.90 mm) vertical rise, and inclined at 5.6°.

$$\text{Error} = (\text{magnification})\left(\frac{D_0}{D}\right)^2\left(\frac{s_2}{s_1}\right) = (12)\left(\frac{0.15}{1.50}\right)^2\left(\frac{0.85}{1.00}\right)$$

$$= 0.102 \text{ in./in. water/12-in. length of inclined leg, or } 10.2\%$$

To correct for error, calibrate the scale along the 12-in. (304.8-mm) inclined leg to read from 0 to 1.102 in. (0 to 28.0 mm) water.

The *differential-liquid manometer,* based on the same principle as the reservoir-type manometer, consists of two vertical limbs of large cross section connected by a horizontal tube of much smaller cross section. A small pressure difference between the two limbs corresponds to a large displacement in the horizontal measuring limb. By using two different immiscible liquids in the two limbs, the motion of the contact surface can be measured; and using equilibrium equations, the pressure difference can be calculated. This device is very accurate but suitable for laboratory use only.

The *magnehelic manometer,* more like an aneroid barometer, consists of a diaphragm held by a spring. A magnet is attached to one side of the diaphragm, and adjacent to the magnet is a helix that is free to turn to maintain a set gap between itself and the magnet. Any pressure difference across the diaphragm displaces the magnet and rotates the helix. The position of the helix is indicated by a pointer on a dial graduated in pressure units. Several instruments are available to measure differential pressures over the different ranges. Magnehelics are widely used in mines today. The magnehelic manometer requires calibration for accurate measurements.

There are several other types of manometers for special types of applications. The *inclined/vertical manometer* combines the sensitivity of an inclined manometer and the broader measuring range of a vertical manometer. The *hook-type manometer* utilizes pointed hooks for locating the liquid surface. Its electronic variant requires completion of an electric circuit to indicate location of the liquid surface. The *micromanometer,* based on the same principle as the hook gage, requires a pointer–mirror combination in locating the liquid level precisely. A micrometer moves the pointer until its mirror image appears to be just touching the pointer. At this time, the pointer is at the surface of the liquid, and an accurate reading can be obtained. Micromanometers are very reliable but difficult to manipulate under fluctuating pressure conditions. They are used underground to some extent for precision surveys.

6.7 VENTILATION SURVEYS

A ventilation survey underground involves the measurement of air velocities, absolute and differential pressures or heads, dry- and wet-bulb temperatures, and airway dimensions at various strategic points in a section, sections, or throughout the whole mine. A mine ventilation survey provides the data for determining air quantities, pressure losses, air specific weights, and airway resistance, which are necessary for ventilation system evaluation and analytical and numerical calculations. Today, these calculations are generally performed on the computer.

In carrying out a ventilation survey, one or more parties are formed, each party with complete instrumentation and instructions to do all the required measurements in one or more areas of the mine. All the raw data, including replicate measurements, are then adjusted, modified, corrected appropriately, and compiled for use in calculations or evaluations. Well-planned and -executed ventilation surveys consider a number of factors, including predicted atmospheric conditions and mine operations. For example, pressure surveys are recommended only when the barometer is steady or the change is regular (Williamson, 1932; Krickovic, 1945/46; Mancha, 1946); and where feasible, surveys should be carried out when the mine is idle.

Accurate ventilation surveys are achieved by (1) using the most suitable instruments, (2) adopting rigorous measurement procedures or modifying procedures that are adequate only for routine ventilation surveys, (3) using one or two properly trained survey groups, (4) judiciously selecting measuring points based on a comprehensive knowledge of the ventilation network, and (5) tying each day's work back to the previous day's work to assure continuity in survey results.

Air Quantity and Temperature Surveys

An air quantity survey involves the measurement of air velocities and cross-sectional areas in various parts of the mine. Velocities are commonly measured with a vane anemometer mounted on an extension rod. Either fixed-point or continuous-traversing methods of measurement are selected, although continuous traversing is more rapid and adequate for routine measurements. The cross-sectional area of regularly shaped airways is easily calculated, and areas of irregularly shaped airways can be obtained by using any one of the methods mentioned previously (Section 6.5). Airflow measurements in horizontal airways are fairly easy to make. In shafts and inclined raises, they may be estimated from measurements of individual flows in airways leading to the shaft or raise.

Air-temperature surveys measure the dry- and wet-bulb temperatures of the mine air, usually with a sling psychrometer. These temperatures, t_d and t_w, together with the barometric pressure p_b measured at the same point, are then used to determine the air specific weight (see Fig. 2.3).

Pressure Surveys

A pressure or head survey measures either (1) the difference in pressure or head between two or more stations by using a manometer or (2) the absolute pressure at each station by using a barometer. Taking manometric measurements is commonly called the *direct method,* indicating the direct observation of the pressure difference between two points in the airway. Taking barometric measurements is called the *indirect method,* since the absolute pressure at each measuring point is adjusted, corrected, and correlated with pressure readings from all the other measuring points in order to obtain the pressure gradient of the mine.

Direct Method   In the direct method, a rubber tubing or hose is laid between the two points whose pressure difference is being measured. A manometer mounted on a tripod and equipped with a leveling bubble is then connected either at one end or at some other convenient point along the tube. The manometer reading is the pressure difference between the two points. This is the commonest and simplest method of obtaining the pressure between two points. Since the pressure difference measured by a manometer should be limited to the static head, care must be taken to prevent velocity and elevation heads from influencing the measurements. To eliminate velocity effects, the ends of the tube should be held or placed normal to the direction of airflow. Sometimes special devices or the static opening of a pitot tube are attached to the open ends of the tube to ensure that only the static component of the pressure is being obtained (Quillian, 1974). The elevation effect due to the different elevations of the two points is caused by the normal change in atmospheric pressure that accompanies differences in elevation. The mean specific weight of the air within and outside the tubing is also different, since flow is inhibited inside the tube. This specific weight-elevation effect may be eliminated by a correction factor that depends on the position of the manometer in relation to the tube (Hinsley, 1948/49; Hemp, 1982). If the elevation difference, however, is not excessive [less than 1000 ft (305 m)], the errors are negligible and may be ignored. Measurements in ventilation pipe when the tube or hose is laid outside the pipe must be corrected for the temperature effects on the specific weight of the air. Care should also be taken to avoid condensation in the tubing, because moisture may plug it.

A simple modification of the direct method for use in shafts has been devised to overcome the above-mentioned problems (Kislig, 1968). The lower end of the tube is fitted with an aspirator bottle containing water. Water flow from the aspirator bottle draws air down the tube. The flow rate is adjusted so that the manometer reads zero, which means that the pressure loss of the air flowing through the tube is balanced by the actual pressure drop of the air in the shaft. From a calibration curve for the tube, the pressure loss for the shaft can be obtained. Since air is also flowing in the tube, errors due to variations in air specific weight, temperature, and humidity are negligible. Application of the direct method to determine pressure drops in mine shafts and across doors is described by Hemp (1982) and Wallace and McPherson (1991).

Example 6.3   The following data are obtained in a trailing-hose survey (Fig. 6.11) in a vertical intake shaft (Hemp, 1982):

| Position | Elevation, ft | Pressure, psi | Temperature, °F | |
			Dry-Bulb	Wet-Bulb
Top	− 1748.70	12.594	59.4	50.0
Bottom	− 4368.70	13.773	67.3	57.0

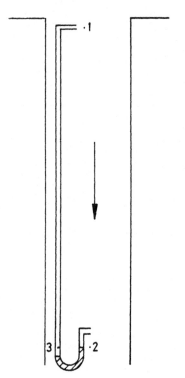

FIGURE 6.11 Gage-and-tube traverse in a shaft.

The manometer is at the bottom of the shaft and reads 1.51 in. water. Calculate the correction to the manometer reading.

Solution: The manometer reading is the difference $(p_3 - p_2)$ in Fig. 6.11. To determine the correction, the mean specific weights of the air in the shaft (column 1–2) $\overline{w}_s$, and the hose (column 1–3) $\overline{w}_h$, must be calculated.

Neglecting velocity head changes and using the appropriate mean specific weights, the head loss in the shaft resulting from flow from 1 to 2 is

$$H_{l_{12}} = (H_{s_1} - H_{s_2}) + (H_{z_1} - H_{z_2})$$

Similarly, noting that $Z_2 = Z_3$, the head loss in the hose due to flow from 1 to 3 is

$$H_{l_{13}} = \frac{144 \, (p_1 - p_3)}{5.2} + \frac{\overline{w}_h(Z_1 - Z_2)}{5.2}$$

$$H_{l_{12}} = \frac{144 \, (p_1 - p_2)}{5.2} + \frac{\overline{w}_s(Z_1 - Z_2)}{5.2}$$

Note that $H_{l_{13}} = 0$ as there is no flow from 1 to 3.

Subtracting $H_{l_{13}}$ from $H_{l_{12}}$ the following expression is obtained for head loss in the shaft:

$$H_{l_{12}} = \left(\frac{144}{5.2}\right)(p_3 - p_2) + \left(\frac{1}{5.2}\right)(Z_1 - Z_2)(\overline{w}_s - \overline{w}_h)$$

In this equation, the manometer reading is represented by the first term on the right, and the correction, by the second term.

Note that $p_3 = p_2 +$ manometer reading (converted to psi) $=$

$$p_3 = (13.773) + \left(\frac{1.51}{27.69}\right) = 13.8275 \text{ psi}$$

Using Eqs. 2.2, 2.3, 2.5, 2.7, and 2.8, the following data are calculated for the air in the shaft:

Position	Specific Humidity, lb/lb	Weight, lb/ft³
Top	0.0072	0.0655
Bottom	0.0082	0.0702

Average specific weight of air in the shaft $\overline{w}_s = \dfrac{0.0655 + 0.0702}{2}$

$$= 0.06785 \text{ lb/ft}^3$$

Conditions at the top of the hose are the same as those at the top of the shaft. Known conditions at the bottom of the hose (point 3) are dry-bulb temperature $= 67.3°F$, barometric pressure $= 13.8275$ psi, and specific humidity $= 0.0072$ lb/lb. Using these data and Eqs. 2.5, 2.7, and 2.8, the specific weight of the air at point 3 is calculated as 0.0705 lb/ft³.

Average specific weight of air in the hose $\overline{w}_h = \dfrac{0.0655 + 0.0705}{2}$

$$= 0.0680 \text{ lb/ft}^3$$

Noting that $(Z_1 - Z_2) = 2620$ ft, the head loss in the shaft due to flow is

$$H_{l_{12}} = \left(\frac{144}{5.2}\right)(13.8275 - 13.773) + \left(\frac{1}{5.2}\right)(2620)(0.06785 - 0.0680)$$

$$= 1.43 \text{ in. water}$$

In this case, the direct method has overestimated the head loss by 5.6%.

Application of the direct method to calculate head losses in a part of a mine ventilation system is illustrated with data from a survey conducted at the Homestake mine (Marks, 1997).

Example 6.4 Figures 6.12 and 6.13, respectively, show an overlay of the 3500 and 3650 levels at the Homestake mine and a simplified longitudinal view of the traverse. The gage-and-tube survey started by dropping the tube down a 16-ft-diameter ventilation raise from junction 3505 to 3603. The tube was then pulled up to a 1000-hp booster fan at junction 3504. At junction 3503, the volume of exhaust air going up an old winze was measured at 323,900 cfm. From junction 3503, the tube was taken down a crosscut containing two air doors to the Ross shaft at junction 3500. From the Ross shaft, the tube was taken through the ramp system that services the Ross Pillar mining area, junctions 3506, 3507, to 3604 on the 3650 level. From 3604, the tube was pulled through an exhaust drift, junctions 3608, 3602, to the starting

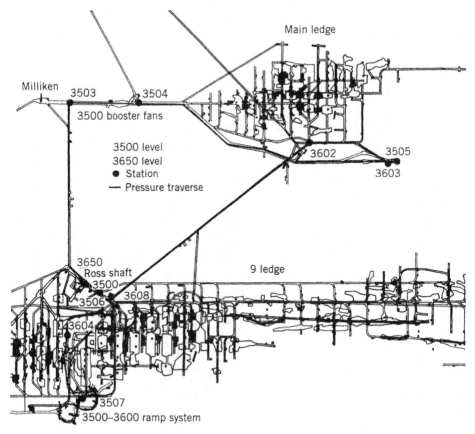

FIGURE 6.12 Plan view of Ross pillar gage-and-tube traverse.

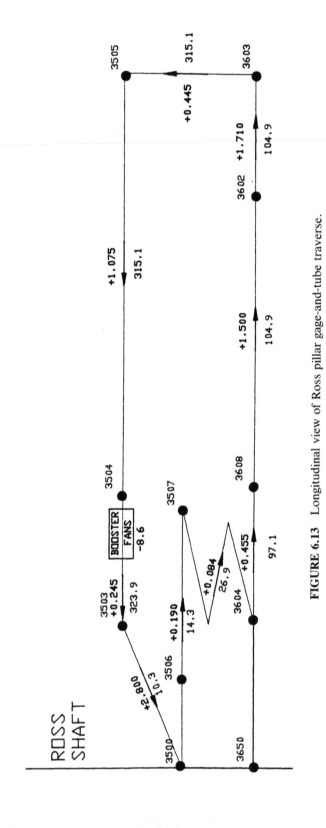

FIGURE 6.13 Longitudinal view of Ross pillar gage-and-tube traverse.

point, junction 3603. Junctions were located wherever airflows changed. The following table summarizes the pressure drops and rises around the loop:

Junctions from–to	Head Losses, in. water Drop (+), Rise (−)
3603–3505	+0.445″
3605–3504	+1.075
Booster Fan	−8.60
Booster-3503	+0.245
3503–3500	+2.80
3500–3507	+0.19
3507–3604	+0.084
3604–3608	+0.455
3608–3602	+1.50
3602–3603	+1.71

Calculate the overall head loss in the circuit and comment on the survey accuracy.

Solution: The direction of traverse for the survey was always in the direction of airflow. The head losses are thus all positive. The only head gain in the loop is due to the booster fan.

The following observations are relevant. The booster fan head is 8.60 in. water. The summation of head losses (+) and rises (−) around any closed mesh or loop must equal zero. The sum of the head losses and head gain is −0.096 in. water. This represents an error of 1.12%, which is very good for a mine survey. Errors under 10% are normally considered satisfactory, although careful surveys under steady-state conditions should do much better.

Where branch resistances and friction factor values are to be calculated, the quantities of air flowing in the branches and the dimensions of the airways including notes on shock-loss conditions should be collected for use with Eq. 5.25.

Indirect Method The indirect method uses a precision aneroid barometer or altimeter. The device indicates absolute static pressure at a point, and the difference in head between two points must be calculated from adjacent readings. Altimeter readings are recorded during the survey in ft (m) elevation and then later converted to head in in. water (Pa).

In conducting an altimeter survey, either of two methods may be used, both requiring two altimeters (Krickovic, 1945/46). In the first, both instruments are taken underground and read simultaneously at adjacent stations. In this method, called the *leapfrogging method,* the upstream instrument becomes the downstream instrument for each successive measurement (Fig.

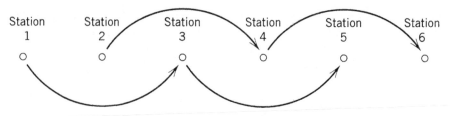

FIGURE 6.14   Leapfrogging method.

6.14). Both instruments are read at the same time with the aid of synchronized watches. Therefore, the effect of atmospheric pressure changes is eliminated. However, time requirements are increased, and the logistics of altimeter station moves add difficulty to its use.

In the second method, termed the *single-base method,* one instrument is used underground in making the traverse, while the second remains on the surface or at some base point underground; readings of the base altimeter are taken regularly at a specified time interval. Alternatively, a continuous-recording barometer may be used as the base instrument and is very convenient (McElroy and Kingery, 1957). Some disagreements on the location of the base altimeter—whether underground (Krickovic, 1945/46) or surface (McElroy and Kingery, 1957)—as well as on the timing of the survey—whether off-shift (McElroy and Kingery, 1957) or during the normal working shift (Krickovic, 1945/46)—are longstanding and have not been resolved.

McElroy and Kingery (1957) have recommended that pressure surveys be conducted independent of quantity surveys since minor changes in mine conditions have a greater effect on absolute pressure measurements than on quantity measurements, and that the quantity survey should be made immediately preceding or following the pressure survey. According to them, for large mines, a pressure survey may take only 1–3 days while the corresponding quantity survey would require 1–3 weeks. For computer-aided analysis of the mine ventilation system, the values for airway resistances R are calculated from measured values of H and Q (Eq. 7.7). The linear and squared relationships of R to H and Q, respectively, indicate that errors in quantity measurement are more serious than errors in pressure measurements. Further, these errors will have a significant effect on the determination of airway resistance. To reduce these errors and discrepancies, it is highly recommended that H and Q be measured at the same time (Luxbacher and Ramani, 1982).

Pressure Survey Planning   The details of planning and conducting the survey are similar for both methods. Selection of points for the pressure survey should be based on a review of a comprehensive line diagram of the mine ventilation system. The line diagram is usually constructed from the mine ventilation map and should include all pertinent information. In addition to facilitating survey planning, the line diagram construction helps to identify problem areas prior to the start of the survey. Stations for pressure measurement should be located at splits, junctions, points of air direction change, and points of major physical change in air courses; across regulators; and at both sides of overcasts. These locations are poor for quantity measurements. Study of the map should be followed by a trip underground, at which time elevations and areas can be measured; and an inspection can be made for possible problems that may be encountered during the survey as well as for changes in station locations, if necessary.

The sequence of stations used in a survey circuit should be located in an orderly manner, proceeding with the air through intakes to the faces and back to the starting point through returns. However, both speed and safety often require that the observer travel intakes between stations on returns and cross from intake to return to make observations at points in the return airways. By traveling slowly in shafts and slopes and allowing time for pressures to equalize when passing through airlocks, the surveyor can minimize the effects of "creep" (i.e., the lag in the registering of changes of pressure). Although two or more readings are taken at a station, primary reliance is placed on the one taken just before the surveyor moves on to the next station; thus the effects of "creep" are, for the most part, obviated by the time required for survey procedures.

Pressure Survey Calculation The Bernoulli equation (Eq. 5.2), rearranged to solve for the head loss or pressure difference H_l between two points, 2 to 1, is as follows:

$$H_{l_{21}} = (H_{s_2} - H_{s_1}) + (H_{v_2} - H_{v_1}) + (H_{z_2} - H_{z_1}) \qquad (6.9)$$

The static heads H_{s_1} and H_{s_2} are the sum of the gage static pressures and the atmospheric pressures at the time of measurements, or

$$H_{s_1} = H_1 + H_{a_1} \qquad H_{s_2} = H_2 + H_{a_2} \qquad (6.10)$$

where H_i is gage static pressure and H_a is atmospheric pressure. Substituting Eq. 6.10 in Eq. 6.9, the equation for the pressure difference between stations becomes

$$H_{l_{21}} = (H_2 - H_1) + (H_{a_2} - H_{a_1}) + (H_{v_2} - H_{v_1}) + (H_{z_2} - H_{z_1})$$

$$(6.11)$$

The last three terms of Eq. 6.11 correspond to the corrections for atmospheric pressure changes, velocity head changes, and elevation head differences between stations, respectively.

In terms of altimeter units, elevations, and temperature measurements, the pressure difference between stations is given by the following expressions (Ramani, 1992a):

$$H_{l_{21}} = - \left[\frac{(p_{A_2} - p_{A_1}) - (p_{B_2} - p_{B_1}) - (Z_2 - Z_1)/DR}{CF} \right] + \frac{V_2^2 - V_1^2}{(4009)^2}$$

(6.12)

$$H_{l_{21}} = - \left[\frac{(p_{A_2} - p_{A_1}) - (p_{B_2} - p_{B_1}) - (Z_2 - Z_1)/DR}{CF} \right] + \frac{w(V_2^2 - V_1^2)}{2g}$$

(6.12a)

where DR and CF are special altimeter factors as defined below; p_A and p_B correspond to roving and base altimeter readings in ft (m); respectively, Z is elevation in ft (m); V is velocity in fpm (m/s); and the subscripts 1 and 2 refer to the trailing and leading stations (or base station and roving station in the single-base method), respectively. Note that in the leapfrogging method, since the readings at the two stations are time synchronized, there is no barometric pressure correction (i.e., $p_{B_2} - p_{B_1} = 0$).

DR or *density ratio* gives the ratio of the specific weights of dry air at 50°F (10°C) and actual mine air at the same barometric pressure, or, from Eq. 6.2,

$$DR = \frac{\left(\dfrac{1.325}{460 + 50} \right) p_b}{\left(\dfrac{1.325}{460 + t_d} \right) (p_b - 0.378 p_v')}$$

(6.13)

$$DR = \frac{\left(\dfrac{p_b}{0.287 (273 + 10)} \right)}{\left(\dfrac{p_b - 0.378 p_v'}{0.287 (273 + t_d)} \right)}$$

(6.13a)

where t_d is the dry-bulb temperature in °F (°C) and p_v' is vapor pressure at dew point in in. Hg (kPa) of the actual mine air. Values for the specific weight ratio for different air temperature and relative humidity ϕ in percent

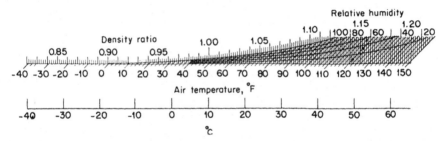

FIGURE 6.15 Ratio of altimeter-scale air specific weight to mine air specific weight. Correction factor for difference of elevation. (After McElroy and Kingery, 1957.)

are shown in Fig. 6.15. These ratios correct the height of the air column between the two stations to equivalent heights of air columns at 50°F (10°C) dry air.

CF or *conversion factor* converts average altimeter readings to pressures in feet of air column equivalent to a pressure of 1 in. (25.4 mm) water (Fig. 6.16). Essentially, the roving altimeter readings at neighboring stations are corrected for changes in (1) base altimeter reading and (2) elevation, taking into account the specific weight difference between the altimeter scale and the mine air. These corrected differences, which are in ft (m) of air, are then converted to in. water (Pa). All calculations are done at a standard, usually 50°F (10°C) dry air. Corrections are negligible for differences in elevation less than about 10–15 ft (3.04–4.56 m). Though using an air specific weight of $w = 0.0500$ lb/ft³ (0.8009 kg/m³) is appropriate, pressure differences due to air-velocity differences are calculated by assuming $w = 0.0750$ lb/ft³ (1.2014 kg/m³), or

$$H_{v_{21}} = \left[\frac{V_2^2 - V_1^2}{(4009)^2}\right]\left[\frac{(w_1 + w_2)/2}{0.0750}\right] \tag{6.14}$$

$$H_{v_{21}} = \frac{w(V_2^2 - V_1^2)}{2}\left[\frac{(w_1 + w_2)/2}{1.2014}\right] \tag{6.14a}$$

Example 6.5 Data from an underground pressure survey shown in Table 6.3 were obtained by the leapfrogging method (Mancha, 1946). Determine the head loss between adjacent stations.

Solution: Example calculations shown are for stations 7 and 0 only. Chart values for DR and CF are first obtained as follows.

1. *Determine DR.* Use psychometric tables (Appendix Table A.2), or Table 6.1 or a psychometric chart (Fig. 2.2) to find relative humidity ϕ.

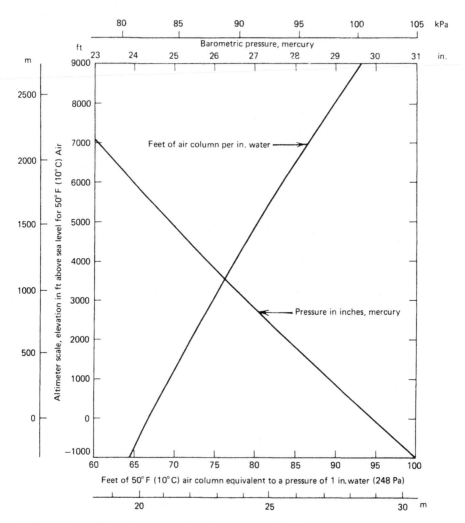

FIGURE 6.16   Chart for converting altimeter readings to heads, in in. mercury, and for converting altimeter differences to equivalent heads, in in. water. The curves are based on an altimeter with an offset of 1000 ft (304.8 m). If the altimeter reading is taken from an instrument without this offset, 1000 ft (304.8 m) must be subtracted from the reading before attempting to read the conversion factor CF from the graph. (After McElroy and Kingery, 1957. Based on Smithsonian Meteorological Table 51.)

TABLE 6.3 Altimeter Survey Data Obtained Using Leapfrogging Method (Example 6.5).

Stations	Elevation, ft		Altimeter Reading, ft		Wet-Bulb t_w, °F		Dry-Bulb t_d, °F		Station Area, ft²		Average Air Velocity, fpm	
	A	B	A	B	A	B	A	B	A	B	A	B
0–7	750	760	800	1073	40	60	45	60	100.0	120.0	1000	833
0–1	750	745	804	834	40	45	45	50	100.0	95.3	1000	1050
1–2	745	748	838	857	45	52	50	55	95.3	94.0	1050	850
2–3	748	750	860	878	52	60	55	60	94.0	95.0	850	632
3–4	750	752	878	948	60	60	60	60	95.0	80.0	632	500
4–5	752	750	945	979	60	60	60	60	80.0	85.0	500	706
5–6	750	755	975	1016	60	60	60	60	85.0	87.0	706	920
6–7	755	760	1015	1090	60	60	60	60	87.0	95.0	920	1055
7–0	760	750	1090	803	60	38	60	40	95.0	100.0	1055	1000

Conversion factors: 1 ft = 0.3048 m, °C = [t(°F) − 32]/1.8, 1 ft² = 0.0929 m², 1 fpm = 0.00508 m/s.

219

For $t_{d_1} = 60°F$ and $t_{w_1} = 60°F$, read $\phi_1 = 100\%$

For $t_{d_2} = 40°F$ and $t_{w_2} = 38°F$, read $\phi_2 = 92\%$

Final average $\phi = \dfrac{\phi_1 + \phi_2}{2} = 96\%$

and average $t_d = \dfrac{t_{d_1} + t_{d_2}}{2} = 50°F$

From Fig. 6.15, DR = 1.00

2. *Determine CF* (Fig. 6.16):

Find average altimeter reading $= \dfrac{1090 + 803}{2} = 946$ ft

Moving horizontally from the altimeter scale until it intersects the appropriate line (ft air column per in. water), CF the air column equivalent to a pressure of 1 in. (25.4 mm) water is found to be 69 ft (25.15 m).

Substituting survey data and the above values in Eq. 6.12:

$$H_{l_{70}} = -\left[\frac{(1090 - 803) - (760 - 750)/1.0}{69}\right] + \frac{(1055)^2 - (1000)^2}{(4009)^2}$$

$$= 4.00 \text{ in. water } (-998 \text{ Pa})$$

The computed head losses between adjacent stations are tabulated in Table 6.4.

TABLE 6.4 Results of Altimeter Survey (Example 6.5)

Stations	Altimeter Difference, ft	Elevation Difference, ft	Head Loss Due to Airflow in. water	Average Air Quantity, cfm
0–7	−273	−10	+4.00	—
0–1	−30	+5	+0.50	100,000
1–2	−19	−3	+0.25	80,000
2–3	−18	−2	+0.25	60,000
3–4	−70	−2	+1.00	40,000
4–5	−34	+2	+0.50	60,000
5–6	−41	−5	+0.50	80,000
6–7	−75	−5	+1.00	100,000
7–0	+287	+10	−4.00	—

Source: Mancha (1946). (By permission from AIME, New York, copyright 1996).
Converson factors: 1 ft = 0.3048 m, 1 in. water = 249.089 Pa, 1 cfm = 0.000472 m³/s.

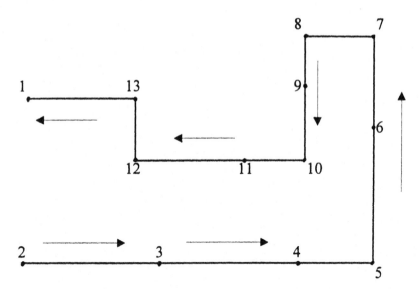

FIGURE 6.17 Fixed-base altimeter survey (Example 6.6).

Example 6.6 Data for this example are taken from McElroy and Kingery (1957). A fixed-base altimeter survey was run from station 1, the surface fan house, following the airflow through the mine, and closed back to station 1, as shown in Fig. 6.17. The altimeter scale was offset 1000 ft. The base altimeter was read every 10 min, and the base altimeter reading at the time of the roving altimeter reading at a station was interpolated.

Separate but simultaneous altimeter readings were made by another crew at stations 12 and 13, respectively, the bottom and top of the return shaft, and at stations 2 and 3, respectively, the top and bottom of the slope. The survey data are summarized in Table 6.5.

Calculate the static and total head losses between the survey stations and the total head at each station. Comment on the fan and closure error. Calculate the head losses in the slope and shaft from the simultaneous altimeter readings.

Solution: Calculation of the head loss between two adjacent stations is accomplished by the application of Eq. 6.12. The two corrections to the differences in the roving altimeter readings are due to differences in the base altimeter readings and differences in elevations. Note that positive differences result in negative corrections, and vice versa. The calculations are best performed in a tabular manner as shown in Table 6.6. As an example, the following calculations pertain to data at stations 4 and 5:

Step 1. Determine the relative humidities and velocity heads at stations 4 and 5:

TABLE 6.5 Altimeter Ventilation Pressure Survey Data (Example 6.6)[a]

Sta.	Location	I/R	Elevation, ft	Time	RAR, ft	Temperature Wet-bulb, °F	Temperature Dry-bulb, °F	Velocity,[b] fpm	BAR, ft
(1)	(2)	(3)	(4)	(5)	(6)	(7)	(8)	(9)	(10)
1	Surface at fan house	Base	482	8:22	1391	49	53	—	1391
2	Rock slope portal	I	496	8:35	1387	50	53	—	1373
3	Rock slope bottom	I	120	8:59	1010	57	64	475	1355
4	Main N by sta. 1925	I	104	9:10	998	57	65	450	1355
5	Main N by sta. 2700	I	57	9:18	952	60	65	390	1350
6	Junction MN and ME	I	−28	9:45	859	62	65	360	1325
7	ME, opposite 2N	I	−27	10:15	848	62	64	230	1311
8	ME opposite 3N	I	−17	10:30	847	63	65	130	1291
9	ME return at 3N	R	−16	10:50	889	70	72	475	1291
10	ME return at MN	R	−24	11:37	894	69	70	350	1276
11	MN at old coal slope	R	114	1:30	1230	70	70	220	1274
12	Bottom, upcast shaft	R	317	1:53	1410	70	70	880	1264
13	Top upcast shaft—fan inlet	R	482	2:45	1610	68	68	1590	1241
1	Surface at fan house	Base	482	2:53	1238	43	50	—	1239
2	Top of rock slope portal	I	496	8:50	909	31	37	—	Simultaneous readings
3	Bottom of slope portal	I	120	8:50	539	47	54	475	readings
12	Bottom of shaft	R	317	9:25	1041	69	69	880	Simultaneous readings
13	Top of shaft	R	482	9:25	1264	69	69	1590	readings

Source: McElroy and Kingery (1957).

[a] *Abbreviations:* RAR, BAR—roving, base altimeter readings; MN, ME—main north, main east.

[b] Velocity negligible, taken as zero in calculations.

Conversion factors: 1 ft = 0.3048 m, $°C = \dfrac{t(°F) - 32}{1.8}$, 1 ft^2 = 0.0929 m^2, 1 fpm = 0.00508 m/s.

TABLE 6.6 Altimeter Ventilation Pressure Survey Calculation (Example 6.6)

Station	φ	H_z, in. water	Diff. elev.	DR	Alt. diff.	Base corr.	Elev. corr.	Head ft. of air	Avg. alt. rdg.	Feet of air per in. water	ΔH_s	ΔH_v	ΔH_t	H_t
												in. water		
(1)	(2)	(3)	(4)	(5)	(6)	(7)	(8)	(9)	(10)	(11)	(12)	(13)	(14)	(15)
1	75	0						0	1389	67.7	0	0	0	0
2	81	0	14	1.008	−4	+18	−14	9	1199	67.4	−0.133	0.014	−0.119	0
3	65	0.014	−376	1.022	−377	+18	368	3	1004	66.8	−0.045	−0.001	−0.046	−0.119
4	61	0.013	−16	1.034	−12	0	15	4	975	66.8	−0.060	−0.003	−0.063	−0.165
5	75	0.010	−47	1.034	−46	+5	45	14	906	66.8	−0.210	0	−0.210	−0.228
6	85	0.010	−85	1.035	−93	+25	82	2	854	66.4	−0.030	−0.01	−0.04	−0.438
7	90	0.0	1	1.038	−11	+14	−1	9	848	66.4	−0.136	0.0	−0.136	−0.478
8	90	0.0	10	1.038	−1	+20	−10	41	868	66.4	−0.618	0.014	−0.604	−0.614
9	91	0.014	1	1.048	42	0	−1	28	892	66.5	−0.421	−0.004	−0.425	−1.218
10	95	0.010	−8	1.054	5	+15	8	179	1049	67.0	−2.672	−0.01	−2.682	−1.643
11	100	0.0	138	1.048	309	+2	−132	24	1307	67.7	−0.355	0.048	−0.307	−4.325
12	100	0.048	203	1.049	207	+10	−193	65	1510	68.2	−0.953	0.110	−0.843	−4.632
13	100	0.158	165	1.046	200	+23	−158	−370	1424	67.9	+5.45	−0.158	5.292	−5.475
1	55	0	0	1.022	−372	+2	0							0.183

Source: McElroy and Kingery (1957).
Conversion factors: 1 ft = 0.3048 m. 1 in. water = 249.089 Pa.

Station 4   Relative humidity from Table 6.1 ($t_d = 65°F$, $t_w = 57°F$)
$\phi = 61\%$

Velocity head $H_{v_4} = \dfrac{450^2}{4009^2} = 0.013$ in. water

Total head $H_{t_4} = -0.165$ in. water (assume given)

Station 5   Relative humidity from Table 6.1 ($t_d = 65°F$, $t_w = 60°F$)
$\phi = 75\%$

Velocity head $H_{v_5} = \dfrac{390^2}{4009^2} = 0.010$ in. water

Step 2. Calculations for conditions between stations 4 and 5:
Elevation difference, $Z_5 - Z_4 = 57 - 104 = -47$ ft
Density ratio DR for the air column between 5 and 4:

Average relative humidity $\phi = \dfrac{61 + 75}{2} = 68\%$

Average dry-bulb temperature $t_d = \dfrac{65 + 65}{2} = 65°F$

From Fig. 6.15, DR = 1.034
Altimeter difference, $p_{A_5} - p_{A_4} = 952 - 998 = -46$ ft
Base correction, $-(p_{B_5} - p_{B_4}) = 5$ ft

Elevation correction, $-\dfrac{(Z_5 - Z_4)}{DR} = \dfrac{47}{1.034} = 45$ ft

Corrected head difference in ft. of air $= (-46 + 5 + 45) = 4$ ft

Average altimeter reading $= \dfrac{p_{A_5} + p_{A_4}}{2} = \dfrac{952 + 998}{2} = 975$ ft

The conversion factor CF for the average altimeter is read from the 1000-ft offset scale from Fig. 6.16:

$$CF = 66.8 \text{ ft per in. water}$$

Difference in static head $H_s = -\left(\dfrac{4}{66.8}\right) = -0.060$ in. water

Difference in velocity heads $= H_{v_5} - H_{v_4} = 0.010 - 0.013 = -0.003$ in. water

Difference in total head $H_t = -0.060 + (-0.003) = -0.063$
Total head at 5 $H_{t_5} =$ total head at 4 $H_{t_4} +$ the difference in total head between 5 and 4 $= -0.165 + (-0.063) = -0.228$ in. water.
The negative difference in total head ($H_{t_5} - H_{t_4}$) indicates that the airflow is from 4 to 5.
Note that total pressure differences from station 1 back to station 1 must add up to zero.

In the example problem, the sum is $+0.183$ in. water in a total head of 5.475 in. water. This is called the closure error and is the result of errors in measurements and the presence of any natural ventilation (Chapter 8).

The calculations for the head losses in the slope (and the shaft) are similar to those in Example 6.5 and are left for the reader to verify. The head loss in the slope is 0.122 in. water, and in the shaft is 0.856 in. water.

Comparison of Pressure Survey Instruments and Methods   The accuracy of the direct method or the manometer survey is highly desirable for precision work involving major changes in the system, a fan installation or modification, shaft or slope head loss determinations, or the determination of reliable experimental values of friction factors. The convenience and speed of the indirect method or the altimeter survey is advantageous for routine checks of the effectiveness of the ventilation system, in determining the location of sources of excessive head loss, or for measurement of differences in head in a system of moderate complexity or length. Altimeters are widely preferred for underground pressure surveys, especially in coal. The imprecision of altimeters is not a limitation, since the calculations are based on pressure differences rather than absolute pressure measurements. Therefore, altimeters used in a survey should indicate the same elevation change for a given change in pressure. A comparison of air measurements by manometer and altimeter for pressure surveys underground is summarized below:

Manometer	Altimeter
Portable, slow, inflexible, cheap	Portable, fast, versatile, costly
No lag, but fluctuates	Subject to creep or lag
Temperature change causes error	Temperature change correctable
No elevation correction	Must correct for elevation
No barometric correction possible	Must correct for barometer
Station interval limited	Station interval unlimited
Reads directly in in. (mm) water	Readings must be converted
Used for accurate measurements	Used for routine surveys

6.8 CONTINUOUS MONITORING AND REMOTE CONTROL OF THE MINE ENVIRONMENT

Monitoring the conditions of the mine environment involves the sampling and analysis of the mine air to ensure the continued health, safety, and productivity of underground personnel. Control of the mine environment, meanwhile, involves the adjustment, modification, alteration, or correction of an existing or incipient undesirable environmental situation in order to attain or maintain a suitable underground condition. Although incremental, cyclical, or random monitoring of the mine environment is satisfactory in some cases, lamentable catastrophes in mines due to inadequate monitoring and control of the mine atmosphere have focused attention on the need for better methods of mine ventilation monitoring and control.

On-line monitoring and control usually imply data collection in the process and its environment through devices that are wired directly to a computer and operate continuously to control the process. The use of methane monitors on face equipment and airflow monitoring at fans, practiced for some time now, has been partially responsible for minimizing major underground disasters, such as fires and explosions. Cyclical checking of the mine air for toxic, noxious, and explosive gases and dust has also reduced injuries and fatalities in underground mining.

Continuous monitoring of airflow and the gases and dust contained therein would go a long way in upgrading the mine environment. Continuous monitoring aids the day-to-day operation and maintenance of the mine system in minimizing hazards and maximizing efficiency, and provides early detection of potential problems well before they develop into major health and safety hazards.

Although a large number of environmental parameters could be monitored, the most suitable are air velocity, differential pressure or head between points in an airway, concentrations of critical gases (methane, oxygen, carbon monoxide, oxides of nitrogen, etc.) and dusts, temperature, and relative humidity. Suitable location of sensors is essential. Sensors placed only in areas where problems are most likely to occur may miss an unsuspected area where problems may arise. A more desirable system would be the division of the entire mine ventilation system into zones in which surveillance is maintained over each zone by a set of sensors (Nilsson, 1995). A principal set of sensors located in the major intake and return airways, including the bleeders, would oversee the minewide environment.

Monitoring of the Mine Atmospheric Environment

Currently, there are two approaches to the monitoring of mine environments. In the first, sensors are sited at various points in the mine to monitor parameters of interest. The sensor output is transmitted to a data-processing center for analysis. In the second approach, known as *tube-bundle sampling,* mine air is drawn through the tubes installed throughout the mine by small vacuum pumps situated at a sample analysis center.

A generalized block diagram of the first approach is shown in Fig. 6.18. Sensors and/or transducers monitor the current values of the parameters of interest. The output of the sensors, which is usually in the analog form, is sent through an amplifier preconditioner to standardize the signals. The amplifier/preconditioner can also be used to convert the analog signals to digital signals. The preconditioner output is passed to a multiplexer, where the signals from all the sensors are combined into one composite signal with a unique frequency. A sample holder may be inserted in the circuit between

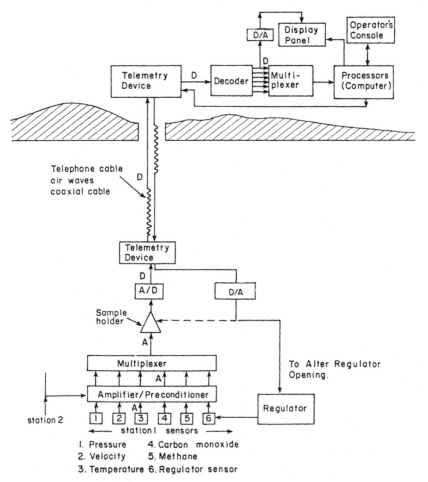

FIGURE 6.18 Block diagram of an automated monitoring and control system (A—analog signal; D—digital signal; A/D—analog-to-digital converter; D/A—digital-to-analog converter). (After Hormozdi, 1979.)

the multiplexer and the analog/digital (A/D) converter. The purpose of the sample holder is to enable sampling from one or more sensors as required rather than on a regular programmed time sequence for each sensor. Also the sample holder may have a small memory unit, whereby the value of a parameter can be stored for a short time. The A/D converter converts the multiplexer signal, if it is in the analog form, to digital form.

The digital signal is modulated in a telemetry device (telemetry station) and is transmitted to a receiving telemetry device. This data transmission can take several forms such as (1) wireless (radio), (2) wire (telephone cables), and (3) coaxial cables. Amplifier, multiplexer, A/D converter, and

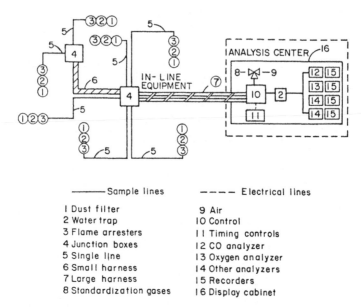

- ─────── Sample lines   ── ── ── Electrical lines

1 Dust filter	9 Air
2 Water trap	10 Control
3 Flame arresters	11 Timing controls
4 Junction boxes	12 CO analyzer
5 Single line	13 Oxygen analyzer
6 Small harness	14 Other analyzers
7 Large harness	15 Recorders
8 Standardization gases	16 Display cabinet

FIGURE 6.19 Schematic diagram of a tube–bundle system; automatic control. (After Fink and Adler, 1975.)

telemetering equipment can be packed in a telemetry station (outstation). The modulated signal is demodulated at the receiving telemetry station and then is transmitted to a decoder at a central control station. At the decoder, the composite signal is converted back to individual signals based on the multiplexer's reference frequency. These signals can be passed through a D/A converter and can be displayed on a process display or can be passed through a multiplexer to a computer for data processing. The decoder and multiplexer functions can be performed in a computer software program. The control signal from the computer, which may be interrupted by the operator, follows a similar sequence of manipulations before resulting in a physical action such as altering a regulator opening.

In the tube-bundle system (Fig. 6.19), the sensors and telemetering devices are replaced by an exhauster pump situated in a remote location (perhaps on the surface). The pump draws a continuous sample of air from selected points underground through small-bore plastic tubing. The sample can be analyzed by laboratory-quality instruments for constituents such as carbon monoxide, methane, oxygen, and other gases.

In either system, an enormous amount of data is continually collected whose reading, processing, interpretation, and storage can be an onerous task without the help of a computer. In computer-based data-processing systems, the computers can be used in both off-line and on-line configurations. Separate off-line computers enable input data and information to be accumulated and provide access to relevant files. Such a facility would be useful for routine analyses and planning. In on-line computers, indicating devices, data storage, data display, and discriminatory and alarm facilities can be provided. Such systems are usually supported by standby computer capability and integrated facilities for operation during maintenance failures and emergencies.

Remote Control of the Mine Environment

A successful mine monitoring system is a prerequisite for the introduction of remote control. For example, changes in airflow quantities and directions can be achieved in a relatively short time by remotely and electrically controlling the regulators and fans. In the case of explosion or fire, the ability to redistribute and control the direction of airflow would be extremely valuable in aiding the safe evacuation of personnel and in controlling the spread or propagation of mine fires. Rustan and Stöckel (1980) describe a three-way valve for use in mine ventilation for controlled distribution of air. Known as the *flexistor*, it is an air valve that works on a refined fluidic principle. By causing a small geometric change inside the valve, the distribution of flow can be altered to any desired proportion between splits in the mine airways (Fig. 6.20).

The monitoring and control system of the ventilation at the Waste Isolation Pilot Plant (WIPP) in Carlsbad, NM, is designed with 15 air-velocity sensors and 8 differential pressure sensors, and for remote control of the air regulators to control the airflow in four main splits called the experimental areas, mining areas, waste storage area, and the waste shaft station (Strever et al., 1995).

Automatic Control

The availability of minewide remote monitoring gives rise to the possibility of adjusting airflow control devices such as fans, regulators, and doors auto-

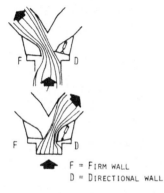

F = FIRM WALL
D = DIRECTIONAL WALL

FIGURE 6.20 Principle of operation of SKEGA flexistor air valve for the control of air distribution to two workplaces. (After Rustan and Stöckel, 1980.)

matically to maintain the desired quantity, quality, and climatic conditions underground. An important requirement for the introduction of automated-control technology is a clear understanding of the cause-and-effect relationship, whereby the monitored factors (such as high methane concentration) can lead to actions (such as opening a regulator, changing the operation of a fan, or turning off power to a cutting machine) that will result in achieving the desired effect (i.e., lowering the methane concentration). In essence, a control policy must be predetermined and programmed into the computer. The policy must be governed by the magnitude of the deviation from the desired value (error), its rate of change, and its immediate past history. To date, there are very few examples of completely computer-controlled ventilation systems.

One such monitoring and control system is the Mine Operating System (MINOS), which has been under development in the United Kingdom since the early 1970s. The principles employed on MINOS were first established on an experimental system at Bagworth Colliery in 1973 (Bexon and Pargeter, 1976). Among the basic components of MINOS is a digital computer linked to a control console and driving underground data-transmission equipment (Thomas and Chandler, 1978). The surface station includes a printer for report generation and optional remote display terminals for use by officials away from the control center (Fig. 6.21). Normally, operation is automatic, with the computer controlling the remote equipment linked to the system. The computer monitors the actions and values, and stores data and events

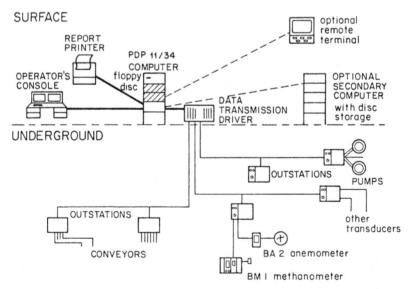

FIGURE 6.21 Basic elements of the MINOS processing system. (After Thomas and Chandler, 1978. By permission from *The Mining Engineer,* London.)

MINE MONITORING AND CONTROL

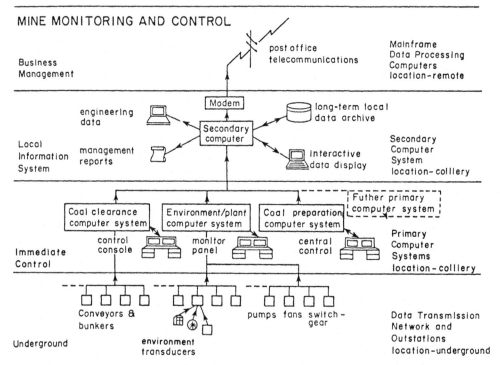

FIGURE 6.22 Distributed computer system for overall mine monitoring and control. (After Bexon and Pargeter, 1976. By permission from West Virginia University, Morgantown.)

in its memory or sends them to a secondary system. Only when anything abnormal occurs is operator action required.

The MINOS computer was installed in 1975 at Brodsworth Main Colliery in the United Kingdom (Bexon and Pargeter, 1976). A tube-bundle sampling system with infrared and other analyzers at the surface was also linked up to the computer system. The analyzers measure methane and carbon monoxide concentrations. Trends in carbon monoxide concentrations are examined by the computer to give early warnings of heating or fires and to distinguish from short peaks due to blasting or changes in barometric pressures.

A fundamental aspect of MINOS philosophy on computer-based colliery systems is the relationship defined between primary systems (e.g., environmental monitoring) and the secondary computer (Fig. 6.22). The secondary computer is designed to promote effective modularity and standardization without placing arbitrary constraints on the primary system. In this manner, primary system functions may be developed or evolved with the minimum self-generated constraints of the secondary system. The secondary computer will provide the common services to all such primary systems that are not considered vital to the effective daily or short-term functions of the latter but that optimize plant operations in the medium and long term.

The typical secondary system compromises a computer; supporting random-access bulk-data storage devices; a printer; various interactive terminals for control and data entry; and a high-resolution graphic or graph-plotting facility, with high-speed data links to the primary systems and, where appropriate, to the area or regional computer facilities. The secondary computer is programmed to be a flexible computational tool, supporting programs written in a variety of high-level languages. Typical secondary system functions may well include some or all of the following: (1) production monitoring, (2) maintenance scheduling, (3) energy utilization, and (4) long-term environmental analysis. Such a system is now fully in operation in the Selby complex of mines (Houghton, 1991).

Automatic monitoring, remote control, and automatic control of the mine atmospheric environment are technically feasible. However, the introduction of a system is very much dependent on the magnitude of the problems and the economics and benefits to be derived from its installation. It is better to plan for an overall monitoring and control system, even though only parts of it may be installed in the beginning. In the development of the total system, the ability to tie the system at a later date to production operations such as longwall and continuous mining and service operations such as haulage and drainage should be considered. In fact, an automated monitoring and control system, when applied to these operations, may become more justifiable than when applied to the atmospheric environment alone.

6.9 ORGANIZATION FOR VENTILATION FUNCTIONS

Organizing for the effective management of ventilation functions must address all aspects of ventilation: planning, designing, staffing, directing, and controlling. Ventilation organizational activities can be viewed as line and staff functions. Ventilation planning and design are staff functions, often located within the corporate engineering departments. Mine ventilation, which is primarily an engineering and service function, may not gain a principal focus in the strategic planning process. However, ventilation planning acquires considerable importance in the implementation of production, health and safety policies, and top-management resource allocation responsibilities, for example, for shafts and slopes. The management of the ventilation system at the mine is a line function, intertwined with the management of the production operations in the mine. The responsibility for maintaining the integrity of the system lies with the mine operating personnel (the mine manager, the mine superintendent, etc.). The mine engineer is generally responsible for carrying out needed changes to the system. This diffusion of the ventilation functions among the various parts in the organization should not lead to a state where ventilation is everybody's business and nobody's responsibility. Whatever the manner in which ventilation functions are organized, there must be a clearcut definition of the responsibility, authority, and accountability for ventilation decisions.

Ventilation Decisions

Ventilation decisions vary extensively with the nature of mining operations such as metal, nonmetal, or coal mines; gassy or nongassy mines; and deep or shallow mines. Other factors influencing decisions are the available time, personnel involvement, scale of operations, mining method, type of equipment, and spatial extent and age of the mine. The decisions themselves can range from those that have primarily local significance such as airflow changes at a single mining face to those having minewide ramifications such as the establishment of corporate policies, the sinking of a new ventilation shaft, or the installation of a new fan (Ramani et al., 1985). Ventilation situations warranting decisions may be of two types: repetitive and highly predictable, and infrequently occurring and less predictable. Ventilation decisions can also be classified as those made under normal mine operating conditions that generally evolve in a more structured environment and those made during mine emergencies. Emergency situations consist of two categories: (1) emergencies or abnormal situations where emergency operating plans have been drawn up and (2) emergencies where the existing plans are not adequate for tackling the situation. Emergencies falling in the second category present extreme difficulties in decisionmaking.

In making these decisions, ventilation personnel have to operate with good data and within a broad set of guidelines. Guidelines are of three types: legal requirements, company procedures, and policies that are based not necessarily on minimum legal requirements, but on research and development findings. Higher levels of management, when making decisions related to ventilation, will evaluate the available options in the context of company policy, any recent research findings, and capital requirements. On the other hand, operating management will be bound by legal requirements and good practices.

All management personnel from section supervisors to higher officials have to rely on a good flow of data and information for decisionmaking. The ventilation information system is a system that has data, information, and models to support ventilation decisionmaking. With the emergence of automatic monitoring and control systems and an increased use of computers, the role of data acquisition, analysis, synthesis, and presentation has become very important. A schematic of a knowledge-based, mine ventilation decision-support system incorporating planning, designing, monitoring, and control functions is shown in Fig. 6.23 (Ramani and Prasad, 1987).

Ventilation Department

The importance of ventilation is well recognized in the mining industry. The need for good planning, proper engineering, and adequate control is also

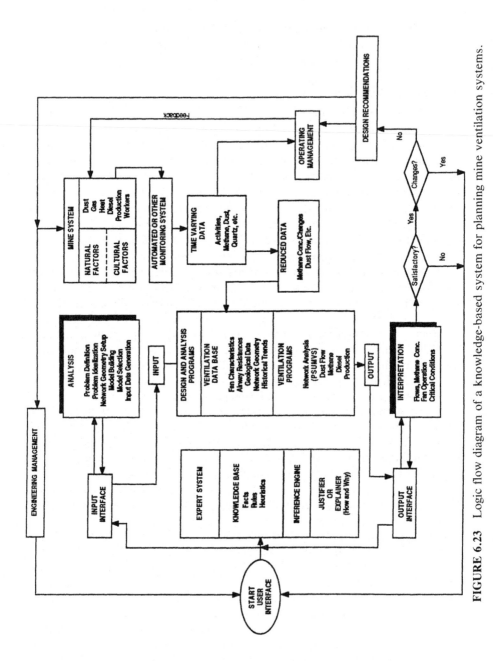

FIGURE 6.23 Logic flow diagram of a knowledge-based system for planning mine ventilation systems.

234

well recognized. However, the organizational structure and lines of authority for design and control of ventilation systems are very diverse. In the United States, mining laws and regulations provide guidelines as to the ventilation standards to be maintained, frequency and kinds of measurements, and maintenance of records (Anon., 1995a). Instructions on organizational structure and lines of authority are not very specific, although statutory responsibilities are defined in terms of certified and authorized persons. For some engineering and design work, an engineer must have been certified by a state or national board of professional registration.

In the United States, mines facing severe gas, spontaneous combustion, heat and humidity, and other similar ventilation problems or having extensive methane drainage or air conditioning programs often have engineering and management positions staffed by personnel with extensive knowledge and experience in these areas of mining (Stevenson, 1980). Both senior and junior engineers as well as supporting technicians are necessary. Where much construction work needs to be done, a construction crew may also be necessary. When the law specifies the organization, as in South Africa, a more formal structure is warranted (LeRoux, 1974; Burrows and Roberts, 1980). In South Africa, the best predictor of ventilation-personnel requirements is mine production; a less reliable predictor is number of underground employees (Fig. 6.24). Typical organizations at large mines in the United States and South Africa are compared in Fig. 6.25.

Staff selection and training in the ventilation department warrants special care. This is particularly true because of the lack of availability of experienced personnel. Assigning ventilation duties only temporarily to engineers,

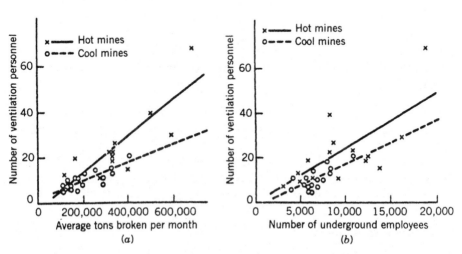

FIGURE 6.24 Number of ventilation personnel in South African gold mines as a function of (*a*) mine production and (*b*) number of underground employees. (After Burrows and Roberts, 1980. By permission from AIME, New York. Copyright 1980.)

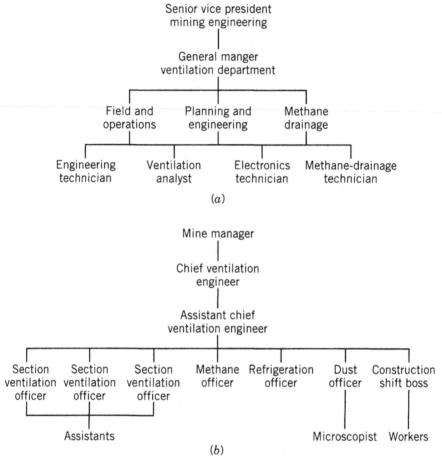

FIGURE 6.25   Organization charts of mine ventilation departments: (*a*) American coal mine; (*b*) South African gold mine. [After (*a*) Stevenson, 1980; (*b*) LeRoux, 1974.]

and then not necessarily to those specially trained for the task, is not conducive to the development of professional expertise. Ventilation personnel should understand both the theoretical and practical sides of their field. Skilled management and well-conceived reward systems are crucial to the success of the department's operation (Rose, 1980; Shaw, 1980).

Ventilation Equipment

The ventilation department must have enough equipment to perform effectively its planning, designing, and operating functions. As a minimum, the equipment available will include the following:

1. A mercury barometer
2. An adjustable inclined-tube manometer
3. Magnehelic differential-pressure gages for different ranges
4. Pitot tubes
5. Two altimeters, capable of reading pressures to the nearest foot (m)
6. Two vane anemometers, including the digital type
7. Smoke tubes
8. A sling psychrometer, a stationary hygrometer, and kata thermometers
9. Instruments for air-quality sampling and monitoring

Item 9 (air-quality sampling and monitoring instruments) includes a wide variety of equipment: aspirator bulb with stain tubes for a number of gases commonly found in the mines, portable electronic instruments for detecting methane, carbon monoxide, nitrogen oxides, oxygen deficiency, etc. Instruments for mine aerosol sampling should include several personal gravimetric dust samplers, at least one Mine Research Establishment (MRE) gravimetric dust sampler, and at least one real-time aerosol sampler such as RAM-1 or a DataRAM (MIE, Inc., Billerica, MA).

Technical literature on the various equipment, including operational instructions and maintenance guidelines, should be maintained in the department. Field notes, ventilation reports, and mine maps must be properly cataloged and indexed for ready reference. Some equipment such as vane anemometers may require periodic calibration.

The ventilation department must have access to personal computers; computers networked to larger systems; mine ventilation–specific software such as pressure survey, heat transfer, and ventilation network programs; and general-purpose software including graphic and word-processing packages.

The broad objective of mine ventilation and air conditioning—to provide a comfortable, safe atmospheric environment for workers—must never be lost from sight. However, its role must be placed in the proper perspective: atmospheric environmental control is only one phase of a broader, more general, environmental-control mission. Mine environmental control can well be integrated into a single department and expanded to include health and safety engineering as well (Ramani, 1992b). A number of progressive mines have already established such organizations.

PROBLEMS

6.1 The smoke-tube method is employed to determine the velocity of a slow-moving airstream in a drift. The observer records the travel time of a smoke cloud as 35 s over a measured distance of 20 ft (6.1 m). Calculate the approximate average air velocity in the drift.

6.2 A pitot tube is set up in a main drift, 8 × 10 ft (2.4 × 3.0 m), and connected to an inclined manometer to read velocity head directly. Nine readings are obtained by equal-area traversing as follows, in in. water (Pa):

0.090 (22.4)   0.086 (21.4)   0.083 (20.7)
0.089 (22.1)   0.085 (21.2)   0.082 (20.4)
0.088 (21.9)   0.084 (20.9)   0.080 (19.9)

Determine the average velocity and quantity of air flowing in the drift if psychrometer readings are 85 and 75°F (29 and 24°C) and the barometer reads 28.6 in. Hg (96.6 kPa). Calculate by averaging heads and then by averaging velocities, and compare the results.

6.3 Calculate the centerline-velocity correction factor for a 7 × 9-ft (2.1 × 2.7-m) drift in which velocity measurements were made by anemometer as follows:

Reading by traverse, 2 min: 770 ft (235 m)
Centerline reading, 1 min: 450 (137 m)

Also determine the quantity of airflow.

6.4 Design an inclined manometer using oil of specific gravity of 0.8 and having a magnification of 5 and a corrected scale range of 2 in. (50.8 mm) water. The tubing bore is 0.1 in. (2.5 m) and the reservoir diameter 0.75 in. (19.1 mm).

6.5 The following data were obtained in a two-instrument altimeter survey of a main split of air. Complete tables of calculated data to determine the overall drop in static head in in. water (Pa). The average wet-bulb temperature was 33°F (0.6°C) and the barometric pressure 28.65 in. Hg (96.7 kPa) underground. Assume no splitting of the air:

Time	Surface Altimeter ft (m)	Surface Temperature, °F (°C)	Station	Mine Altimeter ft (m)	Mine Temperature, °F (°C)	Station Elevation ft (m)
9:00	6500 (1981)	80 (27)	1	5000 (1524)	45 (7)	2240 (683)
9:10	6504 (1982)	80 (27)	2	5029 (1533)	43 (6)	2236 (682)
9:20	6505 (1983)	82 (28)	3	5041 (1536)	39 (4)	2245 (684)
9:30	6505 (1983)	84 (29)	4	5050 (1539)	38 (3)	2251 (686)
9:40	6504 (1982)	86 (30)	5	5067 (1544)	38 (3)	2249 (685)
9:50	6501 (1982)	87 (31)	6	5081 (1549)	36 (2)	2243 (684)
10:00	6499 (1981)	88 (31)	7	5093 (1552)	35 (2)	2248 (685)
10:10	6496 (1980)	91 (33)	8	5099 (1554)	35 (2)	2252 (686)
10:20	6494 (1979)	93 (34)	9	5112 (1558)	34 (1)	2255 (687)
10:30	6488 (1978)	96 (36)	10	5135 (1565)	35 (2)	2262 (689)

6.6 Plot the static head gradient from Problem 6.5.

6.7 The corresponding velocity heads for the stations of Problem 6.5 are as follows, in in. water (Pa):

Station 1	0.14 (34.8)	Station 6	0.21 (52.3)
Station 2	0.23 (57.2)	Station 7	0.09 (22.4)
Station 3	0.21 (52.3)	Station 8	0.15 (37.3)
Station 4	0.08 (19.9)	Station 9	0.20 (49.8)
Station 5	0.17 (42.3)	Station 10	0.12 (29.9)

Plot all the head gradients for the system.

6.8 Calculate the actual head loss between stations in Problem 6.5, using the data given and gradients obtained in Problem 6.7. What is the overall head loss for the entire system? Calculate the mine heads, assuming system discharge at station 10.

CHAPTER 7

Mine Ventilation Circuits and Networks

7.1 RELATIONSHIP BETWEEN HEAD AND QUANTITY

In a given ventilation system or portion thereof, a fixed proportionality exists between the head applied and the quantity of airflow. This means that once the heads have been determined in a system for a given flow, the heads at any other flow can be determined easily. This relationship is of immense value in mine ventilation work, because varying the airflow in a given system and in portions of a system occurs frequently.

From the Atkinson equation for the calculation of head loss in an airway (Eq. 5.25), it will be recalled that parameters K, O, L, L_e, and A are constants for a given airway or duct; therefore,

$$H_l \propto Q^2 \tag{7.1}$$

that is, the head loss varies as the square of the quantity for a given airway. Since the static head of a system is the sum of the series head losses of the system, again

$$H_s \propto Q^2 \tag{7.2}$$

Recalling the relationship for velocity head (Eq. 5.13), and that $Q = VA$ (Eq. 3.10), then

$$H_v \propto Q^2 \tag{7.3}$$

Finally, since the total head of a system is the sum of the static and velocity heads, then

$$H_t \propto Q^2 \tag{7.4}$$

Therefore, *any mine head or head loss varies as the square of the quantity.* Representing difference in head by the symbol H,

$$H \propto Q^2 \tag{7.5}$$

This is a basic law of mine ventilation.

Characteristic Curve

The static head of a system can be plotted against the quantity on graph paper. The word *system* can be interpreted as a portion of the mine, or it can refer to the entire mine if only a single fan is being used. For an entire mine, this graphic representation is called the *mine characteristic curve,* or simply the *mine characteristic*. It is also often helpful to plot the mine total head as a function of the mine quantity.

One point on the curve is determined by measuring or calculating the system head for an actual or assumed airflow. Then additional points are determined from the basic head–quantity relation (Eq. 7.5), which can also be expressed as follows:

$$\frac{H_1}{H_2} = \left(\frac{Q_1}{Q_2}\right)^2$$

or

$$H_2 = H_1 \left(\frac{Q_2}{Q_1}\right)^2 \tag{7.6}$$

Example 7.1   Given a mine with a single fan whose static head is 2 in. water (497.7 Pa) and total head 3 in. water (746.5 Pa) at a quantity of 400,000 cfm (188.8 m³/s), determine and plot the mine characteristic curve.

Solution: Assume quantities and calculate the corresponding heads by Eq. 7.6.

Mine Q cfm (m³/s)	Mine H_s in. water (Pa)	Mine H_t in. water (Pa)
0 (0)	0 (0)	0 (0)
200,000 (94.4)	0.5 (125)	0.8 (200)
400,000 (188.8)	2.0 (497)	3.0 (745)
600,000 (283.2)	4.5 (1120)	6.8 (1690)
800,000 (377.6)	8.0 (1990)	12.0 (2990)

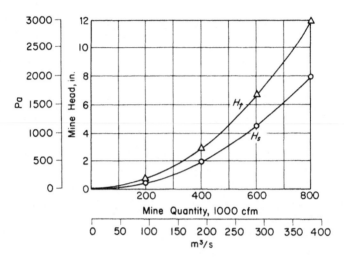

FIGURE 7.1   Mine characteristic curve (see Example 7.1).

The mine characteristic is plotted in Fig. 7.1, which can be used to determine the head corresponding to any given quantity.

Airway Resistance

Since the head loss of an airway is directly proportional to the square of the quantity flowing through it, the head–quantity relationship can be expressed in equation form by including a constant of proportionality. Thus Atkinson's equation (Eq. 5.25) can be written as

$$H_l = RQ^2 \qquad (7.7)$$

where R is the constant of proportionality and is referred to as the resistance of an airway. In effect, the constant terms K, O, L, L_e, and A of Eq. 5.25 are lumped into a single constant, namely, the resistance R, which is expressed as follows:

$$R = \frac{KO(L + L_e)}{5.2\,A^3} \qquad (7.8)$$

$$R = \frac{KO(L + L_e)}{A^3} \qquad (7.8a)$$

in which the units of R are in.·min²/ft⁶ (N·s²/m⁸).

The Atkinson equation states that the head loss for a given airway is equal to the resistance of the airway times the square of the quantity of air flowing through the airway. An analogy can be drawn between the Atkinson equation and Ohm's law, which is the fundamental equation used in electrical-circuit analysis. Ohm's law states that the potential difference or voltage across a resistive device is equal to the resistance of the device times the current flowing through the device. Therefore, head loss is analogous to potential difference, quantity to current, and airway resistance to electrical resistance. As a result, many of the techniques used in the analysis of electrical circuits are applicable to the analysis of ventilation circuits. However, caution must be exercised when applying these techniques, since the Atkinson equation is a square-law relation and Ohm's law, a linear relation.

Equivalent Orifice

A term of long standing in fan selection but in little use today is *equivalent orifice*. It compares the resistance of a system to the resistance of a circular opening in a thin plate through which the same quantity of air is flowing. Using a vena contracta coefficient of 0.65, Murgue obtained (McElroy, 1935)

$$O_e = \frac{3.9 \times 10^{-4} Q}{\sqrt{H_s}} \tag{7.9}$$

$$O_e = \frac{1.2 Q}{\sqrt{H_s}} \tag{7.9a}$$

where O_e is equivalent orifice in ft^2 (m^2), Q is quantity flowing in cfm (m^3/s), and H_s is static head of the system in in. water (Pa).

7.2 KIRCHHOFF'S LAWS

Two fundamental laws governing the behavior of electrical circuits were developed by the German physicist Gustav Robert Kirchhoff (1824–1887). They are extensively applied in ventilation-circuit analysis using the analogy drawn in Section 7.1.

Kirchhoff's First Law

Figure 7.2 is a segment of a ventilation circuit where four airways meet at a common point or a junction. According to Kirchhoff's first law, also known as Kirchhoff's current law (KCL), the quantity of air leaving a junction must equal the quantity of air entering a junction; therefore, from Fig. 7.2,

$$Q_1 + Q_2 = Q_3 + Q_4$$

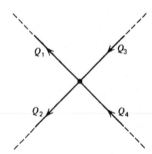

FIGURE 7.2   Application of Kirchhoff's first law (see Example 7.2).

If the quantity of air leaving a junction is defined as being positive, and the quantity of air entering a junction as being negative, the sum of the four quantities in Fig. 7.2 must be zero; thus

$$Q_1 + Q_2 - Q_3 - Q_4 = 0$$

which is the same as the previous equation. Therefore, the following general equation can be used for expressing Kirchhoff's first law:

$$\Sigma Q = 0 \tag{7.10}$$

Example 7.2   From Fig. 7.2, determine the value and the direction of Q_4, given that the following quantities flow in the directions indicated:

$$Q_1 = 200{,}000 \text{ cfm } (94.39 \text{ m}^3/\text{s})$$

$$Q_2 = 300{,}000 \text{ cfm } (141.58 \text{ m}^3/\text{s})$$

$$Q_3 = 900{,}000 \text{ cfm } (424.75 \text{ m}^3/\text{s})$$

Solution: Applying Kirchhoff's first law to the air quantities at the junction yields

$$\Sigma Q = Q_1 + Q_2 - Q_3 - Q_4 = 0$$

$$Q_4 = 200{,}000 + 300{,}000 - 900{,}000 = -400{,}000 \text{ cfm } (-188.78 \text{ m}^3/\text{s})$$

Since Q_4 is negative, it must be in the direction opposite to that indicated in Fig. 7.2. Thus it is in the direction leaving the junction.

Kirchhoff's Second Law

Kirchhoff's second law, also known as *Kirchhoff's voltage law* (KVL), states that *the sum of the pressure drops around any closed path must be equal to zero,* which can be expressed as

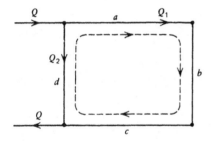

FIGURE 7.3 Application of Kirchhoff's second law.

$$\Sigma H_l = 0 \qquad (7.11)$$

where the summation is over all airways in the closed path and the H_l value for any airway is the algebraic sum of the friction head, the natural ventilation head, and the fan head. Adoption of a consistent sign convention is necessary for correct solution to problems. Consider a closed path consisting of airways a, b, c, and d, as indicated by the dashed line (Fig. 7.3). If one sums the head losses in a clockwise direction around this closed path, the following equation can be written

$$H_l = H_{l_a} + H_{l_b} + H_{l_c} - H_{l_d} = 0$$

where H_{l_a}, H_{l_b}, and H_{l_c} are positive, since the quantity Q_1 flowing through airways a, b, and c is flowing in the clockwise direction. Since Q_2 flows opposite to the direction of summation, there is a head gain in d, or the head loss (H_{l_d}) is negative. This equation can also be expressed in terms of the resistance and quantity for each airway. However, in order to maintain the validity of the sign convention for all cases, the Atkinson equation must be expressed as $H_l = R|Q|Q$, where $|Q|$ is the absolute value of Q (Wang and Hartman, 1967; Wang and Saperstein, 1970). Therefore, the equation is written as

$$\Sigma H_l = R_a|Q_1|Q_1 + R_b|Q_1|Q_1 + R_c|Q_1|Q_1 - R_d|Q_1|Q_1 = 0$$

Kirchhoff's second law must also take into account any pressure sources (fan or natural ventilation) that exist in the closed path. Since a pressure source creates a pressure rise, it must be considered as a negative pressure drop (head loss). Thus a pressure source is assigned a negative value if its airflow is in the direction of summation. It is assigned a positive value if it creates an airflow opposite to the direction of summation.

A simple example is presented to demonstrate the adopted sign convention.

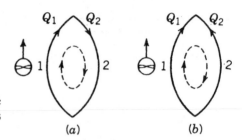

FIGURE 7.4 Demonstration of the sign convention for Kirchhoff's laws (see Example 7.4).

Example 7.3 Figure 7.4a consists of two airways with a fan located in airway 1, causing the air to flow in the indicated direction. Determine the quantities through airways 1 and 2, assuming that the fan is operating at a static head of 1 in. water (248.84 Pa) and the resistances of airways 1 and 2 are 10×10^{-10} and 15×10^{-10} in.·min²/ft⁶ (1.118 and 1.677 N·s²/m⁸), respectively.

Solution: For this simple case, it is apparent that Q_2 is equal to and in the same direction as Q_1. Frequently, however, the direction of an airflow cannot be determined by inspection; thus one must assume a direction of airflow. If the directions indicated by Fig. 7.4a are assumed, and the pressure drops are summed in a clockwise direction, the following expression results:

$$\Sigma H_l = -1 + 10 \times 10^{-10} |Q_1| Q_1 + 15 \times 10^{-10} |Q_2| Q_2 = 0 \quad \text{and} \quad Q_2 = Q_1$$

Substituting Q_1 for Q_2 yields

$$|Q_1| Q_1 = 4 \times 10^8$$

Therefore,

$$Q_1 = 20{,}000 \text{ cfm } (9.439 \text{ m}^3/\text{s}) \quad \text{and} \quad Q_2 = 20{,}000 \text{ cfm } (9.439 \text{ m}^3/\text{s})$$

Since Q_1 and Q_2 are both positive values, their directions must be the same as those indicated in Fig. 7.4a.

For the sake of argument, consider the assumed direction of Q_2 in Fig. 7.4b. Summing the pressure drops in a clockwise direction yields

$$\Sigma H_l = -1 + 10 \times 10^{-10} |Q_1| Q_1 - 15 \times 10^{-10} |Q_2| Q_2 = 0 \quad \text{and} \quad Q_2 = -Q_1$$

Therefore,

$$Q_1 = 20{,}000 \text{ cfm } (9.439 \text{ m}^3/\text{s}) \text{ and } Q_2 = -20{,}000 \text{ cfm } (-9.439 \text{ m}^3/\text{s})$$

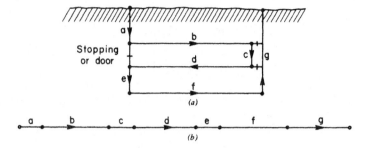

FIGURE 7.5   Mine (*a*) with all airways arranged in series, as demonstrated in schematic (*b*).

Since Q_1 is positive, it is in the same direction as that indicated in Fig. 7.4*b*. However, since Q_2 is negative, it must be opposite to the direction indicated. Therefore, no matter which directions one assumes for the airflows, the correct results will be obtained. However, if the head losses are expressed as RQ^2, substituting $Q_2 = -Q_1$ would result in an incorrect sign and solution.

The application of Kirchhoff's laws to complex networks is discussed in Section 7.7.

7.3   SERIES CIRCUITS

In a ventilation system, two basic combinations of airways are possible: *series* or *parallel*. Both types occur as well as various complex combinations.

A *series circuit* is defined as a circuit whose airways are arranged end to end so that the quantity of air flowing through each airway is the same. An example of a series circuit is shown in Fig. 7.5.

Figure 7.6*a* illustrates a simple series circuit consisting of airways 1, 2, and 3 with resistances R_1, R_2, and R_3 and head losses H_{l_1}, H_{l_2}, and H_{l_3},

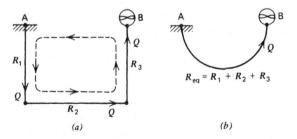

FIGURE 7.6   Airways (*a*) in series and (*b*) equivalent airway.

respectively. The fan static head is H_m. The quantity of air flowing through each airway is the same; thus in general form

$$Q = Q_1 = Q_2 = Q_3 = \cdots \tag{7.12}$$

Applying Kirchhoff's second law to this circuit in a counterclockwise direction results in the following:

$$H_{l_1} + H_{l_2} + H_{l_3} - H_m = 0$$

For this case, the fan head is equal to the total head loss (static head) from points A to B. Since one must often deal with portions of a ventilation circuit that may not involve a fan, the following general expression can be written

$$H_l = H_{l_1} + H_{l_2} + H_{l_3} + \cdots \tag{7.13}$$

This states that the total head loss for a series circuit is equal to the summation of the head losses of the individual airways.

Equation 7.13 can be expressed in terms of the quantity and resistance of each airway by

$$H_l = R_1|Q|Q + R_2|Q|Q + R_3|Q|Q$$

In a series circuit, the quantity and direction of airflow through each airway are the same. Therefore, the preceding equation can be written as follows without disturbing the validity of the adopted sign convention:

$$H_l = R_1Q^2 + R_2Q^2 + R_3Q^2 + \cdots$$

Factoring out Q^2 yields

$$H_l = (R_1 + R_2 + R_3 + \cdots)Q^2 = R_{eq}Q^2$$

where R_{eq} is referred to as the *equivalent resistance* of a series circuit. This equation thus defines the equivalent resistance for a series circuit as the summation of the individual resistances. Therefore, the following general equation for series resistance can be written

$$R_{eq} = \frac{H_l}{Q^2} = R_1 + R_2 + R_3 + \cdots \tag{7.14}$$

In effect, this equation allows the modeling of any number of series airways as a single equivalent airway, as shown in Fig. 7.6b.

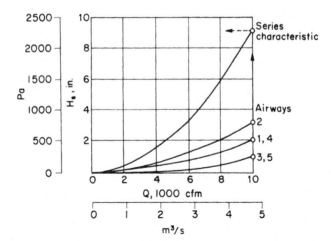

FIGURE 7.7   Characteristic curves of airways in series (see Example 7.4).

Example 7.4   Given five airways in series with the following resistances, all in units of in.·min²/ft⁶ × 10¹⁰ (N·s²/m⁸):

$$R_1 = 200 \ (22.36) \qquad R_3 = 100 \ (11.18) \qquad R_5 = 100 \ (11.18)$$

$$R_2 = 300 \ (33.54) \qquad R_4 = 200 \ (22.36)$$

Find the equivalent resistance and the total head loss of the airways, if 10,000 cfm (4.719 m³/s) flows through them.

Solution:

$$R_{eq} = (200 + 300 + 100 + 200 + 100) \times 10^{-10}$$

$$= 900 \times 10^{-10} \text{ in.·min}^2/\text{ft}^6 \ (100.6 \text{ N·s}^2/\text{m}^8)$$

$$H_l = R_{eq}Q^2 = (900 \times 10^{-10})(10,000)^2 = 9.0 \text{ in. water (2240 Pa)}$$

Series flow calculations may be solved graphically by the use of airway characteristic curves; and, although there is no particular advantage to this approach for simple circuits, curves are useful to visualize conditions with varying airflows. In plotting characteristic curves of ducts in series, the heads are cumulative for a given quantity. Using data from Example 7.4, the curves of Fig. 7.7 result.

7.4 PARALLEL CIRCUITS

Airways are said to be connected in *parallel* when the airways are joined at the same two nodes and the total airflow is divided among them (Fig. 7.8).

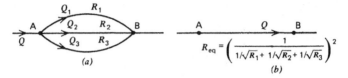

FIGURE 7.8 Parallel airways and equivalent airway.

In mine ventilation, this practice is termed *splitting*, and the parallel branches are referred to as *splits*. There are two forms of splitting. *Natural splitting* occurs when the quantity of air is divided among the parallel branches of its own accord without regulation. *Controlled splitting* occurs when a prescribed quantity of air is made to flow through each parallel branch by means of regulation (see Section 7.7). From Kirchhoff's first law, one can write the general expression

$$Q = Q_1 + Q_2 + Q_3 + \cdots \tag{7.15}$$

Thus when airways are arranged in parallel, the total quantity is the summation of the quantities flowing through the individual airways.

From Kirchhoff's second law, it can also be shown that

$$H_l = H_{l_1} = H_{l_2} = H_{l_3} = \cdots \tag{7.16}$$

which states that the head losses for parallel airways are equal.

Equivalent Resistance for Parallel Circuit

As with series circuits, an equivalent resistance for parallel airways can also be determined. Using the Atkinson equation to express the quantity in each airway as a function of the head loss and the resistance of the airway, and applying Kirchhoff's first law to the flow in Fig. 7.8a, one can write

$$Q = \frac{\sqrt{H_l}}{R_1} + \frac{\sqrt{H_l}}{R_2} + \frac{\sqrt{H_l}}{R_3}$$

where Q is the total quantity and H_l is the head loss of the parallel airways from A to B. The previous equation can now be expressed as

$$Q = \sqrt{H_l}\left(\frac{1}{\sqrt{R_1}} + \frac{1}{\sqrt{R_2}} + \frac{1}{\sqrt{R_3}}\right) = \sqrt{H_l}\left(\frac{1}{\sqrt{R_{eq}}}\right)$$

where R_{eq} is referred to as the equivalent resistance of the parallel circuit.

The general equation for the equivalent resistance of parallel airways can be written as follows:

$$\frac{1}{\sqrt{R_{eq}}} = \frac{1}{\sqrt{R_1}} + \frac{1}{\sqrt{R_2}} + \frac{1}{\sqrt{R_3}} + \cdots \tag{7.17}$$

Equation 7.17 permits the reduction of any number of parallel airways to a single equivalent airway, as shown in Fig. 7.8b.

Example 7.5 Four airways are arranged in parallel with a total quantity of 100,000 cfm (47.19 m³/s) flowing through them. Resistances of the airways are given in the table. Find the head loss for the parallel airways and the quantity of air flowing through each:

Airway	Resistance $R \times 10^{10}$, in.·min²/ft⁶ (N·s²/m⁸)	Quantity Q, cfm (m³/s)
1	23.50 (2.627)	9,540 (4.502)
2	1.35 (0.151)	39,810 (18.788)
3	3.12 (0.349)	26,190 (12.360)
4	3.55 (0.397)	24,550 (11.586)

Solution: Sample calculation using Eqs. 7.7 and 7.17:

$$R_{eq} = \left(\frac{1}{\dfrac{1}{\sqrt{23.5}} + \dfrac{1}{\sqrt{1.35}} + \dfrac{1}{\sqrt{3.12}} + \dfrac{1}{\sqrt{3.55}}} \right)^2 10^{-10}$$

$$= 0.214 \times 10^{-10} \text{ in.·min}^2/\text{ft}^6 \ (0.0239 \text{ N·s}^2/\text{m}^8)$$

$$H_l = R_{eq}Q^2 = (0.214 \times 10^{-10})(100{,}000)^2 = 0.214 \text{ in. (53.25 Pa)}$$

$$Q_1 = \frac{\sqrt{H_l}}{R_1} = \sqrt{\frac{0.214}{23.5 \times 10^{-10}}} = 9540 \text{ cfm (4.502 m}^3/\text{s)}$$

As a check, the sum of the quantities should be approximately 100,000 cfm (47.19 m³/s).

Parallel circuits are commonly employed in mine ventilation because (1) fresh, uncontaminated air is delivered to the workplaces on each split, and (2) the power cost is reduced sharply for a given quantity of air. It is an objective of mine ventilation to provide a separate split of air for each workplace. Where this is not practical or possible, the number of workings per split should be kept to a minimum. Sections in a coal mine and levels in a metal mine are convenient divisions in splitting. Every attempt is made to split the air frequently and as close to the fan installation as possible.

The economics of parallel flow for a given quantity are evident from a consideration of the basic head–quantity and air-power equations. Consider three airways with a given airflow Q, all having the same head loss H_l. In series, the overall static head would be $3H_l$. In parallel, with the same Q, the overall static head is $(\frac{1}{3})^2$ or $1/9$ H_l. The power required (and hence, the power cost) in parallel would be only $(\frac{1}{3})^3$ or $\frac{1}{27}$ that required in series. On the other hand, the power requirement would be the same if the same quantity flowed through each of the splits as in series, but the total quantity would be tripled and the head would be only $\frac{1}{3}$ as large. This is usually a desirable feature in mine ventilation.

Quantity-Divider Rule

The quantity of air flowing through each parallel airway can be determined by knowing the total quantity and the resistance of each airway, without having to calculate the individual airway head losses. Since the head losses for parallel airways are equal, the following can be written

$$R_{eq}Q^2 = R_1Q_1^2 = R_2Q_2^2 = \cdots$$

From this, one can express an individual quantity as a function of the total circuit quantity, the individual airway resistances, and the equivalent resistance; thus

$$Q_1 = Q\sqrt{\frac{R_{eq}}{R_1}}, \quad Q_2 = Q\sqrt{\frac{R_{eq}}{R_2}}, \quad \text{etc.} \tag{7.18}$$

It is apparent from Eq. 7.18 that the airway with the highest value of resistance will have the smallest quantity of air flowing through it.

Example 7.6 Given the four parallel airways of Example 7.5, determine the quantity of air flowing through each airway by using Eq. 7.18.

Solution: Sample calculation:

$$Q_1 = Q\sqrt{\frac{R_{eq}}{R_1}} = 100,000\sqrt{\frac{0.214 \times 10^{-10}}{23.5 \times 10^{-10}}} = 9540 \text{ cfm } (4.502 \text{ m}^3/\text{s})$$

Parallel Airways with Similar Characteristics

Parallel airways with similar characteristics are often encountered in mine ventilation systems. This is particularly true in coal mines where entries are driven parallel to each other, with all entries having the same length and cross-sectional dimensions. The condition of the airways is generally assumed to be the same; therefore, one can use the same friction factor for all airways. With these basic assumptions, a particular equation, applicable to the special case where the dimensions and conditions of parallel airways are the same, can be developed for determining the head loss.

The resistances of all the airways are the same, since their physical parameters are the same. From Eq. 7.17, the equivalent resistance can be expressed as

$$\frac{1}{\sqrt{R_{eq}}} = \frac{N_a}{\sqrt{R}}$$

where N_a is the number of parallel airways and R is the resistance of a single airway. Rearranging this equation yields

$$R_{eq} = \frac{R}{N_a^2} \tag{7.19}$$

This expression for the equivalent resistance is now substituted into the Atkinson equation (Eq. 7.7) to determine the head loss of the airways:

$$H_l = \frac{R}{N_a^2} Q^2 \tag{7.20}$$

From Eq. 7.8, R can be expressed in terms of the physical parameters of the airways and substituted in Eq. 5.25 or 5.25a as

$$H_l = \frac{KO(L + L_e)Q^2}{5.2A^3 N_a^2} \tag{7.21}$$

$$H_l = \frac{KO(L + L_e)Q^2}{A^3 N_a^2} \tag{7.21a}$$

These equations are often expressed as

$$H_l = \frac{K(N_a O)(L + L_e)Q^2}{5.2(N_a A)^3} \tag{7.22}$$

$$H_l = \frac{K(N_a O)(L + L_e)Q^2}{(N_a A)^3} \tag{7.22a}$$

where $N_a O$ is referred to as the *total perimeter* of the parallel airways in ft(m) and $N_a A$ as the *total area* in ft^2 (m^2).

Characteristic Curves of Parallel Airways

Referring to Example 7.5, the same solution can be obtained by plotting the airway characteristic curves (Fig. 7.9). The head loss for one point on the curve of each airway is calculated, assuming a quantity. Then the remainder of each curve is obtained by the head–quantity relation, $H_l \propto Q^2$. After the curves are plotted, the characteristic of the airways in parallel can be found graphically, since the quantities are cumulative for a given head. The quantity flowing in the individual airways is then read along the H_l line that intersects the parallel characteristic at $Q = 100,000$ cfm (47.19 m³/s), as shown in Fig. 7.9.

Pressure Gradients of Parallel Airways

The solutions developed for problems involving natural splitting are predicated on the fact that the pressure drop for the circuit is equal to the head loss (friction plus shock) across any one branch (which is the same across all branches). It should be emphasized that this drop is reflected along the total gradient, as shown in Fig. 7.10.

The head loss across any branch consists of friction loss and shock losses resulting from splitting and junction. In using the equivalent-length method of computing head loss, care should be taken to allow for all the shock losses occurring in a given branch *within* the confines of the split (Section 5.5).

An interesting phenomenon has been observed in parallel flow. The static-head gradients for all branches converge at a common point just prior to junction (Hartman, 1960). This facilitates the plotting of gradients and provides a check on calculations.

The velocity head does not influence the apportionment of air among the

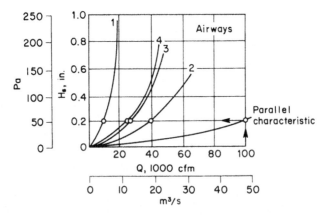

FIGURE 7.9   Characteristic curves of airways in parallel (see Example 7.5).

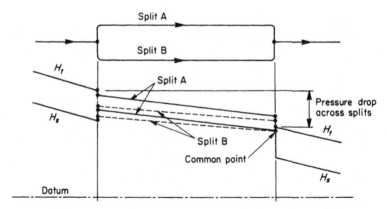

FIGURE 7.10 Pressure gradients for a parallel circuit.

branches in natural splitting; only the head loss determines that. Further-more, there is no relationship between the velocity head prior to splitting and the sum of the velocity heads in the branches. Note (Fig. 7.9) that in plotting the characteristic curve of airways in parallel, airway static-head curves are employed. If the total-head characteristic is desired, the velocity heads in the individual airways are added to the static curve. This need arises when entire mines or major portions of a mine are arranged in parallel and overall heads must be determined.

Controlled Splitting in Parallel Airways

Natural splitting rarely results in a desirable air distribution. Controlled split-ting is often employed to obtain the desired quantity of air in each branch. Controlled splitting in parallel flow is achieved by creating artificial resis-tances in all except one of the branches of the circuit. The branch without an artificial resistance is termed the *free split*. The artificial resistances are in the form of shock losses, which are created by devices called *regulators*. In controlled splitting, the quantities are regulated at will, but not without cost. The artificial resistance created by a regulator dissipates energy, and the overall result is to raise the head and power requirements for the mine. This is reflected in higher power costs to operate the mine fans.

As a control device in mine ventilation, a regulator functions similarly to a damper in home heating systems. In effect, a *regulator* is an orifice that causes alternate contraction and expansion of the air flowing in an airway. It is usually constructed as an opening of adjustable size in a ventilation door. Its size, and hence the amount of shock loss it causes, is varied by a sliding hatch on the door: the larger the opening, the smaller the shock loss. The adjustable feature is desirable to permit changes in airflow after installation because requirements in mine ventilation change frequently as mining activity changes the ventilation network.

To save time in obtaining the desired airflow in all branches of controlled parallel splits, it is advisable to calculate the approximate size of a regulator in advance. In order to determine the size of a regulator, one must first find the shock losses that need to be created in the regulated branches. This procedure involves calculating the head loss for each branch on the basis of its assigned air quantity. The branch with the highest head loss is the free split and needs no regulation. From Kirchhoff's second law, the head losses of parallel airways are equal; therefore, the amount of shock loss that must be created to achieve the assigned air quantities is calculated by subtracting the head loss for each split from the head loss of the free split. This procedure is illustrated in the next example.

Example 7.7 Given the same airways as in Example 7.5, determine the free split and the amount of regulation needed in the other branches to distribute 100,000 cfm (47.19 m^3/s) among the branches in the quantities indicated.

Solution: Calculate the head loss for each branch by the Atkinson equation for its assigned quantity. Find the shock loss needed in each branch by subtracting its head loss from the head loss of the free split.

Airway	Q, cfm (m^3s)	$R \times 10^{10}$, in.·min^2/ft^6 (N·s^2/m^8)	H_l, in. water (Pa)	H_x (shock loss), in. water (Pa)
1	20,000 (9.44)	23.50 (2.625)	0.940 (233.9)	Free split
2	15,000 (7.08)	1.35 (0.150)	0.030 (7.5)	0.940 − 0.030 = 0.019 (226.4)
3	35,000 (16.52)	3.12 (0.349)	0.382 (95.1)	0.940 − 0.382 = 0.558 (139.1)
4	30,000 (14.16)	3.55 (0.397)	0.320 (79.6)	0.940 − 0.320 = 0.620 (154.3)

Note that although the total quantity remains the same, the head loss for parallel airways increases from 0.214 in. water (53.25 Pa) in natural splitting to 0.940 in. water (233.9 Pa) in controlled splitting. This would require an increase of air power in the same proportion.

The approximate size of a regulator can be found from the theoretical shock-loss formula for an assumed, circular, symmetrical orifice (McElroy, 1935):

$$X = \left[\frac{(1/C_c) - N}{N} \right]^2 \qquad (7.23)$$

where X is the shock-loss factor, N is the ratio of the orifice area A_r to the airway area A, and C_c is the coefficient of contraction. The coefficient of contraction is determined from the following equation:

$$C_c = \frac{1}{\sqrt{z - zN^2 + N^2}} \qquad (7.24)$$

Substituting Eq. 7.24 into Eq. 7.23 yields the following expression:

$$N = \sqrt{\frac{z}{X + 2\sqrt{X} + z}} \qquad (7.25)$$

where z is the contraction factor. The area of the orifice A_r is usually called the *area of the regulator*. Since the airway area A is known, the regulator can be readily determined once N has been found:

$$A_r = NA \qquad (7.26)$$

Although the orifice formula assumes a symmetrical opening, the area calculated is sufficiently close in practice to allow the regulator area to be approximated by a rectangular, nonsymmetrical opening.

In order to calculate N by Eq. 7.25, it is first necessary to determine X. This can be done by the basic shock-loss equation (Eq. 5.23):

$$X = \frac{H_x}{H_v}$$

where H_x is the shock loss that needs to be created by the regulator and H_v is velocity head of the airway.

The other factor appearing in Eq. 7.25 is the contraction factor z. This factor varies with the configuration of the regulator. A value of 2.5 (square-edge orifice) is most commonly used in the sizing of mine regulators.

Example 7.8 Given: $Q = 150,000$ cfm (70.79 m³/s), $H_x = 2.25$ in. water (559.9 Pa), and $A = 40$ ft² (3.68 m²). Find the area of the regulator.

Solution:

Using Eq. 3.10: $\qquad V = \dfrac{150,000}{40} = 3750$ fpm (19.05 m/s)

Using Eq. 5.13: $\qquad H_v = 0.0750 \left(\dfrac{3750}{1098}\right)^2 = 0.88$ in. water (219 Pa)

Using Eq. 5.23: $X = \dfrac{2.25}{0.88} = 2.56$

Using Eq. 7.25: $N = \sqrt{\dfrac{2.5}{2.56 + 2\sqrt{2.56} + 2.5}} = 0.55$

Using Eq. 7.26: $A_r = (0.55)(40) = 22 \text{ ft}^2 \ (2.02 \text{ m}^2)$

Charts that permit graphic determination of the size of a regulator have also been devised (Weeks, 1928).

Controlled splitting is very widely used in coal mines and to some extent in metal mines. It should be used in preference to natural splitting, since control over quantity is maintained. In fact, the ideal mine ventilation system utilizes controlled splitting entirely, since natural flow is wasteful and ineffective. Rather than create shock loss in the low-resistance splits, however, it is advisable to consider the alternative of reducing the head loss in the high-resistance (free) split, particularly when the amounts of static pressure to be dissipated are large. This can be accomplished in two ways: (1) by changing the characteristics of the high-resistance airway (reducing obstructions or sinuosity, cleaning up, smooth lining, or increasing the area), or (2) by installing a booster fan in the high-resistance split to effectively reduce the head loss that must be overcome by the main fan. Either or both courses of action may be taken in preference or in addition to the use of regulators; they are discussed in detail in Chapter 11.

7.5 VENTILATION NETWORKS

A mine ventilation system generally consists of a multiple arrangement of airways to and from the surface and workplaces, fans, and control devices, all interconnected to ensure air-quantity and -quality requirements throughout the mine. The term *ventilation network,* analogous to *electrical network,* is often used to describe the mine ventilation system. In considering ventilation networks, it is often useful to have formal definitions that apply. Accordingly, the following network definitions are provided here (Didyk et al., 1977):

Node   A point where two or more segments (lines) or airways meet or intersect. In a ventilation network, a node is called a junction.

Branch (arc)   A connecting line between two nodes. In a ventilation network, the connecting line or branch represents a mine airway or a branch.

Graph   A set of nodes, with certain pairs of the nodes connected by branches.

Connecting graph	A graph in which all the nodes are connected together by branches.
Network	A graph with a flow of some type associated with its branches. A ventilation system is normally a graph with airflows in its branches and is thus properly referred to as a *network*.
Connected network	A network in which the branches connect every node to every other node. The network in Fig. 7.11 is a connected network with seven branches (indicated by encircled numbers) and six nodes (indicated by numbers not circled).
Directed network	A network in which each branch has a sign or direction associated with it. By convention, a positive sign is assigned when traveling from the starting node of a branch and a negative sign otherwise.
Degree of a node	The number of branches that have that node as an endpoint. For example, the degree of node 2 in Fig. 7.11 is 3.
Mesh (cycle)	A connected path through the network in which every node is of degree 2 with respect to the path itself. For example, in Fig. 7.11 the path connecting nodes 2, 4, 5, 1, 6, and 2 (branches 2–4, 4–5, 5–1, 1–6, and 6–2) is a mesh.
Tree (or spanning tree)	A connected graph containing branches that connect all the nodes (i.e., a set of branches spanning the nodes) but create no meshes.
Branch in a tree	A branch that is contained in a spanning tree.

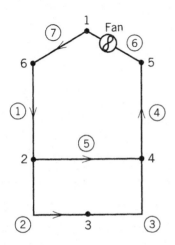

FIGURE 7.11 Connected network with seven branches and six nodes.

Chord (basic branch)	A branch contained in the network but not in a given spanning tree.
Basic mesh	A basic or fundamental mesh in a network is a mesh containing only one chord and the unique path formed by branches in the tree between the two nodes of the chord.
Chord set	A set containing all the chords of a network. There is a unique set of chords for a particular spanning tree.
Mesh base	A set containing all the basic meshes. There is a unique set of basic meshes for a particular spanning tree.
Network degree	For a connected network, the network degree is equal to the number of chords.

These definitions become extremely useful in analyzing certain types of networks as explained in the following paragraphs.

In some instances, the equivalent resistance of the network of airways can be determined by repeatedly applying the laws of series and parallel circuits. In these cases, the airways are arranged in series and parallel and can be combined with other airways to reduce the network to one equivalent airway using the methods developed in Sections 7.3 and 7.4. Such a network hereafter is called a *simple network*. The solution of a simple network does not require knowledge of the network terminology outlined above. Instead, network solutions can be obtained by relatively direct algebraic or graphical methods outlined in Section 7.6.

In other cases, it is impossible to utilize the rules on equivalent resistance of series and parallel circuits to reduce the network to an equivalent airway. Such a network is called a *complex network*. Examples of simple and complex networks are shown in Fig. 7.12. It should be noted that the network definitions listed above become more important in complex networks. In

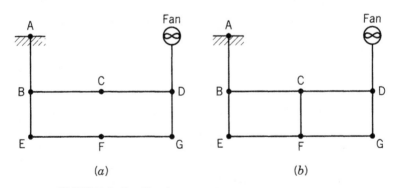

FIGURE 7.12   Simple (*a*) and complex (*b*) networks.

particular, the use of a spanning tree and the identification of chords in the network often are essential to achieving a solution. This topic is discussed in more detail in Section 7.7.

7.6 SOLUTION OF SIMPLE NETWORKS WITH NATURAL SPLITTING

In the solution of simple networks, the principles of combining series or parallel airways outlined in Sections 7.3 and 7.4 are used until the entire network is reduced to an equivalent resistance. This can be performed either algebraically or graphically. Two solutions are illustrated below.

Algebraic Solution

Simple networks may be dealt with algebraically as combinations of series and parallel circuits. The following examples utilize the techniques discussed in previous sections to illustrate the solution procedure.

Example 7.9 For the simple ventilation circuit shown in Fig. 7.13a, the following values of resistance for the individual airways have been determined, with R in units of in.$\cdot$min^2/ft$^6 \times 10^{10}$ (N$\cdot$s^2/m^8) throughout:

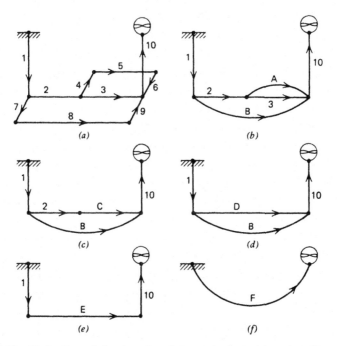

FIGURE 7.13 Reduction of simple network to equivalent airway (see Example 7.9).

$$R_1 = 0.50 \ (0.0559) \qquad R_6 = 1.30 \ (0.1453)$$

$$R_2 = 1.20 \ (0.1342) \qquad R_7 = 0.95 \ (0.1062)$$

$$R_3 = 1.00 \ (0.1118) \qquad R_8 = 1.50 \ (0.1677)$$

$$R_4 = 0.75 \ (0.0838) \qquad R_9 = 1.35 \ (0.1509)$$

$$R_5 = 1.25 \ (0.1399) \qquad R_{10} = 0.40 \ (0.0447)$$

Determine the equivalent resistance for the entire system and the mine static head, given that the fan is exhausting air at the rate of 100,000 cfm (47.19 m³/s).

Solution: The actual airways of Fig. 7.13 are represented by numbers, whereas the equivalent airways are represented by letters.

By inspection, one should recognize that airways 4, 5, and 6 are in series and that they can be combined into a series equivalent resistance, which is illustrated in Fig. 7.13*b* and is referred to as R_A. From Eq. 7.14, the value of R_A is determined to be

$$R_A = R_4 + R_5 + R_6 = (0.75 + 1.25 + 1.30)10^{-10}$$
$$= 3.30 \times 10^{-10} \ \text{in.·min}^2/\text{ft}^6 \ (0.3689 \ \text{N·s}^2/\text{m}^8)$$

Similarly, airways 7, 8, and 9 are combined into a series equivalent resistance R_B where

$$R_B = R_7 + R_8 + R_9 = (0.95 + 1.50 + 1.35)10^{-10}$$
$$= 3.80 \times 10^{-10} \ \text{in.·min}^2/\text{ft}^6 \ (0.4248 \ \text{N·s}^2/\text{m}^8)$$

Figure 7.13*b* reveals that airway 3 is in parallel with airway A; therefore, they are combined into a parallel equivalent resistance R_C as shown in Fig. 7.13*c*. Using Eq. 7.17,

$$R_c = \left(\frac{1}{1/\sqrt{R_A} + 1/\sqrt{R_3}}\right)^2 = \left(\frac{1}{1/\sqrt{3.30} + 1/\sqrt{1.00}}\right)^2 10^{-10}$$
$$= 0.42 \times 10^{-10} \ \text{in.·min}^2/\text{ft}^6 \ (0.0470 \ \text{N·s}^2/\text{m}^8)$$

Airway 2 is now in series with airway C and is combined as shown in Fig. 7.13*d* as airway D, with the following resistance:

$$R_D = R_2 + R_C = (1.20 + 0.42)10^{-10}$$
$$= 1.62 \times 10^{-10} \text{ in.} \cdot \text{min}^2/\text{ft}^6 \ (0.1811 \text{ N} \cdot \text{s}^2/\text{m}^8)$$

Airways B and D are combined into airway E, as indicated in Fig. 7.13e. The value of R_E is given by

$$R_E = \left(\frac{1}{1/\sqrt{R_D} + 1/\sqrt{R_B}}\right)^2 = \left(\frac{1}{1/\sqrt{1.62} + 1/\sqrt{3.8}}\right)^2 10^{-10}$$
$$= 0.59 \times 10^{-10} \text{ in.} \cdot \text{min}^2/\text{ft}^6 \ (0.0660 \text{ N} \cdot \text{s}^2/\text{m}^8)$$

Airways 1, E, and 10 are now in series. By combining these airways into the series equivalent resistance R_F (Fig. 7.13f), one can determine the total equivalent resistance for the entire circuit:

$$R_F = R_1 + R_E + R_{10} = (0.50 + 0.59 + 0.40)10^{-10}$$
$$= 1.49 \times 10^{-10} \text{ in.} \cdot \text{min}^2/\text{ft}^6 \ (0.1666 \text{ N} \cdot \text{s}^2/\text{m}^8)$$

The mine static head is given by

$$\text{Mine } H_s = H_l = Q^2 R_F = (100{,}000)^2 \ (1.49 \times 10^{-10})$$
$$= 1.49 \text{ in. water } (370.8 \text{ Pa})$$

Example 7.10 With the information determined in Example 7.9, find the quantity of air flowing through each airway of Fig. 7.13a.

Solution: The quantity through each airway can be determined by utilizing Fig. 7.13 in the reverse order. The quantity through airway F in Fig. 7.13f was given as 100,000 cfm (47.19 m³/s). Since airway F is the series equivalent of airways 1, E, and 10 of Fig. 7.13e, then

$$Q_1 = Q_E = Q_{10} = 100{,}000 \text{ cfm } (47.19 \text{ m}^3/\text{s})$$

Airway E is the parallel equivalent of airways B and D; therefore, the quantities through airways B and D can be calculated by Eq. 7.18 and Kirchhoff's first law as follows:

$$Q_B = Q_E \frac{\sqrt{R_E}}{\sqrt{R_B}} = 100{,}000 \frac{\sqrt{0.59 \times 10^{-10}}}{\sqrt{380 \times 10^{-10}}} = 39{,}400 \text{ cfm } (18.59 \text{ m}^3/\text{s})$$

$$Q_D = 100{,}000 - 39{,}400 = 60{,}600 \text{ cfm } (28.60 \text{ m}^3/\text{s})$$

The following quantities are determined in a similar fashion:

$$Q_2 = Q_c = Q_D = 60,600 \text{ cfm } (2860 \text{ m}^3/\text{s})$$

$$Q_A = Q_c \frac{\sqrt{R_c}}{\sqrt{R_A}} = 60,600 \frac{\sqrt{0.42 \times 10^{-10}}}{\sqrt{3.30 \times 10^{-10}}} = 21,600 \text{ cfm } (10.19 \text{ m}^3/\text{s})$$

$$Q_3 = 60,600 - 21,600 = 39,000 \text{ cfm } (18.41 \text{ m}^3/\text{s})$$

$$Q_4 = Q_5 = Q_6 = Q_A = 21,600 \text{ cfm } (10.19 \text{ m}^3/\text{s})$$

$$Q_7 = Q_8 = Q_9 = Q_B = 39,400 \text{ cfm } (18.59 \text{ m}^3/\text{s})$$

Graphic Solution

Simple networks are often amenable to graphic solution. For example, the graphic solution to the network of Fig. 7.13a, solved algebraically in Example 7.9, is shown in Fig. 7.14. The procedure is summarized as follows:

1. Combine all series airways into equivalent airways, as in Fig. 7.13b.
2. Plot the characteristic curve for the upper main split (airway D) and the lower main split (airway B).
3. Plot the combined characteristic curve of the main split (airway E).
4. Determine the quantity flowing in each branch of the main split for the mine quantity of 100,000 cfm (47.19 m³/s); read $Q_D = 61,000$ cfm (28.8 m³/s) and $Q_B = 39,000$ cfm (18.4 m³/s).
5. Plot the characteristic curve for the upper secondary split (airway A) and the lower secondary split (airway 3).

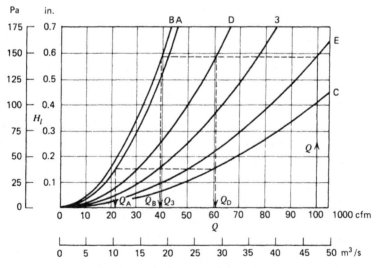

FIGURE 7.14 Graphic solution to simple network of Fig. 7.12 (see Example 7.9).

6. Plot the combined characteristic curve of the secondary split (airway C).

7. Determine the quantity flowing in each branch of the secondary split based on the quantity flowing through D; read Q_A = 22,000 cfm (10.4 m³/s) and Q_3 = 39,000 cfm (18.4 m³/s).

The graphic approach is particularly advantageous for determining the quantities through individual airways for variations in the total quantity.

7.7 ANALYSIS OF COMPLEX NETWORKS

The solution to natural or controlled splitting problems in complex networks is based on combining the engineering characteristics of the flowing air, airways, fans, and other control devices; the generalized relationship of flows and head losses in series and parallel circuits; and the purely structural relationships that exist in the connections of the airways. This combination has enabled the development of very powerful tools to solve airflow problems in complex mine ventilation systems (Scott and Hinsley, 1951; Wang and Hartman, 1967; Didyk et al., 1977).

Some of the network definitions outlined in Section 7.5 are quite useful in solving complex networks. For example, to determine whether an algebraic or an iterative solution is necessary, both spanning trees and chords of a network must be understood. A spanning tree of a network can be created using the following algorithm:

1. Choose any node.

2. Considering only the nodes that are one branch away from the chosen node, arbitrarily connect another node to the chosen node. The result is a connected set of nodes. Proceed to step 3.

3. Considering only the nodes that are one branch away from the connected set of nodes, arbitrarily connect another node to the connected set. Proceed to step 4.

4. If all nodes are in the connected set, then stop. Otherwise, return to step 3.

The graph that results is one spanning tree of the network. An example is shown in Fig. 7.15c. The mine airways in Fig. 7.15a constitute a complex network. The network in reduced form is shown in Fig. 7.15b. One spanning tree is shown in Fig. 7.15c. The branches marked by bold lines constitute the tree, while the other branches are the chords associated with that tree.

If the Hardy Cross iterative technique is to be used to solve the network, the procedure will normally converge in fewer iterations if the minimum-resistance spanning tree is used as the basis of the iterative method. The

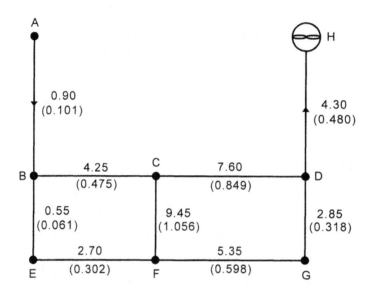

(a)

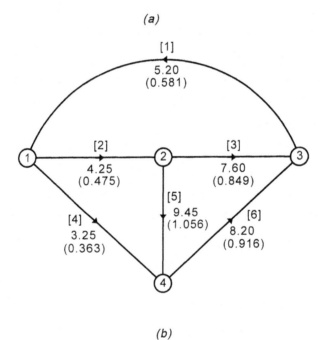

(b)

FIGURE 7.15 Network with natural splitting: (*a*) ventilation system consisting of two shafts and a raise connected on two levels by drifts; (*b*) reduced network (see Example 7.12).

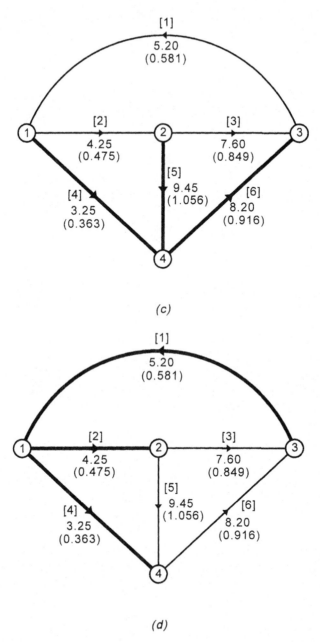

(c)

(d)

FIGURE 7.15 *(Continued)* *(c)* one possible spanning tree of the network (shown in bold lines); *(d)* minimum spanning tree of the network (shown in bold lines).

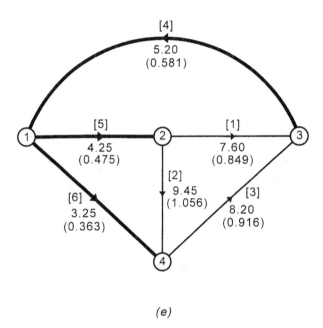

(e)

FIGURE 7.15 *(Continued)* *(e)* rearranged branch numbers (see Example 7.12).

minimum-resistance spanning tree is obtained by first calculating the resistance of each airway in the network. If we consider the resistance of each airway as a "distance," then we can find the minimum-resistance spanning tree by utilizing the following procedure:

1. Choose any node arbitrarily, and then connect it to the closest adjacent node.
2. Find the unconnected node nearest to a connected node, and then connect the two nodes.
3. If all nodes are connected, stop. Otherwise, repeat step 2.

The results of this procedure on the network in Fig. 7.15*b* are shown in Fig. 7.15*d*. The minimum-resistance spanning tree is shown in bold lines. This tree will be identical regardless of the node at which the branch selection is initiated unless ties occur when the final branches are chosen.

Having outlined the methods for choosing an arbitrary spanning tree and the minimum-resistance spanning tree, several properties of networks are outlined here without proof.

1. There exists at least one tree in every connected network.
2. For any given tree of a connected network having N_n nodes and N_b

branches, there are exactly $N_n - 1$ branches in a tree and $N_m = N_b - (N_n - 1)$ chords; therefore, the network degree is equal to N_m.

3. Any fluid flowing through a connected network is in equilibrium if it satisfies the two Kirchhoff laws:

   a. *Kirchhoff's first law:* The algebraic sum of the quantities incident at a node is zero.

   b. *Kirchhoff's second law:* The algebraic sum of the head losses in the branches in a mesh is zero.

4. The quantities flowing in each branch belonging to a chord set are independent, and the quantity of any branch in a tree is a linear combination of the quantities flowing through the chords. The resulting set of quantities satisfies Kirchhoff's first law at every node of a connected network.

5. The set of head losses of the branches of a tree uniquely determines the chord head losses. The pressure drops in the branches satisfying Kirchhoff's second law for a mesh base satisfy this law for any other mesh base in the network and for any other mesh in the network.

Essentially, in a network containing N_n nodes and N_b branches, there will be N_n node equations satisfying the condition stated in property 3a (Kirchhoff's first law). It can be easily shown that only $N_n - 1$ of these equations are independent. Note that there can be many mesh equations, one each for any arbitrarily chosen mesh satisfying the condition stated in property 3b (Kirchhoff's second law). However, if the set of mesh equations is for the $N_b - (N_n - 1)$ basic meshes, then these equations are independent. Also, for a particular spanning tree in this network, if the set of quantities in the chord set is arbitrarily assigned, then according to property 4, all the branch quantities can be calculated. These quantities will satisfy property 3a for any node in the network. Similarly, if the set of head losses across the branches in the tree is arbitrarily assigned, then according to property 5, all the chord head losses can be calculated and these head losses will satisfy property 3b for any mesh in the network.

Mathematical Representation of Ventilation Networks

Given a network with N_b branches and N_n nodes, the variables of the problem are the quantity, head loss, and resistance of each of the N_b branches of the network, for a total of $3N_b$ variables. The relationships among these variables in a balanced network are as follows.

Branch Equations For each branch in the network, an equation of the form $H_l = F(Q)$ exists. These N_b equations are independent. Note that in general, $F(Q_i) = R_i|Q_i|Q_i$.

Node Equations Property 3a provides the relationship that should hold among the quantities Q_j at every node:

$$\sum_{j=1}^{N_b} a_{ij}Q_j = 0 \quad \text{for} \quad i = 1, 2, \ldots, N_n \tag{7.27}$$

where a_{ij} is a constant defined as follows:

$a_{ij} = 0$ if neither of the nodes of branch j is equal to i

$a_{ij} = 1$ if the end node of branch j is equal to i

$a_{ij} = -1$ if the starting node of branch j is equal to i

The matrix $[a_{ij}]$ is often referred to as the *incidence matrix* of the network. Note that if the quantities of Q_i satisfy the equations for $(N_n - 1)$ nodes, they also must satisfy the equation at the N_nth node. In other words, the N_b quantities Q_i are related by $(N_n - 1)$ independent and linear equations. Furthermore, if $N_b - (N_n - 1)$ quantities are given and these quantities belong to a chord set, then the quantity flowing in each branch of the network can be calculated according to property 4. Specifically, the quantities Q_j ($j = 1, \ldots, N_b$) can be expressed as a linear combination of the air quantities W_k ($k = 1, \ldots, N_m$) flowing in the chords:

$$Q_j = \sum_{k=1}^{N_m} b_{kj}W_k \quad \text{for} \quad j = 1, 2, \ldots, N_b \tag{7.28}$$

where b_{kj} is defined as follows:

$b_{kj} = 0$ if branch j is not in the basic mesh corresponding to the chord k

$b_{kj} = 1$ if branch j belongs to the basic mesh corresponding to the chord k, and if, traveling the mesh from the starting node of the chord toward its end node, the starting node of branch j is found first

$b_{kj} = -1$ if branch k belongs to the basic mesh corresponding to the chord k, and if, traveling the mesh from the starting node of the chord toward its end node, the end node of branch j is found first

The matrix $[b_{ij}]$ is called the *fundamental-mesh matrix* of the network. The quantities Q_j calculated as above will satisfy property 3a at every node in the network.

Mesh Equations   Property 3b provides the relationships that should hold among the head losses in the branches of a mesh in the network. As stated before, a great number of mesh equations can be developed. However, there exist only N_m independent mesh equations, and these correspond to a mesh base; that is, the N_b head losses are related by N_m independent equations. The equations for each of the N_m meshes are of the type

$$\sum_{i=1}^{N_b} b_{ki}H_i = 0 \quad \text{for} \quad k = 1, 2, \ldots, N_m \tag{7.29}$$

where b_{ki} is already defined and H_i is the algebraic value of the head loss in branch i. According to property 5, given the head losses in the $(N_m - 1)$ branches in a tree, the head loss for any other branch in the network can be calculated, as these head losses will satisfy property 3b for any mesh in the network.

Solution Approach

The balance between variables and independent equations for any ventilation network is summarized as follows:

Number of Variables		Number of Independent Equations	
Quantities	N_b	Branch equations	N_b
Head losses	N_b	Node equations	$N_n - 1$
Resistances	N_b	Mesh equations	$N_b - (N_n - 1)$
Total	$3\,N_b$		$2\,N_b$

Since there exist more variables than independent equations, the number of variables with initial known values necessary for solving a ventilation network problem is $(3\,N_b - 2\,N_b) = N_b$.

Types of Complex Ventilation Problems

Depending on the type and number of initial data, the difficulty of the ventilation problem to be solved varies. To analyze a complex network more fully, assume that spanning trees and the associated chords can be developed for the network. Also assume that it is possible to determine whether any of the heads or quantities in the airways are known or fixed. Under these conditions, it is possible to determine whether the complex network can be solved algebraically or an iterative procedure is essential.

The algebraic procedure will work if any of the following five conditions are evident (Didyk et al., 1977):

1. The known data values are $N_n - 1$ head losses for the branches in any given tree and N_m air quantities flowing in the chords.

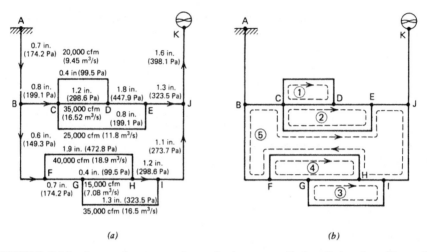

(a)   (b)

FIGURE 7.16   A complex network employing controlled splitting (see Example 7.11).

2. The known data are N_m quantities in the chords and $N_m - 1$ resistances for the branches in a tree.
3. The initial known values are N_k head losses ($N_k < N_m - 1$) in N_k branches of a tree, N_m quantities flowing in the chords, and $N_m - N_k - 1$ resistances for the remaining branches in the tree.
4. The known data points are N_m chord resistances and $N_n - 1$ head losses in the branches of any given tree.
5. The given data are $N_n - 1$ head losses associated with branches in a tree, N_c quantities flowing through the chord ($N_c < N_m$), and $N_m - N_c$ resistances for the remaining chords.

These conditions can be applied to Example 7.11. Note in Fig. 7.16 that the example network has $N_n = 9$ nodes and $N_m = 5$ meshes. The example is thus solvable by algebraic means according to several of the conditions listed above.

Example 7.11   Given the schematic of the complex mine ventilation system shown in Fig. 7.16a, with quantities assigned and head losses calculated as indicated, determine the mine quantity and static head.

Solution:

Upper level: $Q = 20,000 + 35,000 + 25,000 = 80,000$ cfm (37.76 m³/s)

Lower level: $Q = 40,000 + 15,000 + 35,000 = 90,000$ cfm (42.48 m³/s)

Mine $Q = 80,000 + 90,000 + 170,000$ cfm (80.23 m³/s)

For the portion of the network from B to J, Kirchhoff's second law must be satisfied for five meshes, since

$$N_m = N_b - N_j + 1 = 13 - 9 + 1 = 5$$

If the meshes of Fig. 7.16b are chosen, the following equations can be written:

Mesh 1: $0.4 + H_x = 1.2$

$H_x = 0.8$ in. water (199.1 Pa)   (upper branch of CI)

Mesh 2: $1.2 + 1.8 = 0.8 + H_x$

$H_x = 2.2$ in. water (547.4 Pa)   (lower branch of CI)

Mesh 3: $0.4 + 1.2 = 1.3 + H_x$

$H_x = 0.3$ in. water (74.6 Pa)   (lower branch of GI)

Mesh 4: $1.9 = 0.7 + H_x + 0.4$

$H_x = 0.8$ in. water (199.1 Pa)   (branch FG)

Mesh 5: $0.8 + 3.0 + 1.3 = 0.6 + 1.9 + 1.2 + 1.1 + H_x$

$H_x = 0.3$ in. water (74.6 Pa)   (either BF or IJ)

Mine $H_s = 0.7 + 0.8 + 1.2 + 1.8 + 1.3 + 1.6$

$= 7.4$ in. water (1841 Pa)

Other cases will require the use of an iterative method, such as the Hardy Cross method, to achieve a solution. However, in practice, the Hardy Cross method is normally used for solutions of complex networks. This procedure is used because the iterative method works on all complex networks and evaluation of the five conditions above is much more difficult than simply using a computer program that employs the Hardy Cross algorithm. The Hardy Cross procedure has been nearly universally accepted as the method of choice in computer-based ventilation network solvers. The method always converges for a real flow network and can be quite efficient if a favorable spanning tree is chosen. The following develops the method, starting with a generalized network mathematical model.

7.8 GENERALIZED NETWORK MODEL

In order to introduce a more generalized mathematical model for ventilation networks, the head loss between two nodes of a branch is modeled by the following four head elements:

H_i^L head loss caused by the resistance R_i of branch i:

$$H_i^L = R_i|Q_i|Q_i \qquad (7.30)$$

H_i^R head loss caused by a regulator ($H_i^R > 0$) or booster fan ($H_i^R < 0$) in branch i

H_i^F head (pressure gain) of a variable-pressure fan in branch i

H_i^N natural ventilation head (pressure gain), or the head of any independent (constant) pressure source in branch i

In summary, the head loss H_i for branch i is expressed in the form

$$H_i = H_i^L + H_i^R - H_i^F - H_i^N \qquad (7.31)$$

where any term can be a positive, negative, or zero value. Note that Atkinson's equation is used in the first element (Eq. 7.30).

The resistance factor R_i has a nonnegative value and is independent of the direction of a branch. Depending on the application, R_i may include the components for shock losses in the branch and/or at a junction. If desired, the velocity head at the exhaust end of the ventilation system, which is proportional to the square of air quantity, can be converted into an equivalent resistance factor and incorporated in R_i for branch i whose terminal node corresponds to an exit. Fan head H_i^F, which is generally given as a curve and represents a variable-pressure source, can be approximated by a second-degree polynomial. Thus

$$H_i^F = \alpha_i Q_i^2 + \beta_i Q_i + \gamma_i \qquad (7.32)$$

Unless otherwise described, we assume that R_i, H_i^N, α_i, β_i, and γ_i are assigned and known; therefore, they are not considered as variables.

Using these symbols, we rewrite the node equations (Eq. 7.27) and mesh equations (Eq. 7.29) as the following *system equations:*

$$\sum_{j=1}^{N_b} a_{ij}Q_j = 0 \qquad i = 1, \ldots, N_n - 1 \qquad (7.33)$$

$$\sum_{j=1}^{N_b} b_{ij}(H_j^L + H_j^R - H_j^F - H_j^N) = 0 \qquad i = 1, \ldots, N_m \qquad (7.34)$$

This is a nonlinear system of $2N_b$ unknowns (Q_i and H_i^R for each of N_b branches) in N_b equations, where N_b is the number of branches in the network. Theoretically speaking, therefore, if the values for N_b of the $2N_b$ variables are assigned, the system reduces to N_b unknowns in N_b equations, becoming tractable to general solution techniques for a nonlinear system provided that a solution exists. For this type of generalized approach, the reader is referred to the works of Yevdokimov (1969), Jacques (1976), and Wang (1989, 1990).

Two topological conditions that are closely associated with the existence of a unique solution are summarized below. If the network problem of Eqs. 7.33 and 7.34 has a unique solution, then (1) there exists a spanning tree T_1 of the network such that the branches with independent Q_i (known air quantity) are contained in the corresponding cotree (chord set) T_1^* and (2) there exist a spanning tree T_2 of the network such that the branches with dependent H_i^R (unknown head for regulator or booster fan) are contained in the corresponding cotree (chord set) T_2^*. For some network problems, however, T_1 and T_2 can be the same tree.

7.9 NETWORKS WITH FIXED-QUANTITY BRANCHES

A mathematical model of *networks with natural splitting and fixed-quantity branches* is probably the most useful today. The network branches in this model can be grouped into three types: (1) fixed-quantity branches, (2) fan branches, and (3) regular branches.

Fixed-quantity branch: $\quad H_i = H_i^L + H_i^R - H_i^N = R_i|Q_i|Q_i + H_i^R - H_i^N$

$$(7.35)$$

where the value of Q_i is assigned (independent variable) and the value of H_i^R is to be determined (dependent variable).

Fan branch: $\quad H_i = H_i^L - H_i^F - H_i^N = R_i|Q_i|Q_i$

$$- (\alpha_i Q_i^2 + \beta_i Q_i + \gamma_i) - H_i^N \quad (7.36)$$

Regular branch: $\quad H_i = H_i^L - H_i^N = R_i|Q_i|Q_i - H_i^N \quad (7.37)$

Let N_q and N_f be numbers of fixed-quantity branches and fan branches, respectively. For convenience, we label fixed-quantity branches with $1, \ldots, N_q$; fan branches with $N_q + 1, \ldots, N_q + N_f$; and other branches with $N_q + N_f + 1, \ldots, N_b$. Limiting our consideration to the case where $N_q + N_f \leq N_m$ and where fixed-quantity and fan branches are contained in the cotree of a spanning tree, and labeling each of the fundamental meshes with the branch number of the chord containing in it, we put the mesh equations (Eq. 7.34) into the form

$$H_i^R + \sum_{j=1}^{N_b} b_{ij}(R_j|Q_j|Q_j - H_j^N) = 0 \qquad i = 1, \ldots, N_q \qquad (7.38)$$

$$-(\alpha_i Q_i^2 + \beta_i Q_i + \gamma_i) + \sum_{j=1}^{N_b} b_{ij}(R_j|Q_j|Q_j - H_j^N) = 0 \qquad (7.39)$$

$$i = N_q + 1, \ldots, N_q + N_f$$

$$\sum_{j=1}^{N_b} b_{ij}(R_j|Q_j|Q_j - H_j^N) = 0 \qquad i = N_q + N_f + 1, \ldots, N_m \quad (7.40)$$

Thus the problem is to determine the values for N_b unknowns (N_q unknowns in H_i^R and $N_b - N_q$ unknowns in Q_i) using N_b equations (Eqs. 7.33 and 7.38–7.40). In this network problem, the trees T_1 and T_2 defined in Section 7.8 can be the same tree. If $N_q = 0$, then the problem reduces to one of the *natural splitting networks* or *networks with natural splitting*. In practical applications, this type of problem, either with or without fixed-quantity branches, is generally solved by the well-known Hardy Cross method (Cross, 1936; Scott and Hinsley, 1951).

Hardy Cross Method

The Hardy Cross method is an iterative method; that is, an approximate solution is successively improved until the error is acceptably small. The iterative scheme to be discussed below is modified from the original presentation of Hardy Cross and is similar to the Gauss–Seidel method of solving linear equations. It assumes that each equation is a function of only one variable and uses two terms of Taylor's series in the derivation of the improvement formula. It is equivalent to Newton's method of tangents applied to each equation individually.

Recalling that chords are labeled with $1, \ldots, N_m$, we rewrite the node equations (Eq. 7.33) in the form

$$Q_j = \sum_{k=1}^{N_m} b_{kj}Q_k \qquad j = N_m + 1, \ldots, N_b \qquad (7.41)$$

which is equivalent to Eq. 7.28. Substituting this expression for Q_j into Eqs. 7.38 to 7.40, we obtain

$$H_i^R + \sum_{j=1}^{N_b} b_{ij}\left[R_j \left| \sum_{k=1}^{N_m} b_{kj}Q_k \right| \left(\sum_{k=1}^{N_m} b_{kj}Q_k \right) - H_j^N \right] = 0 \qquad i = 1, \ldots, N_q$$

$$(7.42)$$

$$-(\alpha_i Q_i^2 + \beta_i Q_i + \gamma_i) + \sum_{j=1}^{N_b} b_{ij} \left[R_j \left| \sum_{k=1}^{N_m} b_{kj} Q_k \right| \left(\sum_{k=1}^{N_m} b_{kj} Q_k \right) - H_j^N \right] = 0$$

$$i = N_q + 1, \ldots, N_q + N_f \quad (7.43)$$

$$\sum_{j=1}^{N_b} b_{ij} \left[R_j \left| \sum_{k=1}^{N_m} b_{kj} Q_k \right| \left(\sum_{k=1}^{N_m} b_{kj} Q_k \right) - H_j^N \right] = 0$$

$$i = N_q + N_f N_m \quad (7.44)$$

This is a system of N_m unknowns in N_m equations. The unknowns are H_i^R for $i = 1, \ldots, N_q$ and Q_i for $i = N_q + 1, \ldots, N_m$. It is important to point out that the H_i^R appears only in equation i of Eq. 7.42, and that Q_i for $i = N_q + 1, \ldots, N_m$ can be determined independently of Eq. 7.42. For this reason, now we concentrate on the system of $N_m - N_q$ unknowns in $N_m - N_q$ equations: Eqs. 7.43 and 7.44.

Consider that the left-hand side of equation i in Eq. 7.43 or 7.44 is a function of Q_i only and denote it by $f_i(Q_i)$:

$$f_i(Q_i) = -(\alpha_i Q_i^2 + \beta_i Q_i + \gamma_i) + \sum_{j=1}^{N_b} b_{ij} \left[R_j \left| \sum_{k=1}^{N_m} b_{kj} Q_k \right| \left(\sum_{k=1}^{N_m} b_{kj} Q_k \right) - H_j^N \right]$$

$$i = N_q + 1, \ldots, N_q + N_f \quad (7.45)$$

$$f_i(Q_i) = \sum_{j=1}^{N_b} b_{ij} \left[R_j \left| \sum_{k=1}^{N_m} b_{kj} Q_k \right| \left(\sum_{k=1}^{N_m} b_{kj} Q_k \right) - H_j^N \right]$$

$$i = N_q + N_f + 1, \ldots, N_m \quad (7.46)$$

Expanding f_i in Taylor's series and neglecting high-order terms, we have

$$f_i(Q_i + X_i) = f_i(Q_i) + X_i f_i'(Q_i) \quad (7.47)$$

where Q_i and $Q_i + X_i$ denote the current value and an improved value of air quantity, respectively, and $f_i'(Q_i)$ is the derivative of f_i evaluated at Q_i. Setting $f_i(Q_i + X_i) = 0$ (i.e., aiming to eliminate the unbalanced head) and solving for X_i, we obtain the following expression for the correction factor:

$$X_i = -f_i(Q_i)/f_i'(Q_i) \quad (7.48)$$

where

$$f_i'(Q_i) = -(2\alpha_i Q_i + \beta_i) + 2 \sum_{j=1}^{N_b} b_{ij}^2 R_j \left| \sum_{k=1}^{N_m} b_{kj} Q_k \right|$$

$$i = N_q + 1, \ldots, N_q + N_f \quad (7.49)$$

$$f_i'(Q_i) = 2 \sum_{j=1}^{N_b} b_{ij}^2 R_j \left| \sum_{k=1}^{N_m} b_{kj} Q_k \right| \quad i = N_q + N_f + 1, \ldots, N_m$$

$$(7.50)$$

It should be observed that, because of the relationship of Eq. 7.41, the values for f_i and f_i' at the current value of Q_i can be evaluated by

$$f_i = -(\alpha_i Q_i^2 + \beta_i Q_i + \gamma_i) + \sum_{j=1}^{N_b} b_{ij}(R_j|Q_j|Q_j - H_j^N)$$

$$i = N_q + 1, \ldots, N_q + N_f \quad (7.51)$$

$$f_i = \sum_{j=1}^{N_b} b_{ij}(R_j|Q_j|Q_j - H_j^N) \quad i = N_q + N_f + 1, \ldots, N_m$$

$$(7.52)$$

$$f_i' = -(2\alpha_i Q_i + \beta_i) + 2 \sum_{j=1}^{N_b} b_{ij}^2 R_j |Q_j| \quad i = N_q + 1, \ldots, N_q + N_f$$

$$(7.53)$$

$$f_i' = 2 \sum_{j=1}^{N_b} b_{ij}^2 R_j |Q_j| \quad i = N_q + N_f + 1, \ldots, N_m \quad (7.54)$$

The correction factor X_i computed by Eq. 7.48 is then applied to improve not only the air quantity for branch i but also those for the other branches in mesh i:

$$Q_j \leftarrow Q_j + b_{ij}X_i \quad j = N_q + 1, \ldots, N_b \quad (7.55)$$

This application of the correction factor to all branches in a mesh maintains the relationship of Eq. 7.41, which satisfies KCL at every node of the network.

If all $Q_i(i = 1, \ldots, N_b)$ are known, the values for head losses to be adjusted by the regulator or booster fan in a fixed-quantity branch are easily computed by Eq. 7.38, or

$$H_i^R = -\sum_{j=1}^{N_b} b_{ij}(R_j|Q_j|Q_j - H_j^N) \qquad i = 1, \ldots, N_q \qquad (7.56)$$

Assuming that the maximum number of iterations N_x and the tolerance ε, in cfm, are given, the algorithm for the solution of networks with natural splitting and fixed-quantity branches may be summarized as follows:

1. Assign initial values to Q_i for $i = N_q + 1, \ldots, N_m$.
2. For $j = N_m + 1, \ldots, N_b$, assign Q_j values:

$$Q_j \leftarrow \sum_{k=1}^{N_m} b_{kj}Q_k$$

3. For $I = 1, \ldots, N_x$, perform the following:
   3.1. For $i = N_q + 1, \ldots, N_m$, assign f_i, f_i', X_i, and Q_j values:
   3.1.1. If $i \le N_q + N_f$, then

$$f_i \leftarrow -(\alpha_i Q_i^2 + \beta_i Q_i + \gamma_i) + \sum_{j=1}^{N_b} b_{ij}(R_j|Q_j|Q_j - H_j^N)$$

$$f_i' \leftarrow -(2\alpha_i Q_i + \beta_i) + 2\sum_{j=1}^{N_b} b_{ij}^2 R_j|Q_j|$$

otherwise

$$f_i \leftarrow \sum_{j=1}^{N_b} b_{ij}(R_j|Q_j|Q_j - H_j^N)$$

$$f_i' \leftarrow 2\sum_{j=1}^{N_b} b_{ij}^2 R_j|Q_j|$$

   3.1.2. $X_i \leftarrow -\dfrac{f_i}{f_i'}$
   3.1.3. For $j = 1, \ldots, N_b$, assign Q_j values:

$$Q_j \leftarrow Q_j + b_{ij}X_i$$

   3.2. If $X_i \le \varepsilon$ for $i = N_q + 1, \ldots, N_m$, go to 4.
4. If $N_q \le 0$, then go to 5. Otherwise, for $i = 1, \ldots, N_q$ let

$$H_i^R \leftarrow - \sum_{j=1}^{N_b} b_{ij}(R_j|Q_j|Q_j - H_j^N)$$

5. Stop.

Note that the iteration is performed only for the meshes that do not contain a fixed-quantity branch. As described earlier, a negative value of H_i^R represents the head of a fan. If the fan is not allowed in the fixed-quantity branch, then the computed result with a negative H_i^R is not a valid network solution.

7.10 SIMPLIFIED APPLICATION OF THE HARDY CROSS ALGORITHM

The generalized model and the Hardy Cross algorithm are somewhat complicated because of the need for wide applicability. However, the Hardy Cross procedure can be performed easily in many situations. To illustrate an application, we refer the reader to the complex network discussed previously and shown in Fig. 7.15.

Example 7.12 The two-level mine network shown in Fig. 7.15a is to be solved for the quantity and head loss in each branch. Assume that the resistances of the airways given in Fig. 7.15b are in in.·min²/ft⁶ × 10¹⁰ (N·s²/m⁸). To simplify the problem, assume that the natural ventilation is negligible and that the fan is exhausting at a constant head of 8.0 in. water (1991 Pa). Perform five iterations or continue until all meshes satisfy the condition that $|X_i| \leq \varepsilon$ where $\varepsilon = 25$ cfm (0.012 m³/s).

Solution: To use the Hardy Cross method to achieve a solution, a spanning tree of the network is needed. The minimum-resistance spanning tree shown in Fig. 7.15d will be used. For such a spanning tree, the number of chords or fundamental meshes is

$$N_m = N_b - N_n + 1 = 6 - 4 + 1 = 3$$

To follow the equations and algorithm given in Section 7.9, branch numbers are reassigned as shown in Fig. 7.15e so that branches 1, 2, and 3 are chords and branches 4, 5, and 6 are contained in the spanning tree. Figure 7.17 shows the corresponding set of fundamental meshes, which can be represented by the following fundamental mesh matrix:

$$B = \begin{array}{c} \\ 1 \\ 2 \\ 3 \end{array} \begin{array}{cccccc} 1 & 2 & 3 & 4 & 5 & 6 \\ \left[\begin{array}{cccccc} 1 & 0 & 0 & 1 & 1 & 0 \\ 0 & 1 & 0 & 0 & 1 & -1 \\ 0 & 0 & 1 & 1 & 0 & 1 \end{array} \right] \end{array}$$

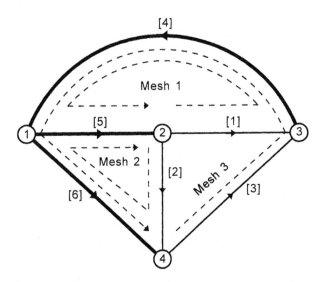

FIGURE 7.17   Network of Example 7.12 showing the minimum-resistance spanning tree, the chords, and the fundamental meshes.

For simplicity, the constant fan head will be treated as natural ventilation pressure; therefore, $N_f = 0$ and $H_4^N = 8.0$ in. water (1.99 kPa). Furthermore, since there is no fixed-quantity branch, $N_q = 0$.

Step 1   Arbitrarily assign $Q_1 = 30,000$ cfm (14.2 m³/s), $Q_2 = 20,000$ cfm (9.4 m³/s), and $Q_3 = 40,000$ cfm (18.9 m³/s).

Step 2

$$Q_4 = Q_1 + Q_3 = 30,000 + 40,000 = 70,000 \text{ cfm (33.0 m}^3/\text{s)}$$

$$Q_5 = Q_1 + Q_2 = 30,000 + 20,000 = 50,000 \text{ cfm (23.6 m}^3/\text{s)}$$

$$Q_6 = -Q_2 + Q_3 = -20,000 + 40,000 = 20,000 \text{ cfm (9.4 m}^3/\text{s)}$$

Step 3

Mesh 1: $X_1 = -\dfrac{f_1}{f'_1} = -\dfrac{R_1|Q_1|Q_1 + R_4|Q_4|Q_4 + R_5|Q_5|Q_5 - H_4^N}{2[R_1|Q_1| + R_4|Q_4| + R_5|Q_5|]}$

$= -\dfrac{[(7.60)(30,000)^2 + (5.20)(70,000)^2 + (4.25)(50,000)^2]\,(10^{-10}) - 8.00}{2[(7.60)(30,000) + (5.20)(70,000) + (4.25)(50,000)]\,(10^{-10})}$

$= 23,030$ cfm (10.9 m³/s)

$$Q_1 \leftarrow Q_1 + X_1 = 30{,}000 + 23{,}030 = 53{,}030 \text{ cfm } (25.0 \text{ m}^3/\text{s})$$

$$Q_4 \leftarrow Q_4 + X_1 = 70{,}000 + 23{,}030 = 93{,}030 \text{ cfm } (43.9 \text{ m}^3/\text{s})$$

$$Q_5 \leftarrow Q_5 + X_1 = 50{,}000 + 23{,}030 = 73{,}030 \text{ cfm } (34.5 \text{ m}^3/\text{s})$$

Mesh 2: $\quad X_2 = -\dfrac{f_2}{f_2'} = -\dfrac{R_2|Q_2|Q_2 + R_5|Q_5|Q_5 - R_6|Q_6|Q_6}{2[R_2|Q_2| + R_5|Q_5| + R_6|Q_6|]}$

$$= -\frac{[(9.45)(20{,}000)^2 + (4.25)(73{,}030)^2 - (3.25)(20{,}000)^2] \,(10^{-10})}{2[(9.45)(20{,}000) + (4.25)(73{,}030) + (3.25)(20{,}000)] \,(10^{-10})}$$

$$= -22{,}278 \text{ cfm } (-10.5 \text{ m}^3/\text{s})$$

$$Q_2 \leftarrow Q_2 + X_2 = 20{,}000 + (-22{,}278) = -2{,}278 \text{ cfm } (-1.1 \text{ m}^3/\text{s})$$

$$Q_5 \leftarrow Q_5 + X_2 = 73{,}030 + (-22{,}278) = 50{,}752 \text{ cfm } (24.0 \text{ m}^3/\text{s})$$

$$Q_6 \leftarrow Q_6 - X_2 = 20{,}000 - (-22{,}278) = 42{,}278 \text{ cfm } (20.0 \text{ m}^3/\text{s})$$

Mesh 3: $\quad X_3 = -\dfrac{f_3}{f_3'} = -\dfrac{R_3|Q_3|Q_3 + R_4|Q_4|Q_4 + R_6|Q_6|Q_6 - H_4^N}{2[R_3|Q_3| + R_4|Q_4| + R_6|Q_6|]}$

$$= -\frac{[(8.20)(40{,}000)^2 + (5.20)(93{,}030)^2 + (3.25)(42{,}278)^2] \,(10^{-10}) - 8.00}{2[(8.20)(40{,}000) + (5.20)(93{,}030) + (3.25)(42{,}278)] \,(10^{-10})}$$

$$= 8{,}464 \text{ cfm } (4.0 \text{ m}^3/\text{s})$$

$$Q_3 \leftarrow Q_3 + X_3 = 40{,}000 + 8{,}464 = 48{,}464 \text{ cfm } (22.9 \text{ m}^3/\text{s})$$

$$Q_4 \leftarrow Q_4 + X_3 = 93{,}030 + 8{,}464 = 101{,}494 \text{ cfm } (47.9 \text{ m}^3/\text{s})$$

$$Q_6 \leftarrow Q_6 + X_3 = 42{,}278 + 8{,}464 = 50{,}742 \text{ cfm } (23.9 \text{ m}^3/\text{s})$$

Since the values of $|X_i|$ are all greater than 25 cfm (0.012 m³/s) and only one iteration has been completed, perform an additional iteration.

The data calculated in four iterations are presented in Table 7.1. At the end of the fourth iteration, all $|X_i|$ values are less than 25 cfm (0.012 m³/s). Therefore, we terminate the iteration process and calculate heads for the ventilation system of Fig. 7.15. The results are listed in Table 7.2. As an example of checking the network solution, let us consider node C and mesh A–B–C–D–H. Using the data in Table 7.2, we have

$$46{,}568 + 3982 - 50{,}550 = 0 \text{ cfm } (0 \text{ m}^3/\text{s})$$

for the summation of quantities at the node, and

TABLE 7.1 Data Calculated in the Hardy Cross Method for Example 7.12

					Quantities in cfm[a]			
Iteration	i	X_i	Q_1	Q_2	Q_3	Q_4	Q_5	Q_6
			30,000	20,000	40,000	70,000	50,000	20,000
1	1	23,030	53,030			93,030	73,030	
	2	−22,278		−2,278			50,752	42,278
	3	8,464			48,464	101,494		50,742
2	1	−2,566	50,464			98,928	48,186	
	2	−1,854		−4,132			46,332	52,596
	3	396			48,860	99,324		52,992
3	1	102	50,566			99,426	46,434	
	2	152		−3,980			46,586	52,840
	3	−25			48,835	99,401		52,815
4	1	−16	50,550			99,385	46,570	
	2	−2		−3,982			46,568	52,817
	3	7			48,842	99,392		52,824

[a] Conversion factor: 1 cfm $= 0.47195 \times 10^{-3}$ m³/s.

TABLE 7.2 Solution of Example 7.12

Branch	A–B	B–C	C–D	B–E	E–F
Q, cfm	99,392	46,568	50,550	52,824	52,824
(m³/s)	(46.91)	(21.98)	(23.86)	(24.93)	(24.93)
R, 10^{-10} in.·min²/ft⁶	0.90	4.25	7.60	0.55	2.70
(N·s²/m⁸)	(0.10)	(0.47)	(0.85)	(0.06)	(0.30)
H^L, in.	0.889	0.922	1.942	0.153	0.753
(Pa)	(221)	(229)	(483)	(38)	(187)
H, in.	0.889	0.922	1.942	0.153	0.753
(Pa)	(221)	(229)	(483)	(38)	(187)

Branch	F–C	F–G	G–D	D–H
Q, cfm	3,982	48,842	48,842	99,392
(m³/s)	(1.88)	(23.05)	(23.05)	(46.91)
R, 10^{-10} in.·min²/ft⁶	9.45	5.35	2.85	4.30
(N·s²/m⁸)	(1.06)	(0.60)	(0.32)	(0.48)
H^L, in.	0.015	1.276	0.680	4.248
(Pa)	(4)	(318)	(169)	(1,057)
H, in.	0.015	1.276	0.680	−3.752
(Pa)	(4)	(318)	(169)	(−934)

$$0.889 \, + \, 0.922 \, + \, 1.942 \, - \, 3.752 \, = \, 0.001 \text{ in. water (0.2 Pa)}$$

for the summation of head losses around the mesh.

The preceding solution utilizes the Hardy Cross algorithm for a network with natural splitting and no fixed-quantity branches. The same algorithm can be used for networks with fixed-quantity branches by assigning all the known quantities to the chords in the network. The spanning tree must be chosen from among the non-fixed quantity branches. When the fundamental meshes have been chosen, only those meshes that do not involve a fixed-quantity branch are adjusted during the iterations as no adjustment can be made in a mesh with a fixed-quantity branch.

The information previously provided in Section 7.7 and the procedure outlined above indicates clearly that the number of fixed branches in a network is limited to the number of fundamental meshes. At that point, the network is solvable by algebraic means, and the algorithm is not essential.

7.11 ADDITIONAL NETWORK INSIGHTS

There are other applications of graph and network theory to mine ventilation problems. The following are some examples where the engineer's ventilation analysis capabilities can be aided through use of various network properties.

Complex Networks with Controlled Splitting

In this type of network model, the air quantities for N_m branches corresponding to a set of chords are given, and those for the tree branches are uniquely determined by Eq. 7.41 such that KCL is satisfied at every node. Therefore, all branches in the network are considered as fixed-quantity branches. The values for R_j and H_j^N ($j = 1, \ldots, N_b$) are given, and those for H_j^L are computed by Eq. 7.30. For convenience, the regulator head is limited to a nonnegative value (i.e., $H_j^R \geq 0$), and the fan head is denoted by H_j^F, which is not to be considered as a function of air quantity although it denoted the head of a variable-pressure fan in Sections 7.8 and 7.9. In addition, we assume without loss of generality that the directions of branches are chosen to coincide with the airflows such that all Q_j values are positive. By considering the air power in addition to KVL, a generalized formulation for the networks with controlled splitting can be presented in the following form (Wang and Pana, 1971). Minimize

$$Z = \sum_{j=1}^{N_b} Q_j H_j^F = \sum_{j=1}^{N_b} Q_j (H_j^L + H_j^R - H_j^N) \tag{7.57}$$

subject to

$$\sum_{j=1}^{N_b} b_{ij}(H_j^L + H_j^R - H_j^N - H_j^F) = 0 \qquad i = 1, \ldots, N_m \qquad (7.58)$$

and

$$H_j^R \geqslant 0, \qquad H_j^F \geqslant 0 \qquad j = 1, \ldots, N_b \qquad (7.59)$$

Equation 7.57 is the linear objective function that is to be minimized. Equation 7.58 defines the linear constraints of the system, and Eq. 7.59 defines the nonnegativity conditions. The problem formulated with these types of equations is known as the *linear programming* (LP) problem and can be analyzed by standard LP techniques. In LP terms, any solution that satisfies Eqs. 7.58 and 7.59 is a *feasible solution*. An *optimal solution* is a feasible solution that minimizes Eq. 7.57. In general, there exist multiple optimal solutions to the ventilation problem. If no limitations are placed on the use of fans, then the optimal solution may always consist of no more than N_m positive values of H_j^F. If fans and/or regulators are not allowed in some branches, the corresponding H_j^F and/or H_j^R may be deleted from the equations. In the following discussion, we first consider the special case where only a main fan is installed as an exhauster or blower. For simplicity, all values for H_j^N are assumed to be negligible.

Because only a main fan is involved, the objective is to determine the head of the main fan and the locations and heads of the regulators. By numbering the branch with the fan as branch 1 and choosing it as a chord contained in mesh 1, the problem can be stated as follows: Minimize

$$Z = Q_1 H_1^F = \sum_{j=1}^{N_b} Q_j(H_j^L + H_j^R) \qquad (7.60)$$

subject to

$$\sum_{j=1}^{N_b} b_{ij}(H_j^L + H_j^R) = 0 \qquad i = 2, \ldots, N_m \qquad (7.61)$$

and

$$H_j^R \geq 0 \qquad j = 1, \ldots, N_b \qquad (7.62)$$

Since Q_1 is a constant, Eq. 7.60 can be replaced by

$$Z' = H_1^F \qquad (7.63)$$

Employing nodal head losses, we can also state the problem as follows. Minimize

$$H_1^F = P_{N_n} - P_1 \tag{7.64}$$

subject to

$$P_j - P_i - H_{ij}^R = H_{ij}^L \qquad \text{for all branches } (i,j) \text{ except branch } (N_n,1) \tag{7.65}$$

and

$$H_{ij}^R \geq 0 \qquad \text{for all branches } (i,j) \text{ except branch } (N_n,1) \tag{7.66}$$

where P_k is the nodal head loss between node k and node 1 (the reference node), in in. water (Pa). The double subscript ij is employed to denote the branch whose initial and final nodes are i and j, respectively. In addition, the nodes corresponding to the entrance and exit of the ventilation system are numbered with 1 and N_n, and a return dummy branch $(N_n, 1)$ having zero resistance is added to the network. The main fan is assumed to be located in this dummy branch.

Critical-Path Approach

The formulation for the problem with a main fan just discussed is similar to that of the *critical-path method* (CPM) for project scheduling. The analogy between the free split in a ventilation circuit and a critical path in a scheduling network was recognized by Owili-Eger (1973). The procedures were further developed by Wang (1982a) and Wang and Mutmansky (1982). The basic property of a network with controlled splitting is that the free split through the network is identified with the critical path, and this fact will allow a ventilation engineer to analyze a number of ventilation problems, including the use of multiple entrances and exits, the generation of alternative network solutions, and the optimization of power consumption using booster fans rather than regulators. The references cited above can supply the ventilation engineer with further information. In addition, standard CPM references can be used to obtain properties and methods for working with critical paths in networks.

Cutset Operations

A *cutset* is defined as any minimal set of branches in a network that, when cut or removed, will divide the network into exactly two distinct subnetworks. Many cutsets often exist for a given ventilation setup. For example,

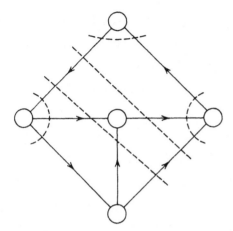

FIGURE 7.18 Some examples of cutsets.

five cutsets are shown for the network in Fig. 7.18. Each set of branches intersected by a broken line is a cutset. However, many additional cutsets can be found in the network. Cutsets have mathematical properties that make them extremely useful in analyzing ventilation problems. Because the directed branches in the cutset follow certain mathematical rules, the use of the mathematical relationships allow a number of different types of analyses to take place. These include the finding of alternative solutions from any known solution, the exchange of booster fans for regulators, the moving of regulators to less objectionable airways, the moving of a fan from one location to another, and the replacement of a single fan with multiple fans. The primary advantage of cutset analysis in all of these operations is that all of the analyses can be performed by straightforward arithmetic operations, and new network solutions are not necessary. More information on the advantages of cutsets and the procedures used can be found in Wang (1982a, 1982b) and Wang and Mutmansky (1982).

Multiple Fans

A system containing multiple main fans, excluding those in series and parallel installations, is always accompanied by multiple entrances and/or exits. In solving a problem involving multiple entrances and/or exits, we may first combine all entrances or exits into a single node and obtain temporary solutions by the longest-path algorithm, assuming a main fan in the return dummy branch. Then a cutset operation is performed to remove the fan in the return branch and, at the same time, add either blowers in all branches whose initial nodes are node 1 (the entrance) or exhaust fans in all branches whose final nodes are node N_n (the exit). In order to minimize the total air power consumed by the ventilation network, the cutset operation should be performed on the longest-path-tree solution corresponding to the forward-pass procedure or backward-pass procedure, depending on whether a blower or exhaust system is desired. See Wang (1982a, 1982b; Wang and Mutmansky, 1982) for more detailed information on this problem.

PROBLEMS

7.1 The following airways in sedimentary rock have characteristics as indicated:

Airway	Size, ft (m)	$L + L_e$, ft (m)	Surface
a	4 × 20 (1.22 × 6.10)	470 (143.3)	Bare, straight, clean
b	6 × 12 (1.83 × 3.66)	1550 (472.4)	Smooth lined, slightly sinuous, slightly obstructed
c	5 × 15 (1.52 × 4.57)	1370 (417.6)	Timbered, moderately curved, clean

Assuming that the airways are arranged in series with a total quantity of 120,000 cfm (56.63 m³/s) flowing, determine the equivalent resistance and the total static head. Neglecting velocity head, compute the air power.

7.2 Plot the characteristic curves for the individual airways and the characteristic curve for the series combination of the airways in Problem 7.1. Plot a total of four points for each curve. Determine the total static head from the curves and compare with the answer obtained in Problem 7.1.

7.3 If the airways of Problem 7.1 are arranged in parallel as natural splits with a total quantity of 120,000 cfm (56.63 m³/s) of air flowing, compute the equivalent resistance, the quantity of air that will flow in each airway, and the static head across the circuit. Neglecting velocity head, compute the air power and compare with Problem 7.1.

7.4 Plot the characteristic curves for the individual airways and the characteristic curve for the parallel combination of the airways in Problem 7.1. Plot a total of four points for each curve. Determine from the curves the quantity of air that will flow in each airway if the total quantity is 120,000 cfm (56.63 m³/s), and check the answers with Problem 7.3.

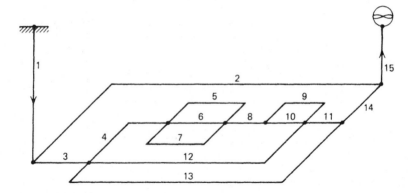

FIGURE 7.19 Mine ventilation schematic for Problem 7.5.

7.5 Assume the resistance for each airway of Fig. 7.19 to be 5.0×10^{-10} in.·min^2/ft^6 (0.559 N·s^2/m^8) and the total (mine) quantity to be 100,000 cfm (47.19 m^3/s). Determine the total equivalent resistance, the total (mine) static head, and the quantity of air flowing through each airway, assuming natural splitting.

7.6 It has been determined that a set of parallel intake entries in a coal mine is required to deliver 150,000 cfm (70.79 m^3/s). The cross section of each entry is 15 × 5 ft (4.57 × 1.52 m). All of the entries have the same physical characteristics: straight, clean, and minimum surface irregularities. Determine the number of entries required to limit the velocity in each airway to 600 fpm (3.048 m/s). Determine the head loss of these entries for a distance of 0.5 mi (0.80465 km).

7.7 Given that the entries of Problem 7.1 are arranged in parallel as controlled splits, determine the free split and the amounts of regulation necessary in the other splits to secure the airflows indicated:

Airway	Quantity cfm (m^3/s)
a	20,000 (9.44)
b	60,000 (28.32)
c	40,000 (18.88)

What is the static head across the circuit? Neglecting velocity head, compute the air power and compare with Problems 7.1 and 7.3.

7.8 Compute the size (area) of the regulators required in Problem 7.7.

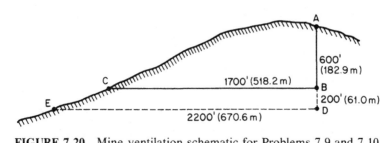

FIGURE 7.20 Mine ventilation schematic for Problems 7.9 and 7.10.

7.9 An adit BC connects to a shaft AB, as shown in Fig. 7.20. It is decided to drive a lower-level DE and deepen the shaft BD. Adits are 5×7 ft (1.52×2.13 m) and the shaft 6×6 ft (1.83×1.83 m). The friction factor for all airways is 60×10^{-10} lb·min²/ft⁴ (0.01113 kg/m³). Assuming that a fan can be installed to exhaust 75,000 cfm (35.40 m³/s) at the collar of the shaft A, determine the quantity that will flow on each level and the mine heads. Solve algebraically.

7.10 Solve Problem 7.9 graphically.

7.11 An exhaust-ventilation system is shown in Fig. 7.21. The head losses

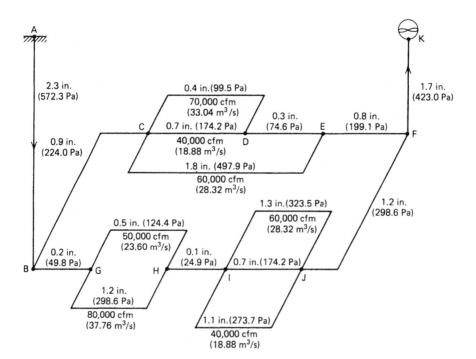

FIGURE 7.21 Mine ventilation schematic for Problem 7.11.

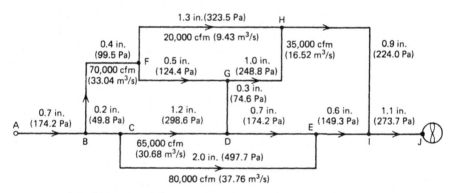

FIGURE 7.22 Mine ventilation schematic for Problem 7.12.

given are based on the assigned air quantities, and the shaft dimensions are 10×7.5 ft (3.05×2.29m). Determine the quantity of air the fan must provide. Locate regulators where needed, specifying the amount of regulation at each. Find the mine heads and the air power.

7.12 The assigned air quantities and their associated head losses are indicated in Fig. 7.22. Determine the quantity and direction of airflow in DG and the quantity exhausted by the fan. Locate regulators where needed, specifying the amount of regulation at each, and find the mine static head.

7.13 Given the ventilation system shown in Fig. 7.23, with resistances and assigned quantities as indicated, locate regulators where needed, specifying the amount of regulation at each.

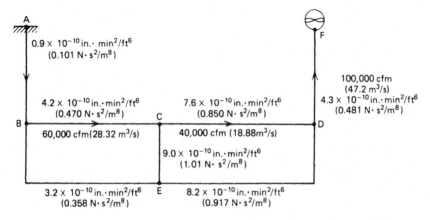

FIGURE 7.23 Mine ventilation schematic for Problem 7.13.

7.14 Given the exhaust ventilation system of Fig. 7.15a, consisting of two shafts and a raise connected on two levels by drifts. Resistance factors, in 10^{-10} in.·min^2/ft^6 (N·s^2/m^8), for each branch are indicated. The head of the exhaust fan can be approximated by

$$H^F = -(2)(10)^{-10}Q^2 - (2)(10)^{-5}Q + 12$$

$$(H^F = -0.22344Q^2 - 10.545Q + 2986)$$

In the raise, the airflow is directed from F to C at a rate of 10,000 cfm (4.72 m^3/s), aided by a booster fan. Solve the network problem applying the Hardy Cross method. Perform four iterations using the following initial values:

Branch	B–C	B–E–F	C–D	F–C	F–G–D	D–H–A–B
Quantity cfm	30,000	50,000	40,000	10,000	40,000	80,000
(m^3/s)	(14.16)	(23.60)	(18.88)	(4.72)	(18.88)	(37.76)

7.15 Refer to the exhaust ventilation system shown in Fig. 7.15a. Using the initial values for air quantities given in Problem 7.14 as the required quantities, solve the network as a controlled-splitting problem. Assume no booster fans.

CHAPTER 8

Natural Ventilation

Airflow through mine openings cannot exist unless energy losses in the system are overcome by energy gains in the system. There are two forces that supply energy to the mine ventilation system: *natural* and *mechanical* (artificial). Mechanical ventilation is considered in the next chapter. The only natural force that can create and maintain airflow in mines is thermal energy exchange with the strata, which occurs as air passes through the mine as a result of the difference in the temperature of the strata and the air. This energy exchange is sometimes sufficient to overcome the frictional and shock losses. As a result, airflow is both induced and maintained.

The common chimney, or stack, effect is known to all. Warm air rises through a chimney, creating a circulating air current from the room to the outside. A similar phenomenon occurs in mines, where, as a result of difference in elevation and difference in temperature of the workings, warm air rises through the workings creating air circulation through the mine. Furnaces were once commonly employed to initiate and stengthen natural ventilation. Other natural forces have been utilized in the past to create airflow, including surface winds and falling water (Fig. 1.1). Although the energy source for mine ventilation today is nearly always mechanical, usually a fan, it is sometimes supplemented by natural ventilation. This chapter is concerned with the cause, characteristics, effects, quantification, and measurement of natural ventilation.

8.1 CHARACTERISTICS OF NATURAL VENTILATION

Source

Natural ventilation depends on the difference in elevation of the surface and the mine workings and the difference in air temperature inside and outside the mine. Ordinarily, the greater this difference, the stronger the natural ventilation. In a mine ventilated by only natural ventilation, greater differences result in higher-heat energy transfer, greater pressure (head) difference, and larger airflow.

Rather than depth, it is the intensity of rock heat and the difference in elevation of the various openings into the mine that determine the strength of the natural ventilation. The principal sources of mine heat (Section 16.2) are autocompression, the wall rock, and electromechanical equipment. Mines in mountains or hilly terrain frequently have strong natural ventilation because of the difference in elevation of the openings. Mines that are both deep and hot also have strong natural ventilation, stronger in mines in cold climates than in ones in warm climates.

Surface temperature variations between summer and winter extremes are common in many mining areas of the world. However, mine temperatures vary little, except in shafts and near portals; the effect of diurnal and seasonal changes tends to be damped out by the strata, which supply heat to the ventilating air or extract heat from it. During the summer months, for example, the noonday hot air is cooled during its downward movement. On the other hand, the cooler midnight air may be warmed. The moisture content of the intake air fluctuates noticeably with the seasons, whereas that of the exhaust air remains comparatively stable, revealing the near-total tempering effect of mine workings (Bruzewski and Aughenbaugh, 1977). The amount of natural ventilation generated by thermal energy is therefore quite variable.

The direction of airflow due to natural ventilation is seldom constant, particularly in shallow mines [those less than 1500 ft (450 m)]. If the temperature difference causing airflow decreases to zero, air movement ceases. If the surface temperature changes from being less than the mine temperature to above the mine temperature, the direction of airflow will also reverse. This can occur not only seasonally but even daily. While the effect of daily atmospheric variation may be of little importance, the effect of seasonal changes can be appreciable (Biswas, 1966). Natural ventilation is usually strongest in winter, weakest in summer, and subject to at least semiannual reversal in the spring and fall.

It may be concluded then that natural ventilation fluctuates, is unstable, and is unreliable. Natural ventilation should never be relied upon solely for mine air conditioning. Natural ventilation may support or act against the mechanical energy source in the ventilation system. Wherever possible, natural ventilation when it is strongest should be utilized to work with the fan (i.e., cause flow in the same direction as the fan). Clearly, control over natural ventilation effects must be exercised.

Direction of Flow

In analyzing the direction of airflow resulting from natural ventilation in simple circuits, the following procedure and observations should be noted:

1. Visualize columns of air of equal height between two horizontal datum lines in order to compare the pressure difference between points in the circuit. This balances the change in elevation between intake and discharge. The columns should extend between the elevations of the highest and lowest points in the mine.
2. Consider surface temperatures in winter to be colder than those in the mine, and vice versa in summer, unless information to the contrary is available.

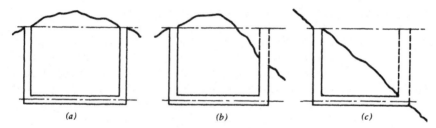

FIGURE 8.1   Determining the direction of flow in basic natural ventilation systems.

3. The colder column of air (the one having the lower average temperature) is the heavier one and will displace the warmer, lighter column.
4. The direction of airflow in the mine will be from the heavier column toward the lighter column.

Consider the simple schematics of three mines, each showing two shafts and the connecting airway between them (Fig. 8.1). In each mine, the two horizontal datum lines that are selected for delineating the top and bottom of the two air columns are of the same height and extend between the highest and lowest points in the mine. In the table that follows, the flow directions are indicated for the two main seasons.

Schematic	Fig. 8.1a	Fig. 8.1b	Fig. 8.1c
Need to induce	Yes	No	No
Direction, winter	Either	Right to left	Right to left
Direction, summer	None	Left to right	Left to right

Note that when the difference in elevation between the two openings at the surface is small (Fig. 8.1a), flow may have to be induced (by a fan, a large and rapidly moving object, or other means). But once induced, flow will continue as long as a favorable temperature difference exists between the mine and the surface, as assumed for the mines in Fig. 8.1a in winter. The airflow may even increase if flow improves the differential. However, flow stops in the summer even if induced because of the assumed unfavorable temperature difference between the mine and the surface.

Even when no elevation difference exists and the air columns are purely imaginary, natural ventilation pressure may be generated. Vehicular tunnels with little or no elevation difference between the two ends often have some natural ventilation. This is due not to the action of wind but to the difference in temperature on the surface at the two portals, especially in areas of high relief, caused by sun and shade conditions and weather changes.

8.2 QUANTIFICATION OF NATURAL VENTILATION

There are two approaches to quantify natural ventilation effects. A simple and, in most cases, a satisfactory approach is to view the cause of natural ventilation as the *density difference* between two columns of air of equal height, the heavier intake air column sinking and the lighter air column rising, as explained before, and then to calculate the *natural ventilation pressure* (NVP) that results. The *thermodynamic approach,* on the other hand, explicitly considers the heat energy exchange in the mine, and applies the equations developed in Section 5.8 to calculate the *natural ventilation energy* (NVE). In all cases, it is necessary to conduct a pressure, quantity, or temperature survey around the mine circuit to obtain the data for estimating natural ventilation (NVP or NVE).

Analysis by Density-Difference-Based Methods

There are several methods to calculate the natural ventilation pressure that are based on the difference in air specific weight between two air columns of equal height. Using available data, the specific weight of air can be calculated using Fig. 2.3 or the procedures outlined in Section 6.3.

Method 1   This method is based on the fact that air specific weight increases progressively, but not linearly, toward the bottom of a column (Weeks, 1926). Consider a column of dry air having a unit cross-sectional area A and a height L, and assume an incremental height dL. Let the pressure on the base of this increment be dp. Then the incremental force exerted is $A\, dp$, which is equal to the weight ($wA\, dL$) of the incremental volume $A\, dL$, where w is specific weight of the air. Noting that specific weight for dry air is (p/RT), from Eq. 1.6, the following relationship is evident:

$$A\, dp = wA\, dL = \left(\frac{p}{RT}\right) A\, dL$$

Therefore,

$$dp = \frac{p}{RT}\, dL$$

Rearranging for integration:

$$\int_{p_1}^{p_2} \frac{dp}{p} = \int_0^L \frac{dL}{R\overline{T}}$$

Integrating, these results:

$$\ln\left(\frac{p_2}{p_1}\right) = \frac{L}{R\overline{T}} \tag{8.1}$$

in which p_1 and p_2 are absolute pressures at the top and bottom of the column, respectively, and $\overline{T}$ is the average absolute temperature between the top and bottom of the column. To calculate the natural ventilation pressure, Eq. 8.1 is solved for each column of air. The pressure difference at the bottom of the two columns is the *natural ventilation head* H_n in in. water:

$$H_n = 13.6 \, (p_2 - p_3) \tag{8.2}$$

and in Pa:

$$H_n = (p_2 - p_3) \tag{8.2a}$$

where p_2 and p_3 are the absolute pressures at the bottom of the two columns expressed here in in. of mercury (Pa). Omission of the effect of water vapor introduces some error in this method.

Method 2 Logically, the difference in the specific weight between two columns of air can be approximated and used to find the natural pressure. Since $p = wL$, the head H_n in in. water,

$$H_n = \frac{L}{5.2} \, (\overline{w}_d - \overline{w}_u) \tag{8.3}$$

and in Pa,

$$H_n = gL(\overline{w}_d - \overline{w}_u) \tag{8.3a}$$

where $\overline{w}_d$ and $\overline{w}_u$ are, respectively, the average specific weights of the downcast and upcast columns in lb/ft^3 (kg/m^3). The values of $\overline{w}_d$ and $\overline{w}_u$ should be obtained by averaging the specific weights rather than temperatures. Temperatures and barometric pressures are measured at various points throughout each column, and the corresponding air specific weights are determined and used to find an average specific weight.

Method 3 On the basis of the difference in dry-bulb temperatures, the natural pressure in in. water can also be calculated approximately as (Rees, 1950)

$$H_n = \left(\frac{\overline{T}_u - \overline{T}_d}{5.2\overline{T}} \right) wL \tag{8.4}$$

and in Pa,

$$H_n = \left(\frac{\overline{T}_u - \overline{T}_d}{\overline{T}}\right) wgL \qquad (8.4a)$$

where $\overline{T}$ is the average absolute temperature $= (\overline{T}_u + \overline{T}_d)/2$, and w the air specific weight at the desired point of reference (fan location or other).

Method 4   Avoiding computation of specific weights, and neglecting moisture in the air, the natural head can be found approximately by mean dry-bulb temperatures (McElroy, 1935):

$$H_n = 0.255\overline{p}_b L \left(\frac{1}{\overline{T}_d} - \frac{1}{\overline{T}_u}\right) \qquad (8.5)$$

$$H_n = 3.484\overline{p}_b L \left(\frac{1}{\overline{T}_d} - \frac{1}{\overline{T}_u}\right) \qquad (8.5a)$$

where $\overline{p}_b$ is the atmospheric pressure in inches of mercury (Pa) at the midpoint of the column, $\overline{T}_d$ is average absolute temperature in the downcast shaft, and $\overline{T}_u$ is the average temperature in the upcast.

Method 5   A rough rule of thumb can be used to estimate the natural pressure (McElroy, 1935). At sea level, the head is 0.03 in. water per 10°F difference in average temperature of the two columns per 100 ft difference in elevation of the surface and the mine workings (7.5 Pa per 5.5°C difference per 30.5 m difference), or

$$H_n = 0.03 \text{ in.}/10°\text{F}/100 \text{ ft} \qquad (8.6)$$

and in Pa:

$$H_n = 44 \text{ Pa}/10°\text{C}/100 \text{ m} \qquad (8.6a)$$

The accuracy in calculating the natural ventilation head from any of the preceding methods is greatly dependent on the correct determination of the temperatures and specific weights. However, the simplifying assumptions used in the derivation of all the preceding formulations are the major limitations to expecting great accuracy in the calculated natural ventilation pressure. In view of the continual change in NVP in an actual mine, any of the five methods seems sufficiently accurate. Bruce (1986) presents data from operating mines that indicate that the second method may be preferable for routine calculations. Since temperature and density changes are highest in shafts and inclined workings, for best results temperatures and densities should be obtained at frequent intervals throughout the depth of the shafts and inclined workings.

Analysis by Thermodynamics

Reference has already been made to the analogy between the mine ventilation system and a heat engine (Section 5.8). Using a thermodynamic approach, the NVP may be determined by two methods. Assume a mine ventilation system consisting of two shafts and connecting workings, such as shown in Fig. 8.1. The processes occurring as air flows through the mine are adiabatic in the shafts, expansion at constant or reduced pressure in the workings, and isobaric cooling in the atmosphere.

Method 1   Hinsley (1965) has developed an equation for the calculation of the natural ventilation pressure that requires only the measurement of the surface pressure and temperatures at the top and bottom of the upcast and downcast shafts. Termed the *virtual temperature* method, this approach allows the natural ventilation head to be calculated in in. water as follows:

$$H_n = \frac{p_1}{5.2}\left[\left(\frac{T_2}{T_1}\right)^{L/R\Delta t_d} - \left(\frac{T_3}{T_4}\right)^{L/R\Delta t_u}\right] \tag{8.7}$$

and in Pa:

$$H_n = p_1\left[\left(\frac{T_2}{T_1}\right)^{L/R\Delta t_d} - \left(\frac{T_3}{T_4}\right)^{L/R\Delta t_u}\right] \tag{8.7a}$$

where p_1 and T_1 are the absolute pressure [in psf (lb/ft^2) or Pa] and temperature, respectively, at the top of the downcast shaft; T_2 is the temperature at the bottom of the downcast shaft; T_3 is the temperature at the bottom of the upcast shaft; and T_4 is the temperature at the top of the upcast shaft (all temperatures are dry-bulb, in absolute units). The shafts are assumed to have equal depths L in ft (m), R is the universal gas constant for air, and Δt_d and Δt_u are the temperature changes in °F (°C) in the downcast and upcast shafts, respectively. If water-vapor effects are to be included, which will make the results more accurate, the method requires that both wet- and dry-bulb temperatures be measured and then converted to virtual temperatures (Hess, 1959). The calculation of virtual temperatures requires that pressure determinations be made at all temperature-measurement stations. This procedure takes into account the decreased specific weight of moist air when compared with dry air of the same temperature, which then satisfies the above equation of state for dry air. For further explanation, the interested reader should refer to Hess (1959) and Hinsley (1965).

Method 2   A more detailed analysis of the system can be performed by plotting the indicator diagram (*p–v* diagram) for the system with knowledge of the temperature and pressure data at the cardinal points (top and bottom of the intake and return shafts) of the system (Hall, 1967, 1981; McPherson, 1993). For an ideal fluid with no heat or friction loss, the diagram resembles a Joule cycle (see Fig. 8.2*a*). The actual cycle for a mine with natural ventilation only is shown in Fig. 8.2*b*. The pressure difference between points 2

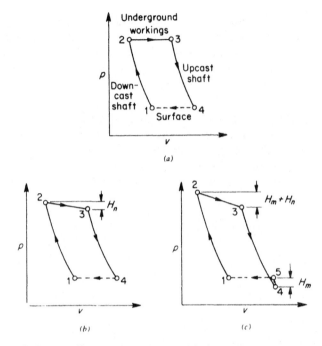

FIGURE 8.2 Indicator diagram for mine ventilation system: (*a*) theoretical cycle; (*b*) actual cycle, natural ventilation only; (*c*) actual cycle, exhaust fan + natural ventilation.

and 3 is equal to the natural ventilation head H_n, neglecting head loss in the shafts. The diagram for a mine with fan plus natural ventilation is shown in Fig. 8.2*c*. The natural ventilation head is equal to the pressure difference between points 2 and 3 minus the fan head H_m (measured between points 4 and 5).

In the actual case, the shaft loss must also be considered. In operating mines, in addition to friction and shock losses, there may be heat additions in the shaft. There may also be heat loss (e.g., due to evaporation of moisture). The processes, instead of being truly adiabatic, may be polytropic. The change in the p–v diagram as a result of these processes would be an altered shape of the process curves. However, the area enclosed by the indicator diagram still represents the natural ventilation energy. For example, a lower temperature at the surface of the intake shaft and greater depths of shafts will, respectively, increase the width and the height of the indicator diagram, leading to larger NVE.

To calculate the natural ventilation pressure, the natural ventilation energy must be multiplied by the specific weight. The specific weight is not constant throughout the system. Therefore, under compressible flow assumptions, the value for NVP is also different at different locations (intake shaft, workings, return shafts) in the mine.

Measurement of Natural Ventilation Pressure

Data collected during ventilation pressure surveys should be examined for the presence of natural ventilation, its magnitude and direction. In a pressure survey conducted from the surface control (usually the fan house) through the mine, across the fan, and back to the surface control, the total head loss around the circuit, in the absence of errors and natural ventilation, would be zero. The existence of a positive or negative closure, in the absence of errors, is an indication of the natural ventilation pressure (Bruce, 1986).

Whether this closure error is numerically equal to the NVP can be readily checked by calculating the natural ventilation by any one or more of the methods described previously with the data collected during the survey. For example, Bruce (1986) determined the error for a pressure survey in an underground metal mine as 0.71 in. water (176.85 Pa). The barometric pressure, dry- and wet-bulb temperatures, and elevations at the cardinal points A, B, and C were measured during the survey (Fig. 8.3). These were used to calculate the relevant specific weights and the NVP by density-difference methods 2 and 5. [Note that an air column of height 101 ft (30.78 m) must be added at A to bring the intake air column to the height of the return air column BC]. The calculated results for H_n by methods 2 and 5 were 0.88 and 0.64 in. water (219.20 and 159.41 Pa), respectively.

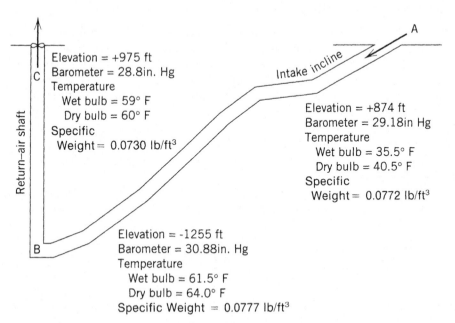

FIGURE 8.3 Conditions causing natural draft pressure in a metal mine. (Conversion factors: 1 ft = 0.3048 m, 1 in. Hg = 3.3768 kPa, °C = $\frac{5}{9}$(°F − 32°), 1 lb/ft³ = 16.018 kg/m³)

It is apparent that the closure error represents the NVP. However, it is important to note that the measured closure error (0.71 in. water) and the two calculated values of NVP (0.88 and 0.64 in. water) are all different, reflecting the earlier comments on measurement errors and formula approximations.

The magnitude of the natural ventilation head in an existing mine can be measured by shutting down the fan, if any, and closing off the flow temporarily with a brattice or bulkhead across a main (series) opening at any point in the system. The head loss across this stopping is measured (by manometer or altimeter), and this is equal to the NVP. For greatest accuracy, readings should be taken rapidly to minimize temperature changes in the mine. The stopping should be erected on a level rather than in a shaft to avoid air-temperature changes during measurement.

Characteristic Curve of Natural Ventilation Pressure

In mine ventilation, a characteristic curve is a plot of head vs. quantity (Section 7.1). Natural ventilation pressure is a function of the temperature difference and the height of the air columns considered, and is not affected by the quantity of air flowing in the circuit. Therefore, the characteristic curve of natural ventilation is a straight line parallel to the quantity axis.

Determining Quantity of Natural Flow

In a mine ventilated by mechanical ventilation, natural ventilation, when present, affects the performance of the mechanical ventilation device. As stated in Section 8.1, natural ventilation can either aid or act against the mechanical ventilation device. In either case, the quantity of air as well as the mine and fan heads will be different from those without the natural ventilation pressure.

In a mine ventilated by natural ventilation only, the quantity of airflow that results in the mine can be calculated knowing the NVP and the mine resistance. The resistance of the mine circuit is calculated using Eq. 7.8 (see also Example 6.7).

Assuming a simplistic single-airway mine, disregarding velocity head at the discharge of mine, and expressing H_n in in. water, the quantity of airflow can be found from Eq. 5.20:

$$Q = \sqrt{\frac{5.2\,H_n A^3}{KO(L + L_e)}} \tag{8.8}$$

or using appropriate SI units, with the head expressed in Pa, from Eq. 5.20a:

$$Q = \sqrt{\frac{H_n A^3}{KO(L + L_e)}} \tag{8.8a}$$

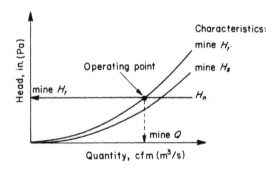

FIGURE 8.4 Characteristic curves of mine and natural ventilation and operating point of the system.

The friction factor for use with the above equation must be corrected for the specific weight conditions in the mine.

For a graphic solution, knowing the mine resistance, the mine static-head characteristic can be developed following the procedure outlined in Section 7.1 and plotted on the same graph as that of the natural ventilation characteristic. The intersection point of the two curves is the operating point, and the quantity of air that will flow in the mine can be read off directly below it on the abscissa (Fig. 8.4). If the velocity head at the discharge of the mine is to be considered, the mine total head characteristic must be developed, again following the procedure outlined in Section 7.1. In the latter case, as can be seen in Fig. 8.4, the quantity of air is less. Calculation of NVP and quantity is demonstrated in the following.

Example 8.1 Given the mine schematic shown in Fig. 8.5 and the following data:

$$\text{Shafts } L = 8000 \text{ ft } (2438 \text{ m})$$

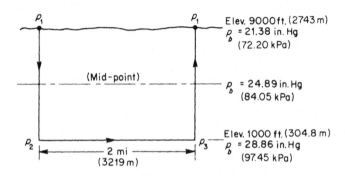

FIGURE 8.5 Mine ventilation schematic (see Example 8.1).

Airways $K = 100 \times 10^{-10}$ lb·min²/ft⁴ (0.01855 kg/m³)

(at $w = 0.0750$ lb/ft³ or 1.2014 kg/m)

Airway size $= 10 \times 20$ ft (3.048 $\times$ 6.096 m)

Beginning at the top of the left-hand shaft in Fig. 8.5 and following counter-clockwise through the airways, dry-bulb temperatures were measured as follows:

$$t_1 = 25°F\ (-3.9°C)$$

$$t_2 = 55°F\ (12.8°C)$$

$$t_3 = 110°F\ (43.3°C)$$

$$t_4 = 90°F\ (32.2°C)$$

Assuming a direction of flow, calculate the natural ventilation pressure by each of the five methods, and find the quantity of flow. Disregard shock loss and velocity head. Assume saturated air. Also calculate the NVP using Eq. 8.7.

Solution: The air in the left-hand column is heavier (lower average tempera-ture), and therefore flow will be from the left-hand column to the right-hand column, as shown in Fig. 8.5. Find the absolute temperatures:

$$T_1 = 460 + 25 = 485°R$$

$$T_2 = 460 + 55 = 515°R$$

$$T_3 = 460 + 110 = 570°R$$

$$T_4 = 460 + 90 = 550°R$$

The mean dry-bulb temperatures can then be calculated:

$$\overline{T}_d = \frac{485 + 515}{2} = 500°R$$

$$\overline{T}_u = \frac{570 + 550}{2} = 560°R$$

$$\overline{T} = \frac{500 + 560}{2} = 530°R$$

Find air specific weights (using Fig. 2.3 and Appendix Table A.1). For $t_d = t_w = 40°F$ and $p_b = 24.89$ in. Hg, read $\overline{w}_d = 0.0660$ lb/ft³. For $t_d = t_w =$

$100°F$ and $p_b = 24.89$ in. Hg, read $\overline{w}_u = 0.0580$ lb/ft^3. Calculate the natural ventilation head by the various methods.

Density difference:

1. By method 1 (Eqs. 8.1 and 8.2):

$$\ln\left(\frac{p_2}{p_1}\right) = \frac{8000}{(53.35)(500)} = 0.2999, \qquad p_2 = 1.3497p_1$$

$$\ln\left(\frac{p_3}{p_1}\right) = \frac{8000}{(53.35)(560)} = 0.2678, \qquad p_3 = 1.3071p_1$$

$$H_n = (13.6)(1.3497 - 1.3071)(21.38) = 12.39 \text{ in. water (3.08 kPa)}$$

2. By method 2 (Eq. 8.3):

$$H_n = \frac{8000}{5.2}(0.0660 - 0.0580) = 12.31 \text{ in. water (3.06 kPa)}$$

3. By method 3 (Eq. 8.4), based on w of downcast shaft:

$$H_n = \frac{(560 - 500)}{(5.2)(530)}(0.0660)(8000) = 11.49 \text{ in. water (2.86 kPa)}$$

4. By method 4 (Eq. 8.5):

$$H_n = (0.255)(24.89)(8000)\left(\frac{1}{500} - \frac{1}{560}\right) = 10.88 \text{ in. water (2.71 kPa)}$$

5. By method 5 (Eq. 8.6), based on w of downcast shaft:

$$H_n = (0.03)\left(\frac{100 - 40}{10}\right)\left(\frac{8000}{100}\right)\left(\frac{0.0660}{0.075}\right) = 12.67 \text{ in. water (3.15 kPa)}$$

There is good correlation among the methods for this example. Select the result of method 2 for calculating Q using Eq. 8.7.

$$Q = \sqrt{\frac{(5.2)(12.31)(200)^3}{(100 \times 10^{-10})(60)(26,560)}} = 179,300 \text{ cfm (84.6 m}^3\text{/s)}$$

at standard air specific weight. If Q is desired at a particular location (shaft collar, level, etc.), correct K for the specific weight at that point.

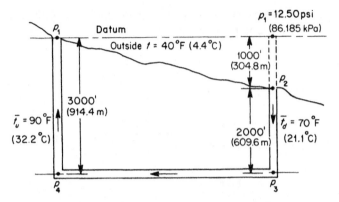

FIGURE 8.6   Mine ventilation schematic (see Example 8.2).

Thermodynamics:

6. By method 1 (Eq. 8.7), based on p at top of downcast shaft:

$$H_n = (13.6)(21.38)\left[\left(\frac{515}{485}\right)^{8000/[(53.35)(55-25)]} - \left(\frac{570}{550}\right)^{8000/[(53.35)(110-90)]}\right]$$

$$= 12.42 \text{ in. water (3.09 kPa)}$$

A further example demonstrates the procedure when a composite column is involved.

Example 8.2   Given the mine schematic shown in Fig. 8.6. Calculate the NVP by method 1 (density difference).

Solution: Write Eq. 8.1 for each part of the right-hand column. Right-hand column, air:

$$\ln\left(\frac{p_2}{p_1}\right) = \frac{1000}{(53.35)(460 + 40)} = 0.03749, \qquad p_2 = 1.0382p_1$$

Right-hand column, shaft:

$$\ln\left(\frac{p_3}{p_2}\right) = \frac{2000}{(53.35)(460 + 70)} = 0.07073, \qquad p_2 = 0.9317p_3$$

$$1.0382p_1 = 0.9317p_3, \qquad p_3 = 1.1143p_1$$

Left-hand column, shaft:

$$\ln\!\left(\frac{p_4}{p_1}\right) = \frac{3000}{(53.35)(460 + 90)} = 0.1022, \qquad p_4 = 1.1077p_1$$

Then using Eq. 8.2,

$$H_n = \frac{144}{5.2}\,(1.1143 - 1.1077)(12.50) = 2.28 \text{ in. } (567.4 \text{ Pa})$$

In applying other methods (Eqs. 8.3 to 8.6) with a composite column, values of temperature or specific weight are obtained by weighting each section of the column according to its length.

To demonstrate the detailed thermodynamic approach (method 2), the following numerical example is presented. The data are taken from Hall (1967) to illustrate calculation procedures for a mine ventilated by natural ventilation only.

Example 8.3 A mine is 10,000 ft. (3048 m) deep, and the shafts are connected by horizontal workings (Fig. 8.7). Dry air is circulating through the mine, exchanging heat adiabatically in the shafts ($n = 1.4$), and at constant pressure in the workings and the atmosphere. Heat is added only in the workings. Horizontal workings are assumed to be level with no friction and no change in the kinetic energy terms. The surface temperature is 40°F (4.44°C), and the surface rock temperature is 65°F (18.33°C). The geothermal gradient is 5°F/1000 ft (9.11°C/1000 m). The surface air pressure is 14.696 psia (101.325 kPa). The air at the bottom of the upcast shaft is at virgin-rock temperature. (Refer to Chapter 6, Section 6.2, or Chapter 16 for definitions of surface rock temperature, virgin rock temperature, and geothermal gradient).

Imagine that the top of the return shaft is blocked by a stopping, and that an ideal turbine is mounted in the stopping with the high–pressure air in the shaft entering the turbine through a small opening in the stopping. The air passing through the turbine follows a reversible adiabatic expansion to atmospheric pressure, doing work, but with negligible changes in kinetic or potential energies.

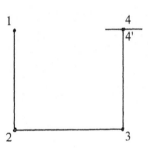

FIGURE 8.7 Mine ventilation schematic (see Example 8.3).

1. Plot the indicator diagram for the airflow through the mine.
2. Determine the work done per pound of air from the indicator diagram, and compare it with that calculated from the turbine input and output.
3. Calculate the heat input into the system, the thermal efficiency, and the heat rejected to the atmosphere.
4. Determine the natural ventilation pressure as measured at the turbine.
5. Determine the natural ventilation air power if the quantity of air entering the downcast shaft is 200,000 cfm.
6. Determine the natural ventilation energy without the turbine in the system.

Solution: To plot the indicator diagram for the process, it is necessary to find the state conditions at the top of intake shaft, bottom of the intake shaft, bottom of the return shaft, intake to the turbine, and top of the return shaft, which are, respectively, points 1, 2, 3, 4', and 4 in Fig. 8.7. The following equations are relevant to calculate the conditions at $i = 1, 2, 4'$, and 4:

$$v_i = \frac{RT_i}{p_i}, \qquad T_{i+1} = T_i + \left(\frac{Z_{i+1} - Z_i}{778 \times 0.24}\right), \qquad p_{i+1} = p_i \left(\frac{T_{i+1}}{T_i}\right)^{n/(n-1)}$$

Note that $p_4 = p_1 = 14.696$ psia

$$p_3 = p_2$$

$$t_3 = 65 + \left(\frac{10,000}{1000}\right)(5) = 115°F$$

Point 1 (top of intake shaft):

$$T_1 = 460 + 40 = 500°R$$

$$p_1 = 14.696 \text{ psia}$$

$$v_1 = \frac{(53.35)(500)}{(144)(14.696)} = 12.605 \text{ ft}^3/\text{lb}$$

Point 2 (bottom of intake shaft):

$$T_2 = 500 + \frac{(10,000)}{(778)(0.24)} = 553.56°R$$

$$p_2 = 14.696 \left(\frac{553.56}{500.0}\right)^{1.4/0.4} = 20.984 \text{ psia}$$

$$v_2 = \frac{(53.35)(553.56)}{(144)(20.984)} = 9.773 \text{ ft}^3/\text{lb}$$

Point 3 (bottom of return shaft):
  As noted before:

$$T_3 = 460 + 115 = 575°R$$

$$p_3 = p_2 = 20.984 \text{ psia}$$

$$v_3 = \frac{(53.35)(575)}{(144)(20.984)} = 10.152 \text{ ft}^3/\text{lb}$$

Point 4' (intake to the turbine):

$$T_{4'} = 575 - \frac{10,000}{(778)(0.24)} = 521.44°R$$

$$p_{4'} = 20.984 \left(\frac{521.44}{575.00}\right)^{1.4/0.4} = 14.903 \text{ psia}$$

$$v_{4'} = \frac{(53.35)(521.44)}{(144)(14.903)} = 12.963 \text{ ft}^3/\text{lb}$$

Point 4 (top of the return shaft)

Note that $\dfrac{p_4}{p_{4'}} = \left(\dfrac{T_4}{T_{4'}}\right)^{n/(n-1)}$

and $p_4 = p_1 = 14.696 \text{ psia}$

$$T_4 = 521.44 \left(\frac{14.696}{14.903}\right)^{0.4/1.4} = 519.36°R$$

$$v_4 = \frac{(53.35)(519.36)}{(144)(14.696)} = 13.093 \text{ ft}^3/\text{lb}$$

Point	1	2	3	4'	4
p, psia (kPa)	14.696	20.984	20.984	14.903	14.696
	(101.325)	(144.680)	(144.680)	(102.753)	(101.325)
v, ft³/lb (m³/kg)	12.605	9.773	10.152	12.963	13.093
	(0.787)	(0.610)	(0.634)	(0.809)	(0.817)
t, °F (°C)	40	93.56	115	61.44	59.36
	(4.44)	(34.2)	(46.11)	(16.356)	(15.20)

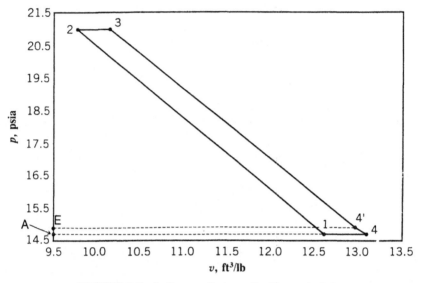

FIGURE 8.8 Indicator diagram for Example 8.3.

1. Indicator diagram (see Fig. 8.8).

2. Work done by a lb of air from the indicator diagram is 12341. The area can be approximated as follows. Approximate width of the enclosed area in the x axis =

$$\frac{(13.093 - 12.605) + (10.152 - 9.773)}{2} \approx 0.433 \text{ ft}^3/\text{lb}$$

the height of the enclosed area along the y-axis = $20.984 - 14.696 = 6.288$ psi.

Work done from the indicator diagram = $(144)(6.288)(0.433) \cong 392.07$ ft·lb/lb (1171.90 J/kg). Since turbine work is the only work done in the cycle, the area of the indicator diagram should equal the turbine work. Work done in the turbine from the indicator diagram is area 4′4AE4′ This area can be approximated as follows:

$$(144)(14.903 - 14.696)\frac{(12.963 + 13.093)}{2} \cong 388.34 \text{ ft·lb/lb (1160.75 J/kg)}$$

The values 392.07 ft · lb/lb and 388.34 ft · lb/lb for work done in the cycle are in close agreement, with the small difference attributed to independent rounding off errors in the numerical calculations and the assumptions of a straight line for the processes. Carrying the calculations to a larger number of significant digits will reduce the errors due to rounding off. Applying Eq. 5.33 to the process from 4' to 4, it is noted that

$$\int_{4'}^{4} v \, dp + H_{l_{4'4}} = (h_4 - h_{4'}) - \dot{q}_{4'4}$$

where $q_{4'4}$ is the heat input (or loss) during flow throughout the turbine. There is no energy loss in the turbine (i.e., $H_{l_{4'4}} = 0.0$), and there is no heat input or heat loss during flow through the turbine, $\dot{q}_{4'4} = 0$. Also applying Eq. 2.10 for the relationship between enthalpy and temperature,

$$\int_{4'}^{4} v \, dp = (h_4 - h_{4'}) = c_p \, (t_4 - t_{4'}) = 0.24(59.36 - 61.44)$$

$$= -0.50 \text{ Btu/lb} \, (-1163.00 \text{ J/kg})$$

Noting that 1 Btu = 778 ft·lb, the work done by the air on the turbine = $(0.24)(778)(59.36 - 61.44) = -388.37$ ft·lb/lb (-1160.84 J/kg). In subsequent calculations, the value used for natural ventilation energy will be 388.37 ft·lb/lb.

3. Heat input to the system occurs only between 2 and 3. As before, applying Eq. 5.33 to the process from 2 to 3 and noting that there is no frictional energy loss ($H_{l_{23}} = 0$) and no flow work ($\int v \, dp = 0$) (constant pressure expansion), then

$$\int_{2}^{3} v \, dp + H_{l_{23}} = (h_3 - h_2) - \dot{q}_{23} = 0$$

and

$$\dot{q}_{23} = (h_3 - h_2) = c_p \, (t_3 - t_2) = 0.24(115 - 93.56)$$
$$= 5.15 \text{ Btu/lb} \, (11,979 \text{ J/kg})$$

In mechanical energy terms, this is equivalent to $(0.24)(778)(115 - 93.56)$ = 4003.28 ft·lb/lb (11,966 J/kg)

$$\text{Thermal efficiency} = \frac{t_2 - t_1}{t_2} = \frac{(93.56 - 40)}{553.56} = 0.097$$

Heat rejected to the atmosphere $\int_{4}^{1} v \, dp + H_{l_{41}} = (h_1 - h_4) - \dot{q}_{41}$

As before, the quantities $H_{l_{41}}$ and $\int_{4}^{1} v \, dp$ are zeros, and

$$\dot{q}_{41} = (h_1 - h_4) = c_p (t_1 - t_4) = 0.24(40 - 59.36)$$
$$= -4.65 \text{ Btu/lb} (-10{,}816 \text{ J/kg}) = (0.24)(778)(40 - 59.36)$$
$$= -3614.90 \text{ ft·lb/lb} (-10{,}805 \text{ J/kg})$$

Note that the sum of the work done on the turbine by the air and the heat rejected to the atmosphere equals the heat input to the system.

4. Natural ventilation pressure at the turbine.

$$H_n = [\text{natural ventilation energy}] \times [\text{air specific weight at the turbine}]$$

Noting that the specific volume at the turbine input is 12.963 ft³/lb:

$$H_n = \left(\frac{1}{5.2}\right)(388.37)\left(\frac{1}{12.963}\right) = 5.762 \text{ in. water (1.435 kPa)}$$

5. Air power generated by natural ventilation.

The mass flow rate through the mine

$$G = \frac{[\text{quantity flow rate at the top of the intake shaft}]}{[\text{specific volume at the top of the intake shaft}]} = \frac{200{,}000}{12.605}$$

$$= 15{,}866.72 \text{ lb/min}$$

$$= 264.45 \text{ lb/h (119.95 kg/h)}$$

The power input from natural ventilation pressure is

$$\frac{(15{,}866.72)(388.37)}{(33{,}000)} = 186.73 \text{ hp (139.24 kW)}$$

6. Note that if there is no turbine in the system, the conditions at 1, 2, and 3 are the same. However, the adiabatic process would have continued to 4 with the following results:

$$T_4 = 575 - \frac{(10{,}000)}{(778)(0.24)} = 521.44°R$$

$$p_4 = p_1 = 14.696 \text{ psia}$$

$$v_4 = \frac{(53.35)(521.44)}{(144)(14.696)} = 13.146 \text{ ft}^3/\text{lb}$$

The natural ventilation energy

$$\frac{[(13.146 - 12.605) + (10.152 - 9.773)](20.984 - 14.696)(144)}{2}$$

$$= 416.5171 \text{ ft·lb/lb } (1244.97 \text{ J/kg})$$

Heat input to the system and heat loss to atmosphere are the same:

$$\text{Heat input} = c_p(t_3 - t_2) = 5.15 \text{ Btu/lb } (11979 \text{ J/kg})$$
$$\text{Heat loss} = c_p(t_1 - t_4) = 0.24 \ (40 - 61.44)$$
$$= -5.15 \text{ Btu/lb } (11,979 \text{ J/kg})$$

Note that the mechanical energy produced by the turbine [388.37 ft·lb/lb (1160.84 J/kg)] is at the expense of the flow energy. Consequently, less air will flow through the system with a turbine.

Hall (1967) continues to build on this example with several variations on the location of the turbine, friction and shock losses, and heat and moisture transfer. The interested reader is referred to Hall's book for further details.

Data for the next example are adapted from McPherson (1993) to illustrate the thermodynamic approach to mine ventilation analysis with a fan in the system.

Example 8.4 The following pressure-temperature survey data are from a mine whose shafts are 4000 ft (1219.2 m) deep. Assume that dry air is moving through the mine. The quantity measured at the fan inlet is 270,000 cfm (127.44 m³/s).

Station No. and Location	Absolute Pressure psia (kPa)	Temperature °F (°C)
1. Top of intake shaft	14.73 (101.560)	50.27 (10.15)
2. Bottom of intake shaft	16.89 (116.453)	71.71 (22.06)
3. Bottom of return shaft	16.53 (113.970)	121.71 (49.84)
4. Top of return shaft	14.39 (99.216)	100.27 (37.93)
5. Fan outlet	14.73 (101.560)	105.89 (41.05)

1. Plot the indicator diagram for the mine ventilation system.
2. Calculate the natural ventilation energy and the natural ventilation pressure at the fan inlet.
3. Determine the input energy to the fan, the fan output, the polytropic efficiency of the fan, and the fan air power.
4. Calculate the heat additions to the air in the mine and the total heat rejected to the atmosphere.

Solution:

1. *Indicator Diagram*   Using the equation $v = RT/p$, the specific volumes at locations 1–5 are calculated. For example,

$$v_1 = \frac{(53.35)(460 + 50.27)}{(144)(14.73)} = 12.834 \text{ ft}^3/\text{lb} \ (0.801 \text{ m}^3/\text{kg})$$

$$w_1 = \frac{1}{v_1} = \frac{1}{12.834} = 0.0779 \text{ lb/ft}^3 \ (1.248 \text{ kg/m}^3)$$

Similar calculations are made for locations 2–5 and are summarized below:

Station	1	2	3	4	5
p, psia (kPa)	14.73	16.89	16.53	14.39	14.73
	(101.560)	(116.453)	(113.90)	(99.216)	(101.560)
v, ft^3/lb (m^3/kg)	12.834	11.663	13.039	14.425	14.233
	(0.801)	(0.728)	(0.814)	(0.901)	(0.889)
w, lb/ft^3 (kg/m^3)	0.0779	0.0858	0.0767	0.0693	0.0703
	(1.248)	(1.374)	(1.229)	(1.110)	(1.125)

The index of the polytropic process n between stations 1 and 2 is calculated using Eq. 5.35:

$$n = \frac{1}{1 - \dfrac{\ln(T_2/T_1)}{\ln(p_2/p_1)}} = \frac{1}{1 - \dfrac{\ln(531.71/510.27)}{\ln(16.89/14.73)}} = 1.431$$

Similar calculations are made for flow between other stations:

Intake shaft (1–2) = 1.431, workings (2–3) = 0.193,

return shaft (3–4) = 1.372, fan (4–5) = 1.749, atmosphere (5–1) = 0.0.

The indicator diagram for the mine ventilation system is shown in Fig. 8.9. The NVE from the indicator diagram (area 12351) is 359 ft·lb/lb (1073 J/kg).

2. *Natural Ventilation Energy*   Frictional energy losses in the intake shaft, workings, and the return shaft are calculated using Eq. (5.33) and, assuming or knowing the following values:

Intake shaft (1–2): $V_1 = V_2$; $Z_1 = 4000$ ft, $Z_2 = 0$ ft; $\dot{W}_{12} = 0$;

$n = 1.431$

Workings (2–3): $V_2 = V_3$; $Z_2 = Z_3$; $\dot{W}_{23} = 0$; $n = 0.193$

Return shaft (3–4): $V_3 = V_4$; $Z_3 = 0$ ft; $Z_4 = 4000$ ft; $\dot{W}_{34} = 0$;

$n = 1.372$

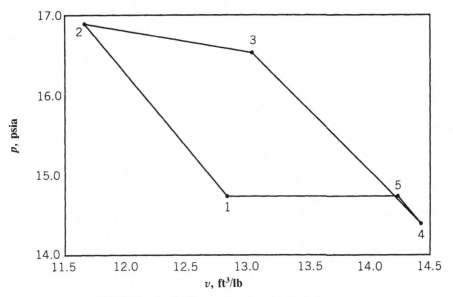

FIGURE 8.9 Indicator diagram for Example 8.4.

For the intake shaft, the energy-balance equation (5.33) reduces to

$$(Z_1 - Z_2) = \int_1^2 v \, dp + H_{l_{12}}$$

Using Eq. 5.37 to evaluate the integral $\int v \, dp$:

$$(4000) = \left(\frac{1.431}{0.431}\right) (53.35) (71.71 - 50.27) + H_{l_{12}}$$

$$H_{l_{12}} = 4000 - 3798 = 202 \text{ ft·lb/lb (604 J/kg)}$$

Similarly,

$$H_{l_{23}} = 638 \text{ ft·lb/lb (1907 J/kg)}$$

$$H_{l_{34}} = 219 \text{ ft·lb/lb (655 J/kg)}$$

Total energy loss in the intake shaft, workings, and return shaft = 202 + 638 + 219 = 1059 ft·lb/lb (3165 J/kg). The mechanical energy supplied to the air to overcome this frictional energy loss is the flow work done by the fan:

$$\text{Fan flow work} = \int_4^5 v \, dp = \frac{1.749}{0.749} \, (53.35) \, (105.89 - 100.27)$$

$$= 700 \text{ ft·lb/lb (2092 J/kg)}$$

(a) Therefore, natural ventilation energy = $1059 - 700 = 359$ ft·lb/lb (1073 J/kg)

(b) Natural ventilation pressure at fan intake

$$= \frac{(359)(0.0693)}{5.2} = 4.78 \text{ in. water (1.191 kPa)}$$

3. *Fan Calculations* For the fan branch, $V_4 = V_5$, $Z_4 = Z_5$, and $\dot{q}_{45} = 0$. From Eq. 5.33, the following relationship is established:

$$\dot{W}_{45} = \int_4^5 v \, dp + H_{l_{45}} = h_5 - h_4$$

Note that

$$h_5 - h_4 = c_p(t_5 - t_4) = 0.24 \, (105.89 - 100.27) = 1.349 \text{ Btu/lb}$$

$$= (1.349)(778) = 1049.37 \text{ ft·lb/lb (3137 J/kg)}$$

The mechanical energy input to the ventilation system ($\dot{W}_{45}$) is 1049.37 ft·lb/lb (3137 J/kg)

$$H_{l_{45}} = \dot{W}_{45} - \int_4^5 v \, dp = 1049.37 - 700 = 349.37 \text{ ft·lb/lb (1044 J/kg)}$$

Useful fan output = $\int_4^5 v \, dp = 700$ ft·lb/lb (2092 J/kg)
Input to fan $\dot{W}_{45} = 1049.37$ ft·lb/lb (3137 J/kg)
Polytropic efficiency of the fan = $(700/1049.37) \, 100 = 66.72\%$
Mass flow of air in the system = $(270,000)(0.0693) = 18,711$ lb/min = 311.85 lb/h (141.45 kg/h)

$$\text{Fan air power} = \frac{(18,711)(700)}{(33,000)} = 397 \text{ hp (296 kW)}$$

$$\text{Fan input power} = \frac{(18,711)(1049.37)}{33,000} = 595 \text{ hp (444 kW)}$$

$$\text{Natural ventilation power} = \frac{(18,711)(359)}{33,000} = 204 \text{ hp (152 kW)}$$

4. *Heat Addition/Rejection*

Note that $\dot{q}_{51}$ = heat rejected to atmosphere

$\dot{q}_{23}$ = heat added at the workings

For the atmospheric branch, assume $V_5 = V_1$, and $\dot{W}_{51} = 0$. From Eq. 5.33, we have

$$0 = h_1 - h_5 - \dot{q}_{51}$$

$$\dot{q}_{51} = (h_1 - h_5) = c_p(t_1 - t_5) = 0.24 \, (50.27 - 105.89) \text{ Btu/lb}$$

$$= (0.24)(778)(50.27 - 105.89) \text{ ft} \cdot \text{lb/lb}$$

$$= -10{,}385.37 \text{ ft·lb/lb } (31{,}042 \text{ J/kg})$$

For the workings $V_2 = V_3$, $Z_2 = Z_3$, $\dot{W}_{23} = 0$

$$0 = h_3 - h_2 - \dot{q}_{23}$$

$$\dot{q}_{23} = (h_3 - h_2)$$

$$= c_p(t_3 - t_2)$$

$$= 0.24 \, (121.71 - 71.71) \text{ Btu/lb} = (0.24)(778)(121.71 - 71.71) \text{ ft} \cdot \text{lb/lb}$$

$$= 9336 \text{ ft·lb/lb } (27{,}905 \text{ J/kg})$$

Note that the sum of the heat input from the workings and the mechanical energy input to the fan ($\dot{W}_{45}$) equals the heat rejected to the atmosphere.

PROBLEMS

8.1 Determine the probable direction of flow, if any, due to natural ventilation during the season indicated in each of the mine schematics shown in Fig. 8.10. Will flow have to be induced, and will it continue, once started?

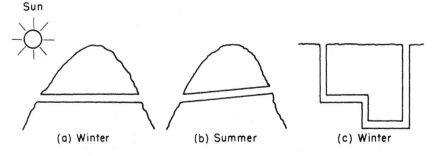

Sun

(a) Winter   (b) Summer   (c) Winter

FIGURE 8.10 Mine ventilation systems (see Problem 8.1).

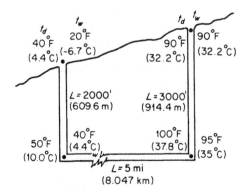

FIGURE 8.11 Mine temperature survey (see Problem 8.2).

8.2 A temperature survey of a mine is made to investigate the natural ventilation. The data obtained are indicated (Fig. 8.11). The friction factor is 100×10^{-10} lb·min^2/ft^4 (0.01855 kg/m^3), all airways are 10×10 ft (3.05×3.05 m), and the elevation of the collar of the higher shaft is 5000 ft (1524 m). Disregarding the shock losses and the velocity head, compute the natural ventilation head by Eqs. 8.2–8.6, and the quantity of flow and air power on the basis of the most reliable calculated head. Indicate the direction of flow and sketch the characteristic curves of the system.

8.3 The mine shown in Fig. 8.12 has warmer temperatures than the air outside.
   (a) Indicate the probable direction of natural ventilation. Does it have to be induced? Will it continue once started?
   (b) How could the natural pressure be measured by manometer? By altimeter, without closing off the flow?
   (c) If the natural pressure is found to be 2 in. water (498 Pa) and the resulting airflow is 40,000 cfm (18.88 m^3/s), plot the characteristic curves of the system (mine and natural ventilation, neglecting velocity head).

8.4 A mine consists of two shafts connected by a drift with the following dimensions and characteristics:

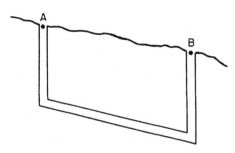

FIGURE 8.12 Mine ventilation system (see Problem 8.3).

	Size		$L + L_e$		K lb·min^2/ft^4	
Airway	ft	(m)	ft	(m)	× 10^{10}	(kg/m^3)
Shaft a	6 × 12	1.83 × 3.66	1080	329.2	65	0.01206
Drift b	7 × 9	2.13 × 2.74	1210	368.8	130	0.02412
Shaft c	5 × 8	1.52 × 2.44	715	217.9	80	0.01484

Direction of flow is a → b → c. The natural pressure is calculated to be 1.5 in. water (373 Pa). Using a graphic solution, plot the characteristic curves of the system and determine the quantity of flow. Consider velocity head.

8.5 A temperature survey is made of a mine that has one upcast and one downcast shaft, both 6000 ft (1829 m) deep. The underground workings are horizontal, and the pressure at the downcast shaft collar was measured as 24.23 in. Hg (81.82 kPa). Dry-bulb temperatures were recorded as follows:

Top of downcast shaft	42°F (5.5°C)
Bottom of downcast shaft	66°F (18.9°C)
Bottom of upcast shaft	84°F (28.9°C)
Top of upcast shaft	72°F (22.2°C)

Calculate the natural ventilation head using Eqs. 8.2 and 8.7.

CHAPTER 9

Air-Moving Equipment

9.1 HISTORY

Miners have traditionally demonstrated ingenuity in the performance of their work underground. Ventilation has been no exception in this regard. Dating back to Agricola's time (sixteenth century) and before, miners attempted to devise ways of harnessing natural forces to provide airflow and later substituting human- and animal-powered devices to induce air movement.

Agricola (1556) describes three early methods used to force air into a mine: deflectors, fans, and bellows (see Fig. 1.1). Deflectors mounted on shafts caught the wind and attempted to divert it into the mine. Crude paddle wheels, sometimes enclosed (the forerunner of the centrifugal fan), and bellows were operated by human, animal, wind, or water power and were usually connected by wooden pipe to underground workings. Natural forces other than wind and thermal energy (natural ventilation) that were utilized were falling water in the downcast shaft and fire in the upcast shaft.

Mine ventilation continued without substantial change for the next two centuries, although furnaces and steam jets were introduced in Great Britain as exhaust aids in upcast shafts. It was the Industrial Revolution at the close of the eighteenth century, bringing mechanization to industry in general and mining in particular, that revolutionized mine ventilation. Just as the earliest application of the Watt steam engine was to drive a mine pump, so other efforts were soon made to operate mine fans by steam. The first successful mechanical ventilation system was reportedly installed in 1807 at the Hebburn colliery in England (Schweisheimer, 1952).

Fan development proceeded rapidly after that date. The Struve air pump, capable of handling up to 10,000 cfm (4.72 m³/s), was introduced in Wales. Mechanically powered centrifugal fans were built and operated by Nasmyth and Brunton in the early 1800s, but to Guibal belongs the credit for the first centrifugal fan with a spiral casing (Dilworth, 1939).

In a much improved form, the centrifugal fan continues in use for mine ventilation today. However, it has been largely supplanted for primary ventilation by the more compact, efficient, axial-flow fan. The disk fan, the first axial-flow type, was introduced in mining early in this century, although present-day axial-flow fans with aerodynamically improved airfoil blades were a development of the 1930s. While most fans installed today are of the axial-flow type, centrifugal fans are still in operation.

9.2 CLASSIFICATION OF AIR-MOVING EQUIPMENT

Mechanical ventilation or *air-moving devices* include all powered machines used to induce airflow in mine openings or ducts. Fans are the most important and most common of these machines, but compressors and injectors also have application to ventilation.

Basically, a *fan* is an air pump, a machine that creates a pressure difference in a duct or airway and thus causes airflow. In a continuous process, the air pump or pressure source receives air at some intake pressure and discharges it at a higher pressure (head). The fan is a converter of energy (from mechanical to fluid), supplying pressure to overcome the head losses in airflow, and as such, must be considered as an energy source in the Bernoulli equation for fluid flow (Section 5.1). The prime mover may be an electric motor, internal-combustion engine, or compressed-air turbine.

Compressors for ventilation use may be thought of as high-pressure fans (operating pressures are in excess of 1 psi or 6.9 kPa), since they also perform as air pumps in the ventilation system. *Injectors* utilize the kinetic energy of compressed air to entrain ambient air, imparting to it mainly kinetic energy.

A classification of mechanical ventilation equipment is given in Table 9.1. There are two major categories of fans, namely, radial flow (centrifugal) and axial flow. In *centrifugal fans,* the air is drawn into a rotating impeller and discharged radially into an expanding scroll casing. Multiblade fans have impellers that resemble somewhat the rotor of a squirrel-cage induction motor, consisting of many small, curved blades. Fans in this class are further identified on the basis of blade curvature.

TABLE 9.1   Classification of Mechanical Ventilation Equipment Used in Mines

Fans	Centrifugal	Radial tip
		Forward curved
		Backward curved
	Axial flow	Tube-axial
		Vane-axial
Compressors	Centrifugal	
	Axial flow	
	Positive displacement	Rotary
		Axial flow
Injectors	Cylindrical	
	Conical	
	Venturi	

Axial-flow fans are classified in two types, both consisting of an impeller in a cylindrical housing mounting disk- or airfoil-shaped blades to cause airflow through the fan in an axial direction. The difference in the two types is in the provision of stationary guide vanes in the vane-axial type to straighten the airflow and recover some rotative energy imparted to the air by the blade motion. Axial-flow mine fans up to 240 in. (6.1 m) in diameter and 5000 hp (3730 kW) are in use.

Centrifugal and *axial-flow compressors* resemble fans of the corresponding classes. They handle smaller volumes of air at much higher pressure than fans. *Positive blowers* have two rotating impellers that mesh in such a way as to cause displacement of a nearly fixed volume of air (in all other mechanical ventilation devices, the volume of air discharged varies with the pressure).

Injectors direct a jet of compressed air into the open end of a short pipe section or ventilation duct, entraining the ambient air and inducing an airflow. The name and characteristics of the injector depend on the shape of the pipe or duct inlet.

It may be concluded from this discussion that there are three essential components to any mechanical ventilation device: (1) power source or mechanical drive, (2) impeller or jet, and (3) housing or casing. In the case of a fan (or compressor), these components are assembled in a conventional rotative-machine design, such as that employed in a pump.

9.3   CENTRIFUGAL FANS

As air pumps, all fans operate on surprisingly similar theory. The rotating wheel, rotor, or impeller does work on the air, imparting to it both static and kinetic energy in varying proportions depending on the fan type. It is a matter of conversion of rotative dynamic energy to radial and/or axial-flow dynamic energy.

Mining engineers have little use for a knowledge of fan theory from the standpoint of designing equipment. But they should have an acquaintance with theory and various design factors in order to understand fan performance and to permit the correct selection of fans for mine ventilation purposes. In discussing fan theory and design, it is desirable to assume that compression effects on the air are negligible. This assumption is generally made in considering fan performance and application as well.

Two separate and independent actions produce pressure (head) in a centrifugal fan: centrifugal force due to the rotation of the air, and kinetic energy imparted as the air leaves the tip of the impeller blades. The amount of kinetic energy developed depends mainly on the tangential velocity (tip speed) of the blades V_t, whereas the centrifugal (static) energy is a function of the increase in the tangential velocity of the air V_a entering and leaving the impeller. The work done on the air by the fan is proportional to the tangential velocity, no work being required to produce the radial velocity.

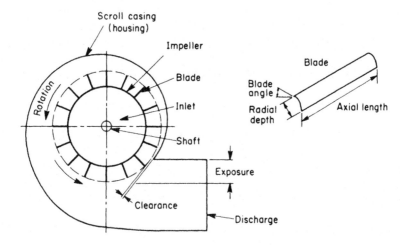

FIGURE 9.1 Construction of a centrifugal fan.

Design Variables of Centrifugal Fans

The distinctive operating characteristics and performance of a centrifugal fan are mainly functions of the effect of design variables on the magnitude and relative proportion of kinetic and static energy developed by the fan. Reference to Fig. 9.1 is helpful in understanding terminology.

Blade Curvature This is the most important variable in fan design. The vector relations of the various components of air velocity for the three principal shapes of blade—radial-tip, forward-curved, and backward-curved—are shown in Fig. 9.2, in which V is absolute velocity of the air, V_t is tangential velocity of the blade, V_a is tangential velocity of the air, V_r is radial velocity of the air, U is velocity of the air relative to the blade, and θ is blade angle in degrees.

Consider a fan with a straight blade, one parallel to the impeller radius r, such as a radial-tip multiblade fan. The work imparted by the rotating im-

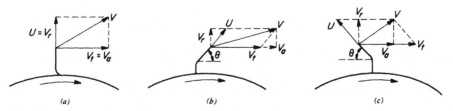

FIGURE 9.2 Velocity relations for multiblade centrifugal fans of different blade shapes: (*a*) radial tip; (*b*) forward-curved; (*c*) backward-curved. For nomenclature, see Section 9.3.

peller to the air per unit time is equal to torque (the change in momentum of the air, or mass times velocity) times angular velocity of the impeller ω in radians per unit time. Since this work imparts kinetic energy to the air, the expression for specific energy of the air, or work per unit weight, may be written

$$\text{Kinetic energy} = \frac{\omega}{g}(r_2 V_{a_2} - r_1 V_{a_1}) = \frac{1}{g}(V_{t_2} V_{a_2} - V_{t_1} V_{a_1})$$

where subscripts 1 and 2 refer to the positions at entrance and exit of the impeller, respectively. If radial entry is assumed, that is, $V_{a_1} = 0$, it can be reduced to

$$\text{Kinetic energy} = \frac{V_t V_a}{g} \tag{9.1}$$

In a perfect machine, this kinetic energy will be recoverable entirely as potential energy, equal to the height of the column of air for a unit weight:

$$\text{Potential energy} = H' \tag{9.2}$$

Then an expression for the theoretical total head H_t developed by the fan can be obtained by equating Eqs. 9.1 and 9.2 and converting the height of the air column into in. water (Pa):

$$H_t = \frac{wH'}{5.2} = \frac{wV_t V_a}{5.2g} \tag{9.3}$$

$$H_t = \rho V_t V_a \tag{9.3a}$$

or

$$H_t = \frac{wV_t^2}{5.2g} \tag{9.4}$$

$$H_t = \rho V_t^2 \tag{9.4a}$$

where $w(\rho)$ is air specific weight in lb/ft^3 (kg/m^3). These relationships are true because $V_a = V_t$ for a radial-tip fan.

The forced-vortex theory states that this head is divided equally between static and velocity pressure; but since a fan operates as a combination of forced and free vortices, the division is actually not equal (Baumeister, 1935). A substantial conversion of velocity head to static head can be made in the scroll casing of the fan. Theoretically, if no flow losses occurred, the head H_t

would remain constant, regardless of the air quantity Q handled. In practice, however, H_t falls off rather rapidly as Q increases. The actual head will always be less than the theoretical head.

With forward- and backward-curved multiblade fans, composition of vectors is necessary to compute the tangential velocity of the air. Referring to Eq. 9.3 and the diagrams in Fig. 9.2, the theoretical total head produced is

Forward-curved:

$$H_t = \frac{wV_t(V_t + V_r \cot \theta)}{5.2g} \qquad (9.5)$$

$$H_t = \rho V_t(V_t + V_r \cot \theta) \qquad (9.5a)$$

Backward-curved:

$$H_t = \frac{wV_t(V_t - V_r \cot \theta)}{5.2g} \qquad (9.6)$$

$$H_t = \rho V_t(V_t - V_r \cot \theta) \qquad (9.6a)$$

where the tangential velocity of the air $V_a = V_t \pm V_r \cot \theta$. Note that the theoretical head is greatest for a forward-curved fan, and that all factors other than θ are constant. This means, however, that a larger percentage of the velocity head must be converted to static head in the casing to utilize the head effectively. A backward-curved fan must operate at a higher speed to produce the same pressure, but a greater proportion of it is static and hence in a more usable form.

The radial and tangential blade velocities in fps (m/s) can be readily computed for any fan:

$$V_t = \frac{\pi D n}{60} \qquad (9.7)$$

$$V_r = \frac{Q}{60A} \qquad (9.8)$$

$$V_r = \frac{Q}{A} \qquad (9.8a)$$

where n is fan speed in rpm, D is diameter of impeller plus blade in ft (m), Q is quantity of air delivered in cfm (m³/s), and A is area of the fan housing measured at the periphery of the impeller in ft² (m²). The following numerical example will demonstrate the use of these formulas.

Example 9.1   Calculate the theoretical head for a forward-curved fan under the following conditions:

Q = 25,000 cfm (11.80 m³/s) b = 2 ft (0.61 m), width

n = 300 rpm θ = 50°

D = 6 ft (1.83 m) w = 0.0750 lb/ft³ (1.20kg/m³)

Solution: Use Eqs. 9.7, 9.8, and 9.5:

$$V_t = \pi(6)(300)/(60) = 94.2 \text{ fps}$$

$$A = \pi(6)^2 = 28.3 \text{ ft}^2$$

$$V_r = \frac{25,000}{(60)(28.3)} = 14.7 \text{ fps}$$

$$H_t = (0.0750)(94.2)\frac{94.2 + (14.7)(0.839)}{(5.2)(32.2)}$$

$$= 4.50 \text{ in. water (1.12 kPa)}$$

In actual design, the blade angle varies from a maximum of 60° for forward-curved fans to 45° for backward-curved fans.

Number of Blades   Increasing the number of blades yields an increase in both the head and quantity of air discharged. However, an increase above a certain number increases the friction losses and decreases the area available for flow. This dictates a compromise. In practice, the following list of number of blades represents the usual range for each type of centrifugal fan (Madison, 1949):

Radial-tip	10–20
Forward-curved	32–66
Backward-curved	14–24

Inlet   The size, shape, and number of inlets to the fan impeller are varied, within limits. There is an optimum inlet diameter, based on minimizing the entrance losses at the inlet and the impact losses against the blades. Generally, a large inlet is provided in multiblade fans and a small one in steel-plate fans, since blade size and number are factors. In shape, the inlet is usually formed to reduce entrance losses. A bell-mouth construction is preferable, although a conical one is nearly as good, and a cylindrical one is common (but little improvement over a plain inlet). The use of a double inlet is effective in increasing the delivery of a fan, because the entrance velocity is halved and the inlet losses quartered. However, because of drive and motor connections to the impeller shaft, this is not always a practical consideration; but a double inlet should be employed if at all possible. When a fan is connected to a duct in the exhaust or booster position, it may not be feasible to provide two inlets.

Impeller Diameter   This dimension usually determines the size rating of similar fans. The diameter varies the quantity as the square but has no effect on head if the tip speed is kept constant (Section 9.8).

Fan Width   Centrifugal fans are referred to as single or double width, depending on the construction of the impeller. A double-width fan has two impellers mounted side by side and rigidly coupled on the same shaft. This has the same effect as operating fans in parallel and produces approximately twice the quantity of a single-width fan.

Scroll   The casing of a centrifugal fan is designed to ensure an efficient conversion of velocity to static head. It is constructed as a gradual expansion duct, and a volute curve is approximated in designing the scroll. In practice, several circular arcs are generally used.

Guide Vanes   Guide vanes at the intake of a fan and diffuser vanes in the discharge casing are sometimes employed in large centrifugal fans to reduce airflow losses. Their expense is justified when the increase in fan efficiency is sufficiently large. Movable inlet vanes, acting as shutters, provide a convenient means of varying fan output and are sometimes used in combination with variable-speed drives as a very economical method of controlling the output quantity.

9.4   AXIAL-FLOW FANS

The principal action of an axial-flow fan in producing pressure is to impart a tangential acceleration to the air as it passes through the impeller of the fan. Any centrifugal force generated is small and practically negligible when the fan operates at rated conditions.

The rotative energy must be converted into linear-flow energy and static head as the air leaves the impeller. This recovery is difficult to accomplish but essential for high efficiency. Guide vanes in the diffuser casing following the impeller are most effective in converting the rotative energy, and most large axial-flow fans are of the vane-axial type. Prerotation vanes at the fan intake have also been tried; they rotate the air in the direction opposite to that of the impeller, the net effect being that the air leaves the impeller with little rotation. Shock losses are high, however. Contrarotation of impellers in series is still another method of straightening out the airflow.

The total head imparted to the air by an axial-flow fan, theoretically, can be found from the following equation (Morris and Hinsley, 1951):

$$H_t = \frac{wV_t(V_1 - V_2)}{5.2\ g} \tag{9.9}$$

$$H_t = \rho V_t(V_1 - V_2) \tag{9.9a}$$

where V_t is the tangential velocity of the blade and $(V_1 - V_2)$ is the change in the rotational component of the absolute velocity (whirl velocity). The value of $(V_1 - V_2)$ varies along a radius. The theoretical head obtained by Eq. 9.9 will be the same in each portion of fluid only if $(V_1 - V_2)$ is inversely proportional to the radius.

Aerodynamic Stall of Axial-Flow Fans

Axial-flow fans exhibit a more or less pronounced trough in the output head–quantity relation. This contraflexure point is called *aerodynamic stall*. In the stall condition, axial-flow-fan operation is very unstable. There is considerable throb and pulsation in the airflow, and in rare cases, fans have been damaged by the fluctuating operating conditions. Vibration and noise increases are customary. Aerodynamic stall is caused by flow separation at the boundary of the blade (Wallis, 1961). Normal operation of an axial-flow fan takes place in the range beyond the stall point. The possibility of experiencing a stall increases if a fan is oversized, if the mine resistance increases significantly, or if the fan is operating improperly. To prevent the occurrence of stall, a monitoring system for the head and flow quantity can be employed. When the operating point of the fan approaches the stall zone, an alarm is sounded and the blade angle is automatically reduced (Anon., 1978). Forward-curved centrifugal fans also exhibit aerodynamic stall, but not as pronouncedly.

Design Variables of Axial-Flow Fans

As with the centrifugal fan, there are design variables that affect the performance of the axial-flow fan. Many of these are identified in Fig. 9.3. A large axial-flow fan employed for mine ventilation is shown in Fig. 9.4.

Angle or Pitch of Blades   Modern axial-flow fans are usually provided with variable-pitch blades. The angle of attack of the airfoil can be adjusted to attain maximum efficiency at the quantity of air desired. Some axial-flow fans are provided with a special feature that allows adjustment of the blade angle without stopping the fan. The quantity of air discharged varies directly with the pitch, and the head varies a lesser amount (Section 9.8). This is the

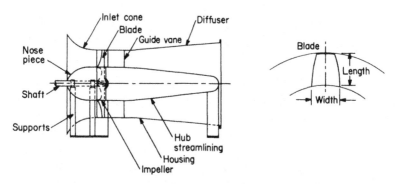

FIGURE 9.3 Construction of an axial-flow fan. (After Madison, 1949. By permission from Buffalo Forge Co., Buffalo, NY.)

most important design variable of an axial-flow fan, because it provides a high degree of flexibility under changing system needs. Operation at a low-pitch setting is not as efficient but can lessen or eliminate aerodynamic stall in an axial-flow fan.

Number of Blades   The number of blades in commercial axial-flow fans varies from 2 to 24 (Madison, 1949). In large mine fans, 6 to 18 blades are

FIGURE 9.4 Axial-flow fans, model 8HU117, dual installation. (By permission from Jeffrey Mining Machinery Division, Dresser Industries, Inc., Columbus. OH).

used. The fan head varies with the number of blades, but above an optimum number, airflow and noise characteristics are adversely affected (Morris and Hinsley, 1951).

Number of Stages   Higher heads can be obtained from impellers in series. Multistage fans are essentially fans in series, and the pressure produced varies as the number of stages. One and sometimes two stages are employed in commercial axial-flow fans.

Ratio of Hub to Impeller Diameter   This ratio effects mainly pressure characteristics. For many blades or high-pressure work, the hub to impeller ratio is large, often 80% (Madison, 1949).

Housing   The inlet and discharge of a main fan are carefully designed to minimize shock loss. A diffuser is employed to aid in the conversion of velocity to static head. Streamlining of the hub is a further efficiency measure.

9.5   FAN CHARACTERISTICS

Departure of Performance from Theoretical

As mentioned earlier, the theoretical relation between head and quantity produced by a fan is a straight line. For example, centrifugal fans having no mechanical losses and handling ideal gases would exhibit the characteristics shown in Fig. 9.5a. These are *fan characteristic curves,* graphic plots of head versus quantity. The curves in Fig. 9.5a are theoretical, based on Eqs. 9.4, 9.5, and 9.6. As would be expected, the curve of the radial-tip fan yields

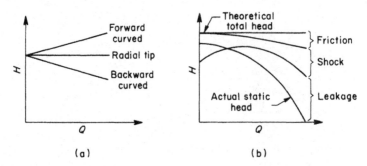

FIGURE 9.5   Departure of centrifugal-fan performance from the theoretical: (*a*) theoretical characteristic curves based on Eqs. 9.4–9.6; (*b*) comparison of theoretical and actual characteristic curves of a radial–tip fan, indicating sources of loss. (After Baumeister, 1935.)

a horizontal relation (constant H_t with increasing Q), whereas those of the forward- and backward-curved fans rise and fall, respectively. The explanation is afforded by the theoretical-pressure (head) equations given earlier. A radial-tip fan depends only on tangential velocity (constant regardless of quantity) in creating pressure, whereas a curved-blade fan depends on a radial velocity (variable with quantity) as well. These curves are plotted for constant speed.

Compare the theoretical with the actual H–Q performance curve of a radial-tip centrifugal fan shown in Fig. 9.5b. Fluid losses account for most of the disparity and leakage for the remainder. The effect of leakage is to shift the vertical axis to the right, or the curve to the left; the shift is determined by the quantity of leakage. Other fans deviate from their theoretical curves in a similar manner and for the same reasons.

Definition of Performance Characteristics

The most important measures of fan performance, the head and quantity delivered, have already been mentioned. Two heads are employed, however—*fan static head* and *fan total head*—and care must be taken to distinguish between them. The static head is more commonly used, but, particularly in large fans, the velocity and total heads must be considered, too. The fan total head is the rise in total head from the fan inlet to the fan outlet, and it is also equal to the fan static head plus the velocity head at the fan outlet:

$$\text{Fan } H_t = \text{outlet } H_t - \text{inlet } H_t \qquad (9.10)$$

Also

$$\text{Fan } H_t = \text{fan } H_s + \text{fan } H_v \qquad (9.11)$$

Solving for fan H_s,

$$\text{Fan } H_s = \text{fan } H_t - \text{fan } H_v \qquad (9.12)$$

These definitions of fan heads are applicable regardless of whether a fan is installed as a blower, booster, or exhauster (Section 10.2).

The air power P_a corresponding to the quantity of flow at any given head can be readily computed from Eq. 5.2 If it is based on fan H_s, it is termed the *static air power P_{as}*, and if based on fan H_t, the *total air power*. The P_a is actually a measure of the output power delivered by the fan to the air. The input mechanical power to drive the fan is termed the *brake power P_m*. It must be measured by test at the same time as the head and quantity are determined. All power is measured in units of hp (kW).

The ratio of the output and input powers is termed the mechanical efficiency, or simply *efficiency* η, of the fan and expressed as a percentage:

$$\eta = \left(\frac{P_a}{P_m}\right) \times 100 \qquad (9.13)$$

If the air power is static, the efficiency is termed the *static efficiency* η_s, and if the air power is total, the efficiency is termed the *total efficiency*. Static air power and efficiencies are commonly employed in comparing fan performances, because the fan static head is more frequently used than the fan total head. They are, in a strict sense, meaningless terms and must therefore be used with caution. Static efficiency is considered a safer rating to use when there is doubt regarding recovery of velocity head. Total powers and efficiencies are really more significant for comparison purposes, however.

The ratio of total air power P_a to the input power P_i supplied to the motor driving the fan is termed the *overall efficiency* η_o, which can be represented as a product of fan total efficiency η, drive efficiency η_d, and motor efficiency η_m, all expressed as decimals. Thus

$$\eta_o = \frac{P_a}{P_i} = \eta\eta_d\eta_m \qquad (9.14)$$

The overall efficiency has no significance in the theoretical discussion of fans, but it should be taken into account in the economical evaluation of a ventilation system involving fans.

Fan Characteristic Curves

The performance of fans can be expressed by data tables or graphs. By far the most useful and informative are graphs, which plot characteristic curves of fan performance in terms of the principal operating variables: head, quantity, power, and efficiency.

The customary form of these curves plots all these variables against quantity. Included may by any or all of the following: static head, total head, brake power, static efficiency, and/or total efficiency. They appear as shown in Fig. 9.6. The most valuable (and essential for the solution of ventilation problems) is the head-quantity characteristic. Curves are plotted for a given fan diameter, constant speed, and standard air specific weight. Characteristic curves can also be plotted in other forms relating other variables, including fan speed, diameter, and sound-power level. Calculations involved in plotting complete curves, such as those shown in Fig. 9.6, are illustrated in the following example.

Example 9.2 Given an axial-flow fan operating under the following conditions:

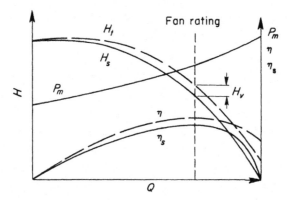

FIGURE 9.6 Characteristic curves of a fan. The fan rating is located through the peak of the total efficiency curve.

H_s = 3.84 in. water (956 Pa) D = 6 ft (1.83 m)

Q = 140,000 cfm (66.07 m³/s) n = 1750 rpm

P_m = 140 bhp (104.4 kW) w = 0.0750 lb/ft³ (1.20 kg/m³)

Calculate fan H_t, η_s, and η.

Solution: Find h_v and P_a, then employ Eqs. 9.11 and 9.13 to obtain H_t and η:

$$A = \frac{\pi}{4} (6)^2 = 28.3 \text{ ft}^2$$

$$V = \frac{140,000}{28.3} = 4950 \text{ fpm}$$

$$H_v = 0.075 \left(\frac{4950}{1098}\right)^2 = 1.52 \text{ in. water (0.38 kPa)}$$

$$H_t = 3.84 + 1.52 = 5.36 \text{ in. water (1.33 kPa)}$$

$$P_{as} = \frac{(3.84)(140,000)}{6346} = 84.8 \text{ hp (63.2 kW)}$$

$$P_a = \frac{(5.36)(140,000)}{6346} = 118.2 \text{ hp (88.1 kW)}$$

$$\eta_s = \left(\frac{84.8}{140}\right)(100) = 60.6\%$$

$$\eta = \left(\frac{118.2}{140}\right)(100) = 84.5\%$$

Since four or more operating points are determined in plotting characteristic curves, all calculations should be tabulated for brevity.

The *fan rating* is the head, quantity, power, and efficiency to be expected when a fan is operating at peak efficiency. It may correspond to the highest point on the static- or total-efficiency curves. The fan rating based on total efficiency is shown in Fig. 9.6. A fan is ordinarily recommended by the fan manufacturer for duty at a point on its characteristic near the rating.

9.6 FAN NOISE

Sound is any source that produces a pressure wave in the audible range, and *noise* is unwanted sound. The sound generated by a mine fan is undesirable and therefore regarded as noise. Fan noise has two major components: (1) a broad band, caused by air turbulence (vortex shedding), and (2) the blade passage (discrete tone with harmonics), due to rotation of the wheel with a discrete number of blades (Smith et al., 1974). Sound levels produced are affected not only by the fan's operational parameters (speed, head, quantity, and efficiency) but also by the geometry of the ductwork or airway in which the fan is installed. Although it may not always be practical in mining applications, fan noise can be reduced considerably by installing silencers at the fan inlet and/or outlet, placing a vibration break in the fan-duct connection, and by lagging the fan and ducting with sound-absorbing materials. The principles involved in fan noise attenuation are illustrated in Fig. 9.7.

Sound-Power Level

Sound-power level is the sound power, which is the total acoustical-power output of a sound source, expressed in decibels (dB) with a reference power of 10^{-12} watts (W). Mathematically, the sound-power level is defined as

$$L_w = 10 \log \left(\frac{P_w}{P_{w_r}} \right) \tag{9.15}$$

where L_w is sound-power level in dB, P_w is sound power in W, and P_{w_r} is reference sound power in W. For example, the sound-power levels for 10^{-7} W, 10^7 W, and 2×10^7 W sound powers are 50, 190, and 193 dB, respectively. Note that the sound-power level increases 3 dB with doubling of the sound power, since $10 \log 2 = 3$. *Sound-pressure level,* similar to the sound-power level, is expressed in dB with a reference pressure of 2×10^{-4} microbars or dyn/cm^2 (2×10^{-5} Pa):

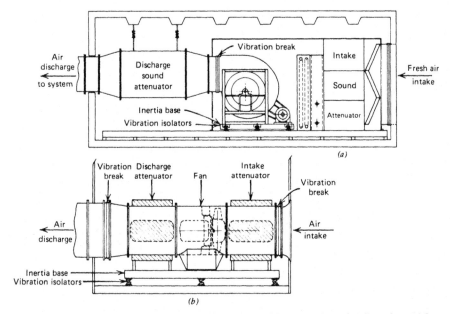

FIGURE 9.7 Sound attenuation for (*a*) centrifugal fan and (*b*) axial-flow fan. (After Graham, 1975. Reprinted with permission of the American Society of Heating, Refrigerating & Air Conditioning Engineers, Inc., Atlanta, GA.)

$$L_p = 10 \log \left(\frac{p}{p_r}\right)^2 \tag{9.16}$$

or

$$L_p = 20 \log \left(\frac{p}{p_r}\right) \tag{9.17}$$

where L_p is sound-pressure level in dB, p is root-mean-square (RMS) sound pressure in microbars, and p_r is reference sound pressure in microbars.

For a given source, sound power is independent of the environment, whereas sound pressure is not. Therefore, the sound rating of fans is based on the sound-power level rather than the sound-pressure level. Fan manufacturers generally provide the sound-power levels in the standard eight *octave bands*. The mean and limit frequencies for the octave bands are listed in Table 9.2. Note that the upper limit of frequency is twice the lower limit in each band. The *mean frequency,* also called *center frequency,* is the geometric mean of the lower and upper frequencies.

The total sound power of a combination of sounds is simply equal to the

TABLE 9.2   Octave-Band Standards

| | Frequency, Hz | | |
Band Number	From	To	Mean
1	45	90	63
2	90	180	125
3	180	355	250
4	355	710	500
5	710	1400	1000
6	1400	2800	2000
7	2800	5600	4000
8	5600	11200	8000

sum of the individual sound powers. But the total sound-power level is not the sum of the individual sound-power levels. The total sound-power level L_{w_t} in dB is given by

$$L_{w_t} = 10 \log \sum_{i=1}^{N} \log^{-1} \frac{L_{w_i}}{10} \tag{9.18}$$

where L_{w_i} is sound-power level for sound i in dB, and N is number of sounds. Applying Eq. 9.18, the sound-power levels given in all eight octave bands can be reduced to a single number.

Noise Characteristics of Fans

The sound-power-level curves for a backward-curved centrifugal fan and a vane-axial fan are shown in Fig. 9.8 together with other performance characteristics. The overall shape of the curve for the vane-axial is slightly different from that for the centrifugal fan. Also shown in Fig. 9.8 is the *specific sound-power level,* which is defined as the sound-power level that would be generated by a fan of specific size (Eq. 9.24) operating at its specific speed (Eq. 9.23) with a unit head and a unit quantity (Section 9.8). Note that the sound-power levels for both fans have a minimum at the peak efficiency. The spectrum for a centrifugal fan may be approximated in most cases by subtracting 9, 7, 5, 7, 10, 13, 20, and 29 dB from the overall level to obtain the sound-power levels in the first through eighth octave bands, respectively, provided the blade-passing frequency falls in the third band. For a high-pressure vane-axial fan, subtract 18, 15, 8, 5, 5, 8, 18, and 27 dB, respectively (Jorgensen, 1970).

9.7   COMPARISON OF FAN PERFORMANCE

A summary of the distinguishing characteristics of the principal types of fans used in mine ventilation is presented it Table 9.3. Comparisons are made on

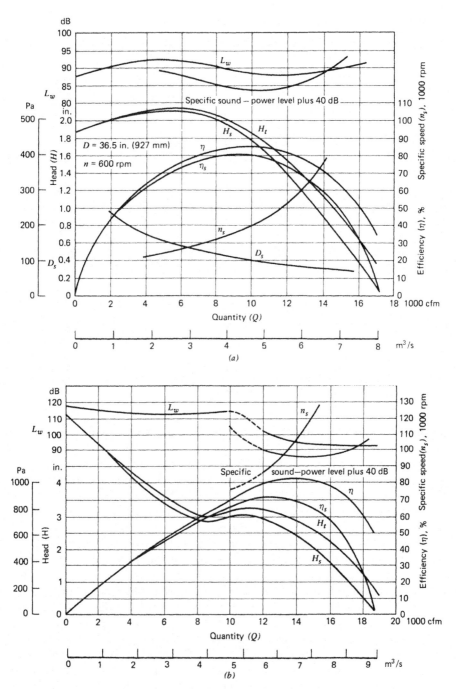

FIGURE 9.8 Typical characteristic curves for (*a*) backward-curved fan and (*b*) vane-axial fan. (After Jorgenson, 1970. By permission from Buffalo Forge Co., Buffalo, NY.)

TABLE 9.3   Comparison of Fan Performance

Fan	Head	Quantity	Power	Efficiency	Speed	Remarks
Centrifugal						
Forward-curved[a]	Medium	Highest	High	Medium (60–70%)	Medium	Inflection point, quiet, bulky, minimum variance in H (constant pressure)
Radial tip, reverse curve	High	Medium	Medium	Medium (50–60%)	Medium	Compromise characteristics
Backward-curved[a]	Highest	Medium	Medium, falling	High (60–80%)	High	Nonoverloading, quietest, minimum variance in Q (constant quantity)
Axial flow						
Tube-axial	Lowest	Medium	Low, falling	Medium (50–60%)	Highest	Small fans, auxiliary blowers, pronounced inflection point
Vane-axial[a]	Low	High	Low, falling	Highest (70–85%)	Highest	Noisy, pronounced inflection point, hard to overload, suited for direct electric-motor drive, flexible

[a] Fans in most common use for mine ventilation.

a relative basis, since the wide range of absolute values within any given class obviates their use.

The most detailed information regarding the performance of a fan can be gained from a study of its characteristic curves. A summary comparison of these curves is presented in Fig. 9.9, in which characteristics are plotted for the same fan rating. The characteristics are plotted on a percentage scale based on rated Q, H_s, and P_m as 100%.

Examination of Fig. 9.9 reveals many interesting points. For example, the static-head curves indicate why the backward-curved centrifugal fan is often termed a "constant quantity" fan and the forward-curved a "constant pressure" fan. Within limits, their curves do approach linearity in the operating range. The power curve of the backward-curved fan demonstrates the nonoverloading characteristic of this fan: Neither a short circuit nor a blocked discharge can damage the fan motor. A falling power characteristic at free delivery also distinguishes the vane-axial fan. The inflection points in the characteristic curves of the forward-curved and vane-axial fans show the regions in which aerodynamic stall occurs. Note that the stall point is more serious in the vane-axial fan because it shifts the operating range farther to the right.

The preference for one type of fan over another for a particular mine

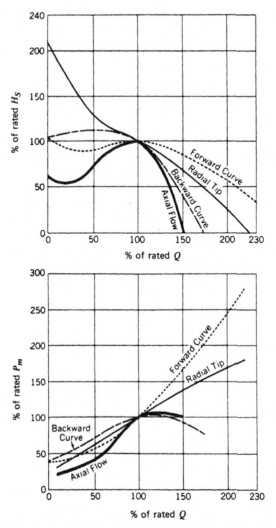

FIGURE 9.9 Comparative characteristic curve for principal types of fans. (After Anon., 1978. By permission from Babcock & Wilcox Co., Barberton, OH.)

ventilation task is largely a matter of the individual requirements in each case. In addition to head and quantity requirements, consideration should be given to fan efficiency, power, speed, cost, flexibility, size, noise, and reliability. In general, centrifugal fans are more serviceable and quiet, but axial-flow fans are cheaper, more compact, and more flexible. Both attain high efficiencies. Vane-axial fans undoubtedly have the edge over all other types for main-fan duty. Forward and backward-curved fans are the most popular of the centrifugal class for mine use. As indicated in Table 9.3, the higher head is generated with the centrifugal fan.

Because of the variability encountered in the characteristics of fans of different types and manufacturers, selection of fans for main ventilation service should always be based on characteristic curves (Section 10.6). These can be obtained from the manufacturers on request.

9.8 FAN LAWS

The behavior of a fan under changing head-quantity conditions is predictable from its characteristic curves. However, there are certain variables other than the flow and resistance of the system that exert a considerable effect on fan performance (as measured by fan head, quantity, power, and efficiency). These variables are fan rotational speed n, fan size (diameter) D, air specific weight w, and, in the case of axial-flow fans, the blade pitch. By dimensional analysis it can be demonstrated that for a series of geometrically similar fans (homologous series), and for a particular point of operation on the head-quantity characteristic, the following three relationships are valid (Berry, 1963; Osborne, 1977):

Quantity:

$$Q \propto nD^3, \quad Q = k_q nD^3, \quad \text{or} \quad Q_2 = \frac{n_2}{n_1} \left(\frac{D_2}{D_1}\right)^3 Q_1 \qquad (9.19)$$

Head:

$$H \propto n^2 D^2 w, \quad H = k_h n^2 D^2 w, \quad \text{or} \quad H_2 = \left(\frac{n_2}{n_1}\right)^2 \left(\frac{D_2}{D_1}\right)^2 \frac{w_2}{w_1} H_1 \qquad (9.20)$$

Power:

$$P_m \propto n^3 D^5 w, \quad P_m = k_p n^3 D^5 w, \quad \text{or} \quad P_{m_2} = \left(\frac{n_2}{n_1}\right)^3 \left(\frac{D_2}{D_1}\right)^5 \frac{w_2}{w_1} P_{m_1} \qquad (9.21)$$

where k_q, k_h, and k_p are numerical constants; and subscripts 1 and 2 refer to conditions 1 and 2, respectively. Because they are not significant in practice, the effects of the Reynolds number are not included in these equations. In Eq. 9.20, either H_t or H_s may be used; in Eq. 9.21, either P_a or P_m may be used. Finally, applying Eqs. 9.19 and 9.20 and the basic Atkinson head–quantity relation, we have

Resistance:

$$R = k_r wD^{-4}, \quad R \propto wD^{-4}, \quad \text{or} \quad R_2 = \frac{w_2}{w_1} \left(\frac{D_2}{D_1}\right)^{-4} R_1 \qquad (9.22)$$

where $k_r = k_h/k_q^2$ is a constant. This implies that the associated duct or mine resistance is assumed to vary with D^{-4} and w.

These relationships, or combinations of them, are useful in expressing or comparing fan performances. For example, the expression for the *specific speed* n_s, which is defined as the speed at which a fan of specific size would run to develop a unit head at a unit quantity, can be obtained by eliminating D from Eqs. 9.19 and 9.20:

$$n_s = \frac{nQ^{1/2}}{H^{3/4}}$$ (9.23)

with w constant. Similarly, by eliminating n, we obtain the expression for the *specific size* or diameter D_s:

$$D_s = \frac{DH^{1/4}}{Q^{1/2}}$$ (9.24)

with w constant. In addition to Eqs 9.23 and 9.24, the following expressions are also employed by some manufacturers to facilitate fan selection:

Specific speed n_s

$$n_s = \frac{Dn}{H^{1/2}}$$ (9.25)

Specific volume Q_s

$$Q_s = \frac{Q}{D^2 H^{1/2}}$$ (9.26)

with w constant in both expressions.

In the use as well as the selection of fans, it is desirable to be able to predict the performance (H, Q, P_m, and η) under new operating conditions of speed, diameter, and density. The so-called *fan laws*, based on Eqs. 9.19–9.21, express mathematically the relationship of these independent and dependent variables. If the fan characteristic curves are known for a given set of conditions, they may be determined quite simply for any new conditions by these relations. The fan laws apply to *all* types of fans, regardless of location with respect to the system (blower, exhaust, or booster). They are summarized in Table 9.4. Note that fan efficiency remains constant, since both air and brake power vary proportionally, and Dn, which is kept constant in law 2, is equivalent to blade-tip speed V_t.

Speed Change

To enable a fan to keep pace with changing resistance conditions in a mine, it may be necessary to change the operating characteristics of the fan. The

TABLE 9.4   Fan Laws

Variance in Performance Characteristics	Law 1, with Speed Change, n (D and w constant)	Law 2, with Size Change, D (w and Dn constant)	Law 3, with Specific Weight Change, w (n and D constant)
Quantity, Q	Directly	As square	Constant
Head, H_s or H_t	As square	Constant	Directly
Power, P_a or P_m	As cube	As square	Directly
Efficiency, η	Constant	Constant	Constant

condition that can be varied the most easily in centrifugal fans is rotational speed. Both centrifugal and axial-flow fans can be controlled in this way, although varying the pitch is the preferred method of control with a vane-axial fan. Because a speed change causes a cubic variation in the power, it is ordinarily not practical to increase fan speed by more than a few percent without providing a more powerful prime mover. Rotational speed is also limited by blade-tip speed, for structural reasons. The following example illustrates the application of law 1.

Example 9.3   Given the fan characteristic curves of Fig. 9.10 at present speed $n_1 = 800$ rpm, plot the characteristic at new speed $n_2 = 1600$ rpm.

Solution: Use the following relationships, based on the first fan law, selecting four or more points on each of the original curves for conversion to the new condition.

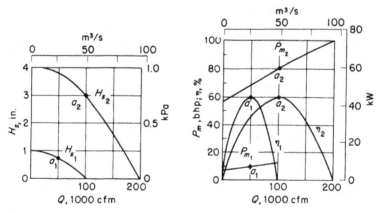

FIGURE 9.10   Fan characteristics at new speed (Example 9.3). Condition 1: $n_1 = 800$ rpm; condition 2: $n_2 = 1600$ rpm.

(a) $Q_2 = \dfrac{n_2}{n_1} Q_1$

(b) $H_{s_2} = \left(\dfrac{n_2}{n_1}\right)^2 H_{s_1}$ (9.27)

(c) $P_{m_2} = \left(\dfrac{n_2}{n_1}\right)^3 P_{m_1}$

(d) $\eta_2 = \eta_1$

For example, at point a_1, $Q_1 = 50{,}000$ cfm (23.60 m³/s), $H_{s_1} = 0.75$ in. water (187 Pa), $P_{m_1} = 10$ bhp (7.5 kW), and $\eta_1 = 59\%$.
   At point a_2

$$Q_2 = (1600/800)(50{,}000) = 100{,}000 \text{ cfm } (47.20 \text{ m}^3/\text{s})$$

$$H_{s_2} = (1600/800)^2(0.75) = 3.0 \text{ in. water } (747 \text{ Pa})$$

$$P_{m_2} = (1600/800)^3(10) = 80 \text{ bhp } (59.7 \text{ kW})$$

$$\eta_2 = 59\%$$

The curves for the new condition are plotted in Fig. 9.10.

Size Change

In predicting the performance of fans of various sizes, similar in all dimensions to a prototype, a manufacturer makes use of the second fan law, which relates the behavior of fans to the ratio of their diameters. Prospective purchasers also find this law useful in aiding them to select the correct size of fan for an intended duty, if they have curves for one fan of a series.

   As law 2 is stated in Table 9.4, a constant tip speed (tangential blade velocity V_t) is assumed for the fans under comparison. To maintain V_t constant, the rotational speed n must vary inversely with the fan size D. In applying the second fan law, it is necessary to compute a required rotational speed for the new fan, inversely proportional to the diameter (or any other fan dimension):

$$\frac{n_2}{n_1} = \frac{D_1}{D_2}$$ (9.28)

The effect of a size change is demonstrated by the following example.

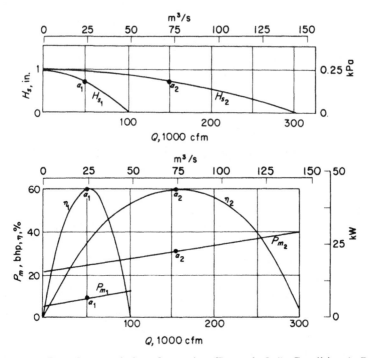

FIGURE 9.11   Fan characteristics of new size (Example 9.4). Condition 1: $D_1 = 48$ in. (1.22 m); condition 2: $D_2 = 84$ in. (2.13 m).

Example 9.4   Given the fan curves of Fig. 9.11 (the same as for Example 9.3) for size $D_1 = 48$ in. (1.22 m) at $n_1 = 800$ rpm, plot the characteristics for a similar fan of size $D_2 = 84$ in. (2.13 m), and find n_2 if the tip speed is constant.

Solution: As in Example 9.3, select four or more points on each of the original curves, converting them to the new condition by the following relations, based on the second fan law.

$$\text{(a)}\quad Q_2 = \left(\frac{D_2}{D_1}\right)^2 Q_1$$

$$\text{(b)}\quad H_{s_2} = H_{s_1} \qquad\qquad (9.29)$$

$$\text{(c)}\quad P_{m_2} = \left(\frac{D_2}{D_1}\right)^2 P_{m_1}$$

$$\text{(d)}\quad \eta_2 = \eta_1$$

Using point a_1 as in Example 9.3, calculate the conditions at point a_2:

$$Q_2 = \left(\frac{84}{48}\right)^2 (50{,}000) = 153{,}000 \text{ cfm } (72.21 \text{ m}^3/\text{s})$$

$$H_{s_2} = 0.75 \text{ in. water } (187 \text{ Pa})$$

$$P_{m_2} = \left(\frac{84}{48}\right)^2 (10) = 30.6 \text{ bhp } (22.8 \text{ kW})$$

$$\eta_2 = 59\%$$

To maintain the tip speed constant, the rotational speed will have to be adjusted. Applying Eq. 9.28,

$$n_2 = \left(\frac{48}{84}\right) (800) = 457 \text{ rpm}$$

Air-Specific-Weight Change

A change in air specific weight is frequently encountered in mine ventilation, making the third fan law the most commonly used law. Air-specific-weight change accounts for much of the discrepancy observed in performance when a fan is moved to a different elevation in the mine. It is often responsible for day-to-day fluctuations in fan characteristics. Specific weight change affects the characteristic curve as demonstrated in the following example.

Example 9.5   Given the fan curves of Fig. 9.12 (the same as for Example 9.3) for air specific weight $w_1 = 0.0500 \text{ lb/ft}^3 (0.80 \text{ kg/m}^3)$, plot the characteristics at new specific weight $w_2 = 0.0750 \text{ lb/ft}^3 (1.20 \text{ kg/m}^3)$.

Solution: Proceed as in Examples 9.3 and 9.4, employing the following relations, based on law 3:

$$\text{(a)} \quad Q_2 = Q_1$$

$$\text{(b)} \quad H_{s_2} = \frac{w_2}{w_1} H_{s_1} \qquad\qquad (9.30)$$

$$\text{(c)} \quad P_{m_2} = \frac{w_2}{w_1} P_{m_1}$$

$$\text{(d)} \quad \eta_2 = \eta_1$$

Conditions for point a_2:

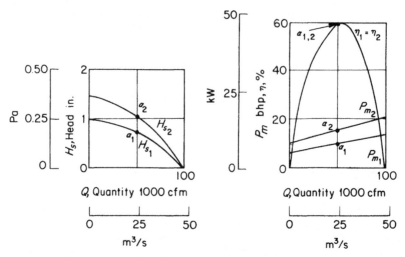

FIGURE 9.12   Fan characteristics for new air specific weight (Example 9.5). Condition 1: $w_1 = 0.0500$ lb/ft^3 (0.801 kg/m^3); condition 2: $w_2 = 0.0750$ lb/ft^3 (1.20 kg/m^3).

$$Q_2 = 50,000 \text{ cfm } (23.60 \text{ m}^3/\text{s})$$

$$H_{s_1} = (0.075/0.050)(0.75) = 1.12 \text{ in. water (279 Pa)}$$

$$P_{m_2} = (0.075/0.050)(10) = 15 \text{ bhp (11.2 kW)}$$

$$\eta_2 = 59\%$$

Combined Change

On occasion, particularly in fan testing (Section 9.9) or in a shift of operating points (Section 10.1), it is necessary to determine fan characteristics under completely different operating conditions, involving simultaneous changes in two or more of the variables—speed, size, and specific weight. Plotting new fan curves can be most easily carried out by using Eqs. 9.19–9.21.

Pitch Change in Axial-flow Fans

Changing the angle of pitch of the blades is the principal means employed in varying the performance of an axial-flow fan. Large mine fans of the vane-axial type are ordinarily variable-pitch fans; the angle of inclination of the blades can be changed by adjusting the pitch setting over a relatively narrow range (10–30°). Pitch settings are limited to safe operating conditions of the fan.

There is no mathematical relation by which the effect of a change in pitch on operating characteristics can be calculated, and reliance must be placed

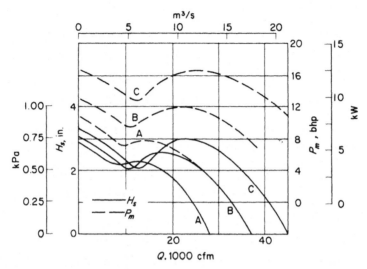

FIGURE 9.13 Typical characteristic curves of variable-pitch vane-axial fan, showing effect of pitch on fan performance. Blade setting: *A*, low; *B*, medium; *C*, high (12° range in pitch).

on experimental data. Tests conducted by Morris and Hinsley (1951) demonstrated that the quantity varies almost proportionally, and the head directly, to a fractional exponent with a change in pitch. Users of fans depend on characteristic curves supplied by the manufacturers to predict performance. Typical curves for a variable-pitch vane-axial fan are shown in Fig. 9.13. Head, quantity, and power at any operating condition can readily be determined.

9.9 FAN TESTING

Procedures for the laboratory testing of fans have been carefully developed over the years. The procedure that is accepted as standard in research and in the fan manufacturing industry is AMCA (Air Movement and Control Association, Inc.) Standard 210-85, *Laboratory Methods of Testing Fans for Rating* (Anon., 1985a).

In brief, standard testing is often carried out with the fan connected to a duct section equipped with flow straighteners and a symmetrical-type damper or regulator mounted at the discharge end. As the damper setting is varied over eight positions to modify duct resistance, readings of the velocity and total heads are taken with a pitot tube by equal-area traversing. A dynamometer measures power consumption during the test, and a tachometer records speed. Heads, quantity, power, and efficiency are calculated from these measurements for each operating condition, and the data are plotted as characteristic curves after conversion to standard air specific weight and rated speed by the fan laws. The result is an accurate determination of fan performance.

Unfortunately, the same precision-type laboratory testing cannot be carried out in the field with fans installed and operating in mine ventilation systems. The mining engineer may never have the need or opportunity to perform a standard fan test but occasionally has to conduct a fan test in the field. The usual purpose of field testing is to determine the operating condition of a fan preparatory to making a change in the system or to evaluate the efficiency of the fan. Objectives of such a test are the measurement of the fan quantity, heads, brake power, and efficiency at the operating condition.

The test procedure must be modified to fit the field conditions. A nonoperating day may have to be selected if the fan cannot be closed down temporarily during working time. Since most fans are driven electrically, the power input to the fan is easily determined by measuring the electrical input to the motor and multiplying this by the motor-drive efficiency to obtain fan power input. Shutting down the fan briefly will allow a wattmeter to be connected in the motor circuit. Measurement of speed is made by a tachometer. Fan heads, velocity, and quantity (Section 9.5 and 9.5.2) are determined by pitot tube and manometer measurements and a uniform cross section in the fan duct(s), and the air power is calculated from these data. The fan efficiency can then be calculated from the power input and output.

9.10   OTHER AIR-MOVING EQUIPMENT

Compressors and *injectors* are mechanical devices employed for ventilation in addition to fans. Their use in mines is restricted almost entirely to auxiliary ventilation. Compressed-air injectors (termed jets, bazookas, and ejectors in mines), either alone or attached to short lengths of ventilation pipe or tubing, find application particularly in metal mines for ventilating dead-end workings, whereas compressors are used with vent tubing to ventilate long tunnels and development headings offering a high resistance.

Only centrifugal, axial-flow, and positive-displacement compressors—those having applications for mine ventilation—are discussed here. These classes of compressors operate in the low-pressure (1–10 psig, or 6.9–68.9 kPa gage) and moderate- to high-quantity range (1000–50,000 cfm, or 0.47–23.60 m^3/s).

Centrifugal Compressors

For pressures above the usual range of fans [0.5–1 psi (3.4–6.9 kPa)], centrifugal compressors have been used. They deliver air at pressure head up to 3–4 psig (85–110 in. water) (2–27 kPa). Compression effects are no longer negligible in this range and must be considered in machine performance as well as in fluid transmission (Section 5.7 and Eq. 5.26).

If the performance (head and quantity) of a compressor is known in either the blowing or exhausting position, its performance when the location is reversed may readily be obtained (Weeks, 1926). This is constant at a given speed and inlet quantity.

Since the air passing through a compressor undergoes compression, the quantity at discharge is reduced in inverse proportion to the changes in absolute pressure. A compressor exhausting air from a working place thus effectively creates less air movement than one blowing air into a working place, although each delivers or handles the same quantity of air. An example illustrates pressure-quantity calculations with a compressor.

Example 9.6   Given a compressor installed as a blower

$$\text{Atmospheric pressure } p_b = 14 \text{ psi } (96.5 \text{ kPa})$$

$$\text{Discharge pressure } p_d = 3 \text{ psig } = 17 \text{ psia } (117.2 \text{ kPa})$$

$$\text{Inlet quantity } Q_i = 6000 \text{ cfm } (2.83 \text{ m}^3/\text{s})$$

determine its performance as an exhauster.

Solution:

$$\text{Compression ratio (as blower)} = \frac{p_d}{p_b} = 17/14 = 1.214$$

$$\text{Discharge pressure (as exhauster) } p_d = p_b = 14 \text{ psi } (96.5 \text{ kPa})$$

$$\text{Inlet pressure } p_i = 14/1.214 = 11.52 \text{ psia}$$

$$= -2.48 \text{ psig } (-17.1 \text{ kPa gage})$$

$$\text{Discharge quantity } Q_d = 6000/1.214 = 4940 \text{ cfm } (2.33 \text{ m}^3/\text{s})$$

where Q_d is the quantity of air removed from the workplace when the compressor is exhausting.

The selection of a compressor for a particular duty is carried out in the same manner as for a fan. Centrifugal compressors have much the same performance characteristics as centrifugal fans; their characteristic curves are similar in appearance.

Axial-Flow Compressors

Compressors of the axial-flow type are characterized by substantially constant quantity at variable pressure, whereas those of the centrifugal type deliver practically constant pressure over a considerable range of quantity. Although multistage units are available, axial-flow compressors used in mine and tunnel ventilation, like centrifugals, are generally single-stage. The performance of an axial-flow compressor, when operating as part of a blower or exhaust system, is analyzed in the same way as that of a centrifugal compressor (Example 9.6).

Positive-Displacement Compressors

A positive- or fixed-displacement compressor delivers a nearly constant quantity of air regardless of pressure drop or resistance in the duct it is supplying. The rotary and axial-flow types are examples that have had some use in mine auxiliary ventilation. In the rotary type, rotating lobes or gears on separate shafts turn in opposite directions, within a casing, mesh, and trap a certain quantity of air with each revolution. This is forced out the discharge side of the compressor. The axial-flow type has screwlike impellers that propel the air axially and raise the efficiency of the compressors. Positive-displacement compressors operate at pressures up to 10 psig or 275 in. water (69 kPa gage).

Because some allowance for clearance must be made at the points of contact of the lobes, the volume of air delivered per revolution is less than the theoretical delivery (equal to the displacement or volume of the compressing chamber) because of leakage (Weeks, 1926). This leakage is termed the *slip* and varies with the operating pressure. It is expressed in terms of a slip speed, determined by operating the compressor at constant pressure and no delivery. The actual capacity of the compressor is then equal to the running speed minus the slip speed times the displacement per revolution. The volumetric efficiency η_v is the capacity Q_c divided by the displacement Q_d per unit of time. The power in hp to operate the compressor equals the displacement in cfm times the gage pressure in psf divided by 33,000 times the mechanical efficiency expressed as a decimal (or in SI units, kW = m^3/s × kPa ÷ 0.01%). Typical calculations involving positive-displacement compressors are illustrated in the example below.

Example 9.7   Given a positive-displacement compressor operating at these conditions,

> Displacement = 6.0 ft^3/rev (0.170 m^3/rev)
>
> Slip speed = 58 rpm at 5 psig (34.5 kPa gage)
>
> Running speed = 500 rpm
>
> Mechanical efficiency = 65%

find the capacity, displacement, volumetric efficiency, and power.

Solution:

$$Q_c = (6.0)(500 - 58) = 2650 \text{ cfm } (1.25 \text{ m}^3/\text{s})$$

$$Q_d = (6.0)(500) = 3000 \text{ cfm } (1.42 \text{ m}^3/\text{s})$$

$$\eta_v = \left(\frac{Q_c}{Q_d}\right)(100) = \frac{(2650)(100)}{3000} = 88.3\%$$

$$P_m = \frac{pQ_d}{(33,000\eta)} = \frac{(144)(5)(3000)}{(33,000)(0.65)} = 101 \text{ bhp } (75.3 \text{ kW})$$

Since the characteristic curve of a positive-displacement compressor approaches a vertical line (constant quantity), the operating point of a given machine depends entirely on the characteristic curve of the duct to which the compressor is connected. The main requirement to determine is the rating of the motor needed to drive the compressor, in terms of power input. The operating pressure must be within the structural strength of the compressor, and care must always be exercised to prevent blockage of the system and subsequent destruction of the compressor.

Compressed-Air Injectors

For localized or temporary auxiliary-ventilation service in noncoal mines, injectors operated by compressed air are still occasionally used. Their capacity (head and quantity) is low. Aside from these limitations, the principal drawback to injectors is their inherently low efficiency, which, excluding losses in compression, seldom exceeds 10–15%. They are consequently many times more costly to operate per cfm (m³/s) of air delivered than electricity-powered fans. However, their first cost is low, maintenance is almost nil, installation is simple, and they are able to operate under conditions where fans would not. Injectors have application where a supply of compressed air is already available but electricity is not.

Injectors are classified according to design of the housing on the intake of the vent pipe as cylindrical, conical, or venturi (Fig. 9.14). The latter is

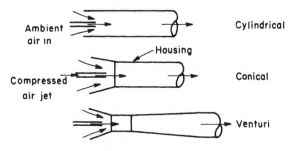

FIGURE 9.14 Compressed-air injectors.

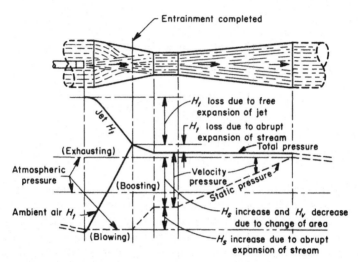

FIGURE 9.15 Pressure gradients for a venturi injector. (After McElroy, 1945.)

most efficient and most commonly used. Injectors can be purchased as separate units consisting of a casing and a nozzle for connection to the compressed-air supply, or they can be custom made. McElroy (1945) discusses the theory and design of injectors in detail. Important design variables are nozzle diameter [0.2–0.4 in (5.1–10.2 mm)] and air pressure [70–90 psig (483–620 kPa gage)].

The principle of operation of injectors is conversion of kinetic energy to

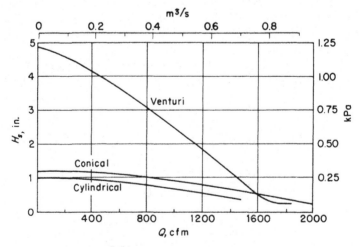

FIGURE 9.16 Typical characteristic curves of principal types of compressed-air injectors. (After Weeks, 1926).

static energy, and transfer of kinetic energy from a small volume of compressed air to a large volume of ambient air. The action depends on utilizing the energy of the nozzle discharge to overcome a shock loss due to expansion and to create a flow of the ambient air into the pipe or housing. Pressure gradients for a venturi injector are shown in Fig. 9.15.

The performance of an injector exhibits a falling head-quantity curve, similar to that of fans Fig. 9.16. Because of their low static head [up to 5 in. water (1.24 kPa)], injectors are limited to relatively short vent-pipe installations or to use without vent pipe. They may also serve as boosters in the high-resistance branch in parallel splitting.

PROBLEMS

9.1 Calculate the theoretical total head produced by a 4-ft (1.22-m) radial-tip centrifugal fan operating at 30,000 cfm (14.16 m³/s) and 450 rpm. Air specific weight is 0.0650 lb/ft³ (1.04 kg/m³), and the casing width is 16 in. (406 mm).

9.2 Plot the theoretical curve for the fan of Problem 9.1 operating from 0 to 40,000 cfm (18.88 m³/s) at standard air specific weight.

9.3 Calculate the theoretical total head produced by a 54-in. (1372-mm) backward-curved centrifugal fan operating at 20,000 cfm (9.44 m³/s) and 600 rpm. Air specific weight is 0.0700 lb/ft³ (1.12 kg/m³), the casing width is 24 in. (610 mm), and the blade angle is 55°.

9.4 Plot the theoretical characteristic curve for the fan of Problem 9.3 operating from 0 to 80,000 cfm (37.76 m³/s) at standard air specific weight.

9.5 The following data are obtained during a test of a vane-axial mine fan:

Static Head		Quantity		Power Input	
in.	(kPa)	cfm	(m³/s)	bhp	(kW)
8.00	(1.99)	0	(0)	37.6	(28.0)
7.65	(1.90)	5,000	(2.36)	35.5	(26.5)
6.00	(1.49)	12,500	(5.90)	24.0	(17.9)
6.65	(1.65)	16,000	(7.55)	25.5	(19.0)
6.10	(1.52)	20,000	(9.44)	25.0	(18.6)
3.70	(0.92)	25,000	(11.80)	21.0	(15.7)
0	(0)	28,600	(13.50)	13.0	(9.7)

The fan has a diameter of 38 in. (965 mm) and a speed of 1750 rpm, and the air specific weight is 0.0750 lb/ft³ (1.20 kg/m³). Plot characteristic curves of the fan for quantity versus static and total heads, power input, static and total air power, and static and total efficiency. Specify the fan rating. Tabulate all data used in plotting curves.

9.6 Given the data of Problem 9.5, plot characteristic curves (quantity vs. static head, power input, and total efficiency) for each of these separate new conditions:
   (a)   Diameter change to 60 in. (1524 mm), tip speed constant
   (b)   Air specific weight change to 0.0675 lb/ft^3 (1.08 kg/m^3)
   (c)   Speed change to 970 rpm

9.7 Assume that the changes listed in Problem 9.6 occur at the same time, and plot the characteristic curves for the new combined conditions.

9.8 A centrifugal compressor installed in an exhaust system discharges 2500 cfm (1.18 m^3/s) at a compression ratio of 1.2. Determine its operating characteristics (pressure and quantity) in the blower position if atmospheric pressure is 14.5 psi (100 kPa).

9.9 Determine the displacement, capacity, volumetric efficiency, and power input for a positive-displacement compressor operating at 450 rpm and 75% mechanical efficiency. Its displacement per revolution is 5.5 ft^3 (0.156 m^3) and slip speed 60 rpm at 4 psig (27.6 kPa gage).

CHAPTER 10

Fan Application to Mines

10.1 APPLICATION OF THE FAN TO THE SYSTEM

Operating Point

A *ventilation system* is composed of a fan or fans and a set of connected ducts. In a mine ventilation system, mine openings comprise the ducts. Assuming a single fan, at a given air specific weight with the fan operating at constant speed, there is only one head and quantity of airflow that can result. This is an equilibrium condition and is known as the *operating point* of the system or the fan, or both. In a broader sense, the operating point is an equilibrium condition specified by the values for certain ventilation parameters and that, mathematically speaking, can be represented by a point in a coordinate system, including a multidimensional system. The discussion in the first three sections of this chapter, however, is limited to the operating point for a single-fan system having a single discharge or outlet.

Since the head generated by the fan (pressure gain) must just balance the mine head (pressure loss), the operating point can be determined from the intersection of the fan and mine $H–Q$ characteristic curves. The head and quantity of flow are read at the intersection, and if the other fan performance characteristics are plotted, the power and efficiency as well can be determined.

It is common practice to employ the static-head–quantity characteristic curve in locating the operating point, as shown in Fig. 10.1*a*. Velocity heads are disregarded, hence the fan head and efficiency curves plotted are both static characteristics. This procedure is satisfactory as long as (1) the velocity heads are equal, (2) the velocity heads are so small as to be negligible, or (3) only an approximation is being sought.

In a main-fan installation involving high quantities and velocities, however, a sizable error can result from neglecting velocity head. Usually a determined effort is made to utilize the fan H_v; if it exceeds the H_v of the mine opening to which connected, a portion may be converted to static head (Section 10.3). Therefore, when (1) a main fan and (2) different areas are involved, the operating point is determined by plotting the recoverable portion of the fan total head or the static head of the fan and its conversion duct. The intersection with the mine total or static head characteristic, re-

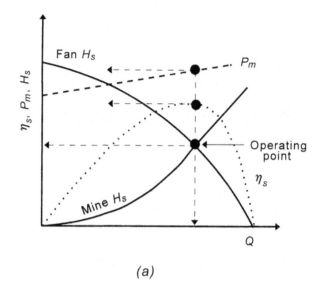

(a)

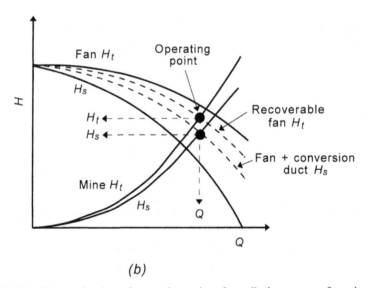

(b)

FIGURE 10.1 Determination of operating point of ventilation system from intersection of characteristic curves of fan and mine: (*a*) use of static head curves for system in which fan H_v = mine H_v; (*b*) use of total head curves for system in which fan H_v ≠ mine H_v. (Fan P_m and η curves omitted to avoid overcrowding.)

spectively, locates the operating point. These two alternative procedures are illustrated in Fig. 10.1*b*. Note the error that would result from using the intersection of the static-head curves. A total efficiency curve based on the recoverable total fan head can then be plotted.

In an exhaust system, the fan H_v and mine H_v may be made equal by design. In a blower or booster system, the two are never equal, except by coincidence. Then the operating point should be determined at the intersection of the total head curves (see Fig. 10.6).

Most fans are designed to operate safely, satisfactorily, and economically over only a certain portion of their characteristic curve. This is known as the *operating range,* lying immediately adjacent to the rating point on the $H-Q$ curve. Fan performance data provided by manufacturers in the form of tables or curves are restricted to this portion of the characteristic, so that the use of the fan will be confined to the operating range. An examination of the curves shown in Fig. 9.9 reveals that the operating range for all types of fans lies generally more to the right than the left of the rating. The portion of the curve to the left of the peak head is unstable and should be avoided, particularly if aerodynamic stall occurs.

It should be emphasized that the operating point remains unchanged only as long as the operating conditions are unchanged. If mine resistance, natural pressure, fan speed, or air specific weight varies, then the equilibrium of mine-fan operation is disturbed. The operating point will shift to the intersection of the new curves.

Changing Conditions

A practical problem that arises and requires plotting of characteristic curves and the application of the fan laws for solution concerns determination of the fan operating condition to supply air at a specified head, quantity, and air specific weight. This may involve a change in fan speed or size, or both, or blade pitch (if an axial-flow fan). Mathematically, the situation is one in which a fan curve must be passed through an operating point (or its equivalent, that the point must be made to fall on a fan curve). Since this operating point must also lie on a mine characteristic, the problem can be solved by plotting and transposing fan and mine curves to obtain an intersection.

Practical problems often arise that require application and understanding of the fan laws, mine characteristic curves, and fan characteristic curves for solution. The simplest case would be a change in fan speed.

Example 10.1 Given the fan curve shown in Fig. 10.2, assume the fan to be operating at a constant speed of 1000 rpm to exhaust 140,000 cfm (66 m³/s) at 3.5 in. water (871 Pa). Determine the speed at which this fan would exhaust 175,000 cfm (83 m³/s), and determine the new operating point.

Solution: From the given operating point, plot the mine characteristic curve, as shown in Fig. 10.2, using the equation $H = RQ^2$. From the mine curve,

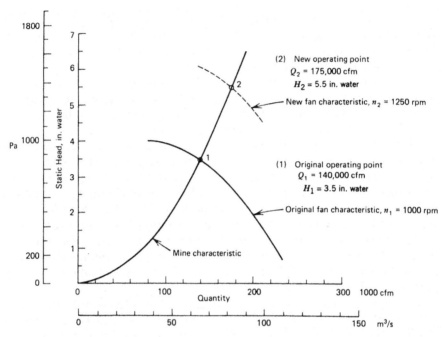

FIGURE 10.2 Determination of new fan speed to provide desired flow (Example 10.1).

it can be seen that to induce an airflow of 175,000 cfm (83 m³/s) requires a head of 5.5 in. water (1369 Pa). Therefore, the new operating point must be $Q_2 = 175,000$ cfm (83 m³/s) at a head of $H_2 = 5.5$ in. water (1369 Pa). The given operating point is $Q_1 = 140,000$ cfm (66 m³/s) at $H_1 = 3.5$ in. water (871 Pa).

Based on the first fan law (Eq. 9.27), the fan speed can be calculated:

$$n_2 = \frac{175,000}{140,000} (1000) = 1250 \text{ rpm}$$

As a check, find the new fan head:

$$H_2 = \left(\frac{1250}{1000}\right)^2 (3.5) = 5.5 \text{ in. water (1369 Pa)}$$

which agrees with the value previously determined.

Care must be taken when a speed change is being considered to ensure that the maximum recommended tip speed of the fan blades is not exceeded.

This tip speed is usually limited to 25,000 to 30,000 fpm (127–152 m/s), depending on the manufacturer's recommendation.

A more complicated situation arises if changes in both air specific weight and fan speed and size are involved. Assume a fan of given size and speed, operating at a given air specific weight, and let it be required to find the speed at which a similar fan of stated size should operate to deliver air at the specified conditions. Two approaches are possible. In the first, a new fan characteristic is computed and plotted for the desired fan size and air specific weight. From the intersection of this curve and the mine characteristic curve drawn through the desired pint, a speed change can be computed. In the second method, the conditions at the desired point are transposed to the fan conditions, a mine characteristic curve is drawn, and a speed change is computed. The latter method may be harder to visualize but is generally quicker.

Example 10.2 Determine the speed at which a 72-in. (1.83-m) fan should operate to deliver 90,000 cfm (42 m³/s) at 2 in. water (498 Pa) and 0.0850 lb/ft³ (1.36 kg/m³). The characteristic curve for a similar fan of 48 in. (1.22 m) diameter operating at 1500 rpm and 0.0750 lb/ft³ (1.20 kg/m³) is shown in Figs. 10.3 and 10.4.

Solution:
Method a (Fig. 10.3). Plot the characteristic curve of the 72-in. (1.83-m) fan at 0.0850 lb/ft³ (1.36 kg/m³) and 1500 rpm, using Eqs. 9.19 and 9.20. For any point on the curve:

$$Q' = \left(\frac{1500}{1500}\right)\left(\frac{72}{48}\right)^3 Q = 3.375Q$$

$$H' = \left(\frac{1500}{1500}\right)\left(\frac{72}{48}\right)^2 \left(\frac{0.0850}{0.0750}\right) H = 2.55\, H$$

Plot the mine characteristic curve through 90,000 cfm (42 m³/s) and 2 in. water (498 Pa) (point 2). Read at the intersection of the fan and mine curves (point 1):

$$Q_1 = 152,000 \text{ cfm } (71.7 \text{ m}^3/\text{s})$$

Then find the new fan speed by Eq. 9.27:

$$n_2 = \frac{90,000}{152,000}\,(1500) = 888 \text{ rpm}$$

Method b (Fig. 10.4): Let x be the desired speed for the 72-in. (1.83-m) fan. Transpose the desired operating point—90,000 cfm (42 m³/s), 2.0 in. water (498 Pa), 0.0850 lb/ft³ (1.36 kg/m³), and x rpm—to the operating point for the 48-in. (1.22-m) fan at 0.0750 lb/ft³ (1.20 kg/m³) and x rpm. Using Eqs. 9.19 and 9.20, we obtain

$$Q_2 = \left(\frac{48}{72}\right)^3 (90,000) = 26,667 \text{ cfm (12.6 m}^3\text{/s)}$$

and

$$H_2 = \left(\frac{48}{72}\right)^2 \left(\frac{0.0750}{0.0850}\right) (2.0) = 0.784 \text{ in. water (195 Pa)}$$

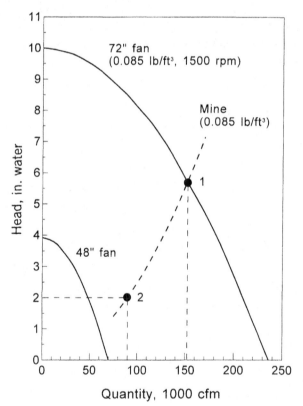

FIGURE 10.3 Determination of speed of similar fan to provide desired flow (Example 10.2): method a. (*Conversion factors:* 1 in. water = 248.84 Pa, 1000 cfm = 0.47195 m³/s.)

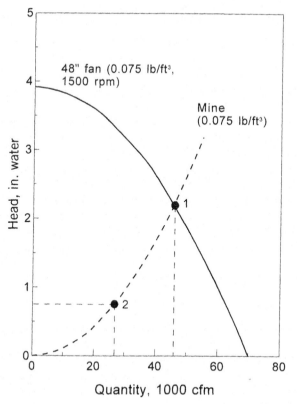

FIGURE 10.4 Determination of speed of similar fan to provide desired flow (Example 10.2): method b. (*Conversion factors:* 1 in. water = 248.84 Pa, 1000 cfm = 0.47195 m³/s.)

Plot the transposed point (point 2) and pass through it a mine characteristic curve, intersecting the fan curve at point 1. Read

$$Q_1 = 45,100 \text{ cfm } (21.3 \text{ m}^3/\text{s})$$

Then find the desired fan speed by Eq. 9.27:

$$n_2 = \left(\frac{26,667}{45,100}\right)(1500) = 887 \text{ rpm}$$

A problem involving change of pitch rather than speed is solved in a manner similar to that of Examples 10.1 and 10.2, except that fan characteristic curves for a complete range of pitch rather than calculations by the first fan law are employed.

10.2   FAN HEADS AND HEAD GRADIENTS

In considering the entire ventilation system, mine plus fan head gradients are an invaluable aid. The gradients employed previously for understanding mine heads and losses in airflow (Section 5.3) are equally useful in representing the head relations throughout the system and in determining the fan heads. Gradients are particularly helpful in visualizing the relation between mine and fan heads with the fan located in any of the three possible positions with respect to the system as a whole: blower, exhaust, or booster. Coupled with plots of the characteristic curves of the system, they can be used to solve for the heads and quantity in the most complicated system, including one equipped with an evasé duct at the discharge (Section 10.3).

Fan heads are the static, velocity, and total heads that the fan must develop to deliver a given quantity of air to the mine. They occur on the system characteristic curves, along with the mine heads, at the operating point (Section 9.5). On a plot of the pressure gradients, the fan heads are read at the fan location.

Gradients for Basic Systems

In Fig. 10.5, pressure gradients are plotted for the three basic types of ventilation systems (McElroy, 1935). The mine is assumed to be the same in each case (the total head loss is constant), and for simplicity, no area changes occur (airway area = fan area). Fan and mine are indicated, together with the manometer reading H when the gage is connected across the fan to read difference in static head.

Neglecting minor differences due to shock loss at entrance and discharge, the mine heads are the same in all three cases, since H_l and H_v do not vary. At the fan in each system, $H_t = H_s + H_v$ (as predicted by Eq. 9.11). Note also, however, that the fan and mine total heads are equal in each case:

$$\text{fan } H_t = \text{mine } H_t \qquad\qquad (10.1)$$

This is a fundamental premise for *every* ventilation system, although seeming disparity in the case of a fan with an evasé or a tapered duct connection will require some qualification. The equality in each case between the static and velocity heads of the mine and fan does not apply if area changes occur within the system, although the total heads remain equal.

Blower System   The fan and mine heads both fall above the datum line but are read at opposite ends of the system. Note that the fan static head, unlike the mine static head, is read to the static gradient.

Exhaust System   The fan and mine heads coincide in their location (and in their magnitude, if no area change occurs at the fan). Like the mine static

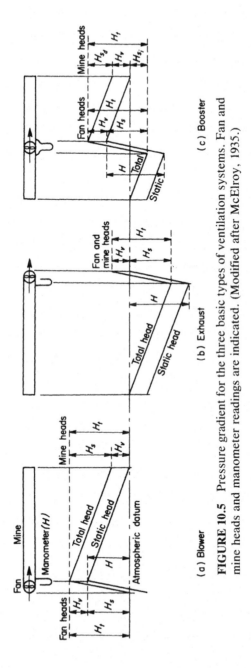

FIGURE 10.5 Pressure gradient for the three basic types of ventilation systems. Fan and mine heads and manometer readings are indicated. (Modified after McElroy, 1935.)

363

head, the fan static head is read to the total-head gradient, here lying below the datum and negative in value. The sign has no significance in specifying the fan rating; in fact, if minor differences in shock losses at entrance and discharge are neglected, a fan delivers the same quantity of air at the same head, regardless of location in the system.

Booster System   As in the blower system, the fan and mine heads are separated. Since this is a hybrid system, the heads lie partly above and partly below the datum. The fan static head is read from the total gradient on the intake side to the static gradient on the discharge side. The fan velocity head is read on the discharge side of the fan, a general rule for any system.

Manometer Readings

In a mine installation, the fan heads can be determined by manometer measurement. If the connections to the manometer are always made for static readings (no velocity component), the leg of the manometers so connected will read from the atmospheric datum to the static-head gradient, regardless of the system. By definition, a manometer with static connection can read only the gage-static pressure of the system above or below atmospheric pressure.

In a blower system, with fan and mine the same size, the manometer reading H is

$$H = \text{fan } H_s \tag{10.2}$$

whereas in an exhaust system

$$H = \text{fan } H_t = \text{mine } H_t \tag{10.3}$$

If no area changes occur in the mine at the fan, Eq. 10.3 also applies to a booster system. To obtain the fan static head in an exhaust or a booster system, a pitot tube attached to the manometer is used on the intake side of the fan (that leg of the manometer then records the differential to the total-head gradient).

Gradients with Area Changes in the System

When changes occur in the area of airways in a mine ventilation system, the velocity head will vary at different points in the system (Figs. 5.5, 5.7, and 5.8). Since this situation prevails in mines, the mine and fan static and velocity heads are rarely equal. Only in an exhaust system are the velocity heads of the fan and mine measured at the same point, and if their areas are unequal, or if an evasé is used, the heads will again be dissimilar. The one irrefutable equivalency is expressed by Eq. 10.1.

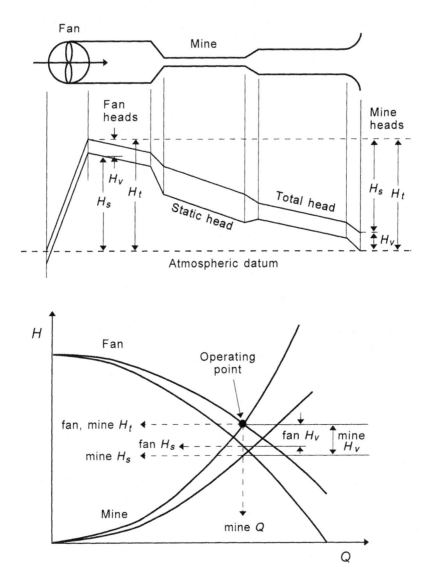

FIGURE 10.6 Plotting both pressure gradients (*above*) and characteristic curves (*below*) to depict head relationships in a mine ventilation system.

Consider the blower system shown in Fig. 10.6. Because of area changes in the mine, the velocity heads (and, likewise, the static heads) at the fan and at the mine discharge are different. These are readily apparent from the pressure gradients and characteristic curves of Fig. 10.6.

A special case is presented by the booster fan in which the areas on the

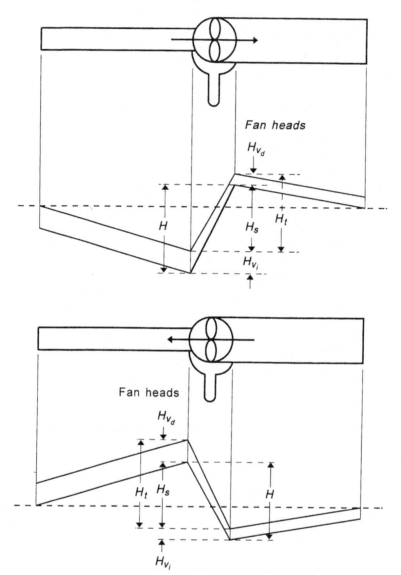

FIGURE 10.7 Pressure gradients for booster systems with fans having different intake and discharge areas, quantity constant.

intake and discharge sides of the fan are not equal (Fig. 10.7). Care must be taken to measure the velocity head on the discharge side of the fan and to correct the manometer reading H for the change in velocity head before equating it to fan total head:

$$\text{Fan } H_t = H_t + (H_{v_d} - H_{v_i}) \tag{10.4}$$

The gradients in Fig. 10.7 are plotted for the same quantity Q in each system; therefore, the fan static heads are equal, but because of the variance in H_v, the total heads are not equal.

10.3 FAN EVASÉS AND DIFFUSERS

Unless the mine opening to which the fan is connected is intentionally constructed so, the fan and mine do not have the same area. A connecting duct, tapering toward the fan since it generally is the smaller, accomplishes this transition in area and incidentally converts static head into velocity head (or vice versa).

An intentional conversion of velocity head into static head is usually made on the discharge side of a main fan. This may be of considerable benefit, since it effectively reduces the mine velocity head, and energy loss, and hence yields a power saving.

Advantage is taken of this interchangeablity of static and velocity heads in attempting (1) to minimize shock losses where section changes occur, particularly at blower fans and (2) to recover some of the velocity head at exhaust fans that normally is discharged to the atmosphere and wasted. In either event, use is made of gradual expansion conversion ducts, called *evasés* on exhaust fans and *diffusers* on blower fans, to convert velocity head to static head and to recover a considerable portion of it. A main exhaust fan equipped with evasé is pictured in Fig. 9.4.

The savings in velocity head, power, and power cost that can be realized are limited by two factors: (1) the minimum, final velocity that must be maintained to keep the air in motion, and (2) the efficiency of recovery that can be made in the conversion duct.

Because of the limitation of the first factor, the saving to be realized from recovery of velocity head is theoretically greater with an exhaust fan than with a blower. This is because the discharge velocity from an exhauster, which is the final velocity, can approach zero with a sufficient length of conversion duct. On the other hand, a blower fan must always discharge into a duct or an airway, and the final velocity head is determined by the duct dimensions, usually fixed or at least limited. An opportunity for savings in a blower system exists in the installation of an evasé at the discharge of the mine. It may happen, however, that the mine opening is larger than the fan evasé discharge; obviously, a blower arrangement may be more conserving of head and power requirements than an exhaust system.

The second factor depends on the slope of the conversion duct, its shape, and its symmetry. Symmetrical, conical ducts, expanding gradually, are the most efficient (Jorgensen, 1983). With ducts of this type and varying area

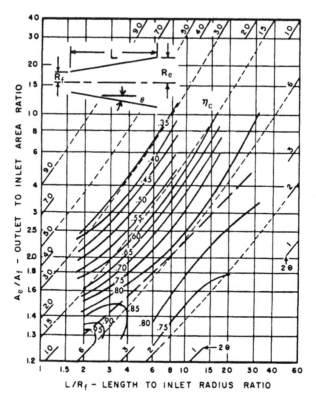

FIGURE 10.8   Recovery efficiencies of conical conversion ducts. (After Jorgensen, 1983. By permission from Buffalo Forge Co., Buffalo, NY.)

ratios, the conversion efficiencies shown in Fig. 10.8 can be expected. Note that negligible benefit is obtained from a conversion duct whose angle (2θ) is 30° or greater. Because of the rapid falloff in efficiency, duct angles exceeding 15–20° are generally not economical. The efficiencies given in Fig. 10.8 are also approximately correct for ducts of rectangular cross section.

In view of the effect of both governing factors, it may be concluded that conversion of the velocity head can be achieved most efficiently by using a gradual slope, a long duct, and the largest possible ratio of areas. On the other hand, the construction cost of the duct increases with its length. It is possible to calculate the optimum length of duct for the least overall cost (capital plus operating) in a given situation, since the power cost decreases and the cost of construction increases with duct length.

Mathematically, the recovery H_r that can be obtained in an evasé or diffuser is a function of the conversion efficiency η_c and the difference in velocity heads H_{v_f} and H_{v_e} existing at the two ends of the duct. The total possible gain $= H_{v_f} - H_{v_e}$. Therefore the recovery

$$H_r = \eta_c(H_{v_f} - H_{v_e}) \tag{10.5}$$

The conversion loss H_l is the difference between the possible regain and the recovery. In terms of duct areas A_f and A_e and quantity of airflow, the recovery

$$H_r = \frac{\eta_c w(V_f^2 - V_e^2)}{(1098)^2} = \frac{\eta_c wQ^2(1/A_f^2 - 1/A_e^2)}{(1098)^2} \tag{10.6}$$

The possible power saving P_r can be found as follows:

$$P_r = \frac{H_rQ}{6346\eta_o} \tag{10.7}$$

where η_o is the overall fan efficiency, from which the saving in power cost can be computed.

In practice, the "savings" with a conversion duct can be realized in either of two ways: (1) in a larger quantity of airflow at the same fan speed and about the same power demand or (2) in the same quantity of airflow at a lower fan speed and a lower power demand.

Plotting both characteristic curves and head gradients is essential in solving problems involving evasés or diffusers. Since most fan manufacturers specify the attachment of a conversion duct to their fans, the usual approach to the problem is to assume that the evasé is part of the fan. This assumption was made in the discussion of Section 10.1 and is demonstrated by the example below.

Example 10.3 From the system characteristic curves plotted in Fig. 10.9b, determine the operating conditions when an evasé is attached to the exhaust fan. The head gradients for the system with the fan alone are plotted in Fig. 10.9a.

Solution: The quantity Q_1 flowing at condition 1 is located from operating point 1. To find the quantity Q_2 that results from the addition of an evasé, compute the recovery H_r at condition 1 by Eq. 10.5. Lay off H_r at operating point 1 to find a point on the fan + evasé H_s curve. At the intersection of this curve with the mine H_s curve, locate operating point 2 for condition 2, fan equipped with evasé, shown in Fig. 10.9a. Manometer readings are noted in each case.

Note that with the larger quantity obtained from the addition of an evasé, lower fan heads but higher mine heads result. They differ by the amount of the recovery. The shift of the operating point to the right is also accompanied by a change in the power requirements. If a power saving (Eqs. 10.6 and 10.7) is preferred to a quantity increase, then the fan speed (or pitch) can

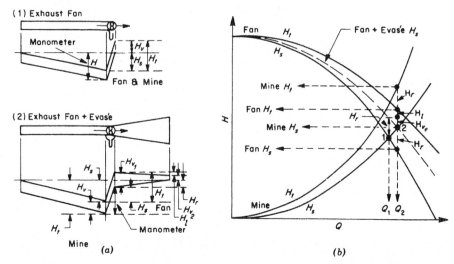

FIGURE 10.9 Determination of quantity and head for an exhaust system equipped with evasé: (*a*) pressure gradients; (*b*) characteristic curves.

be reduced so that the fan + evasé H_s curve passes through operating point 1. The new speed and fan curves can be calculated by Eq. 9.27. Thus $Q_2 = Q_1$, but both heads and power are reduced.

The pressure relations in a ventilation system equipped with evasé are apparently in disagreement with Eq. 10.1, since the mine H_t does not equal the fan H_t. However, the difference is H_r, the amount of velocity head converted to static head and recovered in the conversion duct. This is effective in reducing both the fan H_s and H_t and produces an apparent discrepancy in the total heads. The manometer static reading across a fan equipped with evasé, as in an exhaust system, measures fan total head, if the fan inlet and outlet areas are the same.

The meaning of the term "fan" for which the characteristic curves are given is not always consistent. The evasé or diffuser is considered an integral part of the fan by most manufacturers (e.g., Jeffrey and Joy). Thus the outlet or discharge of the fan refers to the far end of the fan + evasé, and the meaning of the fan heads follows those defined in Section 9.5. In this view, the term fan + evasé H_s is not needed and is replaced by fan H_s, whereas recoverable fan H_t (Fig. 10.1*b*) should be replaced by fan H_t. On the other hand, the outlet *cone* (a short evasé or diffuser that can be added to the outlet of a Joy 1000/2000 series fan) is not considered as an integral part of the fan. In this case, the concept of fan + evasé applies. The manufacturer provides three types of head–quantity curves: (1) fan H_t, (2) fan H_v, and (3) H_c (the evasé outlet velocity head plus head loss in the evasé). As a result, the curves for fan H_s, H_r, and fan + evasé H_s can be constructed by applying the following relationships:

$$\text{Fan } H_s = \text{fan } H_t - \text{fan } H_v \tag{10.8}$$

$$H_r = \text{fan } H_v - H_c \tag{10.9}$$

$$\text{Fan } + \text{ evasé } H_s = \text{fan } H_s + H_r \tag{10.10}$$

In addition, since the outlet area is known, the curve for the velocity head, fan + evasé H_v, can be calculated. Then the head loss in the evasé H_l is given by

$$H_l = H_c - \text{``fan } + \text{ evasé } H_v\text{''} \tag{10.11}$$

It should be noted, however, that

$$\begin{aligned} \text{Recoverable fan } H_t &= \text{fan } + \text{ evasé } H_t \\ &= \text{fan } H_t - H_l \end{aligned} \tag{10.2}$$

Mutmansky et al. (1985) recommended that the fan operating point be determined using the fan total head and the corresponding system/subsystem total head. This approach can be applied to fans in the blower, booster, and exhaust positions, and to both single-fan and multiple-fan networks.

10.4 FANS IN COMBINATION

In a mine ventilation system, two or more fans very often are operating at different locations in the system. When in combination with one another, depending on the arrangement, fans are often said to be operating either in *series or parallel.* It is important to mention, however, that the fans are treated as network branches, and that the terms "series" or "parallel" refer to the topological relationship between two or more branches that represent the fans. To be accurate, therefore, two or more fans are operating either in series or not in series; also two or more fans are operating either in parallel or not in parallel. If two branches that represent the fans are not connected in series or parallel, the principle of series or parallel should not be applied.

Fans in Series

In series operation, heads are cumulative for a given quantity. This means that each fan contributes a portion of the overall head and that each handles the same quantity of air. The combined characteristic is found by summing the individual fan curves, preferably using a pair of dividers to cumulate heads for a given quantity.

Example 10.4   Plot the combined series characteristic curve for fans A and B, whose individual curves are shown in Fig. 10.10a.

Solution: The resultant characteristic is shown in Fig. 10.10b. In plotting, the curve of the fan indicating the larger free delivery (fan B) is laid out first, then the curves of the other (fan A) is plotted above it. Effectively, the same thing is accomplished by using a pair of dividers to cumulate heads for a given quantity. If velocity head is being considered, the standard procedure is first to plot the combined total-head curve from the individual fan total characteristics. Then the combined static-head curve is obtained by subtracting the velocity head of the second fan (downstream fan), here assumed to be fan A.

Note that for a given quantity Q:

$$H_t = H_{t_a} + H_{t_b} \tag{10.13}$$

and

$$H_s = H_t - H_{v_a} \tag{10.14}$$

where the subscripts a and b refer to fans A and B, respectively, and H_s and H_t are the combined heads of the two fans in series.

For mines employing fans in series, either in a blower–exhaust combination or in the booster position, head gradients supplement characteristic curves in analysis. Regardless of the distance that the fans are apart, their characteristic curves can be combined and plotted with the mine characteristic, as shown in Fig. 10.11. From this graph, the mine quantity and heads and the heads on each fan can be determined. Then head gradients for the system can be plotted, from which the heads at any point can be determined. For the mine with booster and exhaust fans shown in Fig. 10.12, head gradients are drawn and mine and fan heads are indicated. The mine total head equals the combined fan total head, which according to Eq. 10.13 is

$$\text{Mine } H_t = \text{combined } H_t = \text{fan } H_{t_b} + H_{t_a}$$

and also as indicated on the gradients:

$$\text{Mine } H_t = \text{mine } H_{s_1} + \text{mine } H_{s_2} + \text{mine } H_v$$

Fans in Parallel

In parallel operation, quantities are cumulative for a given head. It is well to plot first the combined static-head characteristic curve, cumulating quan-

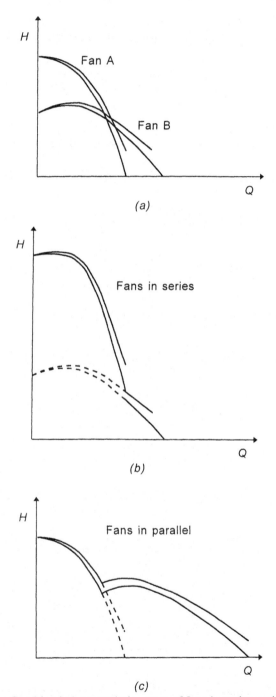

FIGURE 10.10 Combined characteristic curve of fans in series and parallel (Examples 10.4 and 10.5): (*a*) characteristics of individual fans A and B; (*b*) combined characteristics of fans in series; (*c*) combined characteristics of fans in parallel.

tities from the individual static-head fan curves, and then to employ the following equation for the velocity head in plotting the combined total characteristic:

$$\text{Parallel } H_v = \frac{H_{v_a} Q_a + H_{v_b} Q_b}{Q_a + Q_b} \qquad (10.15)$$

where the subscripts a and b refer to fans A and B, respectively. In deriving

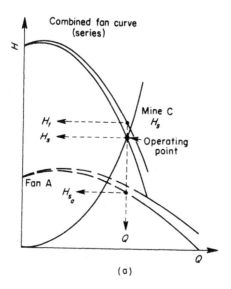

(a)

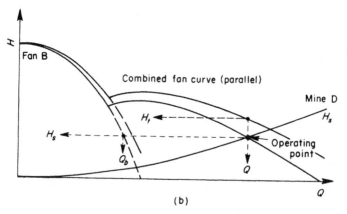

(b)

FIGURE 10.11 Determination of operating point of mine and fans operating in combination: (a) mine C and fans A and B in series; (b) mine D and fans A and B in parallel.

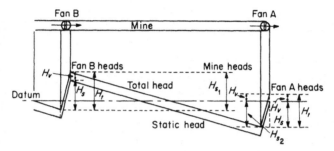

FIGURE 10.12 Head gradients for ventilation system employing multiple fans in series.

Eq. 10.15, the principle of energy conservation has been used. Note that when the two fans are identical, values of velocity head and quantity are equal; thus parallel $H_v = H_{v_a} = H_{v_b}$. The total-head curve of the two fans in parallel is then found by cumulating values of H_s and H_v for constant quantity. The procedure just outlined can be easily extended to a case involving more than two fans.

Example 10.5 Plot the combined parallel characteristic curve for the fans of Example 10.4, whose individual curves are shown in Fig. 10.10a.

Solution: The resultant characteristic is shown in Fig. 10.10c. In plotting, the curve of the fan whose pressure is higher at no-delivery (fan A) is laid out first, then the curve of the other (fan B) is plotted to the right. Again, the use of dividers in effect accomplishes this. Quantities are cumulated for a given static head to obtain the combined static characteristic, and velocity heads are read from the individual fan curves at these quantities. Then the combined velocity head is determined from Eq. 10.15 and added to the static characteristic to find the combined total characteristic.

Note that for a given head, the total quantity

$$Q = Q_a + Q_b \tag{10.16}$$

where Q_a and Q_b are the quantities contributed by the individual fans A and B, respectively.

Characteristics of Fans in Combination

Fans in series are best suited for mines having steep characteristics (high-resistance systems). If the mine curve is too flat to intersect the combined portion of the series curve, then one fan is contributing nothing to the operation, and the quantity of flow will be the same as when using only one fan. In Fig. 10.11a, the operating point of a mine and fans in series is indicated. Operation is satisfactory because the curve of mine C is steep, intersecting the fan curve on the combined portion. Note that the quantity of flow is less than the sum of the individual quantities with each fan operating separately but greater than the quantity produced by either fan alone. The portion of the static head contributed by each fan can be determined from the graph by employing Eq. 10.13.

In practice, a series arrangement may be employed in the later stages of the life of a mine when the mine resistance has become high and the mine characteristic steep. A second fan in series may be installed to supplement the original fan and produce higher heads and quantity to offset the increased mine resistance. Regardless of their location in the system, if two or more fans handle the entire or same quantity of air, they are in series with one another. Another application for fans in series occurs when fans previously acquired must be adapted to a new mine. Sometimes a series combination of two old fans will produce more air at a lower overall cost than a single new fan will. There is usually no difficulty encountered in operating two or more fans in series, provided each is operating on the safe portion of its curve and in the maximum-efficiency range.

Fans in parallel are best suited for mines having flat characteristics (low-resistance systems). If the mine curve is too steep to intersect the combined portion of the parallel curve, then one fan will blow air back through the other. The operating condition of a mine (mine D) and fans in parallel is indicated in Fig. 10.11b. Parallel fan operation for mine D is satisfactory because its characteristic curve is flat, intersecting the combined fan curve well to the right. Note that the total quantity of flow is less than the sum of the individual quantities but greater than the quantity produced by either one alone. The quantity contributed by each fan can be determined from Fig. 10.11b by employing Eq. 10.16.

In Fig. 10.11, mines C and D would not function satisfactorily in combined-fan operation if they were interchanged. The curve of mine C is too steep to intersect the fan parallel characteristic on the combined portion of the curve. Likewise, the curve of mine D is too flat to intersect the fan series characteristic on the combined portion of its curve.

Fans in parallel have practical applications similar to those of fans in series. Infrequently are new fans purchased to operate in parallel (or in series), but in modifications to older systems, multiple-fan installations may be the most economical solution. Two or more fans already on hand may function to advantage in parallel in a new mine or serve adequately in an older mine whose characteristics are modified by the addition of new levels, tending to lower the resistance. Fans in parallel offer some advantage from the safety standpoint also, since in the event of the mechanical failure of one, the other will continue to supply air. However, satisfactory parallel operation is sometimes difficult to obtain with dissimilar fans, because of a

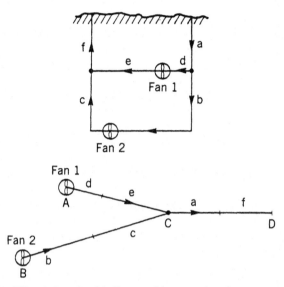

FIGURE 10.13 Mine (*above*) with fans and intervening ducts represented in schematic (*below*).

tendency for the fans to assume disproportionate loads. This may result in overload harm to one or an uneconomical power drain. In selecting fans for parallel duty, those with steep head–quantity characteristics are preferable. Backward-curved and, to some extent, radial-tip centrifugal fans meet this requirement and usually function well in parallel. On the other hand, forward-curved centrifugal and some axial-flow fans are not well suited for parallel operation because of their flat H–Q curves and points of contraflexure. Care must be exercised to ensure that parallel fans operate on the safe portion of their curves, well beyond the stall points.

Fans in Series-Parallel Networks

What has been said to this point regarding the behavior of fans in parallel pertains to their operation with negligible lengths of intervening duct. A more common circumstance is for fans to be installed on alternate airways or alternate levels in a mine. This is known as *parallel operation with intervening ducts*. (Strictly speaking, however, these fans are not operating in parallel.) Very often this arrangement is more satisfactory; in fact, fans may operate well in parallel with intervening ducts and not at all without. Figure 10.13 illustrates how a system with fans in different locations may be represented in schematic with intervening and common ducts.

A problem arising in parallel operation of fans with intervening ducts is to determine the quantities of flow that result (i.e., to solve the associated network problem). The plotting of curves of fans in parallel in the manner of Fig. 10.10c is not applicable now, because head loss occurs in the intervening ducts. In a conventional method for graphic solution (Wang, 1988, 1993), both head-loss and head-gain characteristic curves are plotted on the same $Q–H$ coordinate system; therefore, the basis (i.e., whether head loss or head gain) of each curve has to be identified in the solution process. In the discussion that follows, an alternative approach is taken.

All fan curves, which are usually provided by the fan manufacturer as head gains, will be inverted into head-loss curves before starting series–parallel combinations, and all curves generated from the combination procedures will be expressed in head loss. Furthermore, each fan will be considered as an individual network branch, and its characteristic curve on a head-loss basis is assumed to be a strictly increasing curve extending beyond upper and lower limits of its operating range. This assumption together with the Atkinson equation (5.25), whose graph is also a strictly increasing curve, ensures that the network problem has a unique solution.

As illustrated in Fig. 10.14, the inverted fan curve is obtained by turning

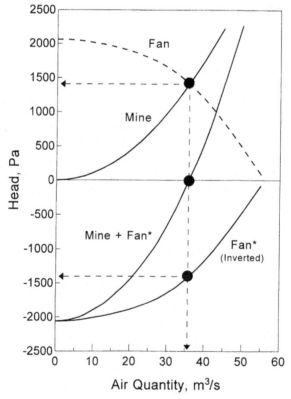

FIGURE 10.14 Determination of operating point of ventilation system with the inverted fan characteristic curve. (*Conversion factors:* 1 in. water = 248.84 Pa, 1 cfm = 0.47195×10^{-3} m³/s.)

the original fan curve 180° around the Q axis; the operating point when the fan is applied to the mine is represented by the Q intercept of the combined characteristic curve for the fan and the mine. In this case, the branches representing the fan and the mine are connected in series; therefore, for a given Q, the H value for the combined curve is obtained by summing the H values for the fan and the mine. Now draw a vertical line through the Q intercept to obtain the operating points for the fan and the mine. Note that fan $H = -(\text{mine } H)$ for the operating points.

The graphic solution to the series-parallel ventilation networks is best illustrated by means of the two-fan network with two meshes, as shown in the closed-form networks of Fig. 10.15, where branches p_1 and p_2 represent fans 1 and 2. There are two types of network problems associated with the two-fan network with two meshes. The problem of fans in parallel with

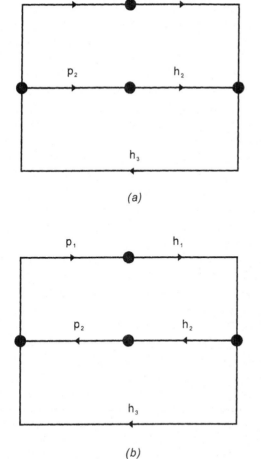

FIGURE 10.15 Two types of problems for the two-mesh ventilation network with two fans: (*a*) type 1 problem; (*b*) type 2 problem. (Reprinted from *Mng. Sci. Tech.*, Vol. 7, No. 1, pp. 31–43, copyright 1988 with kind permission from Elsevier Science—NL, Sara Burgerhartstraat 25, 1055 KV Amsterdam, The Netherlands.)

intervening ducts (Fig. 10.13) is designated type 1; type 2 represents those with a main fan and a booster or underground fan. The networks of two types are different only in the directions of branches h_2 and p_2. Each network consists of five branches: h_1, h_2, h_3, p_1, and p_2. Their characteristic curves are denoted by the same symbols. As illustrated in the following examples, graphical procedures are used in reducing the given network to a single branch for which the operating point is represented by the Q intercept. From this operating point, the operating points for the other branches are traced successively in the reverse order in which the series–parallel reductions are applied. The operating points of series branches and their combined branch have the same Q value; that is, they all lie on a vertical line. On the other hand, the operating points for the parallel branches and their combined branch have the same H value; that is, they all lie on a horizontal line.

Example 10.6   Given the characteristic curves of Fig. 10.16 for the type 1 network shown in Fig. 10.15a. Find the operating points.

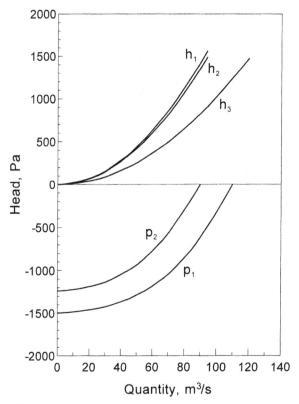

FIGURE 10.16   Characteristic curves for Example 10.6. (Modified after Wang, 1988. Reprinted from *Mng. Sci. Tech.*, Vol. 7, No. 1, pp. 31–43, copyright 1988 with kind permission from Elsevier Science—NL, Sara Burgerhartstraat 25, 1055 KV Amsterdam, The Netherlands.) (*Conversion factors:* 1 in. water = 248.84 Pa, 1 cfm = 0.47195 × 10^{-3} m³/s.)

Solution: As shown in Fig. 10.17, the sequence of reducing the network and determining the operating points is as follows:

1. Perform the series combination of curves p_1 and h_1 to plot curve L_1.
2. Perform the series combination of curves p_2 and h_2 to plot curve L_2.
3. Perform the parallel combination of curves L_1 and L_2 to plot curve L_3.
4. Perform the series combination of curves h_3 and L_3 to plot curve L_4.
5. The operating point of branch L_4 is represented by the intercept of curve L_4 with the Q axis. Draw a vertical line through the operating point of L_4 to intersect curves h_3 and L_3. The intercepts represent the operating points of branches h_3 and L_3, respectively.
6. Draw a horizontal line through the operating point of L_3 to intersect curves L_1 and L_2. The intercepts represent the operating points of branches L_1 and L_2, respectively.
7. Draw a vertical line through the operating point of L_2 to intersect curves p_2 and h_2. The intercepts represent the operating points of branches p_2 and h_2, respectively.
8. Draw a vertical line through the operating point of L_1 to intersect curves p_1 and h_1. The intercepts represent the operating points of branches p_1 and h_1, respectively.

Values for the operating points are summarized as follows:

Branch	Q, cfm (m³/s)	H, in. water (Pa)	
h_1	108,062 (51.0)	1.85	(460)
h_2	81,576 (38.5)	1.00	(250)
h_3	189,639 (89.5)	3.30	(820)
p_1	108,062 (51.0)	-5.14	(-1280)
p_2	81,576 (38.5)	-4.30	(-1070)

Example 10.7 A network for a type 2 problem is shown in Fig. 10.15*b*. Outline the sequence for reducing the network to a single branch.

Solution: Two methods, A and B (Wang, 1988), are outlined in Figs. 10.18 and 10.19, respectively.

In addition to series–parallel combinations, the graphic procedure for reversing the reference direction of a branch (i.e., the branch direction) is needed for some special applications. Figure 10.20 outlines the sequence for obtaining branch L_4 whose characteristic curve is called a subsystem characteristic curve (Section 10.8). Note that the direction of branch L_1 is reversed to yield branch L_2. In this case, the characteristic curve of branch L_2 is obtained by a 180° rotation of that of branch L_1. This means that the two characteristic curves are symmetrical about the origin.

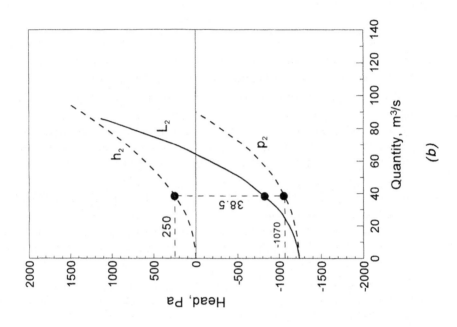

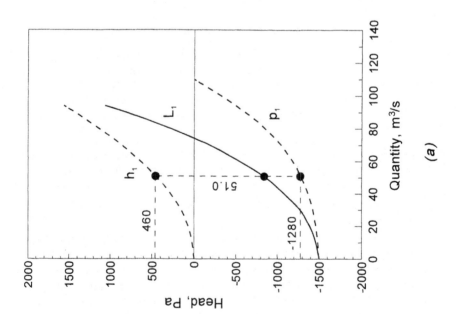

(a)

(b)

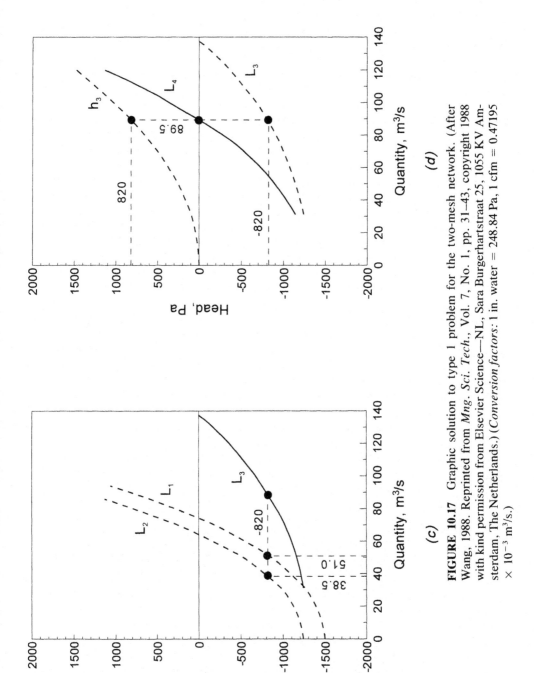

FIGURE 10.17 Graphic solution to type 1 problem for the two-mesh network. (After Wang, 1988. Reprinted from *Mng. Sci. Tech.*, Vol. 7, No. 1, pp. 31–43, copyright 1988 with kind permission from Elsevier Science—NL, Sara Burgerhartstraat 25, 1055 KV Amsterdam, The Netherlands.) (*Conversion factors:* 1 in. water = 248.84 Pa, 1 cfm = 0.47195 × 10⁻³ m³/s.)

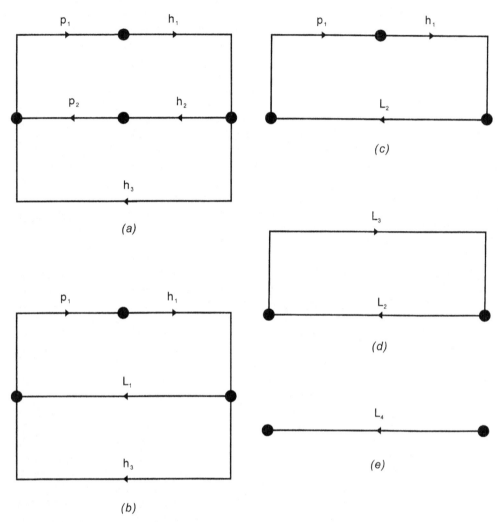

FIGURE 10.18 Method A for type 2 problem for the two-mesh network. (After Wang, 1988. Reprinted from *Mng. Sci. Tech.,* Vol. 7, No. 1, pp. 31–43, copyright 1988 with kind permission from Elsevier Science—NL, Sara Burgerhartstraat 25, 1055 KV Amsterdam, The Netherlands.) (*Conversion factors:* 1 in. water = 248.84 Pa, 1 cfm = 0.47195×10^{-3} m³/s.)

10.5 FANS AND NATURAL VENTILATION

Until now, mine ventilation systems comprised of either a natural or mechanical energy source have been discussed. In the general case, both natural and mechanical ventilation may act together. It is well to understand that natural ventilation is responsible for most of the irregularity in fan performance from season to season and to know how to determine the operating

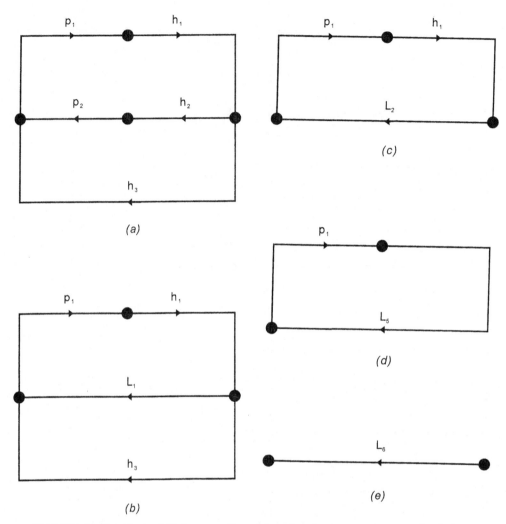

FIGURE 10.19 Method B for type 2 problem for the two-mesh network. (After Wang, 1988. Reprinted from *Mng. Sci. Tech.*, Vol. 7, No. 1, pp. 31–43, copyright 1988 with kind permission from Elsevier Science—NL, Sara Burgerhartstraat Z5, 1055 KV Amsterdam, The Netherlands.)

condition and the combined system characteristics with both forces acting. For simplicity, the natural ventilation head H_n is treated as a fan with a horizontal head–quantity characteristic curve. It is defined based on the head gain. In other words, if the direction of the head gain agrees with the reference direction, it is positive; otherwise, it is negative.

A problem that frequently arises in a mine when both natural and mechanical ventilation are involved is to measure correctly the magnitude of the various heads: What is the actual fan head H_m, the natural ventilation head

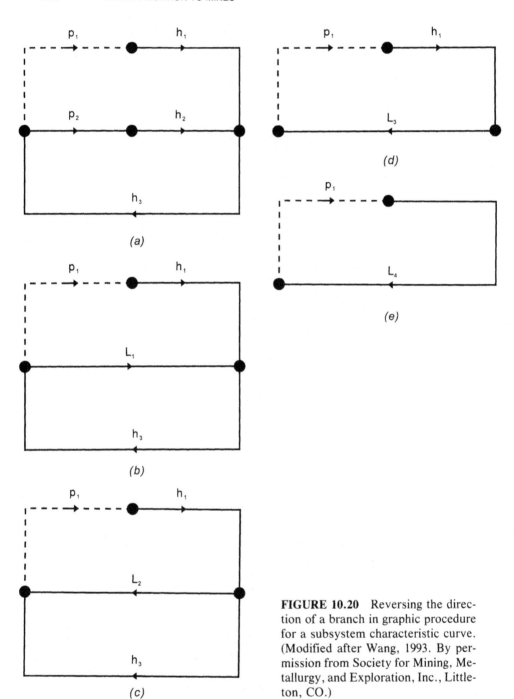

FIGURE 10.20 Reversing the direction of a branch in graphic procedure for a subsystem characteristic curve. (Modified after Wang, 1993. By permission from Society for Mining, Metallurgy, and Exploration, Inc., Littleton, CO.)

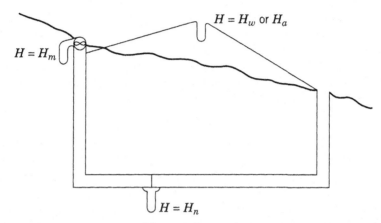

FIGURE 10.21 Manometer readings taken at various points in a mine ventilation system in which both natural and mechanical heads occur.

H_n, and the effective head H_w (natural ventilation with the fan) or H_a (natural ventilation against the fan) for the combined system? Care must be exercised in taking measurements to insure that the various head components are identified. For example, a manometer across a fan always records the actual fan head, while manometer (or altimeter) readings over the entire mine record the effective head of the combined forces. The natural ventilation pressure alone can be measured by shutting down the fan and proceeding as described in Section 8.2. These relationships of the manometer readings H are shown in Fig. 10.21. It is very important to determine H_m when applying the fan laws to achieve a desired change in operating conditions (Weeks, 1926, pp. 112–115). A common mistake is to forget that H_n exists and to assume that H_m is the effective head.

Because natural ventilation exists in nearly all mines (but in widely varying amounts), the mine head and quantity are never constant, even through fan operation is unchanged. This is due to the almost continual fluctuation of natural pressure at a given mine. Hence the presence of natural ventilation usually makes itself readily apparent.

Head gradients of natural ventilation or combined ventilation systems are rarely employed, since the elevation term cannot be disregarded and absolute pressures must be employed. If small, H_n is usually ignored in plotting gradients for combined systems. If large, H_n can be plotted in the gradient as an instantaneous energy source, similar to a fan. Because of fluctuation and reversibility, it may not be worth considering.

10.6 FAN SELECTION

It should become apparent at this point that the problem in providing mechanical ventilation underground is not one of finding a fan to meet the requirements but one of deciding which fan would be best suited and most economical. To ventilate a mine adequately, a desired quantity of airflow at a specified head and specific weight is required, known as the mine head and quantity. This information establishes the operating point for the system. Almost an infinite number of fans are available whose head–quantity characteristics pass through the desired operating point, or can be made to do so by changing speed, pitch, or diameter. The problem: to determine the most satisfactory one.

Just the determination of the operating point for the system (head and quantity requirements) is really not as simple as it sounds, requiring usually extensive calculations. Furthermore, a working mine is not stationary; rather, it is continually changing, advancing here, retreating there. Ventilation requirements are not constant either. Because of changes in the mine layout, production, and working force, the mine resistance changes. Usually this can be forecast within limits, but this fluctuation requires that a fan be selected for a range of application and/or that its performance be variable, also within limits. Since a fan has a useful life of 15–25 years, an attempt is made to select a fan that will be suitable for the range of operating conditions to be encountered in this period. When this is not possible, because of extreme conditions, provision must be made for supplemental ventilation at a later time in the life of the fan. A typical situation is shown in Fig. 10.22. Mine ventilation requirements can be reasonably well predicted at the beginning and end of the lifespan of a fan. Since the resistance of a mine usually increases with age, its characteristic curve steepens. In addition, quantity and/or head requirements may increase with time because of increased production, remoteness of operations, or larger working force. The only way in which it is possible for a fan to satisfy the changing requirements is by an increase in its speed or pitch (if it is an axial-flow type). Of course, these changes must be carried out within the safe and economical operating range

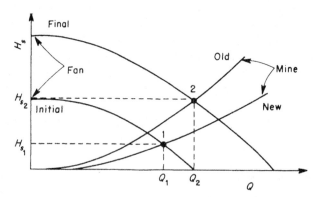

FIGURE 10.22   Operating range required of fan during its life or during life of the mine. Operating points 1 (initial) and 2 (final).

of the fan (i.e., on the steeper portion of its curve and near its peak efficiency) and without overload to its prime mover.

Having fixed the range of conditions to be expected over the life span of a fan, the real problem in selection arises: Which of several "suitable" fans will be "best suited" for the particular application? Assuming that safety, noise-level, and size limitations can be met, the problem resolves almost entirely into an economical one.

As with the selection of the most economical size of air shaft, the annual overall costs (the sum of the capital and operating costs per year) must be compared for the fans in question. The purchase and installation price for each fan depreciated over the expected life must be considered, as well as the power and maintenance charges. The fan having the lowest overall cost per year will be the most economical one to select (and the best suited, all other factors being equal).

Fan Characteristic Curves

Performance curves provide the most convenient and only satisfactory means of selecting a fan. With curves, it is possible to find the fan that will provide a certain head and quantity and to determine how changing conditions will affect fan performance.

Characteristic curves relating the conventional variables (head, quantity, power, and efficiency) are generally employed by the user, but the manufacturer may prepare special curves to assist in the proper choice of a fan. These permit the performance of an entire series of fans to be summarized in a single group of curves. Of the fan constants discussed in Section 9.8, specific speed and specific volume are most commonly used. Typical characteristic curves relating these constants for a series of similar, variable-pitch, axial-flow fans are given in Fig. 10.23. A sample problem illustrates their use.

Example 10.8 Required to deliver 150,000 cfm at 4.0 in. static head. Select the most suitable size of fan of the 8H series, whose characteristics are given in Fig. 10.23.

Solution: Determine the approximate size of fan operating near peak efficiency on the specific volume curve, say, at $Q_s = 2000$ cfm. Use Eq. 9.26.

$$D^2 = \frac{Q}{Q_s H^{1/2}} = \frac{150,000}{(2000)(4.0)^{1/2}} = 37.5$$

$$D = 6.13 \text{ ft}$$

Try a 6-ft fan (8H 72 model):

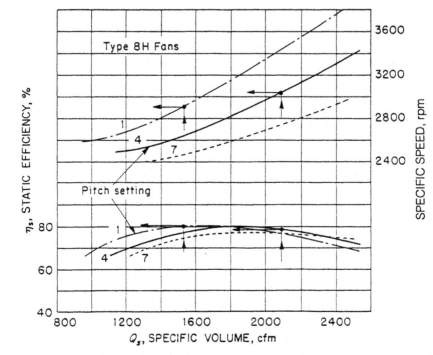

FIGURE 10.23 Characteristic curves for similar axial-flow fans relating specific speed, specific volume, and static efficiency for various pitch settings (Example 10.8). (By permission from Jeffrey Manufacturing Co., Columbus, OH.) (*Conversion factor:* 1 cfm = 0.47195 × 10⁻³ m³/s.)

$$Q_s = \frac{150,000}{(6)^2(4.0)^{1/2}} = 2085 \text{ cfm}$$

From the Q_s–η_s curve, read peak η_s = 79% with a no. 4 blade setting. From the same curve, read n_s = 3020 rpm. Find the speed by Eq. 9.25:

$$n = \frac{(3020)(4.0)^{1/2}}{6} = 1007 \text{ rpm}$$

Calculate the fan brake power, using a modification of Eq. 9.13:

$$P_m = \frac{(4.0)(150,000)}{(6350)(0.78)} = 121 \text{ bhp}$$

Try a 7-ft fan (8H 84 model) for comparison. Calculate Q_s:

$$Q_s = \frac{150,000}{(7)^2(4.0)^{1/2}} = 1530 \text{ cfm}$$

Read $\eta_s = 80\%$ for a no. 1 blade setting. Read $n_s = 2900$ rpm. Find n:

$$n = \frac{2900(4.0)^{1/2}}{7} = 830 \text{ rpm}$$

Calculate P_m:

$$P_m = \frac{(4.0)(150,000)}{(6350)(0.80)} = 118 \text{ bhp}$$

These results indicate that either fan could be used. The higher power cost and lower purchase price of the 8H 72 would have to be weighed against the lower power cost and higher purchase price of the 8H 84 fan.

Calculation of Most Economical Fan

Following the trial selection of several fans that are suitable from all standpoints and supply the desired head and quantity, a final selection is made on the basis of economics. Observe that (1) the most efficient fan is not necessarily the most economical, and (2) the operating costs are proportionately many times more important than the capital costs. If there is little difference in price, a comparison of power requirements will usually indicate which fan will be more economical. The fan supplier should be advised of the method used to make the final selection. One method is to reduce the expected cost of power during the useful life of the equipment to the present value of an annuity sufficient to yield the annual expenditure. The operating cost should be added to the capital cost and expected maintenance cost, and the overall total cost determined. A suitable fan with the lowest overall total cost is then selected. Some fan manufacturers provide computer diskettes containing fan selection software for selecting their products (Huffer, 1995).

Example 10.9   Given the two fans shown below, select the most economical on a present value basis if power costs $0.05/kWh and the fan parameters for the required air quantity are

Fan	Capital Cost, $	Electrical Power, hp (kW)
A	150,000	950 (708.4)
B	180,000	900 (671.1)

Assume that both fans will have a life of 15 years and that the interest rate is 15%.

Solution: To express the power or operating cost in terms of an equivalent present cost, the present-value factor for a series of uniform annual payments is required. For 15% interest and a life of 15 years, the present-value factor (e.g., Grant and Ireson, 1970) is 5.847. The present value of the overall cost is then calculated as

Fan	Present Value of Capital Costs, $	Power Costs, $/yr	Present Value of Power Costs, $	Present Value of Overall Costs, $
A	150,000	310,300	1,814,300	1,964,300
B	180,000	293,900	1,718,400	1,898,400

Thus fan B is selected on the basis of the minimum present value of the overall cost. The same decision would be achieved if the comparison were made on an annual cost basis.

10.7 FAN DRIVE AND CONTROL

Two of the most important external factors that affect the operation of mine fans, especially main installations, are the methods employed to drive and control the fan (Meakin, 1979).

Fan Drives

Where possible, the trend today is to power main fans with direct drives. Vane-axial and backward-curved centrifugal fans, both high-speed units, are always operated by direct drive. An internal drive with the prime mover located in the airstream is preferred where the law permits, both from the standpoint of initial cost and simplified arrangement. Internal drives are available up to 150 hp (112 kW).

For larger fans and where the law requires that the motor be placed out of the airstream, external direct drive is preferred. If the drive is maintained in good alignment, and if the equipment is properly sized, there is very little that can go wrong with an external arrangement.

Direct drive does not necessarily, however, provide the most efficient operation, since the fan often must operate close to the synchronous speed of electric motors where the peak fan efficiency may or may not exist. These speeds are 1800, 1200, 900, 720, 600, 514, and 450 rpm. For auxiliary power at the surface, diesel engines are occasionally used.

Sometimes a belt arrangement is chosen for a main fan, especially if variable control of the fan speed is required. The normal, indirect drive arrangement is by V-belt. Aside from the problem of maintaining matched belts, a V-belt drive provides good power transmission and simplified speed control of the fan.

Fan Control and Regulation

A main-fan installation should provide for control and regulation of the fan output, especially if, as is almost inevitable, a change of duty is anticipated over the life of the fan, motor, and drive. Computerized control of fan operation, although not practiced in this country, is being utilized abroad.

The two major types of fans rely on different mechanisms to vary output. Axial-flow fans (always vane-axial, if main fans) utilize changes in the pitch or setting of the blades to change head and quantity. Small fans do not offer this feature, but all large vane-axial fans do. Formerly, the fan had to be stopped to change the blade settings, but now fans are available on which the pitch can be changed hydraulically while the fan is rotating.

Centrifugal fans formerly had to rely on a speed change to vary output. Now variable dampers, inlet-vane control, or rotor tips can be installed to change head and quantity. With the latest advances in regulating fan performance, both axial-flow and centrifugal fans are capable of delivering approximately the same peak efficiency.

10.8 CHARACTERISTICS OF MULTIPLE-FAN SYSTEMS

The mine characteristic curve (or simply, mine characteristic) has been introduced for single-fan systems. It is a head vs. quantity curve and can be expressed analytically in the form of the Atkinson equation:

$$\text{Mine } H = (\text{mine } R)(\text{mine } Q)^2 \qquad (10.17)$$

where mine R is a constant. It is of considerable help in providing an effective means to display the head–quantity relation. Furthermore, if the fan characteristic curve is given, the operating point of the system can be displayed on the graph. For a multiple-fan system, however, the mine R is difficult to define, and the head–quantity relationship in the analytical form of Eq. 10.17 is not readily available. Instead of a single H–Q curve, a multiple-curve approach referred to as subsystem characteristic curves has been introduced for multiple-fan systems (Wang, 1992).

Conceptually, to construct the mine characteristic curve for a single-fan system, the fan is replaced by a head or quantity source (similar to a voltage or current source). The mine is then driven (ventilated) by this source to generate a series of mine operating points. The plot of data for the operating points yields the mine characteristic curve. Note that fan characteristic data

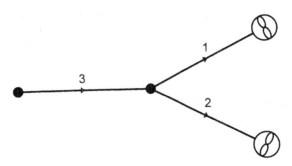

FIGURE 10.24 Example two-fan system, resistance factors (N·s²/m⁸): $R_1 = 0.4$, $R_2 = 0.3$, $R_3 = 0.3$. (After Wang, 1992. By permission from Society for Mining, Metallurgy, and Exploration, Inc., Littleton, CO.)

are not required; that is, the mine characteristic curve is independent of the fan.

Similarly, to construct a subsystem characteristic curve for a multiple-fan system, one fan, say fan k, is replaced by a head or quantity source. The "mine," which is called subsystem k and contains all fans except the kth fan, is then driven by this source to generate a series of subsystem operating points while all other fans remain in operation. If desired, the air quantities for other fans are also recorded with each subsystem operating point. The plot of data for the operating points yields the subsystem characteristic curve with reference to fan k (or referring to as the characteristic curve for subsystem k). In this manner, one can construct a subsystem characteristic curve with reference to any fan in the ventilation system. Note that the characteristic curve for the kth subsystem is independent of the kth fan. Each operating point is equivalent to a solution of the ventilation network with a constant head source or a fixed-quantity branch.

The subsystem characteristic curve just described is the H_k–Q_k type. Another important group of subsystem characteristic curves is the Q_j–Q_k type, where Q_j represents the air quantity of the jth fan in the ventilation system. It should be pointed out that the quantity–quantity curve of any two branches for a single-fan system is a straight line. In a multiple-fan system, however, the quantity–quantity curve is generally not a straight line.

As an example, the subsystem and fan characteristic curves for the two-fan system of Fig. 10.24 are shown in Figs. 10.25–10.29. The subsystem curves are plotted using computational results listed in Table 10.1. In Fig. 10.25, the space curves L_1 and L_2 are called the *solution curves* for subsystems 1 and 2, respectively. They represent the loci of subsystem operating points. Projection of them on vertical plans results in subsystem characteristic curves P_1 and P_2, which are also shown in Figs. 10.26–10.28 with fan curves t_1 and t_2. The subsystem operating points are represented by X_{11}

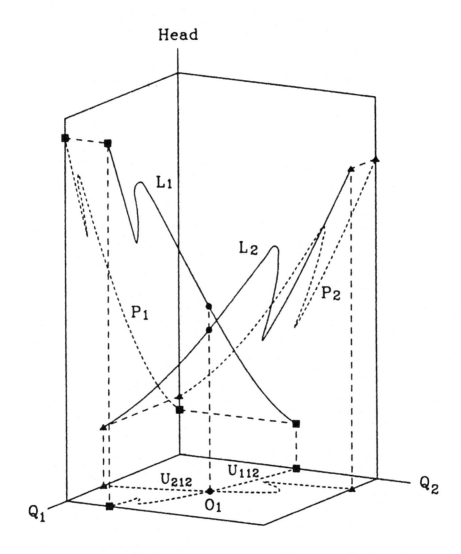

FIGURE 10.25 Solution curves (L_1 and L_2) and subsystem characteristic curves (P_1, P_2, U_{112}, and U_{212}) in a three-dimensional space. (After Wang, 1992. By permission from Society for Mining, Metallurgy, and Exploration, Inc., Littleton, CO.)

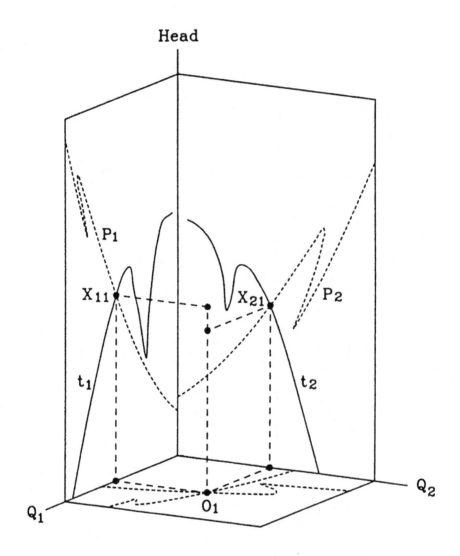

FIGURE 10.26 Subsystem and fan characteristic curves in a three-dimensional space. (After Wang, 1992. By permission from Society for Mining, Metallurgy, and Exploration, Inc., Littleton, CO.)

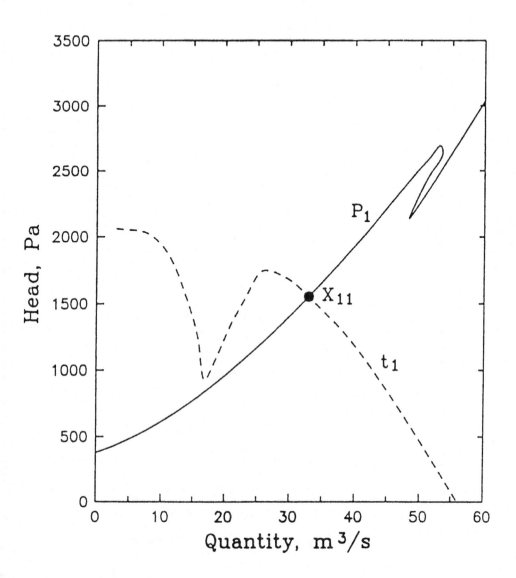

FIGURE 10.27 Subsystem characteristic curve P_1. (After Wang, 1992. By permission from Society for Mining, Metallurgy, and Exploration, Inc., Littleton, CO.) (*Conversion factors:* 1 in. water = 248.84 Pa, 1 cfm = 0.47195×10^{-3} m³/s.)

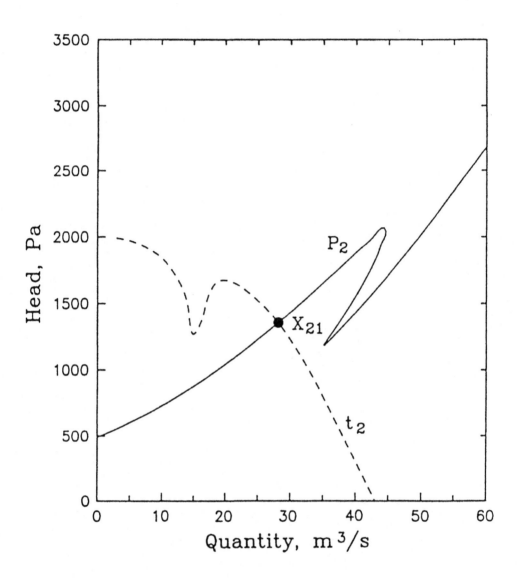

FIGURE 10.28 Subsystem characteristic curve P_2. (After Wang, 1992. By permission from Society for Mining, Metallurgy, and Exploration, Inc., Littleton, CO.) (*Conversion factors:* 1 in. water = 248.84 Pa, 1 cfm = 0.47195 × 10^{-3} m^3/s.)

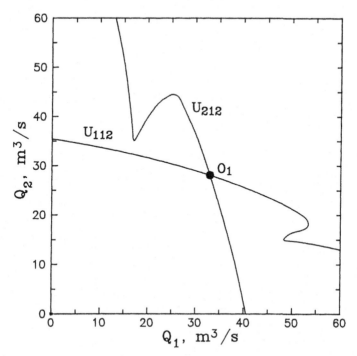

FIGURE 10.29 Subsystem characteristic curves U_{112} and U_{212}. (After Wang, 1992. By permission from Society for Mining, Metallurgy, and Exploration, Inc., Littleton, CO.) (*Conversion factors:* 1 in. water $= 248.84$ Pa, 1 cfm $= 0.47195 \times 10^{-3}$ m³/s.)

and X_{21}. Projection of curves L_1 and L_2 on a horizontal plane yields subsystem characteristics of the quantity vs. quantity type U_{112} and U_{212}, respectively.

As is often done with the mine characteristic curve, the subsystem curves can be employed to display the effect of replacing a fan or changing the fan performance characteristic. If, for example, with replacement of fan 1, the new subsystem operating point, which must stay on the curve P_1 of Fig. 10.27, is Y_{11}. To determine the new operating point for fan 2, read the Q_1 value of Y_{11} from Fig. 10.27; then read the corresponding Q_2 value from curve U_{112} in Fig. 10.29. Next, go to Fig. 10.28 and determine the new operating point Y_{21} on fan curve t_2 at the new value of Q_2. Since the characteristic of fan 1, which is an element of subsystem 2, has been changed, the subsystem characteristic curves P_2 and U_{212} have to be reconstructed.

TABLE 10.1   Data for a Two-Fan System with Two Meshes

Subsystem 1			Subsystem 2			
Q_1 m³/s	Q_2 m³/s	P_1 Pa	Q_1 m³/s	Q_2 m³/s	P_2 Pa	Remarks
0.00	35.43	377	40.51	0.00	492	
2.75	35.00	431	40.00	2.47	543	
26.42	30.00	1234	35.00	21.60	1101	
32.96	28.09	1553	32.96	28.09	1355	Operating point
41.57	25.00	2020	30.00	36.33	1716	
48.43	22.00	2426	28.00	40.71	1914	
52.02	20.00	2639	26.00	44.23	2066	
52.12	17.00	2520	25.00	44.50	2043	
48.31	15.00	2136	23.00	43.46	1892	
55.07	14.00	2644	20.00	39.95	1557	
60.00	13.00	3039	18.00	36.48	1290	
—	—	—	17.00	35.10	1184	
—	—	—	16.80	36.22	1237	
—	—	—	16.00	43.66	1639	
—	—	—	15.00	50.83	2075	
—	—	—	13.11	60.00	2683	

Source: Wang (1992).

(*Conversion factors:* 1 in. water = 248.84 Pa, 1 cfm = 0.47195×10^{-3} m³/s.)

PROBLEMS

10.1   Sketch the approximate shape of the pressure gradients and characteristic curves for a ventilation system consisting of a blower fan connected to a mine with two airways in series. In the first airway, H_l = 1.2 in. water (299 Pa), and H_v = 0.6 in. water (149 Pa). In the second airway, H_l = 0.9 in. water (224 Pa), and H_v = 1.0 in. water (249 Pa). Locate the operating point for the system if 80,000 cfm (38 m³/s) of air is flowing. Determine the mine and fan heads, assuming no area change at the fan connection. What would a manometer across the fan read?

10.2   It is required to supply 35,000 cfm (17 m³/s) of air at 1.5 in. water (373 Pa) static head. Calculate the speed at which the 60–in. (1.5–m) fan in Fig. 10.30 should operate to meet the requirement if the present speed is 720 rpm. Assume standard air specific weight.

10.3   If the air in Problem 10.2 must be supplied at 0.0680 lb/ft³ (1.09 kg/m³) air specific weight, at what speed must the fan of Fig. 10.30 operate if its characteristic is plotted at standard specific weight?

10.4   Determine the speed at which a 48–in. (1.2–m) diameter centrifugal fan should operate to deliver 40,000 cfm (19 m³/s) of air at 4.0 in. water (995 Pa) static head and an air specific weight of 0.050 lb/ft³ (0.80 kg/m³). The characteristic curve for a similar fan of 60–in. (1.5–m) diameter operating at 800 rpm and 0.0750 lb/ft³ (1.20 kg/m³) specific weight is shown in Fig. 10.31.

10.5 The static characteristic curve of a 6–ft (1.8–m) vane-axial fan is shown in Fig. 10.32. It delivers 140,000 cfm (66 m³/s) when connected to a mine in the exhaust position at the collar of a ventilation shaft of the same diameter. To increase the flow of air, an evasé is added to the fan discharge; it has a slope of 10% and a length of 15 ft (4.6 m). Determine the quantity of flow that will result, and plot the new pressure gradients for the system.

10.6 If it is desired to maintain the same airflow in Problem 10.5, at what speed would the fan have to operate after the evasé was installed? Original fan speed was 1250 rpm.

10.7 A mine is to be ventilated by the fan of Fig. 10.30; the point $H_s = 2.0$ in. water (498 Pa) and $Q = 40,000$ cfm (19 m³/s) lies on the mine characteristic. It is calculated that a natural ventilation head of 1.0 in. water (249 Pa) exists with the fan. Determine the effective static head of the system and the quantity of air flow.

10.8 If the natural ventilation is against the fan in Problem 10.7, determine the effective static head and quantity for the system.

10.9 If the speed of the fan in Problem 10.7 is 600 rpm, calculate the required speed to increase the air flow by 50%. Also for Problem 10.8.

10.10 Two fans are available for use underground. Tests determine the following head–quantity relationships for each:

Fan A				Fan B			
H_s		Q		H_s		Q	
in. water	(Pa)	cfm	(m/s)	in. water	(Pa)	cfm	(m³/s)
4.0	(995)	0	(0)	2.5	(622)	0	(0)
3.0	(747)	28,000	(13)	2.9	(722)	15,000	(7)
2.0	(498)	38,000	(18)	2.5	(622)	35,000	(17)
1.0	(249)	45,000	(21)	1.5	(373)	52,000	(25)
0		50,000	(24)	0		70,000	(33)

The fans are to be operated in parallel with intervening and common airways, the characteristics of which are as follows:

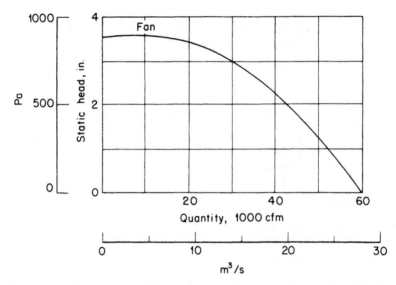

FIGURE 10.30   Fan characteristic curve (see Problems 10.2, 10.3 and 10.7–10.9).

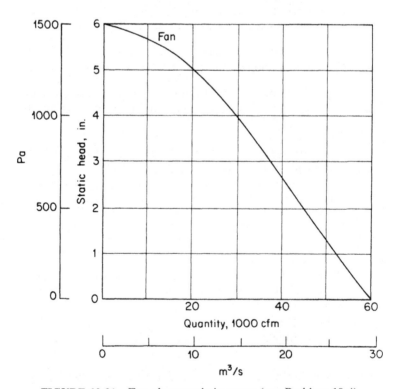

FIGURE 10.31   Fan characteristic curve (see Problem 10.4).

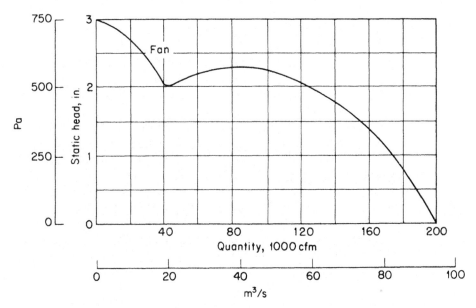

FIGURE 10.32 Fan characteristic curve (see Problems 10.5 and 10.6).

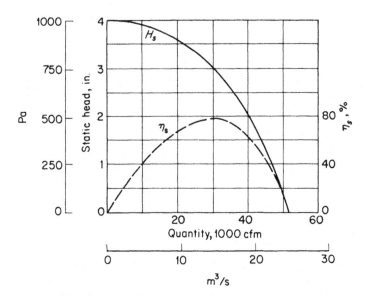

FIGURE 10.33 Fan characteristic curve (see Problem 10.11).

Airway	H_L		Q	
	in. water	(Pa)	cfm	(m³/s)
a	1.0	(249)	25,000	(12)
b	1.0	(249)	30,000	(14)
c	2.0	(498)	60,000	(28)

The common airway is c; fan A is connected to airway a, and fan B is connected to airway b, with airway c connected to the intersection of airways a and b. The fans are exhausting. Disregarding velocity head, determine the quantity of flow in each airway.

10.11 Select a fan operating at peak efficiency to deliver 110,000 cfm (52 m³/s) of air weighing 0.0450 lb/ft³ (0.72 kg/m³) against a mine static head of 7.4 in. water (1841 Pa). Specify the fan diameter and speed. The characteristic curves of a similar fan are plotted in Fig. 10.30 for D = 36 in. (0.9 m), n = 600 rpm, and w = 0.0750 lb/ft³ (1.20 kg/ m³).

CHAPTER 11

Auxiliary Ventilation and Controlled Recirculation

11.1 IMPORTANCE OF FACE AND AUXILIARY VENTILATION

The ventilation of working faces in underground mines is one of the most important tasks of the mine ventilation engineer because it is at the face that the ventilating air performs its most useful functions. Face ventilation can be a normal extension of the main ventilation system. However, the primary system is not ideally suited for ventilation of dead-end openings. Because of this, reliance is placed often on supplemental means of supplying air to working faces. The practice of augmenting the main ventilation system is termed *auxiliary ventilation*.

Two primary applications of auxiliary ventilation in mining are to ventilate dead-end workings and to provide supplemental flow to assist the main ventilation system, as in booster ventilation or in controlled recirculation. Ventilation of dead-end workplaces is the most common application of auxiliary ventilation. The method is employed in both coal and metal mining, for both development and exploitation.

Drifts, raises, shafts, inclines (slopes), and winzes normally require auxiliary ventilation during their development as do stopes with only one entrance. The ventilation of these openings often reverts back to the main ventilation system after their development is complete. In coal mines, the location of the air inlet or exhaust for each working face is required by law to be within 10 ft (3 m) of the face. This can be performed via a fan and duct system or by a line brattice system.

In coal mining, face ventilation is a critical task of the ventilation system that ensures that methane is swept away from the face and that dust is diluted or carried away. The ventilation system must offer sufficient airflow at the face to accomplish both of these objectives. Historically, line brattice installed near the face has been used for this purpose. Because line brattice works by utilization of the airflows supplied by the main fan or fans, it does not qualify strictly as an auxiliary ventilation method. However, its use parallels that of the auxiliary fans and vent pipe or tubing used in some mines, and so it is covered here. Brattice and ventilation ducts are used in very similar ways in coal mines; these methods are outlined and compared.

In addition, other methods that affect the face ventilation system, such as scrubbers, sprayfans, and injectors, are discussed as well.

The final topic considered in this chapter is recirculation of air in a mining system. It is a general principle of mine ventilation that recirculation of air can be utilized as a ventilation strategy only under carefully controlled situations. It is important therefore that recirculation be prevented from occurring accidentally, especially in coal mines where, in this country, it is prohibited by law.

11.2 CHECK CURTAINS AND LINE BRATTICE

Brattice and *curtain* are mining terms that refer to partitions to separate airways into intake and return splits. In coal mines, brattice is commonly used to control the airflow in the working section and to direct the air to each active mining face. Brattice or curtain material is currently manufactured of plastic sheeting with or without fabric reinforcement. It should be resistant to tearing and other abuse, have a low flame-spread index, and be relatively leakproof and inexpensive. For safety purposes, it is desirable to install transparent or translucent curtain material where haulage vehicles or other equipment will regularly pass through the brattice.

The two primary patterns of airflow when using brattice are shown in Fig. 11.1. In Fig. 11.1a, an exhaust pattern of face ventilation is illustrated. Check curtains divert the normal flow of air, and the line brattice directs the air to the face where it is used to remove gas and dust. Workers are usually subjected to less dust in this system as the dust is carried away from the workers and exhausted behind the brattice. Because of the tendency of airflows to seek out the minimum resistance path, however, the air may have a tendency to move around the end of the brattice without adequately sweeping the face. If this is a problem because of gas emission at the face, then the blowing pattern shown in Fig. 11.1b is normally used. In this system, the air is directed to the face through the narrower split with resulting higher velocity and better ability to sweep gases from all corners of the working area. This may, however, subject the workers to higher dust concentrations as the airflow tends to move the dust toward the worker locations.

Brattice is easy to set up and can be changed quickly as the equipment changes faces. The brattice material is ordinarily hung from the roof using roof bolts or spads as anchor points. Because of the nature of brattice material, it may leak badly even if hung with care. It is therefore not normally used for distances of greater than about 200 ft (60 m). In addition, it impedes the passage of workers and equipment. Because of its simplicity, however, it continues to be used extensively in coal mines.

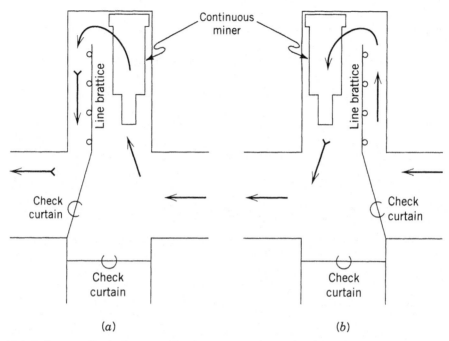

FIGURE 11.1 Basic layouts of exhaust (*a*) and blowing line (*b*) brattice systems for a continuous miner section.

11.3 AUXILIARY FANS AND VENT PIPE OR TUBING

The use of auxiliary fans and ventilation pipe or tubing parallels the use of brattice in many ways, as shown in Fig. 11.2. The check curtains are optional; they improve the amount of air at the face but are not necessary in all systems. An exhausting auxiliary fan system (Fig. 11.2*a*) is advantageous since it exhausts dusty air from the face and removes it from the breathing zone of the face workers. The disadvantage of this system is that the air does not sweep the face particularly well. The blowing-fan system (Fig. 11.2*b*) provides better capability for sweeping any gas out of a working face (where such gases exist), because a blowing arrangement tends to extend its influence far ahead of the end of the ventilation tubing. The reason for this is illustrated in Fig. 11.3. The isovels (velocity contours) indicate that the blowing tubing can effectively direct its airflows for significantly more than 10 ft (3 m) beyond the end of the tubing while the exhausting tubing cannot.

A wide range of fans, both centrifugal and axial-flow, is available to power auxiliary ventilation systems. Internally driven axial-flow fans are popular because of their compactness and adaptability to arrangement in series. This series capability is one primary advantage of auxiliary fans that allows them to be used in many underground workings where the faces to be ventilated are far from the main ventilation air source. Auxiliary fans vary in diameter, power, and weight and are chosen to meet the specific needs of the area to be ventilated. Fans for drifts, raises, and tunnels are normally meant to be mounted or hung above the opening floor, but coal mine auxiliary fans and larger noncoal fans are often mounted on wheels and are sometimes self-propelled to enhance their mobility. While auxiliary fans are quite useful, they generate high levels of noise that can be of concern. Manufacturers offer optional noise-attenuation packages for fans that reduce the noise generated.

Auxiliary fans deliver air to the face through either rigid or flexible ducts that are normally 8–48 in. (200–1220 mm) in diameter. Rigid duct or vent pipe is generally made of galvanized steel, fiberglass, or resin extrusions. Most are circular in cross section, but oval fiberglass duct is available for use in mines where headroom is a problem. Many mines use flexible tubing instead. These are normally made of brattice-type material and are available in steel-ribbed versions for use with exhausting fans and collapsible versions for blowing fans. Collapsible vent tubings are often referred to as "bags" in the mines. Selection of the optimal size of duct is outlined in Section 11.5.

Typically, fans and tubing are more expensive but ensure less leakage than brattice. Always more popular in metal mines, fans and tubing are displacing brattice in coal mines. Both blowing and exhaust auxiliary ventilation systems are in common use with dust- and gas-control characteristics similar

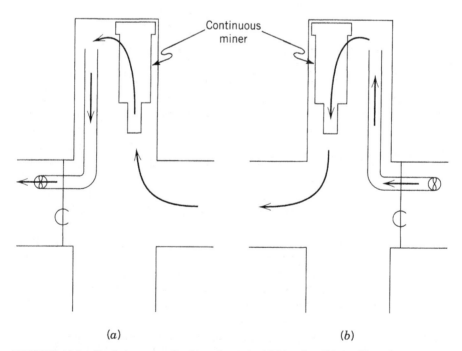

(a)   (b)

FIGURE 11.2 Basic layouts of exhausting (*a*) and blowing (*b*) auxiliary-fan systems for a continuous miner section (check curtains shown are optional).

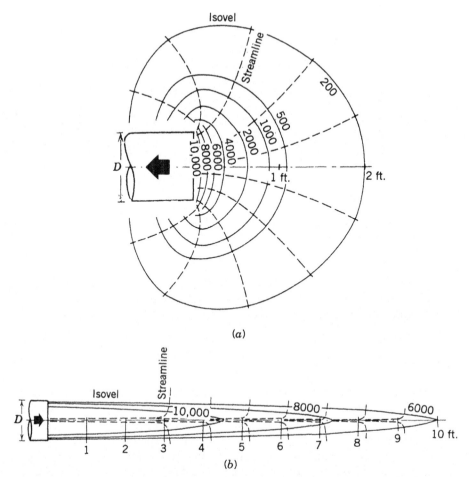

FIGURE 11.3 Comparison of airflow patterns for vent tubing at (*a*) the intake of an exhaust system and (*b*) the discharge of a blower system. Shown are isovels (constant-velocity lines) and streamlines of flow for a 1-ft (0.3-m) duct transporting 7850 cfm (3.70 m³/s) and having an average velocity of 10,000 fpm (50.8 m/s). (After McElroy, 1943; Hartman, 1962.) (*Conversion factors:* 1 ft = 0.3048 m, 1 fpm = 0.005080 m/s, 1 cfm = 0.47195 × 10^{-3} m³/s.)

to brattice systems. One advantage of fan systems is that they supplement the head of the primary fan, resulting in added total flow quantities when compared with brattice face ventilation systems (Wallace et al., 1990).

11.4 OTHER OPTIONS FOR FACE AND AUXILIARY VENTILATION SYSTEMS

A variety of special devices and ventilation arrangements can be utilized in mines to provide adequate face ventilation. These options often implement

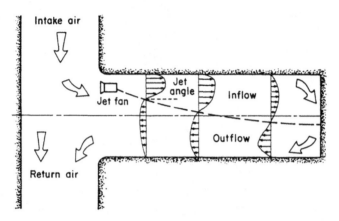

FIGURE 11.4   Diagram illustrating typical jet fan velocity profile when used in a dead-end opening (Goodman et al., 1992).

specialized equipment to enhance the ventilation of the working face or use ventilation concepts in combination to better supply face ventilation.

Jet Fans

Jet fans are used as free-standing air movers to provide for the mixing of fresh air at a given working face. These fans are normally set up without tubing in larger mine openings or in situations where tubing or brattice would interfere with the mining process. The general pattern of airflow that is generated by a jet fan in a dead-end opening is illustrated in Fig. 11.4. The depth of penetration of the air movement can be well over 300 ft (100 m), sufficing for many operations. The flow of air reaching the face under these conditions may be rather unreliable; therefore, this type of system is not consistent with good practice in gassy mines.

Injectors

Injectors are often called *static air movers* because they have no moving parts. An injector is powered by compressed air and is used mostly in noncoal mines as a means of providing good flows of air at the working face. Air movement is provided primarily by the venturi effect discussed in Section 9.10. Injectors are very inefficient for use in routine ventilation of faces or headings, but they can provide a quick and easy method of ventilation for headings after blasting or for occasional face ventilation needs because they are easy to set up and use when compressed air is available.

Diffusers

Diffusers are small fans used to provide fresh air and mixing at the working face. The diffuser is often mounted on face machinery, as shown in Fig. 11.5a. It ensures that gas generated at the face is swept away and thus is helpful with exhaust face ventilation systems where the flow of air to the face is not definitive. However, the use of a diffuser on a continuous miner in coal mines in the United States is not presently permitted by the Mine Safety and Health Administration because of the fear of recirculation at the face.

Scrubbers

Scrubbers are primarily dust control devices and have been described in Section 4.11. To perform their job in dust control, scrubbers often move in

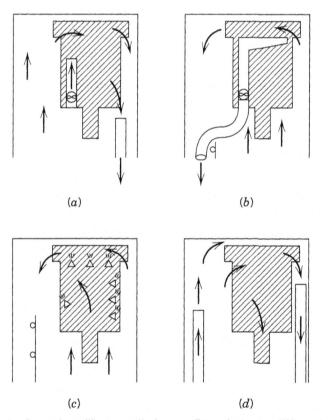

(a)   (b)

(c)   (d)

FIGURE 11.5 Several auxiliary ventilation configurations: (*a*) diffuser fan and exhaust tubing; (*b*) scrubber with exhausting line brattice; (*c*) spray fans and exhaust line brattice; (*d*) push–pull tubing system.

excess of 5000 cfm (2.4 m³/s) through the scrubber ductwork. If the ductwork is designed similar to that in Fig. 11.5b, then the air drawn into the scrubber acts also to ventilate and remove methane from the working face. Thus the scrubber can be an effective means of improving the ventilation in a coal mine. As with a diffuser, however, recirculation circuits can be set up, and thus the quantity of air moved through the scrubber should be regulated to less than that of the quantity flowing into the face area.

Sprayfans

Sprayfans are water sprays that move air by entrainment. These have been described in Section 4.11 as venturi sprays. As shown in Fig. 11.5c, spray fans are often used in conjunction with exhausting ventilation systems to ensure that ventilating air reaches all corners of the working face.

Push–Pull Systems

Push–pull systems use both blowing and exhaust elements, as illustrated in Fig. 11.5d. To bring air to the face or to exhaust it, either tubing or brattice is utilized. This system has been used in coal mines where auger miners are employed or where abnormally wide entries are employed.

Combination Systems

In most of the situations outlined above, the face ventilation plan is designed to provide for dust, gas, or other contaminant removal through the process of directing the air to the working face. Their thrust is toward the current working face only. In some mining situations, it is important to provide for ventilation to the current working place while also providing for environmental control in an adjacent face. Such a situation occurs in many coal mines when roof bolters must work downwind of the continuous mining operation (Fig. 11.6). In this situation, air exhausted from the continuous miner face may contain a high concentration of dust, as illustrated in Fig. 11.6a. This dust will then contaminate the downwind faces. A solution to the problem is depicted in Fig. 11.6b. In this combination plan, an auxiliary fan is used to exhaust air from a continuous miner while a brattice directs air into the bolter location. However, instead of simply exhausting the dust-laden air from the auxiliary fan to the downwind working faces, the tubing discharges through the brattice to the return. This example illustrates the need to consider all combinations of techniques when a face ventilation system is installed. The right combination plan may greatly improve the ventilation of one or more working faces.

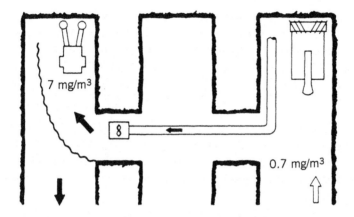

(*a*) Vent tubing without bypass

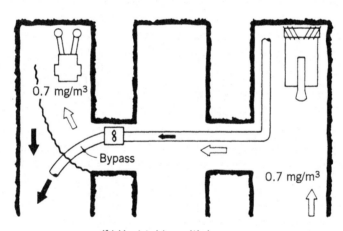

(*b*) Vent tubing with bypass

FIGURE 11.6 Combination of vent tubing [(*a*) without and (*b*) with bypass] and brattice ventilation for reduction of dust at roof bolter face; typical dust concentrations shown (Anon., 1985b).

11.5 EQUIPMENT SELECTION AND DESIGN CONSIDERATIONS

In selecting fans and ducts for an auxiliary ventilation system, the designer must keep in mind a number of factors. The most important are the following:

1. *Compliance with Company Rules and Government Regulations.* A number of considerations may be important here, including noise, state

and federal regulations on auxiliary fans, and whether strata gases may be encountered. Noise associated with a blowing system is generally more of a problem than with an exhaust system. In either system, noise is generally high, and noise-attenuation equipment may be desirable or required to meet company rules or governmental regulations. Noise also affects productivity, and this should be considered in the design procedure.

2. *Efficiency of the System.* An auxiliary fan always requires power, and this power adds to the overall costs of operating the ventilation system. If the fans are to be operated continuously during production, the efficiency of the fans should be considered. It may be justifiable to purchase more costly fans if their efficiency is higher. The type of tubing should also be considered, as the energy efficiency varies significantly with tubing type.

3. *Standardization of Equipment.* A mine that uses auxiliary fans may wish to standardize the duct types and sizes and the models of fans utilized to reduce the inventory of equipment and to allow for easy design of new installations. Standardization also helps to ensure that equipment and spare parts are available when needed.

4. *Interface with Other Ventilation Circuits in the Mine.* The designer must keep in mind the possible effects that auxiliary ventilation systems will have on the other ventilation circuits in the mine. It may be desirable to add doors, stoppings, brattices, or other ventilation control devices to ensure that ventilation is suitable both while the auxiliary fan is operating and when it is not operating.

5. *Leakage, Return Air, and Compressibility Considerations.* When designing longer ventilation pipe or tubing installations, it may be necessary to consider leakage, head losses in the return airflow, and the compressibility effect in the tubing (Section 5.7). These effects are generally ignored in shorter tubing systems but may be significant as the length of tubing increases.

The selection process requires that fans and ducts operate at a reasonable efficiency in the intended application. A characteristic curve of a proposed system can be plotted for various settings of the fan blades so that design information is readily available.

Basic Design Considerations

Chapter 5 deals with the principles of airflow in all mine openings, including ducts. Useful for duct calculations is Example 5.6, based on generic friction factors for ventilation tubing and pipe. It may be advantageous, however, to utilize specific manufacturers' data for head-loss calculations, because their data may be more accurate for their products. Shock-loss approximations for various tubing fittings such as couplings and elbows, should also be consulted and included in the calculation where applicable. Data for some

TABLE 11.1   Equivalent Lengths for Vent Pipe and Tubing

Source	Equivalent Length, ft (m)
Entrance loss[a]	35 (10.7)
Exit loss[a]	100 (30.5)
Bend (generic), 90°	35 (10.7)
Bend (generic), 60°	24.5 (7.5)
Bend (generic), 45°	17.5 (5.3)
Coupling, spiral-reinforced duct	8 (2.4)
Fiberglass duct	
90° round sharp[b] bend	23D (7D)
90° round smooth[c] bend	12.5D (3.8D)
90° oval sharp[b] vertical bend	20D (6.1D)
90° oval smooth[c] vertical bend	10D (3.0D)
90° oval sharp[b] horizontal bend	29D (8.8D)
90° oval smooth[c] horizontal bend	14D (4.3D)
45° bend	50% of 90° value
60° bend	70% of 90° value
Lay-flat tubing	
coupling	6 (2)
90° bend[d]	150 (46)
45° bend	50% of 90° value
60° bend	70% of 90° value

[a] Used only if fan is not located at the point of loss.

[b] Sharp bend duct centerline radius = 0.75D.

[c] Smooth bend: duct centerline radius = 1.25D.

[d] Duct centerline radius = D.

Notes: (1) most values derived from ABC Industries, Inc., laboratory measurements (Anon., 1995c); (2) the table variable D refers to duct diameter.

(*Conversion factor:* 1 ft = 0.3048 m)

of these fittings are presented in Table 11.1, in terms of equivalent lengths of ducts to be used in calculations.

Example 11.1   Determine head versus quantity values for an auxiliary ventilation system that is expected to deliver 6000 cfm (2.832 m³/s) of ventilating air through 250 ft (76.2 m) of new 16-in. (410-mm) spiral-reinforced flexible duct with a blowing fan. Assume that two 90° bends are necessary.

Solution:

1. Calculate the discharge velocity head (Eq. 3.10):

$$A = \frac{\pi}{4} \left(\frac{16}{12} \right)^2 = 1.396 \text{ ft}^2$$

$$V = \frac{6000}{1.396} = 4297 \text{ fpm}$$

Using Eq. 5.14,

$$H_v = \left(\frac{4298}{4009}\right)^2 = 1.15 \text{ in. water}$$

2. Determine the equivalent length of tubing from fan to farthest point, using the actual length of tubing plus equivalent lengths for bends and couplers. Neglect the return airflow head loss.

Item	Unit Equivalent Length, ft	Total Length or Equivalent Length, ft
16-in. spiral-welded steel pipe	—	250
Couplers (25-ft pipe lengths, 10 required)	8	80
90° bend (2 required)	35	70
Total length + equivalent length	—	400
Exit loss	100	100
Total lengths + equivalent length	—	500

3. Multiply the total length by friction loss/100 ft from Appendix Fig. A.3 to obtain the head loss caused by airflow in the tubing. Note that this graph applies to new steel pipe and must be corrected for other materials. Read $H_l = 1.4$ in. water/100 ft. The correction factor for spiral-reinforced flexible duct is 1.50. For 500 ft of flexible duct

$$H_l = 1.4(1.50)\left(\frac{500}{100}\right) = 10.50 \text{ in. water}$$

4. Total head loss of the tubing is

$$H_v = 1.15 \text{ in. water}$$

$$H_l = \underline{10.50}$$

$$H_t = 11.65 \text{ in. water (2.89 kPa)}$$

5. Obtain additional points in order to plot system curves, using the head–quantity relationship (Eq. 6.6):

\multicolumn{2}{c}{Q}		\multicolumn{2}{c}{H_l}		\multicolumn{2}{c}{H_t}
cfm	(m³/s)	in. water	(kPa)	in. water	(kPa)
4000	(1.888)	4.67	(1.16)	5.18	1.29
6000	(2.832)	10.50	(2.61)	11.65	(2.90)
8000	(3.776)	18.67	(4.65)	20.71	(5.15)

These points can be plotted as the tubing characteristic curves on various fan performance curves in order to select a fan that would be suitable for this application. Resistance charts and fan performance curves provided by manufacturers are based on standard conditions. Therefore, designers must modify their calculations if conditions are not reasonably close to standard.

Provision of Head for Return Flow

In an auxiliary ventilation system involving fan and tubing installed in a mine opening, it should be understood that the auxiliary fan must provide sufficient head for the system to supply air to the face through the duct as well as return it to the main ventilation airstream through the opening. Since the quantity of flow is usually small, however, and the size of the opening in relation to the size of the duct is large, the head required for the return airflow is small and can generally be disregarded in calculations of fan head. An example illustrates this point.

Example 11.2   If the auxiliary fan in Example 11.1 must also supply the head to return the air to the main ventilation circuit through an 8 × 10-ft (2.4 × 3.0-m) drift that has a friction factor of 70×10^{-10} lb·min²/ft⁴ (0.013 kg/m³), what return head must be supplied? What percentage of the total fan head does the return head represent?

Solution: The return head loss can be calculated most directly using Eq. 5.20 (or Eq. 5.20a), as follows:

$$H_f = \frac{(70 \times 10^{-10})(36)(250)(6000)^2}{(5.2)(80)^3} = 0.0009 \text{ in. water}$$

Auxiliary fan total head $H_t = 11.65 + 0.0009 = 11.65$ in. water (2.89 kPa). Percentage of head for returning the air through the drift is

$$\frac{0.0009}{11.65} = 0.0001 \quad \text{or} \quad 0.01\%$$

The head required to return the air is thus insignificant.

Compressibility Effects

Unlike primary mine ventilation systems dealing with major mine openings of large cross-sectional areas and relatively low air velocities and head losses, auxiliary fan and vent pipe systems for development openings of great length operate at relatively high pressures. Therefore, compressibility effects cannot be ignored. Instead of the conventional Atkinson equation based on the flow formula for noncompressible fluids, the formula for compressed air (Eq. 5.27) discussed in Section 5.7 must be employed. In this equation, use K_c values of 0.0006 lb²·in.·min²/ft⁷ (4440 kg²/s²·in.⁴) for metal pipe and 0.0007 lb²·in.·min²/ft⁷ (5180 kg²/s²·in.⁴) for flexible tubing as the following demonstrates.

Example 11.3   It is required to circulate 2530 cfm (1.194 m³/s) at the face of an 8 × 10-ft (2.4 × 3.1-m) drift through 5500 ft (1676 m) of 12-in. (300-mm) metal vent pipe. Atmospheric pressure is 14.7 psi (191.35 kPa). Calculate the head loss and quantity handled by the fan (1) in a blower system and (2) in an exhaust system. Disregard leakage.

Solution:

1. As blower system:

$$p_1^2 - (14.7)^2 = \frac{(0.0006)(2530)^2(5500)}{(12)^5} = 84.9$$

$p_1 = 17.35$ psia $= 2.65$ psig $= 73.4$ in. water (18.26 kPa)

At the fan:

$$Q = 2530 \text{ cfm } (1.194 \text{ m}^3/\text{s})$$

2. As exhaust system:

$$(14.7)^2 - p_2^2 = 84.9$$

$p_2 = 11.45$ psia $= -3.25$ psig $= 90.0$ in. water (22.40 kPa)

At the fan:

$$Q = 2530 \left(\frac{14.7}{11.45}\right) = 3250 \text{ cfm } (1.534 \text{ m}^3/\text{s})$$

Head loss in drift (assume $K = 100 \times 10^{-10}$ lb·min^2/ft^4 or 0.01855 kg/m^3):

$$H_l = \frac{(100 \times 10^{-10})(36)(5500)(2530)^2}{(5.2)(80)^3} = 0.0048 \text{ in. water } (1.19 \text{ Pa})$$

which is negligible. As a blower, a fan would be selected to handle 2530 cfm (1.194 m^3/s) at 73.4 in. water (18.26 kPa), and as an exhauster, 3250 cfm (1.534 m^3/s) at 90.0 in. water (22.40 kPa).

The variance in Q is usually disregarded in the head-loss calculation; for greater accuracy, an average value of Q may be used or the calculation repeated for segments of vent pipe, using corrected values of Q. The increased $H–Q$ requirements of an exhaust fan over a blower fan, in addition to the high-velocity dropoff discussed earlier, are factors that should be considered in a comparison of these two alternatives, especially in long installations.

11.6 BOOSTER VENTILATION

The use of booster fans underground to handle only a portion of the total air circulated and to supplement the primary fan is a category of auxiliary ventilation properly termed *booster ventilation*. Booster fans have been widely employed in noncoal mines; but they are not permitted in U.S. coal mines. Where permitted, booster fans may be used effectively on high-resistance splits to reduce the overall head loss across a parallel circuit that the primary fan must overcome. There are three ways in which desired quantities of air may be distributed to branches of a parallel circuit having unequal head losses across all the branches (see Section 7.4): (1) controlled splitting using regulators in the low-resistance split(s) to balance the circuit; (2) controlled splitting using booster fan(s) on the high-resistance split(s) to create a balance; and (3) free splitting achieved by decreasing K, O, or L or by increasing A or the number of airways in the high-resistance split to balance the pressure drops. On the basis of economical operating cost, they are listed in order of increasing preference. An example illustrates the savings that can be expected by comparing the power requirements for each method in a given system.

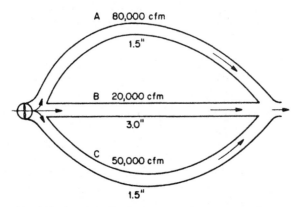

FIGURE 11.7 Parallel circuit in Example 11.4. (*Conversion factors:* 1 in. water = 248.84 Pa, 1 cfm = 0.47195 × 10⁻³ m³/s.)

Example 11.4 Given the parallel circuit of Fig. 11.7 with desired quantities and computed head losses as shown, and assuming a fan efficiency of 80%, determine the power required to ventilate the mine by (1) using regulators, (2) using a booster fan, and (3) reducing the resistance.

Solution:

Mine $Q = 80,000 + 20,000 + 50,000 = 150,000$ cfm (70.79 m³/s)

1. Using regulators in branches A and C:

    Mine $H_s = 3.0$ in. water:

    $$P_m = \frac{(3.0)(150,000)}{(6346)(0.80)} = 88.6 \text{ bhp (66.1 kW)}$$

2. Using booster fan in branch B:

    Booster $H_s = 3.0 - 1.5 = 1.5$ in. water

    $$P_m = \frac{(1.5)(20,000)}{(6346)(0.80)} = 5.9 \text{ bhp (4.4 kW)}$$

    Mine $H_s = 1.5$ in. water

    $$P_m = \frac{(1.5)(150,000)}{(6346)(0.80)} = 44.3 \text{ bhp (33.0 kW)}$$

    Total $P_m = 5.9 + 44.3 = 50.2$ bhp (37.4 kW)

3. Reducing resistance of branch B to achieve a head of 1.5 in. water:

Mine $H_s = 1.5$ in. water

$$P_m = \frac{(1.5)(150,000)}{(6346)(0.80)} = 44.3 \text{ bhp } (33.0 \text{ kW})$$

From the standpoint of operating cost alone, reducing the resistance of the free split is preferred. When overall costs are considered, the booster fan arrangement could be the most economical, regulations permitting. The indiscriminate use of regulators is often a costly solution, particularly when the difference in head across the branches is widely divergent. With the rising cost of electrical energy, prudent mine planning is necessary to avoid what may become future high-resistance splits as the mine develops. In the initial driving of airways, full advantage should be taken of methods that have been developed to optimize the parameters relating to airflow.

11.7 CONTROLLED RECIRCULATION

Recirculation Logic

The reuse or recirculation of air in underground mines is a practice that has been discouraged historically, particularly in coal mines where methane is likely to be found. *Reuse* of air is the practice of taking air from one working face and using it to ventilate another working face or area. While reuse is discouraged, it is often practiced even in coal mines to a limited extent where the exhaust air at a continuous miner may be further utilized to ventilate the roof bolter location (or vice versa). *Recirculation* has been defined as the movement of mine ventilation air past the same point more than once (Jones, 1987). In essence, recirculation is a specific form of reuse where air is used to ventilate the same face or same mining district more than once as it traverses through the mine. *Controlled recirculation* is a term that is used for a recirculation circuit that is purposefully designed and utilized in a controlled fashion to provide some ventilation benefits without adversely affecting other ventilation variables. It is this type of recirculation referred to here.

Recirculation has been prohibited in many countries for coal mining operations because of the fear that ventilating air would be used in a manner that would allow unsafe quantities of methane gas to accumulate. The use of controlled recirculation was suggested in the 1930s (Lawton, 1933), but it was not given serious consideration until the 1960s. At that time, a number of researchers began to investigate more thoroughly the potential benefits and disadvantages of recirculation in mining operations (Bakke et al. 1964; Leach, 1969). Many papers have appeared in the technical literature since that time that outline the benefits of recirculation. The use of controlled recirculation circuits is considered to be beneficial in mines where

1. Mine intake air must be heated because of cold climates.
2. Mine air is refrigerated for reasons of comfort or productivity.
3. Added velocity at the face would result in better turbulent mixing of air and methane at the point of release.
4. Added velocity at the face would more effectively carry away dusts.
5. Working faces are far removed from the mine portals, such as in undersea mining.

Either economic or environmental benefits can be cited for each situation in this list.

To effect a recirculation circuit in an underground mine, it is necessary to have an underground energy source such as a fan. Potential recirculation circuits can therefore occur where booster fans, auxiliary fans, and scrubbers are used. To consider the effects of a recirculation circuit, refer to the two mine-ventilation diagrams shown in Fig. 11.8, where the contaminant is assumed to be a point-source strata gas introduced under steady-state conditions. In Fig. 11.8a, a section or mining district is ventilated by a quantity of air Q_i that contains a decimal proportion of the contaminant x_i. At the working face, a quantity of the contaminant Q_g is introduced into the airstream and increases the level of contamination. As a result of this set of conditions, the following ventilation parameters are found in the mine:

Intake quantity: Q_i

Intake contamination level: x_i

Return quantity: $Q_i + Q_g$

Return contamination level:

$$x_r = \frac{Q_i x_i + Q_g}{Q_i + Q_g} \tag{11.1}$$

These relationships apply to any consistent set of ventilation units.

If additional air is desired in the mining face or district without increasing the amount of air into the district, then a recirculation circuit can be set up by developing a recirculation crosscut as shown in Fig. 11.8b. To provide the design quantity of air Q_r from point B to point A through the recirculation crosscut, and therefore increase the total flow through the working face to a quantity level of $Q_i + Q_r$, a single booster fan may be used at points C, D, or E. In addition, the crosscut flow quantity Q_r can be established by a

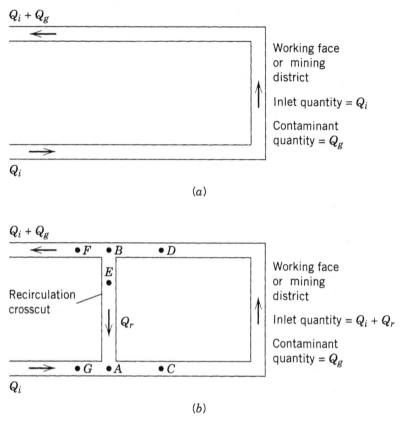

FIGURE 11.8 Basic layout of face or district ventilation flow quantities without (a) and with (b) a recirculation system. Fans may be located at point C, D, or E.

variety of multiple-fan setups. These new fans added for recirculation will affect the operation of the main fans, and the main fans may require adjustment to accommodate the desired recirculation and face quantities. In a recirculation circuit, the proportion or fraction of the air that is returned through the recirculation crosscut is termed the *recirculation factor* RF. In Fig. 11.8b, the recirculation factor is defined as

$$RF = \frac{Q_r}{Q_i + Q_r} \qquad (11.2)$$

When a recirculation circuit is set up in the mine openings as shown in Fig. 11.8b, the effect on the air quality can often be determined rather easily. For example, assuming that the influx of the contaminating gas in Fig. 11.8 remains unchanged for the case of recirculation, then the contamination levels can be calculated based on the newly established air quantities. The quantities and contamination levels around the recirculation loop and in the intake and return are as follows:

Intake (point G) quantity: Q_i

Intake (point G) contamination level: x_i

Point C quantity: $Q_i + Q_r$

Point C contamination level:

$$x_c = \frac{Q_i x_i + Q_r x_r}{Q_i + Q_r} \qquad (11.3)$$

Point D quantity: $Q_i + Q_r + Q_g$

Point E quantity: Q_r

Point F quantity: $Q_i + Q_g$

Contamination levels at points D, E, and F:

$$x_r = \frac{Q_i x_i + Q_r x_r + Q_g}{Q_i + Q_r + Q_g} \qquad (11.4)$$

Note that Eq. 11.4 algebraically reduces to Eq. 11.1. This indicates that recirculation does not change the contamination level in the return airway. Complete step-by-step analyses of these ventilation conditions have been conducted in a number of publications (Jones, 1985; Rao et al., 1989; and McPherson, 1993, pp. 119–129).

It is sometimes more useful to utilize the recirculation factor in recirculation analysis. To introduce this, note that the following is true:

$$1 - RF = \frac{Q_i}{Q_i + Q_r} \qquad (11.5)$$

When Eq. 11.5 is used in Eq. 11.3, the expression for the contamination level at point C becomes

$$x_c = RF \cdot x_r + (1 - RF)x_i \qquad (11.6)$$

Because x_i will always be less than x_r for $Q_g > 0$, this expression establishes the fact that the contamination level at point C is higher with recirculation than without. In addition, comparison of Eq. 11.6 with Eq. 11.1 indicates that the intake contamination level at point C is always less than that in the return. Therefore, the critical airway from a gas contamination standpoint is still the return.

The introduction of contamination into the face intake via the recirculated air quantity is one of the weaknesses of the recirculation concept. Another disadvantage of recirculation occurs when the booster fan or fans used to generate the recirculation quantity are not operating. The air can short-circuit through the recirculation crosscut. In addition, any fire or other carbon monoxide–generation activity between the working face and point B will result in CO in the face area if the recirculation crosscut is not monitored. As a result of these disadvantages and the logic of the contamination process outlined above, the following procedures should be observed in the implementation of a recirculation circuit:

1. A recirculation circuit should be implemented only to increase the intake quantity, not replace it.
2. If contamination of the intake air by means of the recirculation quantity is unacceptable, recirculation should not be used.
3. A door system that closes automatically when the recirculation fans are not operating should be used in the recirculation crosscut. This prevents short-circuiting of the ventilation quantity through the crosscut.
4. A CO monitoring system should be utilized in the recirculation crosscut to terminate the recirculation process and close the recirculation doors if CO is detected. This will keep CO from being swept into the working face or district if it is generated anywhere in the return.

In summary, a recirculation circuit should be used only if it solves more environmental problems than it creates and only if the level of safety is improved as a result of the recirculation circuit.

Fan Requirements

When fans are placed in a mine ventilation circuit to produce a desired recirculation quantity, they become a part of the recirculation subsystem. As a result of the heads developed by the booster fans, the main mine ventilation fans and the distribution of air through the remainder of the mine will be affected. The effect on the remaining part of the mine may be quite small, but it needs to be considered to establish the recirculation circuit without undesirable effects elsewhere. An example of this effect is demonstrated in the following.

Example 11.5 Assume that the working face shown in Fig. 11.8*a* is supplied with 40,000 cfm (18.9 m³/s) of air. If the air at the face is sufficient to dilute the methane in the return but not sufficient for proper turbulent mixing in the working face, calculate the head and air power requirement of a booster fan located at point E in the ventilation crosscut to supply an additional 20,000 cfm (9.5 m³/s) to the face. Also calculate what head from point A to point B across the recirculation crosscut must be generated by the main fan(s).

Assume that 6000 ft (1829 m) of 6 × 20-ft (1.8×6.1-m) entry exists from point A to point B when traveling via the working face, and that the recirculation crosscut is 1000 ft (305 m) long and 6 × 20 ft (1.8 × 6.1 m) in cross section. The friction factor in all entries is 50 × 10⁻¹⁰ lb·min²/ft⁴ (0.0093 kg/m³), and the lengths are already in equivalent length units.

Solution: With 60,000 cfm (28.3 m³/s) flowing in the working face, and using Eq. 5.20, the head loss from point A to point B via the face is

$$H_l = \frac{(50 \times 10^{-10})(52)(6000)(60,000)^2}{(5.2)(120)^3} = 0.625 \text{ in. water}$$

The head loss for Q_r = 20,000 cfm (9.5 m³/s) moving through the recirculation crosscut is

$$H_l = \frac{(50 \times 10^{-10})(52)(1000)(20,000)^2}{(5.2)(120)^3} = 0.011 \text{ in. water}$$

Using Kirchhoff's second law, the pressure drops around the recirculation loop must be balanced out by the fan head. Therefore, the head of the fan must be

$$H_t = 0.625 + 0.011 = 0.636 \text{ in. water (157.6 Pa)}$$

(Note that there is no velocity head loss for the fan.) The total air power to be supplied by the fan to the recirculation circuit is calculated using Eq. 5.26:

$$P_a = \frac{(0.636)(20,000)}{6346} = 2.0 \text{ hp (144 kW)}$$

A booster fan should be selected that supplies 20,000 cfm (9.5 m³/s) at 0.636 in. water (157.6 Pa).

With the booster fan operating in the recirculation crosscut, the difference in head between points A and B is 0.625 in. water (155.5 Pa). Without the booster fan and the recirculation crosscut, the head loss from point A to point B is calculated based on the original 40,000 cfm (18.9 m^3/s) flowing through the working face:

$$H_l = \frac{(50 \times 10^{-10})(52)(6000)(40,000)^2}{(5.2)(120)^3} = 0.278 \text{ in. water (69.1 Pa)}$$

The result indicates that the main fan or fans must be changed to provide 0.625 in. water (155.5 Pa) of head from point A to point B instead of the original 0.278 in. (69.1 Pa), or the design recirculation quantity will not be achieved.

Auxiliary Fan–Scrubber Recirculation

Neither auxiliary fans nor scrubbers are normally designed to produce recirculation of ventilating air at the face. However, they are capable of doing so if the fan that is used is powerful enough to overcome the localized effects of the main ventilation current or if other aspects of their use enhance the tendency toward recirculation. However, this recirculation will not increase the maximum contamination level that occurs in the return. The analysis of the contamination levels is similar to that for recirculation circuits. A number of researchers have outlined the effects of this type of ventilation practice. Kissell and Bielicki (1975) and Campbell (1987) have analyzed the case for scrubbers, and McPherson (1993, pp. 121–124) has done the same for auxiliary fans.

Regulations on Recirculation

During most of recent history in coal mining countries, recirculation has been prohibited in coal mines. However, use of district recirculation is often approved as an exception to normal practice in coal mines in Great Britain. Much of the modern activity in testing actual recirculation circuits has been performed there, mainly in response to the difficult ventilation conditions in undersea mines. The practice of recirculation in coal mines in Great Britain has been judged to be satisfactory by the regulatory authorities (Mitchell, 1990a). This, coupled with theoretical analyses, should encourage more applications of controlled recirculation in the future in coal mines. However, this trend to allow recirculation in coal mines has not yet spread to the United States, where both booster fans and controlled recirculation circuits are presently prohibited.

The use of recirculation circuits in metal and nonmetal mines is more common, with many applications arising because of the extensive addition of heat to or removal of heat from the mine airstream. As these applications are more carefully scrutinized, controlled recirculation may become a more versatile and effective tool for mine ventilation in the future.

PROBLEMS

11.1 Compare results obtained by the Atkinson equation and the compressed-air formula for the pressure drop in 4000 ft (1220 m) of 14-in. (360-mm) steel vent pipe (average condition) connected to a blower fan handling 3000 cfm (1.42 m³/s) of air.

11.2 A development heading is to be supplied auxiliary ventilation to dilute and remove diesel exhaust during the mucking cycle, under the following conditions:

Vent pipe, steel	24 in. (610 mm) diameter
Quantity of air to face	16,000 cfm (7.55 m³/s)
Drift length	825 ft (252 m)
Drift conditions	7 × 9 ft (2.1 × 2.7 m), smooth igneous rock, roof-bolted
Barometric pressure	15.0 psi (103.4 kPa)

(a) Neglecting leakage, calculate the pressure drop in the vent pipe and drift and the quantity of air that the fan must handle in the blower position.

(b) Repeat for the fan in the exhaust position.

11.3 A proposed auxiliary ventilation system, consisting of a 25-hp (18.6-kW) fan exhausting from 16-in. (400-mm)-diameter tubing, was tested on the surface prior to underground installation. A manometer connected to a pitot tube centered in the duct read 3.4 in. water (846 Pa) near the duct intake. Assuming standard conditions, what would be the quantity of airflow within the duct at this point?

11.4 Three airways are arranged in parallel for the following required conditions:

	Quantity		Head Loss	
Airway	cfm	(m³/s)	in. water	(Pa)
A	40,000	(18.88)	4.0	(995)
B	10,000	(4.72)	1.0	(249)
C	75,000	(35.40)	1.0	(249)

Compare the horsepower requirements, neglecting velocity head, (a) using regulators in airways B and C, (b) using a booster fan in airway A, and (c) reducing the head loss of airway A to 1.0 in. water (249 Pa). Fan efficiency is 80%.

11.5 An auxiliary fan with a diameter of 24 in. (0.61 m) is being considered for use in a mine with 24-in. (0.61-m) ventilation duct. The total head generated by the fan is as follows:

Total Head, in. water (kPa)	Quantity, cfm (m³/s)
12.8 (3.185)	6000 (2.83)
11.9 (2.961)	7500 (3.54)
9.6 (2.389)	9000 (4.25)
6.2 (1.543)	10,500 (4.96)
2.0 (0.498)	12,000 (5.66)

What will the fan deliver through the duct if 300 ft (91.4 m) of duct is used and if two 90° bends are placed in the duct? Perform the calculations for the following:
(a) Fiberglass pipe in average condition.
(b) Lay-flat tubing in average condition with couplings every 50 ft (15.2 m).
(c) Spiral-reinforced tubing in average condition with couplings every 25 ft (7.6 m).

11.6 Assume that 30,000 cfm (14.16 m³/s) is being delivered to the workplace shown in Fig. 11.8b. If 200 cfm (0.094 m³/s) of methane is liberated in the workplace as shown, what are the methane concentrations at points G, C, D and F on the diagram. If the door in the airway from B to A is removed and replaced by a booster fan with a resultant 10,000 cfm (4.72 m³/s) of airflow from B to A, what are the methane concentrations at points G, C, D and F, assuming that 40,000 cfm (18.9 m³/s) of air moves to the workplace?

11.7 In Fig. 11.8b assume the following lengths of the airways:

A–B	1000 ft (305 m)
A–Working Face–B	2600 ft (792 m)

Assuming that all the openings are 8 × 20 ft (2.4 × 6.1 m) in cross section and have a K value of 60 × 10⁻¹⁰ lb·min²/ft⁴ (0.011 kg/m³), answer the following questions.
(a) What is the head loss from A to B through the workplace with

the door in place and 30,000 cfm (14.16 m³/s) moving past point C?

(b) What head loss from A to B through the workplace must be maintained if the door is replaced by a booster fan and 10,000 cfm (4.72 m³/s) is flowing from B to A and 30,000 cfm (14.16 m³/s) is flowing from G to A?

CHAPTER 12

Economics of Airflow

Every designed system must of necessity consider the costs associated with the design as well as the performance variables of the system. This is as true for mine ventilation systems as it is for other systems related to the mining enterprise. In designing a ventilation system, two major types of design problems are normally recognized: design of individual airways (or groups of airways) and design of the overall ventilation network.

This chapter deals primarily with the design of individual airways. Sections 12.1–12.3 outline the logic and methods for optimal airway design. The procedures employed for this purpose are based on economics and relatively straightforward. The design of the overall network, however, can be much more complicated. Airway systems design in a mine is often more a function of production and haulage variables; consequently, system layout can be more art than science. Some of the considerations in the design of the overall network are outlined in Section 12.4 and in Chapters 13 and 14. The ventilation engineer should be aware of the need for both individual airway design and network design in attempting to optimize a mine ventilation system.

12.1 BASIS OF ECONOMIC DESIGN

When considering the development of a mine airway, the important economic variables that must be considered are the costs of owning and operating the airway. To aid in the analysis of alternatives, these are generally divided into fixed and variable costs. In addition, costs are generally stated on an annual-cost or a present-value basis so that the choice between alternatives can be made on a comparable set of values. In the present discussion, the annual-cost basis is emphasized in the comparison of alternatives.

Fixed costs are those that are incurred regardless of whether an airway is used. The primary fixed cost is the capital cost associated with the construction or development of the airway. The capital cost would ordinarily include the cost of materials and labor as well as the cost of the auxiliary services and equipment necessary to complete the airway. However, the total capital cost is sometimes reduced to an annual cost by some procedure that distributes the cost over the life of the airway. This is usually done by means of an annual investment recovery charge that pays back the original capital investment plus interest on the capital investment over the lifetime of the opening. In addition to the investment recovery charge, the fixed costs would normally include the taxes, insurance, and maintenance costs, because these would be incurred regardless of whether the airway was in use.

Variable costs are those costs necessary to operate the facility. In the case of an airway, the variable cost normally consists simply of the cost of power to move the prescribed amount of air through the airway.

The final decision about the most economical choice among several options is often made on the basis of the overall cost per year, that is, the sum of the fixed and operating costs. Thus the choice of airways is based upon the overall annual cost expressed as

$$
\begin{array}{l}
\text{Overall} \\
\text{annual} \\
\text{cost}
\end{array}
=
\begin{array}{l}
\text{investment-} \\
\text{recovery} \\
\text{charge}
\end{array}
+
\begin{array}{l}
\text{annual} \\
\text{insurance} \\
\text{and taxes}
\end{array}
+
\begin{array}{l}
\text{annual} \\
\text{maintenance} \\
\text{cost}
\end{array}
+
\begin{array}{l}
\text{annual} \\
\text{power} \\
\text{cost}
\end{array}
$$

$$(12.1)$$

The airway having the lowest overall cost is chosen, provided that no intangible factors are considered in the choice of alternatives. The following sections discuss methods for minimizing the overall airway cost through proper design of the airway.

12.2 EFFECT OF AIRWAY CHARACTERISTICS ON POWER CONSUMPTION

As air flows through a mine opening, it must be supplied with power to overcome the head losses that occur in the airway. In Chapter 5, the air power was shown to be proportional to both the head loss H_l and the air quantity Q. However, the head loss, in turn, is proportional to Q^2, resulting in the power being proportional to the cube of the quantity:

$$P_a \propto Q^3 \tag{12.2}$$

From this relationship, it is apparent that the quantity is extremely important in the determination of air power, and that every attempt should be made to reduce the air quantity to the lowest possible value. This means that no excess air should be provided and leakage should be minimized so that the total air quantity will be minimized. Generally, the ventilation engineer does not have great control over the quantity of air flowing to each face because the quantities are normally dictated by safety or legal requirements. The leakage, however, is often a function of the quality of the separation between the intake and return airways and can be affected by the effort put into constructing stoppings and shaft dividers (Section 13.5).

Once the air quantities have been minimized, control over power consumption and operating cost depends on minimizing the mine static and velocity heads. Although P_a varies with the first power of H_l, the reduction of head loss in mine openings is nonetheless important. This is because there are many ways of reducing the head loss, and some of these head-reduction methods can be implemented without great difficulty.

It was shown in Chapter 5 that the total head loss is directly proportional to K, O, $L + L_e$, and Q^2 but inversely proportional to A^3 (Eq. 5.25). When the quantity Q is controlled, then the only way to decrease head loss is to reduce K, O, L, or L_e or to increase A. In the following paragraphs, these possibilities are discussed in detail.

Size and Shape

The size and shape of an airway are both important in determining the head loss due to friction, and the two variables are closely related. In considering the effects of these variables, it is perhaps best to first isolate the size and concentrate our attention on how size alone controls the head loss. To accomplish this, it is necessary to choose a simple opening such as a circle and study the effect of change in the diameter on the head loss. For a circular airway,

$$H_l \propto \frac{O}{A^3} \propto \frac{\pi D}{(\pi D^2/4)^3} \propto \frac{1}{D^5} \tag{12.3}$$

This tells us that the head loss for a circular opening is inversely proportional to the diameter raised to the fifth power.

Similar results apply to a square with sides of length D and to a rectangle, but only if the sides of these openings are increased in dimension proportionally. Analysis of other opening shapes is not as simple since either the perimeter or the area is a more complex function of any given dimension.

The shape of an airway does not have as large an effect on the head loss as the size. However, the two variables are difficult to separate because the only valid measure of the shape is the hydraulic radius. Recalling that the hydraulic radius $R_h = A/O$, then

$$H_l \propto \frac{1}{R_h} \tag{12.4}$$

for a constant velocity of flow. This relationship tells us that we must minimize the perimeter of the opening for a given cross-sectional area. The optimal shape is provided by a circle. The efficiency of other airway shapes depends on how closely they approximate the circle, with the more rounded shapes being more efficient. McElroy (1935) expressed the relative efficiency of various shapes by calculating the relative head losses of various airway shapes having a constant area. The relative head losses of an array of shapes are presented in Table 12.1.

It is obvious that a return in ventilation power savings can be realized by proper shaping of the airway. However, other considerations may be of more importance in determining the shape of the opening. The greater utility of rectangular openings for accommodating roadways, pipe, cables, ventilation tubing, and haulage vehicles may overshadow the ventilation cost advantage of a more rounded opening. The primary opportunities for power cost savings come when driving openings that are to be used exclusively for ventilation purposes. Ventilation shafts offer obvious possibilities, but other openings occasionally fall into this category. These openings should be driven with the maximum hydraulic radius whenever possible.

With increasing emphasis on rapid excavation technology in recent years, round ventilation openings are more commonplace. In particular, progress has been made in the driving of shafts and raises by both pilot-hole and blind-boring techniques. In sedimentary deposits, a rectangular shape is often chosen for practical reasons. However, a horseshoe-shaped opening is a favorable alternative from both a mechanics and ventilation standpoint when an opening with a flat bottom is desirable.

Character of the Surface

The irregularity or roughness of the surface of an airway, as reflected in the friction factor, may affect the head loss significantly. As shown in Tables 5.1 and 5.2, the friction factor can vary within any category of airway and any degree of sinuosity. Because the head loss is proportional to the friction factor, any reduction in K will produce a like reduction in the head loss. Another way of expressing the effect of a reduced friction factor is to determine the effect on the quantity. This is evident in the relationship

$$Q \propto \frac{1}{K^{1/2}} \qquad (12.5)$$

Thus either a reduced head loss or an increased quantity can be achieved by a reduction in the friction factor.

The more common procedures followed for decreasing the value of K include lining the opening to provide smooth surfaces, changing the support plans to eliminate timbering, better blasting practice, and good housekeeping. The lining of airways is practiced primarily in major airways such as shafts and is usually accomplished with concrete, gunite, or lagging. There is generally a reduction in area as a result of the installation of the lining

TABLE 12.1 Relative Head Loss in Airways of Various Shapes Having a Constant Area

Airway Shape		Dimensional Relationship (b/a)	Relative Head Loss
Circle		—	1.00
Square		—	1.13
Octagon		—	1.03
Rectangle		1.50	1.15
		2.00	1.20
		3.00	1.30
		4.00	1.40
Ovaloid		2.00	1.09
		3.00	1.21
		4.00	1.33
Horseshoe		1.00	1.07
Ellipse		1.25	1.01
		1.50	1.04
		1.75	1.08
		2.00	1.12

material, but this is more than offset by the decrease in friction factor in most cases.

A change in the roof-support plan that eliminates the need for timbers in the airways can have a significant effect on the friction factor. Strong evidence to support this exists in Table 5.1. If the friction factors for timbered airways with average irregularities are compared with the comparable friction factors for sedimentary rock, the friction factors are found to drop an average of over 35% when the timbers are removed. The final decision in the elimination of timber depends, however, on whether a suitable roof-support procedure can be substituted. The practice of roof bolting has positively affected the ability to ventilate a mine, but it cannot be applied in all cases.

Length

Although the length of airways required may be fixed for a predetermined mine layout, it is often possible to reduce the length to a minimum when planning the mine. Ventilation cost should be a major consideration when the mine planner designs the mine. The layout must be accomplished considering all of the subsystems that make up the mine, and this often conflicts with the ventilation objectives. However, minimizing the length of the airways is most easily controlled during the initial planning stage.

A second manner of controlling the length of the airways in a mining layout is in the planning and location of ventilation shafts that are developed after the mine is placed in operation. The closer together the shafts are placed, the shorter the length of the airways traversed from intake to exhaust and the lower the ventilation cost. However, the capital cost of the ventilation shafts also enters into consideration, and a balance must be achieved between capital and operating costs.

Shock Loss

The various sources of shock loss should be investigated when attempting to reduce pressure and power requirements for a ventilation system. Losses due to bends and area changes can be reduced substantially by careful planning or minor revision of the mining system. The basic equation (Eq. 5.23) for shock loss indicates that for a given airflow, the shock-loss factor determines the magnitude of the loss. Straightening airways, installing guide vanes at bends close to the fan, and making area changes more rounded and gradual reduces the shock loss factor, the equivalent length, and hence the head loss in an airway.

Velocity Head

Although not strictly a head loss, the velocity head nevertheless represents a loss to the system at discharge (Section 5.2). In a blower or booster system,

H_v can be reduced by enlarging the airway gradually just prior to discharge. In an exhaust system, a gradual-expansion duct (termed an *evasé*) is generally placed on the discharge of the fan for the same purpose (Section 10.3).

12.3 ECONOMIC DESIGN OF AIRWAYS

One of the best means of lowering the cost of ventilation in a mining operation lies in the design of individual airways. The ventilation engineer most frequently must work with the mining system already in use; but when afforded the opportunity of designing a system, the engineer must exercise care in specifying the size, shape, and surface condition of the airways, particularly the main ones.

To analyze any airway in a ventilation network, the engineer must isolate it and consider it as an independent entity. This can be done in any ventilation network and any airway as long as the engineer knows the quantity of air that the airway is to carry. This variable, the quantity, is of primary concern and must be determined before the proper economic design of the airway can be attempted.

When an airway has been isolated, the method of analysis must utilize the concept of minimizing the overall cost of the airway. A number of approaches have been outlined in the literature to tackle this problem. McElroy (1935) was one of the first to give serious consideration to this problem. He made cost calculations for various types of airways to compare relative overall costs and to provide a means for choosing between the various possible configurations of openings.

Mancha (1950) provided a mathematical approach to the optimization of several opening shapes. This approach requires that the overall cost be expressed mathematically so that the first derivative of the expression can be obtained in order to solve for the minimum cost. He provided solutions for the optimal sizes of round shafts and also for rectangular and elliptical shafts possessing fixed-ratio dimensions.

A method that is more general than the mathematical approach is the graphic one. The method was used by Weeks (1926) and has been used and expanded several times since. The method requires that the fixed and operating costs be plotted graphically to determine the minimum overall cost. Thus it is necessary that the costs be sufficiently well defined that they can be plotted.

Some authors have utilized both the mathematical and the graphic approaches, choosing the method that best fits any particular airway-sizing problem. Pursall (1960), Hartman (1961), and Lambrechts (1974) have all advocated this procedure. Using this method, problems that can be easily solved mathematically are handled in this manner whereas more difficult ones are solved by means of the graphic approach. This procedure is followed here and applied to a number of different airway configurations.

Circular Airways

Having an ideal shape, the circular airway is often used for main airways such as shafts. It is therefore a shape that is of maximum interest. Because the circular shape can be defined by means of a single dimension, the diameter, it is also easy to define the overall cost for some circular shafts mathematically.

Hartman (1961) provided the following solution for the optimum size of a circular opening. The first element of the overall cost to be considered is the operating cost. To obtain the operating cost, the head-loss equation is first written in terms of the diameter D instead of A and O as follows:

$$H_l = 1.25 \frac{K(L + L_e)Q^2}{D^5} \tag{12.6}$$

(Velocity head need not be considered since only a portion of the total system is involved.) Then the input power P_i required is

$$P_i = \frac{P_a}{\eta_o} = \frac{H_l Q}{6346 \, \eta_o} \tag{12.7}$$

where η_o is the overall efficiency of the mine fan; and the operating cost per annum C_o is

$$C_o = c_o P_i \tag{12.8}$$

where c_o is the cost per hp·yr. Continuous operation is normally assumed.

To find the capital cost, first determine the volume of rock Y excavated:

$$Y = \frac{\pi}{4} D^2 L \tag{12.9}$$

Then the annual capital cost C_c is

$$C_c = c_c c_e Y \tag{12.10}$$

where c_e is unit excavation and lining cost and c_c is annual capital-investment recovery factor.

The sum of C_o and C_c is the overall cost per year C. Substituting Eqs. 12.6 and 12.7 into Eq. 12.8, and Eq. 12.9 into Eq. 12.10, the complete expression becomes

$$C = 1.97 \times 10^{-4} \frac{c_o K(L + L_e)Q^3}{\eta_o} D^{-5} + 0.785 \, c_e c_c L D^2 \tag{12.11}$$

Taking the first derivative of C with respect to D and equating it to zero, the equation for minimum cost becomes

$$c_c c_e L D^7 - 6.25 \times 10^{-4} \frac{c_o K (L + L_e) Q^3}{\eta_o} = 0$$

and solving for D,

$$D = \sqrt[7]{6.25 \times 10^{-4} \frac{c_o K (L + L_e) Q^3}{\eta_o c_c c_e L}} \qquad (12.12)$$

This is the general equation for finding the most economical size of circular airway. Notice that if L_e is small enough to omit, then L cancels out of the equation, indicating that the optimum diameter is independent of the length of the airway. The use of Eq. 12.12 can perhaps be best illustrated by means of the following example.

Example 12.1   Given the following data for a circular concrete-lined shaft:

L = 1000 ft (305 m)
K = 30 × 10^{-10} lb·min^2/ft^4 (5.56 × 10^{-3} kg/m^3)
Q = 250,000 cfm (118 m^3/s)
Life = 20 years
Interest on capital = 10%
Taxes, insurance, maintenance = 3%
Excavation cost = $225/yd^3 ($294/m^3)
Power cost = $0.05/kWh
η_o = 65%
L_e = 0 ft

Find the optimal shaft size.

Solution: To determine the annual fixed charge, it is necessary to combine the annual finance charge, taxes, and insurance. The annual finance charge is determined by considering that the total excavation cost is financed at the current rate of interest, 10%, and that the loan must be repaid through annual payments each year over the life of the opening. To determine the annual finance charge, it is necessary to refer to any engineering economy text that contains the proper financial tables. In this reference, the annual finance charge above is found in the tables that provide the capital-recovery factor. For a 20-year life and a 10% interest rate, the capital-recovery factor or annual finance charge is found to be 0.11746. With the taxes, insurance, and maintenance at 3%, then

$$c_e = 0.11746 + 0.03 = 0.14746$$

Note also that the excavation cost c_e must be converted into proper units ($/ft³) for use in the formula. Thus

$$c_e = \frac{\$225/yd^3}{27 \ ft^3/yd^3} = \$8.33/ft^3$$

The annual power cost c_o must be expressed in units of dollars per hp·yr. Thus the power cost per hp·yr based on continual operation is as follows:

$$c_o = (0.05)(0.746)(24)(365) = \$326.75/hp·yr$$

The mathematical solution for the optimal shaft diameter then follows (Eq. 12.12):

$$D = \sqrt[7]{6.25 \times 10^{-4} \frac{(327.75) \ (30 \times 10^{-10}) \ (250,000)^3}{(0.65) \ (8.33) \ (0.14746)}} = 10.3 \ ft \ (3.1 \ m)$$

In some cases, such as the case of a lined concrete shaft, it is important to consider which sizes are standard. Normally the closest standard shaft would be chosen to eliminate the manufacturing cost of nonstandard concrete forms.

The mathematical solution derived above is based on the assumption that the excavation cost is proportional to the excavated volume. This is not always a valid assumption. In studies on the optimal sizing of shafts, the cost function is given as a nonlinear function of the volume (Solow, 1977; Wang et al., 1979). In particular, the total cost per unit volume generally decreases somewhat as shown in Fig. 12.1. The curves in Fig. 12.1 represent the total cost of excavating and lining a round shaft and most likely decrease in unit cost as a result of the economics of scale and increased maneuvering room in a larger shaft. Similarly, the deeper shafts have a lower cost per unit volume because of similar economies of scale.

When the cost of producing an airway is not proportional to the volume excavated, it is generally best to go directly to the graphic solution approach. This method is illustrated in the following example.

Example 12.2 Determine the optimal size of a circular concrete-lined shaft using the following data:

$L = 1000 \ ft \ (305 \ m)$
$K = 40 \times 10^{-10} \ lb·min^2/ft^4 \ (7.42 \times 10^{-3} \ kg/m^3)$
$Q = 500,000 \ cfm \ (236 \ m^3/s)$
Interest on capital $= 12\%$

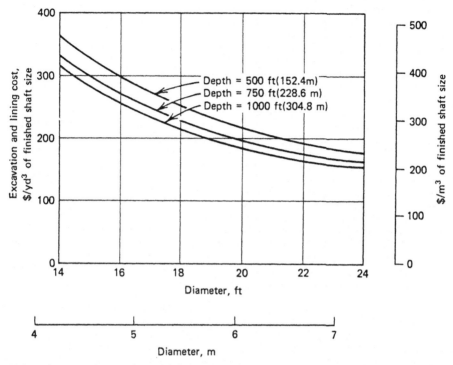

FIGURE 12.1 Excavation and lining costs for round shafts (plotted from data published by Wang et al., 1979).

Taxes, insurance, maintenance = 3%
Power cost = $0.04/kWh
η_o = 65%
L_e = 0 ft
Life = 30 years

Assume that the excavation cost is given by the curve in Fig. 12.1 for a shaft depth of 1000 ft (305 m).

Solution: The capital-recovery factor for a life of 30 years and a 12% interest rate is 0.12414. With the taxes, insurance, and maintenance at a 3% level, the value of c_c becomes 0.15414. To calculate the fixed and operating costs, it is necessary to choose a group of sizes and determine the costs for each size numerically. The sizes chosen for our calculations range from 14 to 24 ft in 2-ft increments. The values of the fixed costs C_c are then determined as follows:

D, ft	Y, ft^3	c_e, \$/yd^3	C_c, \$/yr
14	153,938	315	276,800
16	201,062	256	293,800
18	254,469	213	309,400
20	314,159	183	328,200
22	380,133	164	355,900
24	452,389	154	397,700

A C_c–D plot is shown in Fig. 12.2.

The power cost per hp·yr is then calculated as follows:

$$c_o = (0.04)(0.746)(24)(365) = \$261.40/\text{hp·yr}$$

On the basis of the unit power cost, the operating costs can then be determined to complete the graphic solution:

D, ft	A, ft^2	O, ft	H_l, in.	P_i, hp	C_o, \$/yr
14	153.9	44.0	2.32	280.88	73,400
16	201.1	50.3	1.19	144.07	37,700
18	254.5	56.5	0.66	79.95	20,900
20	314.2	62.8	0.39	47.21	12,300
22	380.1	69.1	0.24	29.31	7,700
24	452.4	75.4	0.16	18.97	5,000

The operating-cost curve is also plotted as shown in Fig. 12.2. The total annual cost curve is then constructed by combining the curves of fixed and operating cost. The overall cost curve in Fig. 12.2 gives us the optimal shaft diameter, which occurs at about 17 ft (5.2 m).

Several facts can be learned by studying this example. First, the overall cost curve is rather flat in the vicinity of the optimum diameter. This makes it possible to choose any diameter close to the optimum without seriously affecting the overall cost. A standard-sized shaft can therefore be selected. It would perhaps be better to choose a standard size larger than the optimum size as a hedge against inflation or the possibility that the air quantity will be increased in the future.

A second feature of the example above is that the optimum shaft size will change slightly for a shaft of a different depth. This variance is due to the change in the cost per cubic yard that occurs as the depth changes in Fig. 12.1. This differs from the result drawn from Eq. 12.12, where we concluded that the optimum diameter is independent of the length if $L_e = 0$. Thus the conclusion that the optimum size of an opening is independent of the length applies only if the cost per unit volume of the opening is not affected by economies of scale.

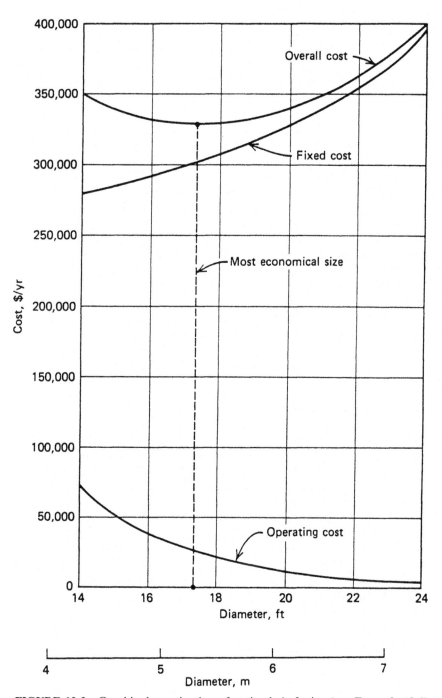

FIGURE 12.2 Graphic determination of optimal shaft size (see Example 12.2).

The procedures discussed above outline the method for using the overall annual cost in evaluating the possible alternatives. These procedures are used when the original cost of the opening is depreciated over the expected life. If the cost of the opening is considered as an expense and is to be charged immediately against the income of the operation, then the present value of all costs over the life of the airway may be the best criterion for judging the alternatives. Papers by Dahl (1976) and Solow (1977) illustrate the use of the present-value concept. These papers also calculate the costs after taxes, thus considering the tax-saving aspect of current expenses and the expected inflation in power costs.

The methods illustrated in Examples 12.1 and 12.2 have outlined the costs in terms of the before-tax values. Any expenses that can be charged against income result in tax savings. Thus the net cost of any type can be considered as the total cost minus the tax saving. However, the use of after-tax costs is important only if there is some difference in the rate of tax saving of one or more of the categories of cost. Otherwise all the costs are reduced proportionally and the optimal opening size remains the same.

The consideration of inflationary effects on power costs is an important refinement that can be added to the calculation of the optimal opening size. To analyze this aspect of the problem, however, a reasonably accurate estimate of the inflation rate of power costs and the general infiltration rate of the overall economy must be obtained. Analysts interested in applying this concept are referred to the publications by Dahl and Solow.

Single Openings With Other Shapes

Openings other than circular often predominate in underground workings and thus are also of importance to the engineer. In analyzing noncircular shapes, it is often desirable to limit consideration to those in which the ratio of the dimensions is held constant. This makes the optimization of size an easy task because the cost becomes a function of only one dimensional variable.

As an example of this process, one can view the work of Mancha (1950) in deriving the optimal size of rectangular and elliptical openings. In a simplified version of determining the optimal size of a rectangle, let us take a rectangle possessing a width b. To keep the ratio of the dimensions constant, let the height of the opening be given by xb, where x is the constant of proportionality between the height and width. With these dimensions, then $O = 2b(1 + x)$ and $A = xb^2$.

Assuming that $L_e = 0$, the head loss for this opening is given by the relationship

$$H_l = \frac{2K(1 + x)LQ_2}{5.2\ x^3 b^5} \tag{12.13}$$

From this equation, it is possible to obtain the annual operating cost. Using the same notation for unit costs that was used in the derivation of the optimum circular-opening size, the annual operating-cost expression is found to be

$$C_o = \frac{2K(1 + x)LQ^3c_o}{33,000 \ x^3b^5\eta_o} \tag{12.14}$$

With the assumption that the excavation cost is proportional to the volume of material excavated, the expression for the annual fixed cost becomes

$$C_c = xb^2L \ c_e c_c \tag{12.15}$$

By summing the annual fixed and operating costs, we obtain an expression for the overall annual cost C. Taking the derivative of this expression with respect to the width b and setting this expression equal to zero, we get the following expression for the optimal width:

$$b = \sqrt[7]{\frac{K(1 + x)Q^3c_o}{6600 \ x^4c_e c_c \eta_o}} \tag{12.16}$$

In both Eqs. 12.12 and 12.16, the optimal dimension of the airway is proportional to the three-sevenths power of the air quantity Q. This is characteristic of the expression for the optimal size of any airway.

This solution is based on the assumption that the excavation cost is proportional to the excavated volume. This assumption may not hold in practice. When the cost of excavating a rectangular opening is not proportional to the volume excavated, it is easier to apply the graphic method to the problem of finding the optimal size of the opening. The practical constraints on the opening size can then be evaluated before the final size is chosen.

Multiple-Opening Sets

A single opening offers the engineer an opportunity to analyze ventilation economics on a relatively simple basis. In some cases, especially in sedimentary deposits such as coal, the airways are made up of multiple openings connected with each other in parallel. These openings, called headings or entries, may be identical in size and shape. The analysis of the ventilation economics of multiple openings is more difficult than that for single openings but will follow a similar procedure of determining the fixed and operating costs and minimizing the overall cost. If the size and shape of each opening are fixed, the process is simplified since only the number of openings used must be determined.

The determination of the proper number of entries can be initiated by

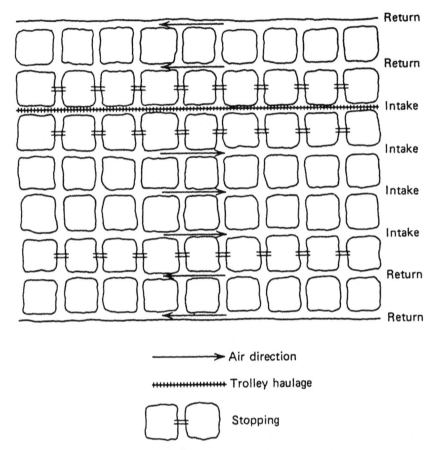

FIGURE 12.3 Typical multiple-entry airways.

referring to Fig. 12.3. In this illustration, the entries are bidirectional with the return air moving in the opposite direction along both sides of the intakes. Ordinarily, both the intake and return airways must be divided into more than one split of air to meet regulatory or practical constraints. In some cases, an additional opening is required for a belt conveyor. Because of limitations on the airflow through a belt heading, the heading may not contribute to the supply of ventilation air. The ventilation process may be further complicated by a requirement that the air velocity on a trolley heading be limited to some maximum value.

In analyzing the fixed cost of multiple headings, several changes in the procedures may be necessary. First, the entries will ordinarily be driven in the coal, ore, or mineral product desired, in which case the overall cost of driving the openings is not of concern. The major fixed-cost element is therefore the differential in the cost of coal obtained developing the entry set and the cost of producing coal from a production section. If the cost of producing tonnage from development openings is less than or equal to that for production tonnage, then there is no economic reason to limit the number of entries in the development set. If the opposite is true, then the applicable cost to be assigned to the ventilation entries is the loss of profit because the cost of development tonnage is higher than the cost of production tonnage. This differential in mineral production cost will be used as the initial "capital cost" of the entry set.

Because the differential in cost occurs at the start of the life of the entry set, it can be treated as a one-time expense. The operating costs, on the other hand, will continue to be incurred throughout the life of the opening. Thus this situation can be analyzed on a present-value basis with both the initial cost and the annual operating costs being discounted to a present value. The overall cost is still the governing decision variable but will be compared on the basis of total present value rather than on an annual-cost basis.

An additional change in the fixed cost is the charge for taxes, insurance, and maintenance. Ordinarily these costs can be ignored for development openings because they are identical to the tax, insurance, and maintenance costs for production openings. The process of analyzing the number of entries in a set can perhaps be best illustrated by means of an example.

Example 12.3 Determine the optimal number of coal intake and return airways in a main entry set using the following operating conditions:

$K = 80 \times 10^{-10}$ lb·min²/ft⁴ $(14.84 \times 10^{-3}$ kg/m³) for intakes
$K = 100 \times 10^{-10}$ lb·min²/ft⁴ $(18.55 \times 10^{-3}$ kg/m³) for returns
Interest rate $= 12\%$
Cost differential $= \$2$/ton $(\$2.20$/t) (t $=$ metric ton)
Coal specific weight $= 80$ lb/ft³ $(1280$ kg/m³)
Life $= 20$ years
Entry size $= 16 \times 5$ ft $(4.9 \times 1.5$ m)
Power cost $= \$0.045$/kWh
Distance from crosscut to crosscut $= 90$ ft $(27.4$ m)
Distance from entry to entry $= 80$ ft $(24.4$ m)
$L = 5000$ ft $(1524$ m)
$L_e = 0$ ft $(0$ m)
$\eta_o = 60\%$
$Q = 200,000$ cfm $(94$ m³/s)

Solution: The differential cost of the first entry is the product of the coal tonnage in one entry times the cost differential per ton:

($2/ton) (16 ft × 5 ft × 5000 ft) (80 lb/ft³)(1 ton/2000 lb) = $32,000

The cost of each succeeding entry is the cost differential that arises from mining one entry plus all the crosscuts that join the additional entry to the other entries. Again, this is a product of the coal tonnage and the cost differential per ton:

($2/ton) (16 ft × 5 ft × 5000 ft) (80 lb/ft³) (1 ton/2000 lb)

+ ($2/ton) (16 ft × 5 ft × 64 ft) (1 crosscut/90 ft) (5000 ft)

(80 lb/ft³) (1 ton/2000 lb) = $54,756

Using these values, the present value of the fixed cost can be determined as follows for an entry set containing from one to seven entries:

Number of Entries	Total Cost Differential, $	Present Value of Cost Differential, $
1	32,000	32,000
2	86,800	86,800
3	141,500	141,500
4	196,300	196,300
5	251,000	251,000
6	305,800	305,800
7	360,500	360,500

Note that the present value of the cost differential is the same as the total cost differential. This occurs because the cost differential is incurred at the start of the airway life.

The head loss can then be calculated using the following formula for parallel openings (Eq. 7.21):

$$H_l = \frac{KO(L + L_e)Q^2}{5.2 \, N_a^2 A^3}$$

where N_a represents the number of entries used. The power costs for the intake airways are then calculated using $c_o = \$294.07/\text{hp·yr}$, $A = 80 \, \text{ft}^2$, and $O = 42 \, \text{ft}$.

To obtain the present value of the operating costs over the life of the airways, it is necessary to find the present-value (present-worth) factor for a uniform series of payments over the lifespan of the airways. For a life of 20 years and an interest rate of 12%, the present-value factor is 7.469. This factor gives the present value of the operating cost when it is multiplied by the annual operating cost. When the calculations are completed, the following information is obtained:

Number of Entries	H_l, in. water	P_i, hp	C_o, \$	Present Value of Operating Costs, \$
1	25.24	1324.95	389,600	2,910,100
2	6.31	331.24	97,400	727,500
3	2.80	147.22	43,300	323,300
4	1.58	82.81	24,400	181,900
5	1.01	53.00	15,600	116,400
6	0.70	36.80	10,800	80,800
7	0.52	27.04	8,000	59,400

Note that no attempt was made to provide for inflation in operating costs in this example. The present values of fixed and operating costs are plotted as a bar graph in Fig. 12.4. The optimal number of entries to carry the intake air is 5. However, this number may have to be increased if the maximum air velocity on the intake airways is exceeded.

Calculating the present value of costs for the return airways indicates that the optimal number is also 5. However, it is often advantageous to have an even number of return airways so that the flow through each side of the return-air system is balanced. Thus it may be advisable to put either two or three airways on each side. In this case, two airways on each side is less costly and is recommended. However, it may also be advisable to consider the extra cost for overcasts if they are used in this system.

Note that the analysis of the number of return entries can safely ignore the crosscuts that connect the intakes and returns. This is because the crosscuts interconnecting the intakes and returns are not optional and must be considered as sunken-cost items. This principle can be easily verified by excluding the cost of these crosscuts in the return-airway calculations. The resulting optimal number of returns will be unaffected by excluding the costs or cost differentials for these openings.

Other considerations may affect the economic decisions on mine airways in some cases. An example is the possibility that the optimal velocity is higher than is practical for dust control or for the comfort of miners. An example can be shown for multiple coal mine airways. Mutmansky and Greer (1984) analyzed the optimal velocity for coal mine entries. Using a 20-year life and a power cost of \$0.045/kWh, it was shown statistically that the optimal velocity in intakes with a friction factor of 50×10^{-10} lb·min^2/ft^4 (0.0093 kg/m^3) is 600–1150 fpm (3.0–5.8 m/s). For return entries with a friction factor of 75×10^{-10} lb·min^2/ft^4 (0.014 kg/m^3), the optimal velocity is 450–1000 fpm (2.3–5.0 m/s) for identical assumptions of power cost and life. These velocities may be too high for mine entries used in the winter near the mine intakes. In addition, research on dusts in mine airways (Shirey et al., 1985) indicates that the optimal velocity for longwall faces is about 500 fpm (2.5 m/s). At higher velocities, the dust concentrations increase and may affect the ability

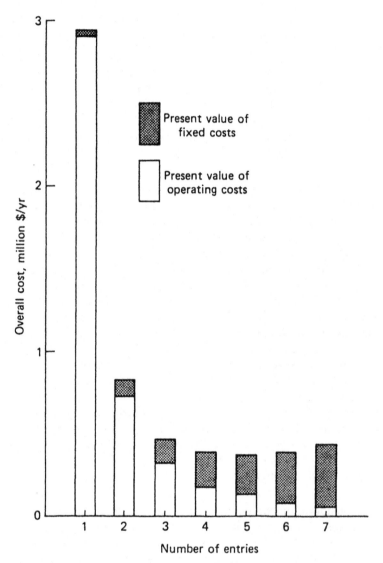

FIGURE 12.4   Overall costs for intake entries of Example 12.3.

to keep in compliance with dust regulations. Thus the velocity designed into the airways may be limited by comfort or dust conditions rather than strictly by the economics of the situation. The mine ventilation engineer needs to consider these other variables in planning the ventilation system.

12.4 ECONOMIC DESIGN OF THE OVERALL NETWORK

The previous sections have discussed the design of individual airways or airway sets. This approach to economic design has a very positive effect on the cost of ventilating a mine. However, it may be more advantageous from an economic standpoint to look at the overall ventilation network. The possibilities for making the overall network more efficient include minimizing the total length of airways in the system, reducing the number of airways, and rearranging the existing airways.

The major factors that constrain the attempt to alter the overall airway system are the form of the mineral deposit and the mining method. The nature of the deposit nearly always restricts the mining method that is applied. When the mining method is chosen, the geometric layout of the production and development openings may be quite rigid, with little or no possibility for change. However, the most opportune time for design and planning of ventilation airways is during the initial layout of the mine. At that time, the airways that supply air can be studied and laid out with minimization of ventilation cost in mind.

A second major opportunity for economic design of the overall system occurs when a ventilation shaft is added to an existing mine. This is especially applicable to coal mines where the lateral extent of the openings and requirements for air are large. At any point during the life of a mine when an additional ventilation shaft is considered, the size and location are of extreme importance. The location, in particular, will affect the overall system since it will affect the total airway length. Thus the number and location of ventilation shafts is one of the most important problems in the engineering of a coal mine ventilation system. The nature of the problem is again one of minimizing the overall sum of fixed and operating costs because more ventilation shafts result in higher fixed costs but lower operating costs.

The methods of analyzing this type of problem have not yet been developed here. Essentially what is required is a method of calculating the cost of operating the overall ventilation system so that any number of different alternative networks can be investigated. To accomplish such a task, a set of procedures for any configuration of openings must be available. Normally these procedures are analyzed by computer so that the computational burden of investigating a number of complex ventilation networks is minimized. The development of a body of theory to solve a complex ventilation network is presented in Chapter 7. Understanding of this chapter is necessary to attack the problems of design of the overall ventilation network.

PROBLEMS

12.1 Determine the most economical size of a circular air shaft by the mathematical method if the following conditions apply:

$Q = 500,000$ cfm (165 m³/s)
Life $= 15$ years
Interest rate $= 12\%$

Taxes, insurance, maintenance = 3%

$K = 40 \times 10^{-10}$ lb·min²/ft⁴ (7.42 × 10⁻³ kg/m³)

Excavation cost = \$190/yd³ (\$249/m³)

Power cost = \$0.04/kWh

$\eta_o = 65\%$

$L = 500$ ft (152 m)

$L_e = 0$ ft (0 m)

12.2 Repeat Problem 12.1 using the graphic method. Plot the fixed cost, the operating cost, and the overall cost curves, and check the results against the mathematical solution.

12.3 Determine the optimal size for a square drift assuming that the following conditions apply:

$K = 70 \times 10^{-10}$ lb·min²/ft⁴ (13.0 × 10⁻³ kg/m³)

Life = 25 years

Interest rate = 15%

Taxes, insurance, maintenance = 2%

$Q = 150,000$ cfm (70.8 m³/s)

Excavation cost = \$25/yd³ (\$32.70/m³)

Power cost = \$0.035/kWh

$\eta_o = 68\%$

$L = 900$ ft (247 m)

$L_e = 0$ ft (0 m)

12.4 Redo Problem 13.3 assuming that the excavation costs are given by the relationship $E = 15.0\ A^{0.35}$, where A is drift cross-sectional area in ft² and E is excavation cost in \$/ft of advance.

12.5 Graphically determine the optimal shaft size for a lined circular shaft for the following parameters:

$L = 500$ ft (152 m)

$K = 40 \times 10^{-10}$ lb·min²/ft⁴ (7.42 × 10⁻³ kg/m³)

$Q = 500,000$ cfm (236 m³/s)

Interest rate = 12%

Taxes, insurance, maintenance = 3%

Power cost = \$0.04/kWh

$\eta_o = 65\%$

$L_e = 0$ ft (0 m)

Life = 30 years

Assume that the excavation cost is given by the curves in Fig. 12.1. Why does the answer differ from that in Example 12.2?

12.6 Determine the optimal shaft size for a lined circular shaft under the following conditions:

$L = 750$ ft (229 m)
$K = 35 \times 10^{-10}$ lb·min^2/ft^4 (6.49 × 10^{-3} kg/s)
$Q = 750,000$ cfm (354 m^3/s)
Interest rate = 10%
Taxes, insurance, maintenance = 4%
Power cost = \$0.05/kWh
$\eta_o = 68\%$
$L_e = 0$ ft (0 m)
Life = 15 years

Assume that the excavation cost is given by the curves in Fig. 12.1.

12.7 Using the present-value concept, calculate the optimal number of intake and return airways using the following data:

$K = 50 \times 10^{-10}$ lb·min^2/ft^4 (9.28 × 10^{-3} kg/m^3) for intakes
$K = 70 \times 10^{-10}$ lb·min^2/ft^4 (13.0 × 10^{-3} kg/m^3) for returns
Interest rate = 10%
Coal specific weight = 85 lb/ft^3 (1361 kg/m^3)
Life = 15 years
Entry size = 18 × 6 ft (5.5 × 1.8 m)
$L = 3000$ ft (914 m)
$L_e = 0$ ft (0 m)
$\eta_o = 65\%$
$Q = 400,000$ cfm (188.78 m^3/s)
Power cost = \$0.04/kWh
Distance from crosscut to crosscut = 100 ft (30.5 m)
Distance from entry to entry = 100 ft (30.5 m)
Cost differential = \$3/ton (\$3.30/t)

Ignore maintenance costs.

12.8 Calculate the most economical number of intakes and returns for a sedimentary mineral deposit subject to the following parameters:

$K = 70 \times 10^{-10}$ lb·min^2/ft^4 (13.0 × 10^{-3} kg/m^3) for intakes

$K = 90 \times 10^{-10}$ lb·min²/ft⁴ (16.7×10^{-3} kg/m³) for returns

Interest rate $= 12\%$

Ore specific weight $= 100$ lb/ft³ (1602 kg/m³)

Life $= 20$ years

Opening size $= 14 \times 6$ ft (4.3×1.8 m)

$L = 2500$ ft (762 m)

$L_e = 0$ ft (0 m)

$\eta_o = 60\%$

$Q = 150{,}000$ cfm (70.4 m³/s)

Power cost $= \$0.035$/kWh

Cost differential $= \$1.50$/ton ($\$1.65$/t)

Distance from crosscut to crosscut $= 80$ ft (24.4 m)

Distance from entry to entry $= 70$ ft (21.3 m)

Ignore maintenance costs.

CHAPTER 13

Coal Mine Ventilation Systems

13.1 INTRODUCTION

The *mine ventilation system* consists of fans, airways, and control devices for air coursing. Quantity control in this system means achieving desired airflows for quality and temperature–humidity control through the optimal selection of openings to surface; the shape, size, and number of airways; location of control devices; and selection and location of fans. The term "optimal" as used here is global and not restricted to ventilation only. The ventilation system adopted for a mine must complement the mining method, and vice versa. Also a well-planned ventilation system offers flexibility in meeting emergencies as well as future needs. Very often, the airflow logistics problem is intertwined with personnel and materials transport as well as with many other safety and productivity considerations (O'Neil and Johnson, 1982; Ramani, 1982, 1992c; Suboleski and Kalasky, 1982). The mine airflow distribution is completely defined by (1) the physical parameters of the airways including shape, area, length, and characteristics of the airway surface; (2) the layout of the mine openings; (3) the pressure sources (e.g., fans) in the system, their location and characteristics; and (4) the interconnections between the airways, mine openings, and pressure sources.

13.2 COAL MINE VERSUS METAL MINE VENTILATION SYSTEMS

The mine ventilation principles discussed in the previous chapters apply equally to coal, metal, and nonmetal mining. Nonmetal mines bear resemblance to both coal and metal mines, depending on the mining methods employed. The differences are largely attributable to two factors: (1) characteristics of the mining system and (2) characteristics of the major atmospheric contaminants generated by mining.

Characteristics of the Mining System

The orientation and the extent of the mining system play a large role in determining ventilation-system design criteria. The mining method, in turn, is prescribed by the ore-deposit geometry, and this is where coal and metal mining often differ greatly. In the United States, underground coal mining is confined to relatively flat-lying seams at shallow to modest depths. Similar conditions are encountered in some limestone, salt, and trona mines, and their ventilation systems are similar to those in coal mines. In cases where metallic ore deposits are also similarly situated, coal and metal mine ventilation systems can be fairly similar. However, the unique characteristics of each deposit are generally more significant in metal mining than in coal mining. Most metallic ore deposits have a relatively large vertical dimension also, and this often yields mining—and therefore ventilation—practices that are not similar to those employed in coal mining.

Characteristics of Contaminants

One useful way to view a mine ventilation system is as a supply system, delivering a necessary operating supply (air) to working areas, and then removing a variety of waste products generated in the mining process. In coal mining, the primary pollutant to be removed is methane gas (Secton 3.3). Indeed, reducing methane concentrations to safe levels is a major criterion in the design of coal mine ventilation, such that when adequate ventilation is achieved for this purpose, all other ventilation requirements (e.g., control of airborne coal dust concentration) have usually been satisfied also.

By comparison, explosive atmospheres are an exception in metal mines. However, a wider range of pollutants is often encountered; the limiting design criterion could be due to diesel exhaust, radioactivity, or heat, as well as toxic-gas emissions from mining and blasting. Thus the nature of the controlling contaminant is a significant difference between coal and metal mining in ventilation system design. With increasing mining depths and the growth of diesel equipment use in coal mining, this distinction may become less significant in the future.

Differences in Design Criteria

Because of the preceding observations, there are several differences in the design criteria used in coal, metal, and nonmetal mines. Further, there are some specific areas where application differences occur (O'Neil and Johnson, 1982; Johnson, 1992a).

Air Reuse   In coal mines, it is not a common practice to ventilate more than one working area on one split of air, whereas it is the usual practice in metal mining. As long as the oxygen content is high enough and the level of contaminants low enough, some portion of the air may be recycled in metal mines. In U.S. coal mines, reuse of air to ventilate working areas is illegal.

Booster Fans   Although underground booster fans are prohibited in U.S. coal mines, they are used in coal and metal mines abroad and in metal mines in this country. In fact, it is not unusual for a large underground metal mine stoping operation to have literally dozens of auxiliary and booster fans operating underground, supplying air to individual stopes.

Piecemeal Mine Development   Coal mining operations tend to follow closely an approved mine development–extraction plan. On the other hand, because of the impact of cutoff grade on ore reserves, the erratic nature of many metallic ore deposits, and the relatively high cost of exploration in delineating such deposits, many metal mines are developed on a piecemeal basis. Thus, although an extraction system is selected on the basis of preproduction reserves, the system employed 10 or 15 years later may be quite different because of newly discovered reserves, low-grade mineral that became ore under improved economic conditions, or a variety of other similar factors. In recent years, such changes have also become frequent in coal mining due to the adaptation of longwall systems in predominantly room and pillar mines.

Cost of Ventilation   The proportion of overall production cost attributed to ventilation is typically lower in metal mining than in coal mining. It is difficult to justify a new shaft for ventilation purposes alone, particularly for deep mines in hard rock. As noted above, the development of mines is often unpredictable, and the ventilation system is adapted to support configurations based largely on operating factors.

The preceding generalizations have notable exceptions—for example, new mine ventilation and air-quality regulations are having a major impact on mining-method design and ventilation costs; on the whole, however, they constitute a reasonably valid characterization of the factors and issues that create major differences between coal mine and metal mine ventilation systems.

13.3 GENERAL CONSIDERATIONS

Coal mine ventilation systems must be designed so as to maintain at all times healthy and safe atmospheric working conditions. Correct estimation of the required air quantity at each of the workings is vital. This quantity should be determined on the basis of number of workers in the workings, kinds of machinery used, and the makeup of gases, dust, heat, and humidity. The required quantities of fresh air at the faces and other places must be increased in proportion to the amount of leakage anticipated in the airways. Leakage estimation may be problematical, but an allowance of 100% of the estimated quantities for faces is common. Quantity estimation is critical as head losses and power are proportional to the square and cube of quantity, respectively.

Small errors in quantity estimates can therefore lead to large errors in head and power estimates and affect the total mine ventilation, health, safety, and production performance.

The number of airways required to course the various quantities through the mine is important as this number influences the velocity of the air in the airway. Air velocities are critical from two points of view. From the safety viewpoint, inadequate air velocities can cause undesirable accumulations and layering of gases; very high velocities, on the other hand, can raise dust clouds. From the economic viewpoint, the head loss in a mine airway is proportional to the square of the velocity, and high velocities will entail large horsepower dissipation. Thus velocities must be carefully chosen so that the safety and economic factors are not compromised. Since velocity is a function of the area of an airway and of the number of airways in parallel, definition of the velocity and quantity requirements will automatically determine the number of airways once the shape and size of an airway are determined by mining-machine and ground-control conditions.

The head loss in individual splits can be calculated once the resistances of the airways and the volumes of air circulating in them are known. The head loss in the individual splits can be accumulated using series or parallel circuit laws as applicable until the total head loss in the system is calculated. Knowing the quantity and head required for a mine, selection of a suitable mine fan can be made. This is not to imply that mine-ventilation-system planning and designing is a simple straightforward process but to emphasize that it is an ordered process with many interactions with other aspects of mine planning and designing (Luxbacher and Ramani, 1980; Suboleski and Kalasky, 1982; Stefanko, 1983). Factors such as ground control, production requirements, and equipment limitations may often have great bearing on the mine design. In fact, mine ventilation planning is quite complicated when data on strata gases, mining methods, and mine openings are not readily available or estimates of these have large variances. Therefore, the final ventilation design selected must be flexible in order to adapt to changing conditions.

13.4 OVERVIEW OF MINE AND VENTILATION SYSTEM DESIGN

The planning and design of the ventilation system as they relate to the overall mine planning problem are illustrated in Fig. 13.1a. At the exploration stage, data must be collected on those specific geologic factors that may affect the ventilation system. Mine design also includes many other geologic data and, as indicated, is affected by rules and regulations on health and safety. The end result is the development of a mining method and a mine infrastructure. Mine ventilation considerations play an important role at this initial design stage. However, a more detailed ventilation analysis is required as more definite plans are made to mine the coal seam (Fig. 13.1b). This analysis

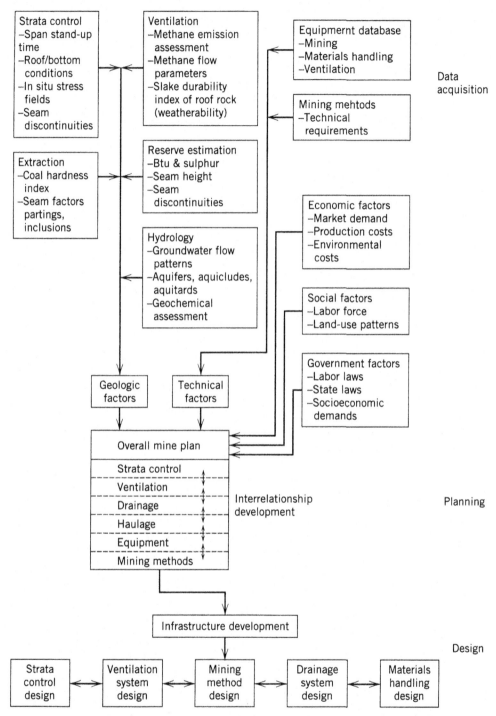

FIGURE 13.1a Mine design and ventilation system design (Luxbacher and Ramani, 1980).

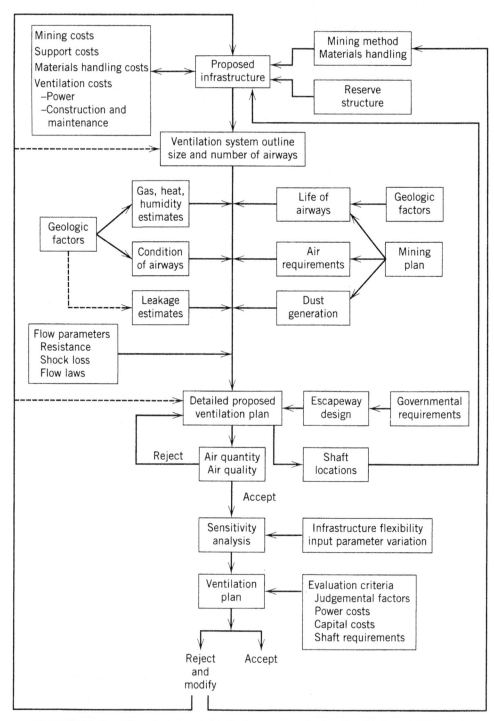

FIGURE 13.1b Mine design and ventilation system design (Luxbacher and Ramani, 1980).

must be very specific with respect to quantity and quality control of the mine air, and should include a sensitivity analysis covering factors such as methane and leakage, which cannot be accurately estimated. Once a suitable plan has been selected, it must be examined with relation to such factors as performance in case of emergencies (e.g., fires and explosions) and adaptability to changing mining plans (e.g., from room and pillar to longwall).

On the basis of such an analysis, the plan may be accepted as presented or modifications may be required in the ventilation plan or even in mine design. Early liaison between mine planning and ventilation staffs can prevent many problems and shorten the length of planning time by revealing the need for additional investigations or new calculations or engineering revisions. In effect, mine development schemes should incorporate full details of ventilation arrangements, and the means of achieving an effective ventilation system, therefore, should be considered in relation to the long-, medium-, and short-range mine projections (Anon., 1971). While safety is the prime consideration, good ventilation planning is essential to maintain or to increase production and productivity levels (Ramani, 1992c).

Long-Range Plan

A long-range plan is defined 10–20 years in advance. The general layout of airways underground should be decided by ventilation considerations as well as by other considerations. The sizes of the underground entries must be sufficient for their varying duties as airways at different stages of development. Since production plans may change with changing geologic and/or market conditions, the ventilation plan must be based on a split system that will give some flexibility to blocking off old or opening new sections as the need arises. The evaluation of specific details related to ventilation should be made as early as possible. Although there may be uncertainty with regard to geologic and market data, decisions regarding development of access to the seams (slopes, shafts, adits, etc.) and method of mining, for instance, cannot be deferred indefinitely. Such decisions as those pertaining to main and submain layouts, cross-sectional areas, and lining of the airways must be made. Since mine openings and fans normally function for the life of a mine, they should be selected carefully, taking into account their life-cycle duties. Such total system considerations can reveal the need for relocation of shafts or the opportunities for eliminating shafts in the preliminary plans. Further, they will amplify the sequencing needs for these critical ventilation-related activities.

Medium-Range Plan

Medium-range plans must show the projected method of working and the general method of ventilation to be used for at least 3–5 years into the future. Projections should be updated minimally once a year to include changes and variations from the plan. These plans must be available to the operational staff to schedule ventilation-related actions in their operational plans. Medium-range plans should be based on underground head/quantity, methane, and temperature surveys except in cases where the workings are entering virgin areas. In the latter case, use of good estimation techniques for parameter values and of extensive sensitivity analysis cannot be overemphasized.

Short-Range Plan

The short-range plan, as the name implies, is for the immediate future, usually a year or so, and is based on the projected long- and medium-range plans. Short-range planning of ventilation is necessary to maintain control of the nature and timing of ventilation construction and activities. It must include at least the following: (1) layout of development work with a time sequence; (2) air quantities; (3) ventilation arrangement at the working faces; (4) schedule for ventilation changes; (5) schedule for the construction of air crossings, doors, and so on; (6) special precautions for firefighting; and (7) emergency escapeways.

13.5 VENTILATION CONTROLS

Success in providing adequate ventilation to the active workings of a mine for any method of mining depends on adequate fan capacities and good air distribution (primary ventilation) and, when the air reaches the face area, good control and distribution of the air at that location (face ventilation). An acceptable system provides various control devices such as stoppings, overcasts, airlocks, and regulators, so arranged that air is coursed in the desired manner in the desired quantities. Control devices in mine ventilation serve to (1) separate the intake and return airstreams in adjacent airways (stoppings), (2) allow the crossing of the intake and return streams without mixing (overcasts), and (3) regulate the flow of air through the various airways in the desired manner when the quantity has to be split between the airways (regulators). For a list of map symbols for control devices, see the List of Map Symbols in the frontmatter of this book.

Stoppings

Stoppings are physical barriers erected between intake and return airways to prevent the air flowing in them from mixing with each other. Stoppings can be temporary or permanent.

Temporary stoppings are often constructed of fire-resistant jute fabric, plastic, rough lumber covered with plastic, or even various types of sheet metal sections. These are extensively used in areas where frequent adjustments to air directions are necessary, such as in the working panels. Where

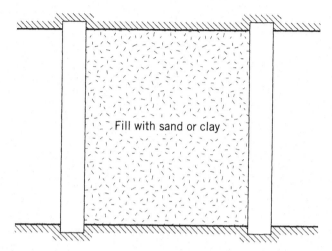

FIGURE 13.2 Well-constructed explosionproof stopping (seal).

the stoppings are constructed of concrete, cinder, or slag blocks, the blocks are usually wedged in place with dry joints. This permits fast erection and even allows for recovery of the blocks.

Permanent stoppings are installed in places where a permanent or a long-term control of flow is needed, such as between the main intakes and returns and neutral entries. Most permanent stoppings generally consist of 8 × 8 × 16-in. (≅200 × 200 × 400-mm) solid or hollow blocks, laid with mortar joints. The entire sides of the stopping are then coated with cement or some type of block coating. The stopping is usually keyed into the roof, floor, and sides. It is common practice to provide a fire-resistant stringer or cushion across the top to prevent cracking due to strata pressure. Since stoppings must be accessible for inspection and repair, it is advisable that they be built where the roof and sides are secured. Additionally, the stopping must be accessible for both inspection and maintenance.

A *seal* is a special stopping used to isolate abandoned workings or as fire bulkheads in coal mines. Although no definitive design criteria have been established, ordinary practice is to construct two solid block stoppings 1–2 ft (0.3–0.6 m) apart, and to fill this void with concrete, earth, or sand. The stoppings should be substantially built so that they are airtight. They must resist the disruptive forces of explosions. For the latter purpose, the stoppings must be built according to specifications of the Coal Mine Health and Safety Act of 1969 as given in Title 30, *Code of Federal Regulations* (CFR) (Anon., 1995a). Shown in Figs. 13.2 and 13.3 are two such constructions; the one in Fig. 13.3 is able to withstand greater pressures than the stopping in Fig. 13.2. The greater the distance between the two masonry walls, the more effective will be the seal.

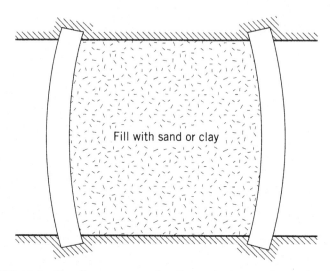

FIGURE 13.3   Excellent construction for explosionproof stopping (seal).

A *door* is simply a hinged or movable partition within a stopping designed to permit the passage of personnel and equipment. Mine doors may be constructed of metal (required for doors between the intake and return) or of lumber covered with tarpaper, plastic, or other sealant material. They serve the same function as stoppings and are frequently used in haulageways. To avoid short circuiting between the return and intakes, main doors should always be arranged in pairs to provide an *airlock*—one door is always closed while the other is open. The distance between doors should be made great enough to accommodate the longest train of cars or equipment passing. Automatic self-opening doors are especially useful along haulageways and equipment travelways. However, doors along the haulageways do not have sills, so that conveyor belting or brattice cloth should be attached to the bottom of the door. When the head loss across the door is high (over 1–2 in. water, or 250–500 Pa) and difficulty is experienced in opening it, a small door that can be operated by hand is usually cut in the main door.

Doors ordinarily serve as stoppings but also as regulators if propped open. Emergency, fireproof doors of metal construction are also located at strategic points for closing in event of fire or explosion to isolate the combustion area. Whatever their use, care must be exercised that doors are left in the position desired. Therefore, personnel should be trained in their proper use. They should be marked in bold letters—**KEEP CLOSED** or **KEEP OPEN**—so that they serve their intended purposes.

Overcasts

Overcasts or *crossings* are air bridges that allow the intake and return airflows to cross one another without mixing. Undercasts have the advantage of not disturbing the roof in the mine airways but are less frequently used because these are below the level of the surrounding openings, and water tends to accumulate in them. Shown in Fig. 13.4 is an ideal overcast with long sweeping curves at both the approach and discharge sides to ensure very low shock losses. The approach is gradually contracted and the discharge gradually expanded, and the cross-sectional area of the overcast is one-half the area of the approaching airway.

The overcast should be properly built with high-quality material. In the area where an overcast is required, the roof is raised to the necessary height. Common practice where continuous miners are used is to ramp up at the spot where the overcast is to be located before returning to bench the coal (Stefanko, 1983). Overcasts in coal mines are of metal or masonry construction. In constructing masonry overcasts, solid blocks set in mortar should be used. The two wing walls must be anchored firmly in the rib. Steel beams are then placed on the wing walls over which the roof is built. Two walls are then built from this roof to the heightened roofline. When well constructed,

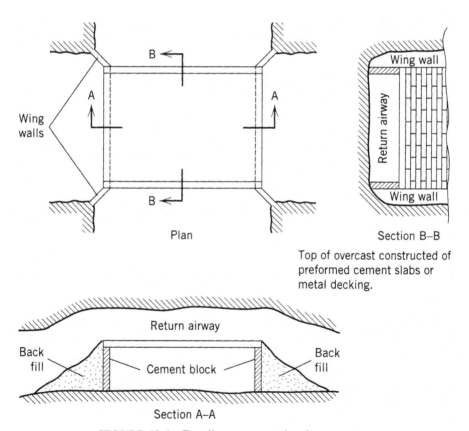

FIGURE 13.4 Excellent construction for overcast.

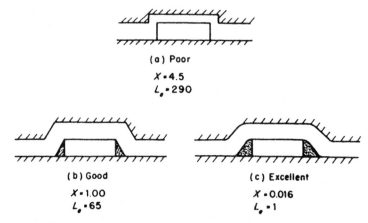

FIGURE 13.5 Comparison of theoretical values of shock-loss factors and equivalent lengths for various overcast constructions, calculated from information in Table 5.3.

resistance to airflow is increased nominally and in the upper airflow only. Shock losses for various types of overcasts are compared in Fig. 13.5. Prefabricated, culvert-type units have also been used extensively. Small-diameter tubing may be used for equalizing head losses or for other low-volume applications where travel is not required. An inexpensive but good-quality overcast, most suitable for temporary use, is illustrated in Fig. 13.6. Here a solid (stopping) wall is built on either side of the airway, and a sufficient number of pipes of large diameter are laid on the top of the walls to carry the air across the main airway. It is possible to transport up to 25,000 cfm (11.8 m³/s) with two 36-in. (914-mm) or three 30-in. (762-mm) pipes. The ends of the pipes should be embedded in the stopping and made airtight. Special overcasts are available to enclose belt conveyors, often featuring special openings for easy access and cleaning of spillage.

Regulators

Regulators are used to control and redistribute the quantity of flow in each split of air. Regulator sizing has been discussed previously (Section 7.7). The regulator is an opening in a stopping in an airway and may be equipped with an adjustable or sliding door (Fig. 13.7). These are known as box regulators and are used almost exclusively in coal mines. Regulators should be easily accessible and located in places where roof and rib conditions are good. They should be kept clear of debris and obstructions. The ideal location for regulators in coal mines is in the return near the beginning of a ventilation split so as not to interfere with haulage and materials transport. In this location, it may be utilized for the life of the working section and after the section is mined out or idled.

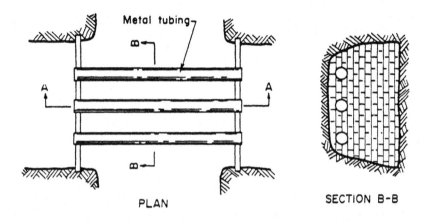

PLAN

SECTION B-B

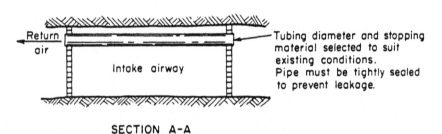

SECTION A-A

FIGURE 13.6 Inexpensive but efficient overcast (Rock et al., 1971).

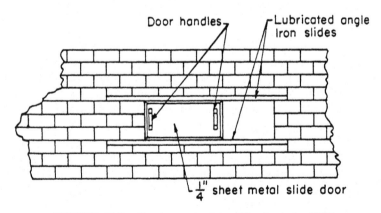

FIGURE 13.7 Box-type regulator (Rock et al., 1971).

Line Brattice

A partition placed in the opening to divide it into intake and return airways is termed a *line brattice*. The use of line brattices in coal mines to advance development headings and rooms beyond the last breakthrough is discussed in Section 11.2.

Fans

Except for auxiliary fans, which may be used in face areas in lieu of line brattice, all fans must be located on the surface in coal mining. Exhaust fans are favored for larger mines because of the temporary retardation of methane gas in the event of a power failure. However, small drift mines often use a blower system, since unidirectional airflow can easily be established in panels simply by ''punching out'' to the surface.

For large mines, multiple main fans are normally required (Chapter 10). The use of multiple fans in the original planning of the ventilation offers versatility and flexibility desirable in large mines (Kingery and Kapsch, 1959). There are strong arguments for utilizing a multiple-fan system.

1. In the event of failure of one fan, the ventilation will be reduced but not stopped; the load can be redistributed among the remaining fans so that production can be continued without interruption.

2. In the event of emergency (fire or explosion), close control over the ventilation system can be obtained with the other fans.

3. Expansion of the system can be carried out by installing an additional fan as needed rather than modifying the original fan.

4. There are limitations on the head–quantity output of a single fan; the use of multiple fans allows several combinations of head and quantity to be obtained.

Considerable experience with multiple-fan ventilation has accrued over the years in coal mining. One large coal mine meets a quantity requirement of 3,500,000 cfm (1652 m^3/s) with a ventilation system composed of four fans with heads of up to 24 in. water (6.0 kPa) (Stevenson, 1988).

There are certain disadvantages to multiple-fan ventilation systems. These include the need for control, necessity of close coordination, and the tendency for fans to assume disproportionate loads. Interventilation of mines is objectionable for the same reasons. It has been recommended that in addition to the measures listed above, duplicate fan and drive units be installed at each fan site to ensure uninterrupted ventilation (Mancha, 1958). A backup power source, such as an automatic-starting diesel engine, is sometimes employed in place of a complete duplicate unit.

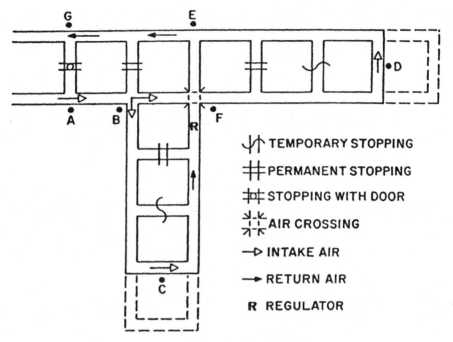

FIGURE 13.8 Ventilation circuit with leakage sources.

Vent Pipe or Tubing

An *auxiliary fan* and *ventilation pipe* are used to supply or remove air from the face of a working by providing an additional duct for airflow. The use of fans and vent pipe in metal and coal mining, similar to the use of the line brattice in ventilating dead ends in coal, is discussed in Sections 11.3 and 11.4.

13.6 LEAKAGE CONSIDERATIONS

Consider a somewhat oversimplified diagram of a part of a mine ventilation system (Fig. 13.8) (Ramani, 1992a). Fresh air to the workings is split at point B. The roles of the stoppings, door, overcast, and regulators are evident. Stoppings are also built in the connecting entries between ABFD and DEG and in the connecting entries between BC and CF. The dotted lines show the mine advance projection; as the mine advances, the temporary stoppings will be replaced by permanent stoppings. The return from the workings at C must cross the intake in ABFD at F; an overcast is built at F to course the return air from C directly to E over the intake air to D. The two splits of air reunite at E. Since access to the return entry may be needed, a door is built into the stopping in the connecting airway AG. A regulator R is placed in the return airway CF to regulate the flow into the workings at C. Since air always flows from a point of higher total head to one of lower total head, the head at every point on the intake side is higher than that at the return side. Consider the head across the stopping in the airway AG. On the side of the stopping toward A, the head is higher than it is on the side of the stopping toward G. Because of the stopping, free flow of air from A to G is not possible. In practice, however, some air will flow through the cracks and crevices in the stopping due to the pressure difference. Such unintended losses of air directly to the return from the intake are known as *leakage*. Leakages occur, as explained before, through stoppings, overcasts, and doors. Leakages have also been known to occur through crushed pillars and improperly packed gob. The leaking air does not help in the ventilation of the working areas. Thus leakage is doubly disadvantageous: it does not improve health and safety conditions, and it increases the cost of getting the quantity of air needed into the main split (Fig. 13.9a). Leakage should be kept to a practical minimum. Leakage estimates are essential for determining the quantity requirements at the fan.

Leakage phenomena are complex. Leakage through stoppings, doors, and regulators depends not only on the pressure across the control device but also on the condition of the device itself. Ground pressures destroy control devices, and fugitive air losses become so great that it is almost impossible to provide adequate air at the face. Generally, leakage is most severe through old stoppings in the outby portion of the circuit. These are also subjected to higher pressure differences than the newer inby stoppings. Therefore, the circuit air quantity diminishes at a decreasing rate as the air progresses from outby to inby in the circuit. For example, between adjoining intake and return airways in coal (and some metal and nonmetal) mines, leakage will occur at

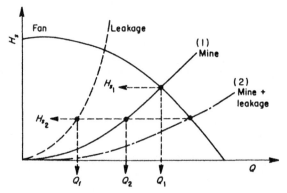

FIGURE 13.9a   Effect of leakage in a simple ventilation circuit. Condition 1, no leakage; condition 2, mine with leakage.

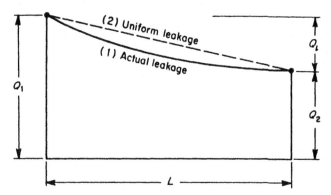

FIGURE 13.9b Comparison of (1) actual leakage rate and (2) uniform leakage rate between adjacent parallel intake and return airways. (After Mancha, 1942. By permission from AIME, New York. Copyright 1942.)

the fastest rate where the pressure differential is highest, ordinarily near the fan. In Fig. 13.9b, line 1 shows the actual decrease in quantity between two points in adjacent entries located a distance L from the portals, and line 2 represents a linear decrease, sometimes assumed as the limiting case. Painting and plastering a stopping decreases leakage, on the average, to around 2 or 3% of the leakage through a mortar-laid, unplastered, and unpainted stopping (Kawenski et al., 1965). The exact amount of leakage, however, is determined by how well the plastering and painting is done.

Kharkar et al. (1974) carried out studies to determine leakage across stoppings under several different airflow conditions. Graphs were plotted for cumulative leakage against the number of stoppings (Fig. 13.10). The graphs show that the rate of air loss is variable over the length of the airway, the largest values being farthest from the working face with three-quarters of the total loss occurring in the first one-half of the air intake.

In ventilation planning, leakages have been handled by a system of rough allowances since they often cannot be measured accurately, particularly for leakage through overcasts, doors, and airlocks. Table 13.1 lists some recommended figures for use in planning (Pursall, 1960). Where volume measurements can be made on either side of the leakage source, these measurements are not only more reliable than visual inspection, but may also serve as a base for estimating leakages.

In the past, an average of less than 20% of the air quantity measured at the fans reached the workplaces. Conditions today are much better in that this figure is anywhere from 40–60%. For new mines, fairly liberal leakage allowances should be assumed in the design, generally 50% or more of the quantity at the inby end of the split.

In allowing for leakage when using the computer, high-resistance splits may be inserted at various points to simulate more realistically the effect of

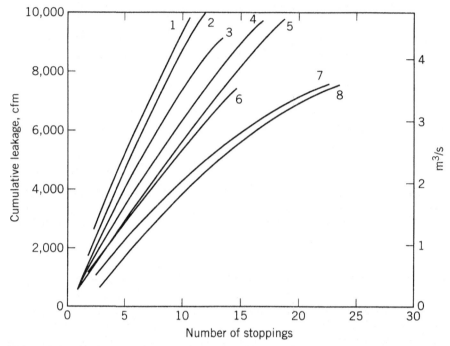

FIGURE 13.10 Results of study to determine leakage vs. number of stoppings in several coal mines: (1) mine 4, section a–$H = 1.7$, $L = 1530$; (2) mine 2, section b–$H = 2.8$, $L = 1520$; (3) mine 2, section b–$H = 2.3$, $L = 630$; (4) mine 3, section a–$H = 1.9$, $L = 1200$; (5) mine 2, section c–$H = 0.51$, $L = 1050$; (6) mine 4, section B–$H = 0.71$, $L = 990$; (7) mine 1, section B–$H = 0.78$, $L = 1360$; (8) mine 1, section a–$H = 1.32$, $L = 3125$ (L–length in feet of roadway considered; H–head loss in inches of water across stoppings. (*Sources:* Kharkar et al., 1974; Hartman et al., 1982).

leakage. For hand calculations, this may be too complicated an approach. Mancha (1950) has proposed that the ratio of the head losses with and without leakage may be considered proportional to the ratio of the quantities at the two points in the circuit, or

$$\frac{H_{l_L}}{H_l} = \frac{Q_1}{Q_2} \tag{13.1}$$

where H_l is head loss without leakage (computed for Q_2) and H_{l_L} is head loss with leakage (based on Q_1 decreasing to Q_2). Although only an approximation, the validity of this assumption has been supported by others (Holdsworth et al., 1951). If uniform leakage is assumed, the calculation for head loss will be based on the average quantity $[(Q_1 + Q_2)/2]$. An example demonstrates the differences in the calculated head losses.

TABLE 13.1 Leakage Estimates in Coal Mines
Leakage across newly formed gobs (longwall faces)

Distance between Intake and Return, ft	Leakage across Gob as Percentage of the Air on the Face, %
150	20
300	10
600	5

No allowance made where solid stowing is done

Separation doors: 3000 cfm
Overcasts: 3000 cfm

Surface leakage: air that leaks through the casing at the top of the upcast shaft; a function of fan head

Fan Head, in. water	Leakage, cfm
5	25,000
10	35,000
15	45,000
20	50,000

Leakage through stoppings
Explosionproof: none
Temporary stoppings: 100–200 cfm dependent on the age, physical construction, maintenance, and head across the stopping.

Source: After Pursall (1960).
Conversion factors: 1 ft = 0.3048 m, 1 cfm = 0.47195 × 10^{-3} m³/s, 1 in. water = 248.84 Pa.

Example 13.1 The main entries in a coal mine consist of four intakes and four returns, each having a friction factor of 217×10^{-10} lb·min²/ft⁴ (0.0403 kg/m³) and a cross section 5 × 20 ft (1.5 × 3.0 m). If 42.8% leakage occurs between two points in the entries 4000 ft (1220 m) apart, calculate the head loss across the intakes (or returns) when 100,000 cfm (47.2 m³/s) of air, exclusive of leakage, is delivered.

Solution: In each airway

$$Q_2 = \frac{100,000}{4} = 25,000 \text{ cfm}$$

Head loss without leakage:

$$H_l = \frac{(217)(10^{-10})(30)(4000)(25,000)^2}{(5.2)(50)^3} = 2.5 \text{ in. water}$$

Original quantity, with leakage:

$$Q_1 = \frac{100,000}{(1 - 0.428)} = 175,000 \text{ cfm}$$

Leakage:

$$Q_l = Q_1 - Q_2 = 175,000 - 4(25,000) = 75,000 \text{ cfm } (35.4 \text{ m}^3/\text{s})$$

Head loss with leakage (Eq. 13.1):

$$H_{l_L} = (2.5) \left(\frac{175,000}{100,000}\right) = 4.38 \text{ in. water } (1.090 \text{ kPa})$$

If a uniform leakage rate were assumed, the average quantity in each airway will be 34,380 cfm (16.2 m³/s), and the head loss will be

$$H_{l_L} = (2.5) \left(\frac{34,380}{25,000}\right)^2 = 4.73 \text{ in. water } (1.177 \text{ kPa})$$

Uniform leakage assumption results in a 8% increase in head loss. The assumption of uniform leakage over a long stretch of airway is not realistic, because head differences (and consequently, leakage) across stoppings in the beginning of the circuit will be much higher than those toward the end of the circuit.

13.7 LAWS AFFECTING MINE VENTILATION

Many features of mine ventilation systems owe their origin to regulations enacted by federal, state, or local governing bodies. Methods of air distribution and the layout of the ventilation system are governed by these regulations. Familiarity with the provisions of mining laws that are applicable to ventilation is essential in the design and/or operation of a mine ventilation system. The advancements in technology (e.g., automatic monitoring system) may permit variations to air distribution systems and must be explored for technical and economic feasibility and for enhancing health and safety (Section 6.8).

   In the United States, at the federal level, the Mining Safety and Health Administration (MSHA) is responsible for the enforcement of health and safety standards defined in *CFR* Title 30 (Anon., 1995c). These regulations specifically outline the quantity and quality requirements expected. These regulations are modified on the basis of new findings, experience, and good practices, and therefore the following primary coal-mine system design criteria affecting ventilation quantity and airways must be verified for their relevance.

(75.302f)   In mines with multiple fans, the system must be designed to prevent air reversal if stoppage of any fan or fans occur.

(75.321a)   Oxygen content must be greater than 19.5% by volume; carbon dioxide must be no more than 0.5% by volume.

(75.321b)   Regardless of other limit values, the following gases must not have concentrations by volume greater than

    (a) CO   2.5%
    (b) H_2 0.80%
    (c) H_2S 0.80%
    (d) H_2S_2 (acetylene)   0.40%
    (e) C_3H_8 (propane)   0.40%
    (f) MAPP (methyl acetylenepropylenepropodiene)   0.30%

(75.322)   Concentrations of noxious or poisonous gases must not exceed the 1972 threshold limit values (TLVs) as specified by the American Conference of Governmental Industrial Hygienists (ACGIH) (Anon., 1972d).

(75.323)   Maximum methane content is 1% in face areas and section return airways, although it may be 2% in return airways that are not the return split from the working section (e.g., bleeders).

(75.325)   Minimum quantity must be 9000 cfm (4.2 m^3/s) at the last open crosscut and 3000 cfm (1.4 m^3/s) at the working face.

(75.326)   Minimum mean air velocity at all working faces is 60 fpm (0.3 m/s).

(75.327b)   Unless special approval is granted, velocity in trolley-haulage entries must be less than 250 fpm (1.25 m/s).

(75.332a)   Each mechanized mining section must be ventilated by a separate split of intake air.

(75.332b)   Does not allow the use of air that has passed through any area not examined under 75.360, 75.361, or 75.364 or where second mining has been done to ventilate any workplace.

(75.334)   Bleeder entries or seals are required in pillar areas. Seals must be explosionproof.

(75.350)   Air in belt-haulage entries must be isolated, the air velocity is limited, and the air may not be used to ventilate active working faces unless needed.

(75.380)   Escapeways; bituminous and lignite mines (primary intake air; alternate escapeways).

(75.381)   Escapeways; anthracite mines.

(75.384)   Longwall and shortwall travelways.

In general, state laws conform to the federal code or have less-stringent regulations, although in a few cases special requirements may exist.

13.8 LOCATION OF FAN AND PRINCIPAL AIRWAYS

Main Openings and Direction of Flow

Of paramount concern in any mine ventilation system is the location of the main openings and the direction of airflow. Regardless of the method of mining employed, at least two openings must be provided, separated by a prescribed distance. It would seem logical to use one of these as the intake airway and the other as the return, and this is frequently done if the openings are shafts; the one serving as the intake is termed the *downcast shaft* and the one serving as the return the *upcast shaft*. The main concern is safety. However, economy and convenience are also considerations to be weighed in making the final decision.

It is desirable to place the principal routes of access and egress on intake air. This is done not only for health reasons, permitting miners to travel to and from work in fresh, uncontaminated air, but also to provide a generally safe exit in an emergency (fire, explosion, etc.). To escape the toxic gases from a fire, miners would usually exit the mine through the primary escapeway. Neither the belt- nor trolley-haulage entry may be used as part of the separate intake escapeway. The main fan must be located on the surface in a fireproof housing and must be offset with a "weak wall" in the line of force to minimize damage from explosions. It is a common practice to put down a single two-compartment air shaft, with the downcast and upcast portions separated by an airtight, concrete curtain wall. Mancha (1950) has demonstrated, however, that two separate air shafts are cheaper than a single two-compartment shaft. Furthermore, the leakage problem is eliminated with separate shafts. It is becoming more common to drill smaller-diameter shafts in an effort to reduce the number of large air shafts required. Where a slope belt is used for haulage from the mine, tubing and a small fan are often used to ventilate the slope separately.

The ideal arrangement of main openings in a mine is to locate the intake airway(s) at or near the center of operations and to ring the active mining area with exhaust airways. In practice, this is never completely or continually realized as the active mining areas are constantly shifting or additional property is acquired. In a mine of great depth, the number of openings that can be afforded is strictly limited by cost considerations. In shallower but large-area coal mines, multiple intake-return shaft combinations are preferred. Consequently the actual layout of main openings in a given situation must be a compromise, approaching the ideal arrangement as nearly as possible.

Main Entries   The main entries that extend from the shafts and slopes carry the major quantity of air to the splits and are usually required to last throughout the life of the mine. These two aspects make their design with regard to number, size, support, and lining material important. Since material (coal and supplies) is usually conveyed through one or more of the entries, limitations such as isolation of the belt and maximum velocity in trolley haulage airways must be considered in determining the number of airways. The air quantity flowing in the airway is a function of the air-velocity range desirable in the airway.

Example 13.2 Given these ventilation and haulage specifications, calculate the number of entries in the mains:

Mainline haulage	belt
Supply haulage	battery-powered scoops
Range of maximum velocity	600–800 fpm (3.05–4.06 m/s)
Minimum quantity to be carried	300,000 cfm (141.6 m³/s)
Seam height	6 ft (1.82 m)
Width of entry	20 ft (6.1 m)

Solution:

(a) In this case, the area of an airway is 6×20 ft $= 120$ ft^2 (11.15 m^2). But it is possible that this entire area may not always be available for airflow since there may be some obstructions in the form of stationary equipment and supplies that are either required or left in the entries. Additionally, due to convergence, the actual height of the entry may be less than 6 ft (1.82 m). Assuming the actual area for airflow at 100 ft^2 (9.29 m^2) and an air velocity of 800 fpm (4.06 m/s), the number of airways needed to convey 300,000 cfm (141.6 m^3/s) will be

$$\frac{300,000}{800 \times 100} = 4 \text{ entries}$$

Based on a velocity of 600 fpm (3.05 m/s), the number of airways needed will be

$$\frac{300,000}{600 \times 100} = 5 \text{ entries}$$

(b) It can be assumed that the main entries will be carrying the entire quantity over only a short distance, and that air from the main entries splits to the submains and panels at frequent intervals. Thus, from the velocity point of view, at least four intakes and four returns (i.e., a total of eight entries) are necessary.

(c) Because a belt entry must be isolated, an additional entry is needed to accommodate the belt, making nine the total number of entries required. If trolley haulage is used for personnel and material transport, then the intake escapeway must be isolated from the trolley entry. Additionally, the velocity in the trolley entry must not exceed 250 fpm (1.27 m/s). This may increase the number of entries in the mains by at least one.

If, instead of a belt, trolley haulage is used for personnel, supplies, and coal transport, then only the intake escapeway must be isolated from the entry in which the trolley operates. However, the velocity restriction still demands that the trolley be isolated and regulated. In this case, however, eight-entry mains may be sufficient.

Shown in Fig. 13.11 is an eight-entry mains development. In this example, the two outside entries on each side are the returns. The No. 3 entry (always counting from the left facing inby) contains the tracks for trolley haulage. The trolley entry is isolated and followed by the three intake entries. One of these is designated as the intake escapeway. Whenever the trolley haulage has to be carried into the splits, the trolley entry needs to be isolated from the other intakes and returns by a set of overcasts. It is also good practice to connect the returns on either side by a set of overcasts every 2000 ft (600 m). These are known as "equalizers." Thus, if there is a fall on the returns blocking any set of overcasts completely, the air can cross over and back without unduly affecting the overall resistance of the returns. The idea of putting the returns adjacent to the solid rib of coal is advantageous, because any gas that bleeds off the solid coal will not be carried to the face. If symmetry with regard to the haulage is required, the trolley or belt can be moved to the fourth or fifth entry. In this case, however, an additional row of stoppings will be required (i.e., two rows of stoppings to isolate the intakes from the return on each side, and two rows to isolate the trolley or belt from the intakes on either side).

There are several other possible layouts for the mains. For example, to minimize the number of stoppings, one side could be the intake and the other the return, as shown in Fig. 13.12. The disadvantage of this layout is the contamination of the intake air from the solid rib bleedoff and worked-out panels. Because of ventilation economics, some mines with low methane emissions use such an arrangement (Stefanko, 1983).

Shown in Fig. 13.13 is a standard set of mains in a high-methane, high-production mine (Stefanko et al., 1977). Sixteen entries divided into three groups are laid out so that, during development, each group has its own belt (B), track (IT), intakes (I), and returns (R). Once development is complete, the outlying groups, which consist of five parallel entries, will be used as returns. The central group, which consists of six entries, will utilize five entries as intakes, and an isolated belt will be placed in the No. 4 entry. The eventual isolation of the three groups into returns and intakes with 250-ft (75-m) barriers between them reduces both leakage and resistance. Furthermore, this type of mains development is suitable for use when large tonnages are being produced during first mining. Fig. 13.14 shows a scheme employed at a mine using eight entries per section with two sections paralleling each

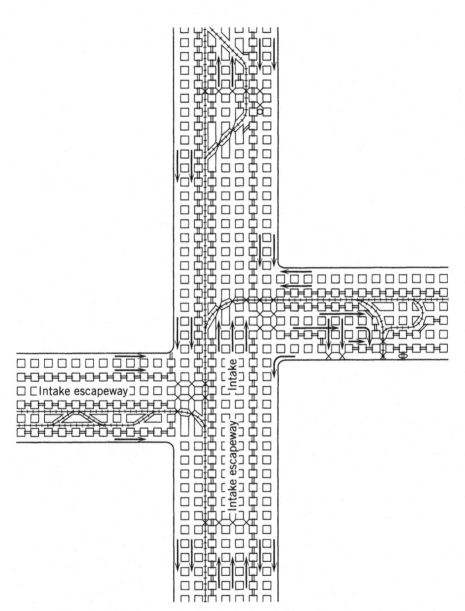

FIGURE 13.11 Eight-entry main development with track haulage.

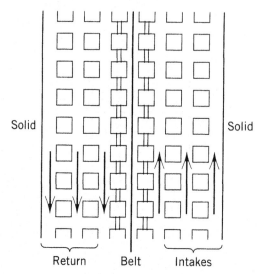

FIGURE 13.12   Mains development with belt haulage to minimize number of stoppings.

other. The two sets are connected at convenient intervals and have the same advantages as listed above.

Submains   Submains or panel entries have a much shorter life than mains and carry much less air. For example, according to the 1969 Coal Mine Health and Safety Act, the minimum volume of air required is only 9000 cfm (4.25 m³/s) at the last open crosscut. In practice, however, the quantity of air required at the last open crosscut is a function of the methane liberation rate and the number of machine units and workers in the section. It is good ventilation practice to place the miner operator on one split of air that will be conveyed directly to the return, and the roof bolter operator (or other machine units) on another split of air. This type of face ventilation, called the *dual-split* or *fishtail arrangement,* is very common in gassy mines as well. The air in the face areas is coursed by line brattice and check curtains. Here leakage losses will be severe. For example, the air quantity on the basis of the preceding considerations at the last open crosscut is, say, 40,000 cfm (18.88 m³/s)(20,000 cfm, or 9.44 m³/s, on each side). The air at the entrance to the panel will have to be much more than this quantity to account for the leakage from the intake into the returns all along the panel length. This can be estimated on the basis of the number of stoppings and the stopping construction. For the purposes of illustration, the air at the panel entrance is assumed to be 60,000 cfm (28.32 m³/s). The air velocity in the submains (or panel entries) is considerably less than that in the mains.

Example 13.3   Given the following ventilation and haulage considerations, calculate the number of panel entries:

1. Maximum velocity: 300 fpm (1.52 m/s)
2. Width of entries: 20 ft (6.1 m)
3. Height of entries: 6 ft (1.8 m)

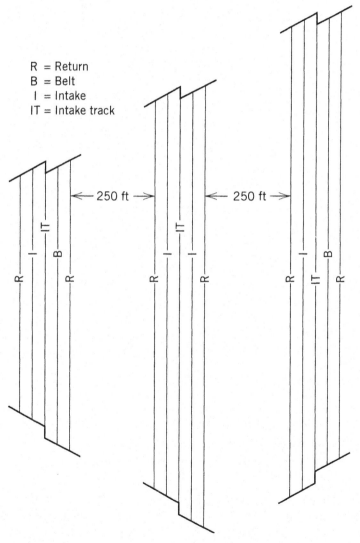

FIGURE 13.13   Mains development for achieving high production and for reducing leakage. (*Conversion factor:* 1 ft = 0.3048 m.)

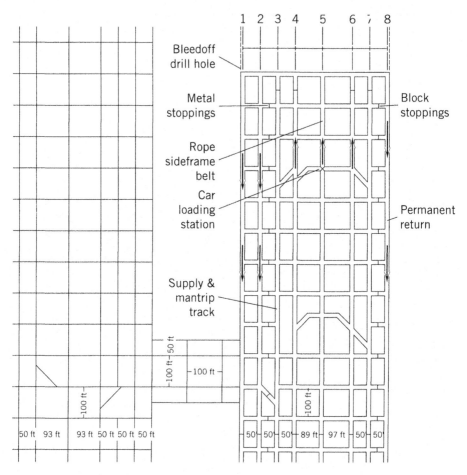

1 2 3 4 5 6 7 8

Bleedoff
drill hole

Metal
stoppings

Block
stoppings

Rope
sideframe
belt

Car
loading
station

Permanent
return

Supply &
mantrip
track

100 ft — 50 ft

100 ft

100 ft

100 ft

50 ft   93 ft   93 ft 50 ft 50 ft 50 ft

50   50   50   89 ft   97 ft   50   50

FIGURE 13.14 Design of mine entries according to pressure-arch theory and reflecting future ventilation expansion (Belton, 1962). (*Conversion factor:* 1 ft = 0.3048 m.)

4. Belt for coal transportation
5. Battery haulage for men and materials
6. Minimum quantity to be carried: 50,000 cfm (23.6 m³/s)

Solution: The cross-sectional area of an airway, although 120 ft² (11.15 m²), is rounded off to 100 ft² (9.29 m²) as before for the same reasons. Using the maximum velocity of 300 fpm (1.52 m/s), it is obvious that one airway can convey 30,000 cfm (14.16 m³/s). Thus two intakes and two returns will be enough from the ventilation point of view. However, a separate entry must

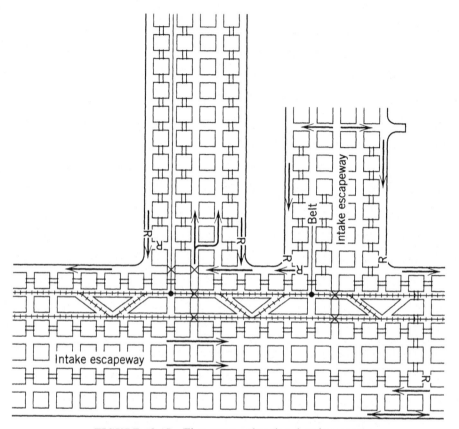

FIGURE 13.15 Five-entry submains development.

be driven for the isolated belt. In summary, a five-entry development is adequate.

Shown in Fig. 13.15 is an example of a five-entry development taking off from a seven-entry submain with trolley haulage. Notice that in the submains, two entries are isolated from the intakes and returns for the trolley. Studies by the U.S. Bureau of Mines (Deul and Kim, 1986) have shown that isolating a large block of coal by developing sets of headings around it to allow bleedoff for at least one year prior to mining can provide safety and production advantages. An example of such a panel development is shown in Fig. 13.16 where several blocks have been isolated before extraction within any one block begins. From an economic point of view, this practice is not most desirable due to the high investment in the mine, the high cost of development coal, and the delay caused in the start of second mining (i.e., pillaring or longwalling). Methane drainage ahead of mining through vertical

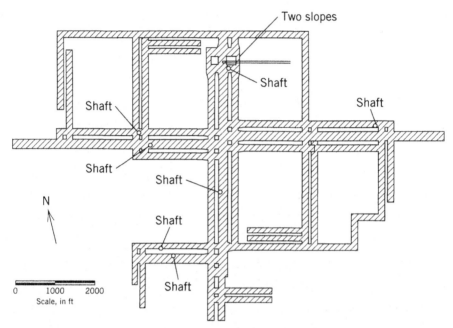

FIGURE 13.16   Mine plan for high-development production and degasification; development after 36 months. (*Conversion factor:* 1 ft = 0.3048 m.)

boreholes from the surface or through horizontal holes from the section is a preferred method (Aul and Ray, 1991; Section 3.7, of this book).

13.9   ROOM AND PILLAR VENTILATION

In the United States, the room and pillar method is predominant. In utilizing the multiple openings found in room and pillar workings, either unidirectional or bidirectional distribution may be employed. In *unidirectional flow,* the air in adjacent openings flows in the same direction and is entirely fresh or exhaust air; whereas in *bidirectional flow,* the air in adjacent openings flows in opposite directions and is partly fresh and partly used air. To maintain separate flow in bidirectional distribution, stoppings must be erected in connecting crosscuts. Even shafts are utilized in this manner, with a concrete curtain wall separating compartments and the two airflows. Leakage is a serious problem in bidirectional flow, and for this reason, there is a trend toward unidirectional flow in horizontal systems. It is particularly desirable in major airways (e.g., main entries). As discussed earlier, unidirectional flow in main entries may be approached by leaving a barrier block between two sets of main entries and using one set as intakes and the other as returns

(Figs. 13.12–13.14). Ordinarily, most systems are a mixture of unidirectional and bidirectional flow.

During development work, when entries are being advanced into virgin territory, a bidirectional system of either one or two splits can be used (Fig. 13.17). In the single-split arrangement, a high-velocity airstream ventilates the faces of all entries in series. A neutral heading, necessary for the installation of a belt conveyor, is easily maintained. In the double-split arrangement, doors or curtains may be eliminated in the haulageways, and two workplaces may be ventilated independently, an advantage in reducing dust exposure. In addition, with the double-split method, the outside ribs, which tend to release more methane, are exposed only to return air. However, a double row of stoppings is required, and the velocity of flow at the face is reduced. A single split is commonly used in mines containing little gas and when rooms are driven from only one side of the entries at a time, whereas a double split is preferable if rooms are driven on both sides simultaneously or if the mine generates much methane. The ventilation system arrangement for a typical four-entry development is shown in Fig. 13.18, and for a five-entry development in Fig. 13.19. The location of the belt and track entries and the cut sequence are also shown in these figures.

A mine completely ventilated by bidirectional flow is shown in Fig. 13.20. Details of the face ventilation in a similar system appear in Fig. 13.21.

Bidirectional flow modified with bleeder entries produces a composite system that has many of the advantages of unidirectional flow. The modified room and pillar method employed in Fig. 13.22 is ventilated by a composite system (Artz, 1951). It is applicable to the use of either continuous or conventional mining equipment, with a separate split of air supplied to each section. The avoidance of doors in major airways can be achieved in a similar system (Kingery and Harris, 1958).

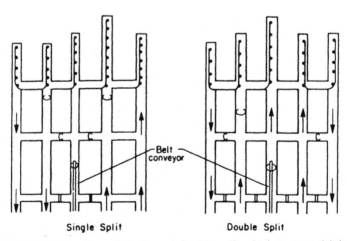

Single Split   Double Split

FIGURE 13.17 Use of single and double splits during entry driving.

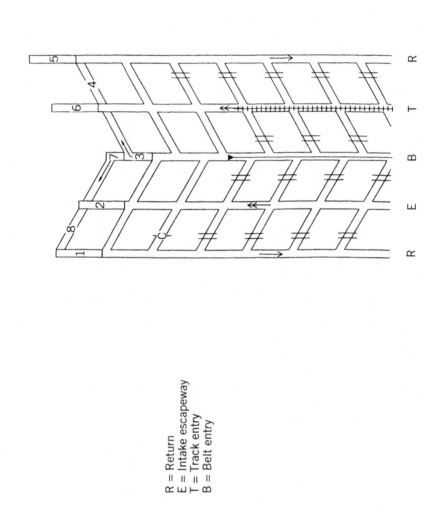

FIGURE 13.19 Typical five-entry main line development.

R = Return
E = Intake escapeway
T = Track entry
B = Belt entry

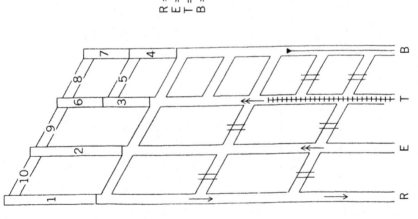

FIGURE 13.18 Typical four-entry longwall development.

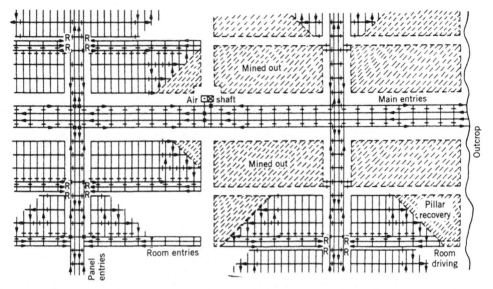

FIGURE 13.20 Bidirectional flow commonly employed in room and pillar mining of coal with conventional mobile-loading equipment.

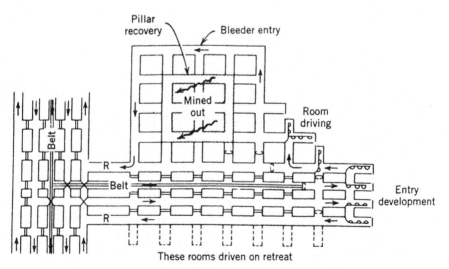

FIGURE 13.21 Details of face ventilation in room and pillar mining of coal. Entries employ bidirectional flow.

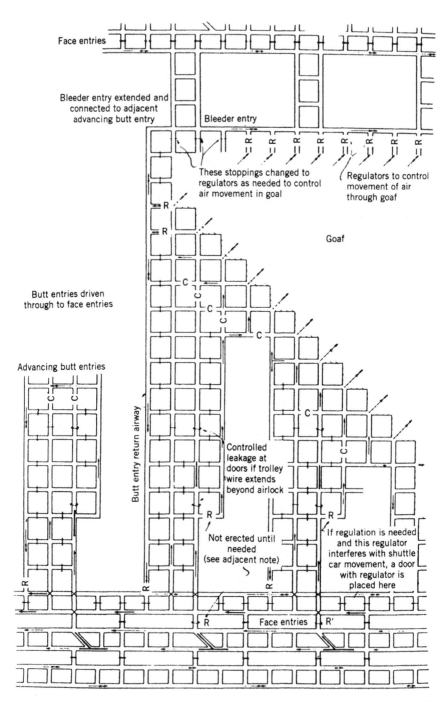

FIGURE 13.22 Composite unidirectional and bidirectional flow with bleeder entries employed in room and pillar mining (After Artz, 1951).

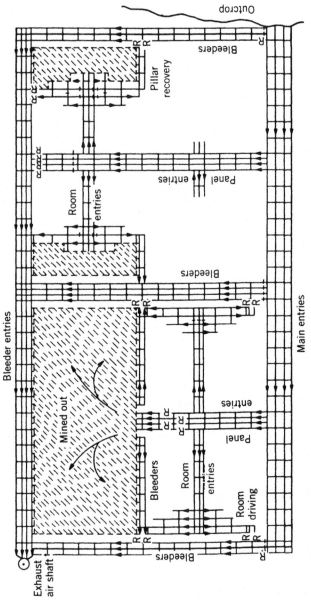

FIGURE 13.23 Unidirectional flow desirable to employ in room and pillar mining of coal, particularly with continuous equipment.

489

For contrast, compare Fig. 13.20 with the ventilation system shown in Fig. 13.23, which utilizes unidirectional flow solely, except during development. This system is designed especially for continuous mining under gaseous conditions (Anon., 1956). All control devices in the major openings are eliminated. Note the arrangement of bleeder entries to attain unidirectional flow.

Some general rules can be cited as a guide to laying out the major and minor features of the ventilation system for room and pillar mining in order to obtain the most satisfactory quantity control:

1. To achieve unidirectional flow, have one surface opening serve as the intake and another as the exhaust; that is, distribute air from one boundary of the deposit across to the opposite boundary. Furthermore, employ bleeder entries leading directly from the active workings to the return airways and connect panels through to adjacent entries. For abandoned, caved areas, methane drainage by bleeders is usually considered preferable to sealing, except where spontaneous combustion is a problem.

2. In the multiheading system utilizing three or more entries and bidirectional flow, employ the inner entries as intakes and outer ones as returns if air is to be taken off on both sides. If there are an even number of entries (excluding neutral airways), divide half and half; if an odd number, make the majority returns (since exhaust air is less dense and occupies a greater volume than air in the intakes, and because of leakage from the gob).

3. Always maintain return airways on the side nearest the gob and intake airways on the side next to the unmined coal.

4. Whether advancing or retreating, direct air to the end of the panel entries and return through the panels; but use separate splits to avoid circulation of air from gob line to room work.

5. Where possible, erect control devices in returns rather than intakes to avoid interfering with travel.

6. During room development, erect curtains across entrances to all except the first and last working rooms and in all but the last crosscuts nearest to face.

7. During pillar recovery, direct air to the working face by taking down curtains on the retreat.

13.10 LONGWALL VENTILATION

Several factors of ventilation system design for longwall mining are unique. Virtually all longwall faces in the United States are of the retreat type. Therefore, the development for longwall mining is analogous to that for room and pillar mining. It is standard practice to maintain a set of bleeder entries at the rear of the panels. Panel entries are driven and ventilation is established as in ordinary room and pillar practice. When these entries are used as the headgate entries, either all entries are placed on intake or one may be left as a return airway. Conversely, as tailgate entries, either all are made return airways, or the entry nearest the solid coal block is left as an intake. The latter has been found to reduce dust and methane at the tailgate in some mines. Extensive cribbing and/or the use of roofbolt trusses are usually necessary to prevent caving of these entries. Vertical drainage holes are frequently used to prevent methane buildup in the bleeder entries from a longwall panel. One element that is very important is the provision of escape routes from the longwall face. In U.S. practice, ventilation planners are moving toward providing escape routes in intake air off both sides of the longwall.

The development for a longwall face consists of two sets of parallel entries with two, three, or four entries in each set (Stevenson, 1988; Fuller, 1989). The most common in the eastern United States is the three- or four-entry development, accounting for over 80% of the longwalls in operation. In the western United States, two-entry developments are widely used. The distance between the two sets (i.e., the face length) is variable and can be anywhere from 600–1500 ft (182.9–457.2 m). These entries may be driven 6000–15,000 ft (1829–4572 m) in length.

The amount of air required in a longwall face is dictated by the methane emission rate during mining and conveying. The velocity of the air at the face is determined by the area available. The total cross-sectional area of the face is a function of the height of coal cut (usually the thickness of the coal seam), and the distance from the face to the gob line [usually 12 ft (3.7 m)]. However, the area occupied by the machine, roof support equipment, and conveyor must be subtracted from the total area to calculate the effective area available for ventilation. This area may be as little as 50% of the total. The net result of such a low area for ventilation is high air velocities at the face. While it is advantageous from the methane point of view to have high velocities, from the dust point of view, velocities over 600 fpm (2.79 m/s) are generally undesirable. Dust control in longwall is a problem, particularly when the face is mined in both directions by the so-called bidirectional cutting method (Chapter 4). In bidirectional cutting, when cutting in the direction of the flow of air, the dust generated by the machine is blown away from the operator. However, when cutting in the reverse direction, the operator and other workers are in the downwind side.

Minimum velocity and quantity across the face are to be specified in the approved ventilation plan (Anon., 1995a). The quantity of air must be at least 30,000 cfm (14.16 m^3/s) unless a lesser quantity is approved. The velocities are to be measured at locations in the face not less than 50 ft (15.2 m) nor more than 100 ft (30.5 m) from the headgate and tailgate. Leakage into the caved area is generally less than 20% of the entering air volume where the caving action is good and shield-type supports are used. In deciding on minimum design quantities, the higher mining rate and the great length of face being exposed must be considered in estimating methane liberation. Assuming a 5-ft (1.5-m) coal seam and a 12-ft (3.66-m) distance from the face to the gob line, the maximum cross-sectional area for airflow is 60 ft^2 (5.57 m^2). Assuming that only 75% of this area is available for ventilation, the remaining 25% being blocked by obstructions such as the machine, conveyor, and supports, the maximum velocity in the face for a flow of 30,000 cfm (14.16 m^3/s) will be about 700 fpm (3.26 m/s). A limit on the face velocity, somewhat on the high side, is 800 fpm (3.05 m/s), although higher velocities are common in highly gassy mines. Thus, if the methane emission rate per ton of coal mined is high, then the emission of methane can be reduced only by mining at a lower rate, by degasifying the seam prior to longwalling by isolating panels and/or by drainage through vertical or horizontal holes, or by water infusion (Aul and Ray, 1991). In modern high-production longwalls, the air quantity traversing the face can be as much as 50,000–70,000 cfm (23.60–33.04 m^3/s).

As an example, consider a gas emission rate of 50 ft^3/ton (1.55 m^3/t) of coal mined and a maximum production rate of of 10 tons/min (9 t/m). The amount of gas liberated is 500 cfm (0.24 m/s), requiring nearly 50,000 cfm (23.60 m^3/s) of fresh air for dilution at the face. An additional 10,000 cfm (4.72 m^3/s) can be assumed to directly enter the gob from the face. Therefore, at the entrance to the face, there must be 60,000 cfm (28.32 m^3/s). Providing for leakage, say, 25,000 cfm (11.80 m^3/s) in conveying the air from the main intakes to the face through the panel entries, and for ventilating the panel belt, say, 5000 cfm (2.36 m^3/s), a total of approximately 90,000 cfm (42.48 m^3/s) must be provided by the main intake.

This quantity can be easily handled by two entries, each with cross-sectional area of 100 ft^2 (9.29 m^2). However, requirements of a neutral belt, bleeding from the solid coal, and entries to serve the adjacent longwall face demand a three- or even a four-entry longwall development.

Advancing Longwall Ventilation Systems

The ventilation system of an advancing longwall face is illustrated in Fig. 13.24 (Dalzell, 1972). In this system, the five-entry panel is advanced by a continuous miner ventilated by 17,000 cfm (8.02 m^3/s) of air, while the longwall face is ventilated by a separate split of 18,000 cfm (8.50 m^3/s). The No. 4 entry in the panel serves as a common intake to the development section and the longwall face. The air is split inby the longwall face to provide separate air currents to each unit. The longwall and the development utilize a common belt in the No. 3 entry of the panel. The belt entry is ventilated by low-velocity air flowing inby to a regulator installed on the third crosscut outby the No. 1 butt development faces, where it enters the No. 1 and No. 2 entries serving as return airways.

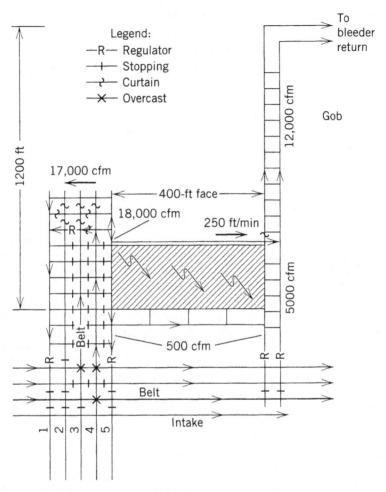

FIGURE 13.24 Advancing longwall face ventilation (Dalzell, 1972). (*Conversion factors:* 1 ft = 0.3048 m, 1 cfm = 0.47195 × 10⁻³ m³/s.)

The bleeder system for the advancing gob is developed by converting into bleeders the No. 5 entry of the active panel (adjacent to the gob) and No. 1 and 2 entries of the previously mined panel which is the tailgate for the advancing longwall. Bleeder flow is controlled by checks and regulators. The return air from the longwall face is directed to the main returns through No. 1 and No. 2 entries of the previously mined panel (5000 cfm or 2.36 m³/s) with the remainder diverted to the bleeder entries [12,000 cfm (5.66 m³/s)], allowing 1000 cfm (0.472 m³/s) leakage through the active gob to the No. 5 entry in the active panel. The advancing longwall system is not commonly practiced in the United States.

Retreat Longwall Ventilation Systems

Double-entry longwall panels are, almost without exception, mined on the retreat. The use of a double entry in the United States must be approved by MSHA. Probably the most common ventilation system design for double-entry longwall panels is a modification of the U-ventilation system. Return air from the face is split at the tailgate corner with the majority of the air flowing out the tailgate return and the smaller portion flowing through the tailgate bleeder (Fig. 13.25). Small leakage volumes also enter the headgate bleeder, although these are normally kept to a minimum. Intake air is routinely supplied to the face through both the belt entry and the headgate intake entry.

In the Y-ventilation system design, intake air is supplied through all the headgate and tailgate entries. Routinely, air is supplied to the face from the tailgate side, which keeps gob gases pushed away from the tailgate corner. Intake air supplied to the headgate corner via the belt entry, and the headgate intake is forced back into the gob. All the air supplied across the face is required to flow through and alongside the gob to return through the first crosscut immediately behind the face, as shown in Fig. 13.26.

If seals are constructed in the tailgate and headgate bleeder connections adjacent to the gob, and successively constructed in the headgate crosscuts as the face retreats, methane drainage system installations drawing through those seals become feasible. A high-pressure differential is applied across the gob with the Y-ventilation system, which may cause difficulty in controlling spontaneous combustion in the gob (Fuller, 1989).

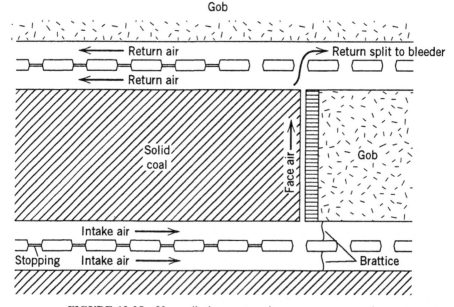

FIGURE 13.25   U-ventilation system in a two-entry panel.

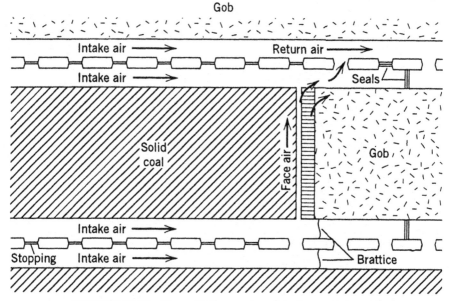

FIGURE 13.26 Y-ventilation system in a two-entry panel.

Almost all ventilation planning for three-or-more-entry longwall panels specifies some variation of the simple U-ventilation system. The most common difference is a return-air split requirement at the tailgate (Fig. 13.27). Belt-entry air is not used to ventilate the face in most cases. A recent trend has been to bring intake air to the face in the tailgate entry adjacent to the panel, improving the air quality at the return side of the face results. A wide variety of modifications to the U-ventilation system is practiced.

Only recently have Y-ventilation systems been proposed as designs for longwall panels of three or more entries. The main advantage being exploited in this case is that the tailgate bleeder connection entries need not be heavily cribbed because the middle entry in the tailgate will remain largely open. It is preferable to keep the headgate corner clear of methane by using the belt entry for face intake, but dust control problems usually preclude this practice. In any case, specific arrangements are largely dictated by local conditions and established practices. For example, two variations of the longwall ventilation system in a highly gassy eastern U.S. mine requiring about 100,000 cfm (47.19 m³/s) in the panel is shown in Fig. 13.28.

13.11 SHORTWALL VENTILATION

A shortwall layout, as shown in Fig. 13.29, is very similar to that of a longwall (Stefanko, 1983). The primary difference is that the shortwall face is generally

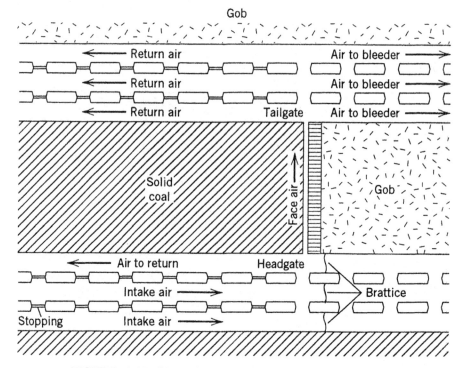

FIGURE 13.27   U-ventilation system in a three-entry panel.

150–250 ft (45.7–76.2 m) wide. The web depth is about 10 ft (3.0 m), as opposed to a maximum of about 30 in. (0.76 m) in a longwall. The minimum operating height of a shortwall face is about 54 in. (1.37 m). Thus the face area open for ventilating air is much larger than that for a longwall.

The shortwall face shown in Fig. 13.29 was located in eastern Kentucky in the Elkhorn No. 3 seam that averages 44 in. (1.12 m) in thickness, ranging from 42–48 in. (1.07–1.22 m). The main entries, which are not shown in the figure, consist of five intakes and four returns in a two-split arrangement with two returns on each side. The submains, represented by 10 left in the figure, consist of three intakes, one controlled belt-track intake, and five returns in a two-split or fishtail arrangement. One Right Flats, off 10 Left Submain, has a similar configuration but consists of only seven entries. The panel entries perpendicular to One Right Flats consist of an intake on the right side (No. 3 entry) with a return in No. 1 entry and a neutral belt-track entry in No. 2. A bleeder system is shown in use with panel production. This ventilation system is very effective in maintaining the methane below the 1% level.

In shortwall, the continuous miner cuts only in one direction. Then it is backed up to the headgate and another cut is started. Therefore, by arranging the ventilation flow from headgate to tailgate, the miner operator and the

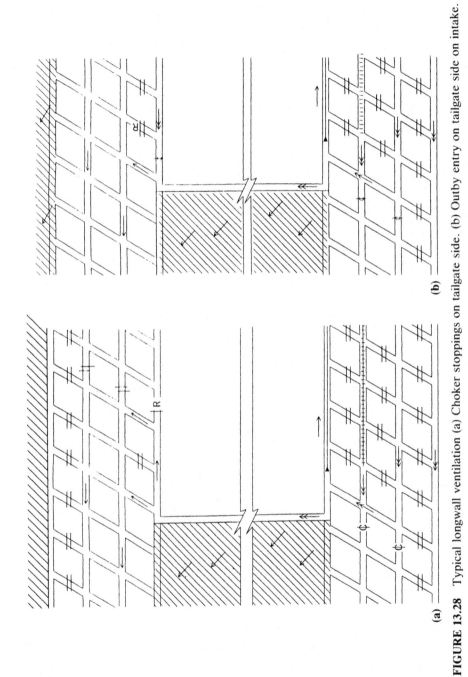

FIGURE 13.28 Typical longwall ventilation (a) Choker stoppings on tailgate side. (b) Outby entry on tailgate side on intake.

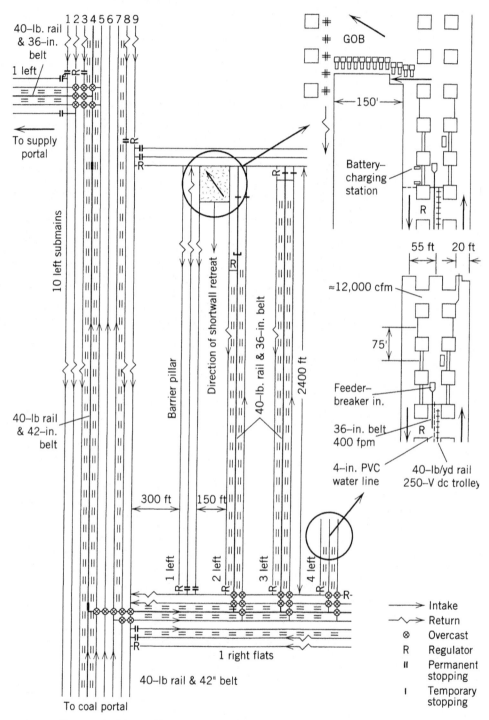

FIGURE 13.29 Shortwall ventilation scheme (Stefanko, 1983). (*Conversion factors:* 1 ft = 3.048 m, 1 cfm = 0.47195 × 10^{-3} m³/s.)

jacksetters are always in fresh air and are not exposed to the dust generated by the machine. Other than this, there is essentially little difference between the ventilation of longwall and shortwall faces.

13.12 BLEEDERS

Methane control during development, although requiring constant and careful attention, is easier than methane control during pillaring. During development, frequent inspections and regular monitoring are possible. However, as soon as conventional pillaring or retreat of longwall is started, it is impossible to inspect the gob areas. In the United States, it is common to ventilate the gob to dilute and carry away the methane gas from the gob area. This is known as bleeding the gob (Kalasky and Krikovic, 1973). It is usually achieved by placing regulators at strategic locations at both panel ends and utilizing whatever permeability has been established in the gob. The location of the regulators should be chosen to provide as much flow across the gob as is needed to sweep away gases and avoid dangerous accumulations in the gob. These regulators should be accessible since, as retreat mining increases, regulator adjustment is needed to compensate for the increased natural resistance of the developing gob. To accomplish "bleeding" an adequate pressure differential from the front to the back of the pillar line, as well as to both sides, is required.

Effective bleeding should remove the methane sufficiently far from the active pillar line or face so that a temporary ventilation interruption or sudden atmospheric pressure drop will not cause a large influx of gas from the gob.

Supplementary degasification through vertical boreholes drilled from the surface ahead of approaching pillaring lines or in the longwall gob is also practiced. However, effective bleeding systems should still be used to reduce the methane problem in the gob. It is common practice to drill shafts, up to 8 ft (2.4 m) diameter, at the back end of a bank of longwalls and use high-pressure exhaust fans [24–48 in. water (6–12 kPa)] to draw upward of 150,000 cfm (70.79 m^3/s) through the gobs.

Bleeders for Room and Pillar Systems

A simple bleeder entry system for room and pillar mining is shown in Fig. 13.30. Here the mining method involves mining panel headings usually advancing from one side of a set of submains to another set, that is, from 6 North to 7 North, and from 7 North to 8 North, as shown. When the panel headings approach the submains, 7 North and 8 North, they are reduced to a single heading that acts as the bleeder. Regulation is provided in the bleeder system as the gob becomes larger in size; regulators in the bleeders are closed to provide greater resistance to flow through the more permeable areas of the gob, and opened to provide airflow through the less permeable areas. If

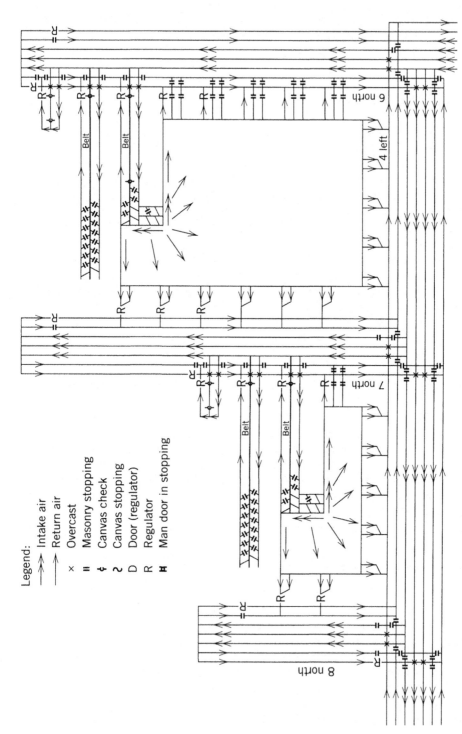

Legend:

→→ Intake air

→→ Return air

→ Overcast

× Masonry stopping

= Canvas check

ᵠ Canvas stopping

∿ Door (regulator)

D Door (regulator)

R Regulator

ⴶ Man door in stopping

FIGURE 13.30 Bleeder system for room and pillar workings (Kalasky and Krikovic, 1973).

the pressure differential across the gob is not adequate, the fan pressure may have to be increased; and, if that proves impractical, vertical boreholes must be drilled from the surface and a high-pressure exhauster installed on the top. One significant advantage of this arrangement is that the bleeders are accessible from the intake airway. This facilitates cleanup of falls and pumping of water accumulations. Otherwise, such conditions would cause restrictions to adequate bleeding.

In the mining plan shown in Fig. 13.31, panel headings are driven to mine the block of coal on the left side of the 6 north submains, and retreated on the adjacent side (from 7 north submains) with intermediate bleeders. In this case, the distance between the submains is double the panel length, minimizing the number of submains required for development. In this method of extraction, the same bleeder headings utilized in mining the first group of panels are also utilized for the second group. This arrangement may be unsatisfactory, particularly in very gassy mines, since it is difficult, if not impossible, to maintain the bleeder headings between the two gobs, especially during the final retreating cycle. In any case, all bleeders in this arrangement must be adequately supported. For pools of water that might block the bleeders, one remedy is to drive panel headings in advance and provide pumping facilities to remove water and, of course, avoid placement of bleeders in the low spot of a seam.

An alternative method of providing bleeders for the mining plan is shown in Fig. 13.32. Here a four-entry system is developed in between the submains with two intakes in the center and a return on either side. The butt drivage at the end is reduced to a single heading (bleeder) and is connected to the nearest return with regulation in the bleeder. This plan provides an effective and flexible bleeder system, approachable from the intake side for cleaning the bleeders and adjusting the regulators.

Bleeders for Longwall Systems

Bleeder entry development for longwall mining calls for careful planning as the gob is tightly packed. A good method consists of providing a set of four or five entries as main bleeder returns with solid barriers on each side. In the system shown in Fig. 13.33, the head–tailgate development for the longwall is connected at the back end by four or five entries that form the backend bleeder. In the design shown in Fig. 13.34, a protective barrier of 200 ft (61 m) is left between the bleeders and the setup entries for the longwall. Extremely gaseous conditions encountered when the immediate roof and overlying strata collapse have, in some cases, necessitated the use of vertical degasification boreholes equipped with exhausters to supplement the underground bleeder system.

The problems involved in bleeding and bleeder system development require critical attention in the planning stage. To avoid ventilating large gob areas only for the purpose of clearing the gobs of methane, an alternative is

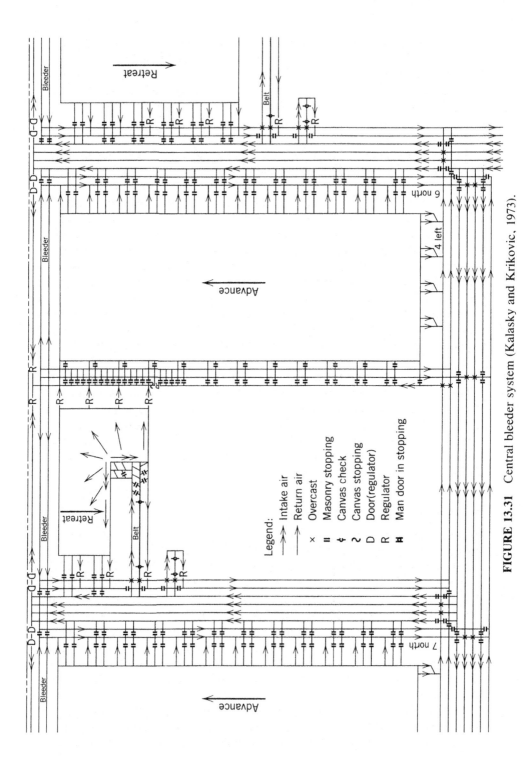

FIGURE 13.31 Central bleeder system (Kalasky and Krikovic, 1973).

502

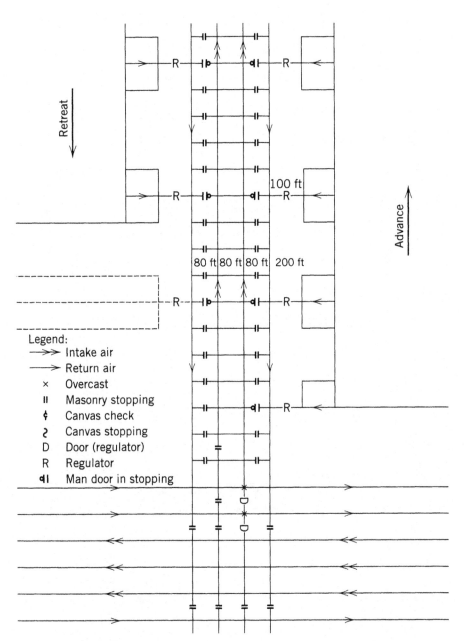

FIGURE 13.32 Better design for a central bleeder (Kalasky and Krikovic, 1973). (*Conversion factor:* 1 ft = 0.3048 m.)

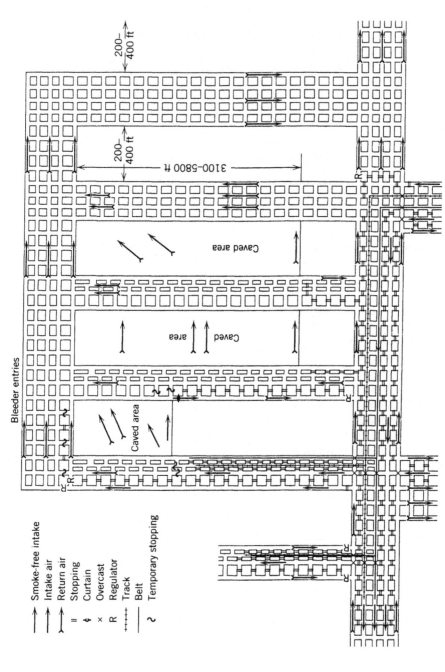

FIGURE 13.33 Bleeder system for longwall workings: case A (Kalasky and Krikovic, 1973). (*Conversion factor:* 1 ft = 0.3048 m.)

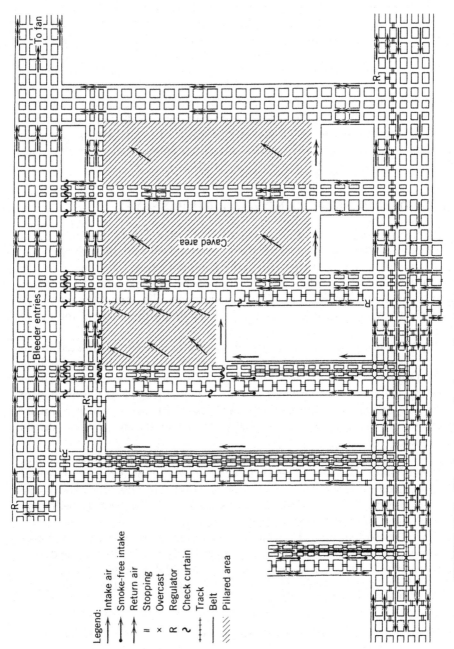

FIGURE 13.34 Bleeder system for longwall workings: case B (Kalasky and Krikovic, 1973).

Legend:
→ Intake air
→ Smoke-free intake
→ Return air
= Stopping
× Overcast
R Regulator
∼ Check curtain
+++ Track
— Belt
/// Pillared area

Bleeder entries

Caved area

To fan

to plan the mine ventilation system to seal off the mined areas. This alternative is gaining increasing attention as the areal extent of mines increases, and bleeders are becoming difficult and costly to maintain.

It is becoming more common today in the United States to seal gob areas using seals of the type described in Section 13.5. In older mines, sealing is cost-prohibitive, because the number of seals required to isolate the gob from active workings is large. In newer mines, or mine extensions, the number of seals can be reduced by laying out the mine workings whereby the number of entries connecting several panels to the mains or submains is kept to a practical minimum. After the panels are mined out, seals can be erected in these connecting entries. The gob is now isolated to prevent spontaneous combustion. In highly gassy mines, coalbed methane drainage can be practiced using vertical degas gob holes (Section 3.7).

13.13   DESIGN OF THE MINE VENTILATION SYSTEM

The selection and layout of the air-distribution plan is only the beginning of the design of the complete ventilation system. Determination of the mine quantity and heads for the system is the main objective, since this fixes the operating point and permits the selection of the fan that is best suited and most economical for the ventilation requirements. The approach employed in finding the mine quantity and heads depends on the stage in the life of the mine.

Ventilation requirements for a projected mine must be estimated in advance; the quantities usually are assigned and the corresponding heads calculated. Ordinarily, several stages in the mine's life need to be evaluated; however, by judiciously examining the mine development plan, the number of cases may be kept to a minimum. Worst-case situations, which normally occur before bringing new intake and/or return shafts into service, should be selected.

The requirements for an operating mine are determined by measurement, and then allowances are made for future revision, expansion, or increased quantity needs in arriving at a new operating point. The advantage of working with an existing mine is that values of resistance, leakage, and so forth, may be measured and typical values may be calculated.

Since the estimation of the ventilation requirements for an undeveloped mine involves the basic procedure used in either type of problem, the design of a projected ventilation system is dealt with in detail here. Determination of modifications for an existing system falls more in the category of routine work for the ventilation engineer.

Determination of Quantity Requirements

The mine quantity required for the entire system is based on the air quantities assigned to the individual workplaces. Quantities are assigned by judgment or experience to provide adequate, but not excessive, ventilation. The minimum quantity of air necessary in a working place is determined by legal requirements (Section 3.2), human needs (Section 3.3), and comfort or work-efficiency standards (Section 16.3).

The quantities of ventilating air are limited by the economic restrictions on velocity. Montgomery (1936) and Peele (1941) listed maximum velocity recommendations for economic use in coal mine entries. A more recent analysis based on both capital and operating costs (Mutmansky and Greer, 1984) provided the following rules of thumb:

1. For intake airways with a friction factor K of about 50×10^{-10} lb·min²/ ft⁴ (0.00928 kg/m³), velocity in the airways should be in the range of 600–1150 fpm (3.0–5.8 m/s).
2. For return entries with a friction factor K of about 75×10^{-10} lb·min²/ ft⁴ (0.0139 kg/m³), velocity in the airways should be in the range of 450–1000 fpm (2.3–5.0 m/s).

When setting velocities for a coal mine ventilation plan, it is necessary to keep other concerns in mind as well as ventilation economics. For example, effective gas and dust control may be more important than the goal of optimizing velocity.

Example 13.4 On the basis of minimum quantity and velocity standards for quality control, how much air should be assigned to a workplace that is 5 × 10 ft (1.5 × 6.1 m) in cross section? On the basis of quality standards, what airflow is required for dilution of a methane emission of 30 cfm (0.014 m³/s)?

Solution:
(a) Specified minima:
  (1) $Q = 9000$ cfm (4.2 m³/s) at last open crosscut or 3000 cfm (1.42 m³/s) at face
  (2) $V = 60$ fpm (0.30 m/s)
    $A = (5)(20) = 100$ ft² (9.3 m²)
    $Q = VA = (60)(100) = 6000$ cfm (2.8 m³/s)
(b) Assuming TLV = 1% for methane and using Eq. 3.8 or 3.9, $Q = 3000$ cfm (1.42 m³/s). Where legal minima govern or the expected conditions are not well known, a safety factor may be applied to the calculated quantities. In this case, the law governs, and the minimum required Q is 9000 cfm.

Determination of Head Requirements

Once quantities have been assigned to the workplaces and determined for all airways throughout the ventilation system, head losses can be computed for each segment of airway of constant quantity and given cross-sectional and surface characteristics (Section 5.5). Then the mine static head can be determined in accordance with the procedure outlined in Section 7.5, employing controlled splitting. If the number of regulators or the amount of regulation is excessive, the reduction of head losses should be considered in place of regulators.

Summary of Procedure in Design of Ventilation Systems

The principal steps in designing a coal mine ventilation system are as follows:

1. Develop the major features of the distribution plan, including arrangement of main airways, flow direction, fan location, and so on (Section 13.8).
2. Develop the minor features, coordinating the mining and ventilation systems (Section 13.4).
3. Represent the ventilation system and its circuits by a schematic, grouping the airways into splits (Section 7.4).
4. Calculate the air quantities required in the workplaces to maintain adequate ventilation (Section 13.7). Allowances for leakage should be made.
5. Distribute the air quantities and calculate the required number of entries using the rules of thumb for maximum velocity (Section 13.8) and the mine quantity required (Section 7.4).
6. Calculate the head loss corresponding to the quantity in each airway of the system (Section 5.5).
7. Determine the head loss over all splits, replace by equivalent series circuits, and find the regulation required in each split and the mine static head (Section 7.4).
8. Calculate the mine velocity head at the discharge of the system (Section 5.5). For an exhaust system, this will depend on the fan selected.
9. Sum the mine static and velocity heads to find the mine total head (Section 5.5).
10. Select a fan based on the basis of the operating conditions (Section 10.6).

13.14 VENTILATION DATA FROM OPERATING MINES

Operating Conditions

The range of ventilation requirements in actual operating mines is approximately as follows (Suboleski and Kalasky, 1982):

Quantity of air
Per person   200–5800 cfm (0.09–2.7 m³/s)
Per ton of coal   7–21 tons (6.3–19 t)
Total   50,000–2,990,000 cfm (24–1410 m³/s)
Static head   2–17 in. water (500–4200 Pa)

Although the total quantity at most large mines is a million cfm (470 m³/s) or less, the largest ventilation requirement at a coal mine approaches 3.0 million cfm (1410 m³/s).

Ventilation Costs

The capital costs for various ventilation plant items and control devices may be approximated as follows (Suboleski and Kalasky, 1982):

Air shaft, bored $775–$1275/ft diameter/ft depth
                          ($8340–$13,725/m diameter/m depth)
Air shaft, sunk $2000–$4350/ft depth ($6560–$14,270/m depth)
Exhaust drift $1000–$2500/ft length ($3280–$8200/m length)
Vertical degasification $75–$100/ft ($245–$330/m)
  (gob) hole installation
Main fans $100–$400/hp ($135–$535/kW)
Auxiliary fans $150–$400/hp ($200–$535/kW)
Vent pipe, metal, 12 in. $10–$20/ft ($35–$65/m)
  (305 mm) diameter
Vent tubing, flexible, 12 $3–$5/ft ($10–$15/m)
  in. (305 mm) diameter
Brattice cloth or plastic $0.75–$1.20/yd² ($0.90–$1.45/m²)
Stopping, temporary $1–$3/ft² ($10–$30/m²)
Stopping, permanent $2.50–$5.00/ft² ($27–$54/m²)
Doors, man, 30 × 30 in. $50–$60
  (762 × 762 mm)
Doors, haulage $5–$7/ft² ($54–$75/m²)
Overcasts (installed) $10,000–$15,000

The operating costs are quickly estimated, once the unit cost of power at the mine is known. Convenient figures to remember are the following: If the unit power cost is $0.01/kWh ($65/hp·yr), the annual power cost to supply 100,000 cfm at 7 in. water total head (47 m³/s at 1.7 kPa) and overall efficiency of fan and drive 70% is about $10,000.

The overall costs for ventilation at a mine include all capital and operating expenses. They are usually expressed on a unit basis, either per ton of mineral product or per quantity of air. Approximate ranges in 1982 were as follows:

Based on quantity of air $0.25–$0.50/100,000 cfm ($0.55–$1.05/100 m³/s)
Based on tonnage of coal $0.75–2.50/ton ($0.85–$2.75/t)

13.15   DESIGN EXAMPLE

To illustrate the complete procedure involved in designing a coal mine ventilation system, the mine shown in Fig. 13.35 is used (Stefanko, 1983). It is
located on a property 1 × 3 mi (1.6 × 4.8 km) in area at a depth of 1000 ft (305
m) in a highly gaseous seam. Mains are driven in the north–south direction [3
mi (4.8 km)] along the center of the property. The property will be mined
on a half-advance, half-retreat arrangement. Panels [2000 ft (610 m) long]
will be driven and mined to the left on advance until the property line is
reached; panels will then be driven on the right on retreat. Only the northern
half of the mine on advance is shown as the southern half is a mirror image
of the northern half. All entries are 16 ft (4.9 m) wide on 100-ft (30.5-m)
centers. It is assumed that six machine units (four on development and two
on pillaring) are required to meet the production tonnage of 2000 tons (1814
t)/shift. Panel belts transfer coal to a main belt.

Procedure

Step 1. Quantity Calculations: Mining Unit   Assuming a gas liberation
rate of 40 ft^3/ton (1.25 m^3/t) of coal mined and a mining rate of 5 tons/min
(4.54 t/min), the amount of gas liberated per minute during cutting is 200 ft^3
(5.66 m^3). The amount of fresh air required to dilute the gas to a methane
content below 1% is 20,000 cfm (9.44 m^3/s).

It is good practice to provide two splits of air in a panel, one for the
continuous miner and another for the roof bolter. In this arrangement, the
dust-laden air from the continuous miner is placed directly in the return.
Further, there is no line brattice through which shuttle cars must tram, reducing leakage that otherwise adversely affects air distribution in the face area.
Thus two splits of air, 20,000 cfm (9.44 m^3/s) for the miner and another 20,000
cfm (9.44 m^3/s) for the other machine units, are utilized in a section.

Belts may not be placed in main intakes or returns but must be isolated
in order to prevent the spreading of a fire, and only that amount of air is to
circulate in them that is necessary to maintain methane concentrations below
1%. The air passing over the belt must be placed directly into the return rather
than being transported to working faces. Since there is a certain amount of
air leakage where the belt passes through the doors near the head end, the
belt-air travel direction is usually into the panel to utilize the leakage. Normally, a temporary stopping will be placed across the belt entry just outby
the belt loading point, with a regulator in the outby crosscut that regulates
the belt air directly to the return.

In the sample calculation, a small quantity of air (5000 cfm or 2.36 m^3/s)
will be split from the main intakes at the head of the belt and dumped directly
into the returns near the tail of the belt.

Step 2. Leakage Considerations   There will be leakage through the stoppings in the panel from intake to return. The quantity of air required at the

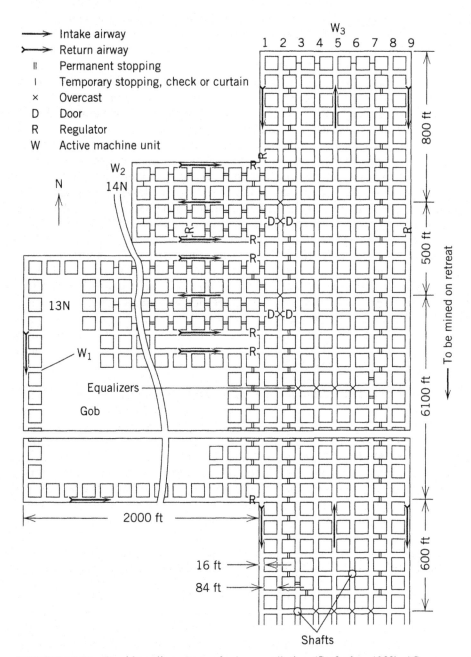

FIGURE 13.35 Double-split system of mine ventilation (Stefanko, 1983). (*Conversion factor:* 1 ft = 0.3048 m.)

entrance to the panel will be higher than the 40,000 cfm (18.88 m^3/s) calculated as required at the last open crosscut. When the panel is fully developed [i.e., 2000 ft (610 m) with 20 crosscuts], there will be two rows of stoppings between the intakes and the returns on either side of the intakes, and one additional row of stoppings for the isolated belt. Assuming a maximum leakage of 1000 cfm (0.472 m^3/s) through the stoppings in each crosscut, a total leakage quantity of 20,000 cfm (9.44 m^3/s) is estimated, making the air required at the panel entrance 65,000 cfm (30.68 m^3/s) [60,000 cfm (28.32 m^3/s) for the unit, plus 5000 cfm (2.36 m^3/s) for the belt].

The fan air quantity must provide for air leakage between the main intakes and returns through the stoppings, overcasts, and main doors. The farthest panel (i.e., the one at the property line) will be approximately 8000 ft (2438 m) or 80 crosscuts away from the fan shaft. Fortunately, control devices in the mains are built more substantially than those in panels because they have to be in service through the life of the mine. Assuming 500 cfm (0.236 m^3/s) leakage of air in the stopping in each crosscut, a total leakage of 40,000 cfm (18.88 m^3/s) exists for a panel. The volumetric efficiency [(air at last open crosscut + belt air)/(air at the fan)] is just under 45%. Thus, with three active panels in the north, the total air required for the panels will be 315,000 cfm (148.66 m^3/s). To this must be added the air required for the main belt, which is estimated at 10,000–15,000 cfm (4.72–7.08 m^3/s), making the total air required for the north side 325,000 cfm (153.38 m^3/s). The air required for the south side will be another 325,000 cfm (153.38 m^3/s), resulting in a total fan quantity of 650,000 cfm (306.76 m^3/s).

Step 3. Head-Loss Calculations: Main Entries The cross-sectional area of an entry is 16 × 6 ft (4.88 × 1.83 m) = 96 ft^2 (8.92 m^2). The maximum velocity permitted in the mains is, say, 800 fpm (4.06 m/s). Then one entry can carry 76,800 cfm (36.25 m^3/s) of air. The number of entries needed to carry 315,000 cfm (148.66 m^3/s) [10,000 cfm (4.72 m^3/s), will be carried by the belt entry] is just over four entries (i.e., 315,000 ÷ 76,800). Because the air from the mains is split into the submains and panels, and leakages occur between the main intakes and returns, in practice, the mains handle progressively lower quantities. On the other hand, the entire 96-ft^2 (89.2-m^2) area of an entry may not be available for airflow. Therefore, for this example, the number of intake entries is set at 5 and the number of return entries at 4; additionally, one isolated belt entry is provided, making a 10-entry main.

Panel Entries Since the panel entries have to carry only 60,000 cfm (28.32m^3/s), a three-entry development—one intake entry, one return, and one isolated belt entry—can suffice. However, for providing a separate intake escapeway and greater production potential, a five-entry development is recommended.

Bleeder The quantity of air to be regulated into the bleeder from the active pillaring section is 20,000 cfm (9.44 m^3/s). The air flows through the gob and joins the main return at a distance of 600 ft (183 m) away from the return shaft.

Leakage For the purpose of head-loss calculations, unless it is definitely known that an inordinate quantity of air is leaking through a particular airway, it is assumed that leakage air is uniformly distributed over the entire airway. Thus the quantity flowing in an airway is the average of the quantity flowing in the beginning and end of the roadway. In the ventilation schematic shown in Fig. 13.36, the North mains have been developed to the property limits. The most critical situation with regard to ventilation will arise when the unit in 14N panel is pillaring, 15N panel is nearing completion of development to the boundary to the west prior to pillaring, and 16N panel (the first panel to the east of the mains) is on development. Examination of the

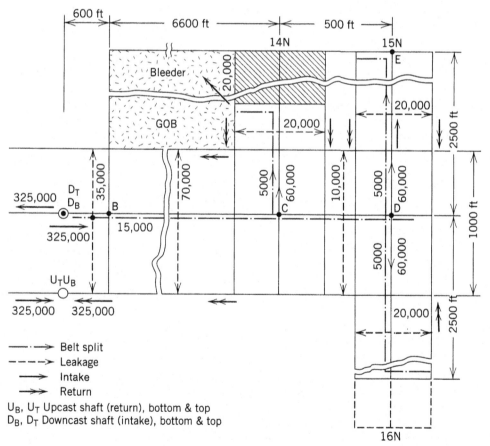

FIGURE 13.36 Ventilation schematic for mine in Fig. 13.35. (*Conversion factors:* 1 ft = 0.3048 m, 1 cfm = 0.47195 × 10⁻³ m³/s.)

ventilation schematic reveals that the free split is the path from the downcast shaft through the 15N panel to the upcast shaft.

Friction Factors   Values for friction factor must always be chosen on the basis of field measurements under conditions that are similar to the mine conditions. For new mine projects in virgin areas, the following general values can be used:

Concrete-lined shaft	15×10^{-10} lb·min^2/ft^4 (0.00278 kg/m^3)
Intake airways	80×10^{-10} lb·min^2/ft^4 (0.0148 kg/m^3)
Return airways	100×10^{-10} lb·min^2/ft^4 (0.0186 kg/m^3)

Shock-Loss Considerations   In the calculation of head loss in each segment of the airway, consideration must be made for shock losses that may be encountered with such conditions as splitting, changes in areas, overcasts, and changes in direction. Shock losses can be minimized by gradual changes in area and direction but can never be eliminated. The longest segment in the example mine is from the bleeder junction (B) to the 14N entrance (C). Considerations for shock loss on the basis of the number of overcasts, doorways, obstructions, and area changes must be made for this segment. The second largest segment is the panel mains in 15N from D to E. Here also, considerations for shock loss on the basis of the number of doorways and obstructions must be made. In the calculations in Table 13.2, therefore, equivalent lengths of 1000 ft (305 m) for the segment BC and 200 ft (61 m) for the segment DE have been assumed.

Total Head Loss   It is convenient to calculate the head loss in each segment of the airways and add the head losses in series to calculate the total head loss. It is also preferable to use a tabular format for the calculations. The formulas used for calculating head loss are Eqs. 7.8, 7.19, and 7.7, in that order.

Table 13.2 shows the calculation of mine resistance for each segment, and Table 13.3 shows the calculation of head loss for each segment. The head loss in the circuit is calculated as 5.68 in. water (1.413 kPa). Adding about 0.32 in. water (79.63 Pa) for losses across the face in 15 N on a line brattice, the fan must develop a static head of nearly 6 in. water (1.493 kPa).

Step 4. Fan Selection   A fan that can develop a static head of 6 in. water (1.493 kPa) and 650,000 cfm (306.77 m^3/s) is needed, and it must be able to handle this quantity with high efficiency. The fan motor must be selected to provide adequate horsepower for the total fan head. The fan must operate in an adequate range of capacities so as to accommodate the changing requirements for quantity and head during the life of the mine.

Using the Jeffrey fan selection chart (see Fig. 13.37), an initial selection of a fan can be made. It can be seen that a 8HU117 fan operating at 710–880

TABLE 13.2 Resistance Calculations for Mine Segments

Mine Segment	Description Figure	Code in Fig. 13.35	Friction Factor, K lb·min²/ft⁴	Distance L, ft	Equivalent Length L_e, ft	Height H, ft	Width W, ft	Perimeter $O = 2(H+W)$ ft	Area $A = H \times W$ ft²	Resistance of One Entry, R_1 = $\dfrac{K(L+L_e)O}{5.2A^3}$ in.·min²/ft⁵	Number of Parallel Entries	Resistance of n Entries $R_T = R_1/n^2$ in.·min²/ft⁶
1	Intake shaft	D₁–D_B	15	1000	0	12	20	64	240	0.0134	1	0.0134
2	Intake shaft bleeder Jct.	D_B–B	80	600	0	6	16	44	96	0.4591	5	0.0184
3	Bleeder Jct.–14N	B–C	80	6600	1000	6	16	44	96	5.8153	5	0.2326
4	15N–15N	C–D	80	500	—	6	16	44	96	0.3826	5	0.0153
5	15N panel intake	D–E	80	2000	200	6	16	44	96	1.6834	2	0.4208
6	15N panel return	E–D	100	2000	200	6	16	44	96	2.1043	2	0.5261
7	15N–14N	D–C	100	500	—	6	16	44	96	0.4782	4	0.0299
8	14N–bleeder Jct.	C–B	100	6600	1000	6	16	44	96	7.2691	4	0.4543
9	Bleeder Jct.	B–U_B	100	600	—	6	16	44	96	0.5739	4	0.0359
10	Upcast shaft	U_B–U_C	15	1000	—	Diameter 18		56.55	294.5	0.0100	1	0.0100

Conversion factors: 1 lb·min²/ft⁴ = 1.855×10^6 kg/m³, 1 ft = 0.3048 m, 1 ft² = 0.0929 m², 1 in.·min²/ft⁶ = 1.117×10^8 N·s²/m⁸.

TABLE 13.3   Head-Loss Calculations for the Mine Segments

| Mine Segment | Quantity in 100,000 cfm | | | | Total Resistance R_T (in.·min^2/ ft^6 × 10^{10}) | Head Loss $H_l = R_T Q_A^2$ in. water |
	Begin Q_B	Leakage Q_L	End $Q_E =$ $Q_B - Q_l$	Average $Q_A =$ $(Q_B + Q_E)/2$		
1. D_I–D_B	6.5	0	6.5	6.5	0.0314	0.5662
2. D_B–B	3.10[a]	0.35	2.75	2.925	0.0184	0.1574
3. B–C	2.75	0.70	2.05	2.40	0.2326	0.1574
4. C–D	1.40[b]	0	1.30	1.35	0.0153	0.0279
5. D–E	0.60	0.20	0.40	0.50	0.4208	0.1052
6. E–D	0.40	0.25[c]	0.65	0.525	0.5261	0.1450
7. D–C	1.45[d]	0.10	1.55	1.50	0.0299	0.0673
8. C–B	2.00[e]	0.70	2.70	2.350	0.4543	0.5089
9. B–U_B	2.90[f]	0.35	3.25	3.075	0.0359	0.3395
10. U_B–U_C	6.5	0	6.5	6.5	0.0100	0.4225

$$\Sigma = 5.6797$$

[a] A split of 15,000 cfm is taken for the main belt.
[b] A split of 65,000 cfm is taken for the 14N section and panel.
[c] The leakage into return also includes the belt return.
[d] This includes return from 15N, 16N and main belt.
[e] Return from 14N is 45,000 cfm.
[f] Return from the bleeder 20,000 cfm.

Conversion factors: 1 cfm = 0.47195×10^{-3} m^3/s, 1 in.·min^{2}ft^6 = 1.117×10^9 N·s^2/m^8, 1 in. water = 248.84 Pa.

rpm can provide the required head and quantity. However, the efficiency of this fan at this high quantity is rather low. To increase the efficiency, a possible combination of two 8HU117 fans may be used, operating in parallel at 710 rpm, each producing 350,000 cfm (165.18 m^3/s) at 6 in. (1.493 kPa) static head. The operating point of each fan is nearly at the middle of its quantity capacity ($\cong$750,000 cfm or 353.96 m^3/s), and the head is within the range of 2–10 in. water (0.498–2.488 kPa). At this operating point, the fan efficiency is well above 80%. The power of the motor (from the curve) at 70% efficiency is 500 hp (373 kW). At 80% efficiency, the fan motor will be sized at 440 hp (328 kW). The static air horsepower at the operating point is approximately 331 hp (247 kW), resulting in an overall static efficiency of approximately 75% [= 331/440 × 100].

Yet another method for initial selection can be the use of charts of specific speeds and specific volumes. Assume a specific volume $Q_s = 2000$ cfm (0.94 m^3/s); then from Eq. 9.26

$$D^2 = \frac{650,000}{(2000)\sqrt{6}} = 132.68 \text{ ft}^2 \text{ (12.33 m}^3\text{)}$$

$$D \cong 11.52 \text{ ft (3.35 m)}$$

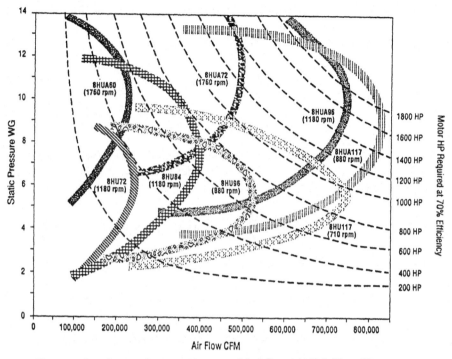

The curves show the approximate operating range of the Jeffrey eight-bladed fans. The area of maximum efficiency (over 80%) occurs at about 1/2 the maximum flow shown on the curves and over approximately the middle two thirds of the pressure range shown at the volume level. The horsepower lines correspond to a fan efficiency of approximately 70%.

FIGURE 13.37 Jeffrey fan selection chart. (*Conversion factors:* 1 cfm = 0.47195 × 10⁻³ m³/s, 1 in. of water = 248.84 Pa, 1 hp = 0.7457 kW.) (By permission of Jeffrey Mining Machine Division, Dresser Industries, Inc., Columbus, OH.)

Choose a fan 12 ft (3.66m)in diameter. The specific volume for a 12-ft (3.66-m) fan is

$$Q_s = \frac{650,000}{(12 \times 12)\sqrt{6}} = 1850 \text{ cfm } (0.87 \text{ m}^3/\text{s})$$

Enter chart (Fig. 10.23) at 1850 cfm (0.87 m³/s), and note that several blade settings (Nos. 1, 4, 7, and 2B–1S) have efficiencies approaching 80% at Q_s ≅ 1850 cfm (0.87 m³/s). The specific speed n_s for the (2B–1S) blade setting is approximately 2500, and for the No. 1 blade setting is approximately 3000. Choosing the lower specific speed, the fan speed can be calculated from Eq. 9.25 as

$$n = \frac{(2500)\sqrt{6}}{12} = 510 \text{ rpm}$$

$$P_m = \frac{65,000 \times 6}{6350 \times 0.8} \cong 768 \text{ bhp } (573 \text{ kW})$$

A 12-ft (3.66-m)-diameter fan connected to a 768-hp (573-kW) motor operating at 510 rpm may be an initial choice.

For a lower Q_s, say, $Q_s = 1500$ cfm (0.71 m³/s),

$$D^2 = \frac{650,000}{(1500)\sqrt{6.0}} = 176.91 \text{ ft}^2 \ (16.43 \text{ m}^2)$$

$$D = 13.50 \text{ ft} \cong 14 \text{ ft} \ (4.27 \text{ m})$$

This is a very large-diameter fan. Assume that two smaller fans will be placed in parallel, to divide the quantity between them equally, the diameter of the fan for the same specific volume (i.e., $Q_s = 1500$ cfm, or 0.71 m³/s),

$$D^2 = \frac{325,000}{(1500)\sqrt{6.0}} = 88.45 \text{ ft}^2 \ (8.22 \text{ m}^2)$$

$$D = 9.35 \text{ ft} \cong 10 \text{ ft} \ (3.05 \text{ m})$$

Assume two 10-ft fans in parallel:

$$Q_s = \frac{325,000}{(10 \times 10)\sqrt{6.0}} = 1327 \cong 1400 \text{ cfm} \ (0.66 \text{ m}^3/\text{s})$$

Enter chart (Fig. 10.23) at 1400 cfm (0.66 m³/s), and note that peak efficiency (80%) occurs at blade setting no. 1 and that the specific speed is 2800:

$$n = \frac{(2800)\sqrt{6.0}}{10} = 685.86 \cong 700 \text{ rpm}$$

$$P_m = \frac{(325,000)(6.0)}{(6350)(0.8)} = 384 \text{ bhp} \ (286.35 \text{ kW})$$

Thus two 10-ft (3.05-m)-diameter fans in parallel, each operating at 700 rpm and connected to a 384-hp (286.35-kW) motor, may be another choice.

Discussion

Each of these alternatives is associated with quantifiable factors such as capital and operating costs, and non-quantifiable but important qualitative factors such as flexibility, availability, standardization, and future mine extensions. The actual selection of a fan will be the result of the judgmental evaluation by the engineer of these factors. Economics in mine ventilation costs must be accomplished by reducing the static and total heads of the mine. The reductions are achieved by having straight, clean, and unobstructed airways and, wherever possible, a number of airways in parallel; by reducing the leakage through stoppings, overcasts, and airlocks; and by selecting and operating fans with high efficiency for the desired mine quantity and head. In addition to the extreme care in the design of the system, frequent mine ventilation surveys and regular maintenance of the components of the ventilation system—airways, control devices, and fans—are required to assure continued safe performance of the ventilation system. Modern automatic monitoring techniques (for gases, quantities, pressures, temperatures, power, etc.) can be effectively applied to the ventilation system to aid managers and engineers in maintaining the integrity of the health and safety aspects of the system (Ramani, 1982; 1992b).

The example just developed is for a relatively small mine. In large mines, it is impractical to take in all the air through one shaft and return it through another, and so a multiple-fan/shaft arrangement becomes necessary. In such a situation, each fan has a zone of influence, and all the zones are interconnected. Difficult ventilation problems may arise at zone interfaces where there may be no airflow (i.e., neutral zones) or air directions may be the reverse of those desired. The areas of influence of each fan vary as a complex function of such factors as the advance rates of the interconnected zones, distances from the main intakes, and the fans and their blade settings (Mishra et al., 1978).

Mine ventilation planning includes provisions for future expansion and exigencies that must be incorporated into the initial design. When planning with computer-oriented, mine ventilation network analysis programs, it is possible to try out various alternatives with regard to factors such as fans, shafts, and mains (Wang, 1982a; Ramani, 1988). Stefanko and Ramani (1972) present the application of a mine ventilation network analysis program to the ventilation planning example discussed in this chapter. The complex calculations involved with multiple-fan/shaft ventilation systems are best carried out on a mine ventilation network analysis computer program (Wang and Hartman, 1967; Wang and Saperstein 1970; Didyk et al., 1977).

PROBLEMS

13.1 Lay out the major features of the ventilation system for a coal mine being worked by the room and pillar method with conventional equipment. One slope and two shafts are available for ventilation use.

13.2 Given the coal mining plan shown in Fig. 13.38, lay out the ventilation system so that each section is on a separate split of air. Assume continuous mining equipment is in use.

13.3 Calculate the mine quantity and heads (static, velocity, and total, to four decimal places); the fan quantity, heads, and power; and the loca-

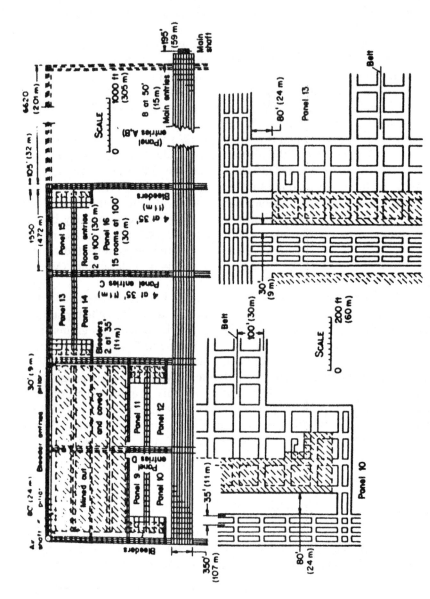

FIGURE 13.38 Horizontal plan view of room and pillar coal mine (see Problems. 13.2 and 13.3).

tion and pressure drop of regulators for the mine of Problem 13.2 (see Fig. 13.38), given the following conditions:

(a) Workings:

Main shaft: depth 600 ft (183 m), two compartments, 10×10 ft (3×3 m) each, $K = 25 \times 10^{-10}$ lb·min²/ft⁴ (0.046 kg/m³).

Air shafts: depth 1100 ft (335 m), 10 ft (3 m) diameter, depth 600 ft (183 m), two compartments, 10×20 ft (3×3 m) each, $K = 25 \times 10^{-10}$ lb·min²/ft⁴ (0.046 kg/m³).

Entries and crosscuts: 6×145 ft (1.8×4.6 m), $K = 50 \times 10^{-10}$ lb·min²/ft⁴ (0.0093 kg/m³).

Rooms and crosscuts: 6×20 ft (1.8×6.1 m), depth 600 ft (183 m), $K = 50 \times 10^{-10}$ lb·min²/ft⁴ (0.0093 kg/m³).

Fan connection to air shaft: equivalent length 100 ft (30 m).

All area changes abrupt, except at shafts.

All right angles sharp, except at shafts.

(b) Airflow requirements:

Consider state and federal mining laws.

Provide minimum face velocity of 125 fpm (0.64 m/s) in workplaces.

Total working sections 16; one continuous miner per place.

Crew: 10 per section.

Belt in room entries only.

(c) Auxiliary ventilation—neglect leakage in line brattice, assume opening divided in half with no change in friction factor.

(d) Evasé duct: angle 6°, length 15 ft (4.6 m)

(e) Fan: axial-flow; read operating efficiency; select the most suitable fan from manufacturer's catalog. Assume 7 in. diameter for evase calculation. Sketch the system characteristic curves and pressure gradients at the fan.

13.4 The schematic diagram of a three-section slope mine is shown in Fig. 13.39. Assume this layout to be the most critical with regard to ventilation planning. In each section, a double split system of ventilation will be used. The air quantity required at the last open crosscut in each split is 13,000 cfm (6.32 m³/s). Face ventilation will use line brattice. The mains, submains, and panels all consist of five entries, with an isolated belt entry in the center, one intake entry on each side of the belt, and one return entry on the outside of each intake entry. All entries will be driven 5 ft (1.5 m) high and 20 ft (6 m) wide. Return airways will be timbered on 5 ft (1.5 m) centers [assume a friction factor $K = 100 \times 10^{-10}$ lb·min²/ft⁴ (0.0186 kg/m³)]. Intake airways will be supported by roof bolts and crossbars [assume a friction factor $K = 70 \times 10^{-10}$ lb·min²/ft⁴ (0.0130 kg/m³)]. For the purposes of mapping the air distribution, assume that the quantity of air entering the mine for each section is twice the air quantities at the last open cross-

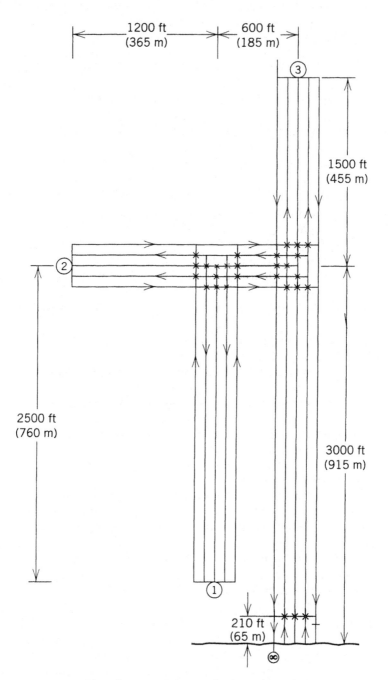

FIGURE 13.39 Line diagram of the ventilation system for Problem 13.4.

cuts and that the quantity of air leaking in a split is proportional to the length of the split. Further assume that 12,000 cfm (5.6 m³/s) will enter the mine through the belt entry and will be distributed so as to provide 4000 cfm (1.87 m³/s) to each section belt. The belt air will be dumped directly in the return in the section. Calculate the mine quantity and head. Show on the diagram the free split and the location of the regulators.

CHAPTER 14

Metal Mine Ventilation Systems

The term *metal mine* as used here refers to both metal and nonmetallic mines including uranium mines. There are several major differences between the ventilation systems in coal and metal mines (Section 13.2). The biggest differences in the United States are the permitted use of underground fans and recirculation or series utilization of air in metal mines (O'Neil and Johnson, 1982; Johnson, 1992a). The planning principles discussed in Chapter 13 often apply to metal mines as well.

14.1 GENERAL CONSIDERATIONS

Air Quantities

In most cases, air quantities are not determined by statute for metal mines, the exception being mines where radioactivity is present. Air quantities where diesels are used are determined from manufacturers' specifications or by testing and certification done on individual units by the Mine Safety and Health Administration (MSHA). It is generally satisfactory to provide 150 cfm/bhp (1.0 m³/s·kW) for each unit (Kenzy and Ramani, 1980). For several units on the same split of air, full quantity requirement for the first unit, 75% of full quantity for the second unit, and 50% of the full quantity for each additional unit are recommended. Otherwise, air quantities are usually determined by practical considerations and experience.

Velocities, wherever possible, should not exceed 1500 fpm (7.6 m/s), both for comfort and for reducing head losses due to friction. This may not always be possible, particularly where a large quantity of air is needed to service an area. Air velocities at a minimum should be kept above 50 fpm (0.25 m/s), as this is about the lowest velocity at which air movement can be felt; however, this requirement may be relaxed in large service bays. In workplaces, air velocities should be 200–400 fpm (1.0–2.0 m/s) in stopes and 400–600 fpm (2.0–3.0 m/s) in drifts. These velocities are recommended for warm mines and may have to be reduced in cold mines.

Overall Mine Layouts

In a shaft mine, intake air should be directed down the main personnel–materials shaft(s) and exhausted by some other shaft or shafts, preferably shafts sunk exclusively for exhaust. In an adit mine, the intake would normally be the adit(s) where personnel enter. Sometimes, a separate adit for exhaust would be used, connected to the workings by raises and winzes. Some mines are close to the surface, and multiple boreholes can be used for ventilation.

From convenience and comfort standpoints, main ventilation openings should be laid out for minimum interference with mining operations and for maximum comfort to personnel. Utilization of a hoisting shaft as a main airway may seriously hamper both the operational and ventilation functions of the shaft. Location of a fan at a hoisting shaft is particularly undesirable, because this requires an airlock in the shaft. In cold climates (e.g., Canada), the usual practice is to nearly neutralize the hoisting shaft, at least during the winter months, to minimize discomfort to the miners and prevent ice formation, damage to water lines, and so forth. This practice requires provision of additional openings to serve as airways. In all cases, the principal access to the mine for workers, whether it is shaft or adit, should be on fresh air.

It is important to consider the direction of flow in the opening and the effect of air temperature and humidity on the opening. Downcast shafts (or drifts) have moisture and freezing problems in wet, cold climates, whereas upcast shafts have condensation and timber decay problems in hot, humid mines. Either of these problems is particularly undesirable in a main hoisting shaft, but in an air shaft having no other use, they are not overly serious. Although mine temperature–humidity control (heating or cooling) affords a possible solution, the separation of hoisting and ventilation functions in shafts appears to be preferable.

Air-Moving Devices

Fans are the main air-moving devices used in metal mines. As opposed to coal mines, there are considerable choices in the location of the main fans as well as in the mode of their operation.

Booster Fans Unlike coal mines, metal mines can usually use booster fans underground either to divert airflow or to give additional air-moving power where needed. Booster fans are often needed in situations where the mine workings are widely scattered. Without boosters, larger main fans will have to be used, and regulation will then be needed in nonremote areas to ensure the needed quantity distribution.

Auxiliary Fans These are relatively small fans used in directing air to individual faces or small areas. Where dead-end faces are to be ventilated, the air is delivered to (or exhausted from) the faces with duct connected to the fan.

In the case of large openings, auxiliary fans are often hung in main airstreams to divert the air into these openings with little or no duct connected to the fan. This is a common technique in room (stope) and pillar systems. A fan used in this manner is termed a "jet fan," as it not only moves the air that is passing through the fan but causes a *vena contracta* that draws in additional air and moves it as well (Dunn et al., 1983).

Compressed-Air Movers These work in the same manner as jet fans, but use a stream of compressed air emanating from an eductor to form the vena contracta. They are often applicable where only small amounts of air are needed and where fans may be impractical. At least one large American mine uses them extensively for secondary development and production ventilation.

Surface versus Underground Location of Main Fan

Neglecting the relevance of mining laws, the advantages of a surface (blower or exhauster) location are as follows:

1. Better, simple, and safer control in event of emergency.
2. Easier access for repairs or in emergency.
3. Simpler and cheaper installation and power supply.
4. Damage to fan by fire or explosion less likely.
5. Recirculation and leakage effects less pronounced.

On the other hand, an underground (booster) location has the following advantages:

1. Stronger and more positive ventilation of deeper and more remote workings.
2. Intake and exhaust shafts are free of airlocks and fan installations, allowing their use for hoisting or haulage.
3. Simplifies modification of ventilation system by addition of fans to new levels and/or remote areas.
4. Handles less air than exhauster (and more than blower) for given quantity at face; of consequence if temperature rise (and specific weight decrease) in mine workings is substantial.

All other factors being equal, *location of the main fan at the surface is usually preferable from the standpoint of its greater safety.* However, the complexity of ventilation networks in multilevel metal mines weighs heavily against the exclusive use of surface fan(s).

Blower versus Exhaust Systems

With a surface location of the main fan, an important consideration is the position of the fan with respect to the system. The following arguments favor an exhaust location of the main fan:

1. Avoids necessity of providing an airlock on main shaft (or drift) if downcast.
2. Rising atmospheric pressure underground in event of failure of fan or power temporarily retards liberation of strata gases (methane, radon, etc.).
3. Greater power savings possible if mine openings are small because evasé discharge on exhaust fan permits higher recovery of velocity head.

On the other hand, a blower location has the following advantages:

1. Fan handles only uncontaminated and noncorrosive air with normal moisture content.
2. Ordinarily handles somewhat less air to deliver same quantity to work-places; of consequence if temperature rise (and specific weight decrease) in mine workings is substantial.
3. Keeps gaseous workings under higher pressure, limiting gas flows.

The issues with regard to blower versus exhaust systems are not as clear-cut as those with surface versus underground fan location. If separate air shafts are provided, this is of no concern, and a blower system may be preferable, particularly if the active mining zone changes rapidly and a number of upcast shafts are in use.

Generally, surface exhaust fans are preferred. However, where the mine contains numerous old open workings, exhaust fans will draw excessive air through areas where it is not needed. Sometimes the best solution is to install main surface exhaust fans supplemented by underground booster fans. If the exhaust shafts are also the hoisting shafts, surface location of the fans is impractical because of the design problem of moving skips through fan plenums.

Multiple-Fan Ventilation

Modern metal mine ventilation practice is to ventilate with multiple fans, usually surface fans in combination with booster fans underground. The terms *intake booster* and *exhaust booster,* respectively, denote that the booster is installed (1) upstream or (2) downstream of the workings. Boosters are generally installed near the downcast shaft to provide positive pressure to the workings. In addition to intake boosters, exhaust boosters may also be installed underground near the exhaust shafts or adits. Exhaust boosters are often needed in mines where the workings have been extended beyond the pressure-generating capabilities of the main exhaust fan.

The airflow pattern in multiple-fan installations is complex. For example, a particularly difficult but not uncommon situation at a western metal mine

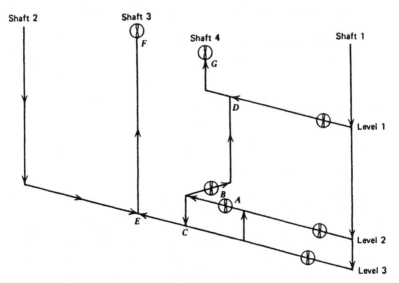

FIGURE 14.1   Ventilation schematic for a typical mature stope mine.

is shown in Fig. 14.1. Here the air downcasts through shaft 1 and is divided among three different levels with an intake booster on each level. Although the intake distribution is simple, the exhaust circuit is fairly complicated. Part of the exhaust from level 3 upcasts and joins the exhaust from level 2; this stream goes into an exhaust booster at A and then splits. Part of it downcasts back to level 3, joins the rest of level 3 exhaust and the downcast air from shaft 2 at point E, and passes out shaft 3 through main exhaust fan F. The rest of the exhaust from fan A passes through a second exhaust booster B, upcasts to level 1, where it is joined by level 1 exhaust at D, and then all this air upcasts through shaft 4 at main exhaust fan G. Similar complex situations are often found in other older mines with extensive workings. However, the safety factor of the backup systems is inherent in multiple-fan installations, as the failure of one or even several fans will result in only a reduction, rather than a complete loss, of ventilation.

Reuse of Ventilation Air

Reuse of air is allowable in most metal mine situations and is preferred for economic reasons. Reuse reduces the air-quantity requirement of the mine and usually simplifies the ventilation network. However, the used air is mixed with fresh air before reuse. In general, there will not be much gas contamination, and oxygen depletion is not a problem. In cases where dust or heat is excessive, filtration or cooling will be needed. South African mines commonly use fume filters to extract nitrous oxide fumes from development blasting when blasting is done on shift and also use baghouse filters at ore transfers located at intake shafts. However, these techniques are seldom employed in the United States.

The danger in air reuse normally arises when it is unintentional or when the magnitude of the circulating loads has been grossly underestimated. Under such conditions, excessive fan loads, increased hazards due to fires, and substandard air quality can develop rapidly.

Leakage

As in coal mines, in metal mines leakage is the most common cause of inefficient distribution of air in mines. While leakage in coal mine ventilation systems can reach 80% of the total volume of air circulated, in metal mines, it averages about only 25%. However, in metal mines, leakages not in excess of 15% are attainable with good ventilation practice.

Coursing of Air

The normal way of coursing air is through *ascentional ventilation,* in which the air is coursed into a stope raise at the lower level and upcasted at the level above. In *descentional ventilation,* fresh air is coursed from an upper level to ventilate a stope below it, and then exhausted to a lower level. Ascentional ventilation is used for two reasons: (1) most metal mines practice some sort of overhand mining method in which ore extraction is commenced at or just above the sill and proceeds upwards with successive cuts; and (2) in most mines, the rock is warmer than the ventilation air, so air is warmed as it rises, and this "chimney effect" aids airflow. Advocates (Bromilow, 1957–58) of descensional ventilation, however, maintain that the dust problem in workplaces is lessened when the air moves in the same direction as the broken ore (so-called *homotropal flow*), and that less heat is added to the air in hot mines when it enters the stopes from an upper level. When airflow and the broken ore move in opposite directions, the ventilation system is called *antitropal flow.*

The design of ventilation for an individual stope is usually simple; however, the handling of air in a multilevel mine requires detailed consideration. An old practice, formerly used in the Coeur d'Alene district of Idaho, was to direct all air down the intake shaft to the lowest level of the mine, course it to the raises, let it rise through all the workings to the top active level, and then direct it to the exhaust shaft(s), that is, to practice ascentional ventilation. The disadvantage of this ventilation layout is that the upper stopes get nothing but exhaust air from the lower stopes. While reuse of air is usually desirable for maximum utilization, there is also generally enough contamination with dust, gas, and heat that some addition of fresh air is necessary at each level. With successive additions of fresh air, the upper levels, where there are usually fewer stopes, will have much more air than they need. This problem can be alleviated by directing some of the used air from lower levels up special exhaust raises or unused old stope raises.

A better practice is to have alternate levels as intake levels, with the in-between levels serving as exhaust levels. However, this method necessitates special exhaust raises at the ends of the exhaust levels if the exhaust levels do not connect directly to the exhaust shaft(s). The disadvantage of the fresh air moving downward rather than upward in some areas is relatively slight and can easily be overcome with small exhaust fans.

14.2 LAWS AFFECTING METAL MINE VENTILATION

Threshhold Limit Values

MSHA regulations governing safety and health standards in underground metal and nonmetal mines are contained in Part 57 (CFR 30), subparts D and G (Anon., 1995a). Since the products mined in metal and nonmetal mines vary widely in terms of their mineralogy, chemical and physical characteristics, and health effects, there are not as many generalized standards (such as those for underground coal mines) applicable to all metal and nonmetal mines. For example, few air-quantity requirements are specified. However, air-quality standards do exist. For example, the minimum oxygen content must be 19.5%. With regard to exposure limits on air contaminants, MSHA has adopted the threshold limit values (TLVs) set forth by the American Conference of Governmental Industrial Hygienists (Section 3.2). The TLV-TWA for radon is 1.0 working-level maximum. In 1989, MSHA issued a proposed rule for air quality, chemical substances, and respiratory protection standards (Anon., 1989b). These standards are under consideration for final adoption. Two factors unique to deep mines are heat and humidity, and these must be addressed adequately for both health and productivity reasons (Johnson, 1992b). As with any planning exercise, information on the most recent ventilation and air-quality standards and new, more effective industry practices must be collected.

Local Regulations

Some states or localities have regulations governing air quantities or velocities. The state of Arizona, for example, requires that "a minimum air speed of 50 fpm (0.25 m/s) shall be maintained in underground drifts where persons are drawing, tapping, or loading rock" (Anon., 1976). Such state or local rules, of course, must also be observed. A set of recommended air velocities that can be used when other considerations do not govern is provided in Section 14.6.

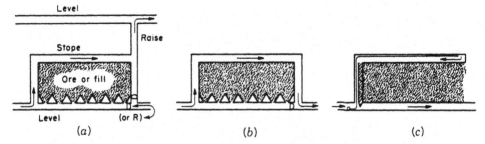

FIGURE 14.2   Ventilation in workplaces for mining methods in a near-vertical plane: (*a*) stopes with interlevel connections; (*b*) stopes with no connections but two access openings; (*c*) stopes with no connections but one access opening.

14.3 VENTILATION FOR VARIOUS MINING METHODS

The physical outline of the ventilation system depends largely on the mining methods in use to extract ore from the mineral deposits. Each type of metal and nonmetal mining method has its own characteristic ventilation system. The term "metal mining method" is usually attributed to extraction methods especially devised for massive or steeply pitching vein deposits. The descriptions of ventilation systems in the following sections for the different metal mining methods are meant to serve as general guidelines and not as hard and fast rules.

Shrinkage, Overhand Cut and Fill, Sublevel, Vertical Crater Retreat, and Square-Set Stoping

All these stoping methods are generally developed with raises between levels at opposite ends of the stopes. Therefore, they can be ventilated with an airstream going in one raise, through the stope, and out the other raise. A common version of this, in which the air is exhausted to the level above, is shown in Fig. 14.2a. Figure 14.3 shows a system for sublevel stoping and

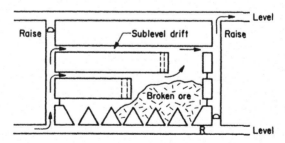

FIGURE 14.3   Ventilation system for sublevel stoping.

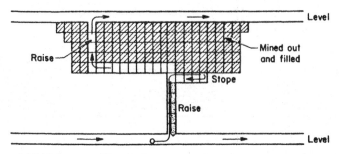

FIGURE 14.4   Ventilation system for square-set stoping.

Fig. 14.4 an arrangement for square-set stoping. Depending on the specific layout, there are many variations in the air distribution system. Some cut and fill mines carry a metal vent pipe up through the fill to within a couple of feet of the top and use this for ventilation rather than the manway–timber raise. An auxiliary fan is often connected to the bottom of this tubing. Other mines prefer to furnish the air raise with flexible tubing, again with an auxiliary fan at the bottom. Either method is preferable to allowing the air to simply upcast the raise, as the air will not pick up as much dust.

Top Slicing, Underhand Cut and Fill Stoping, and Sublevel Caving

All these systems generally involve ventilation of dead-end headings after initial development is complete. Auxiliary fans are therefore generally required. Often the air can be directed from the main level into the headings through tubing, then exhausted up a raise and into the exhaust stream if a raise is kept open between levels. If only a portion of the raise is kept open, the air is directed back through the open part of the same raise through which it entered via tubing. However, recirculation is difficult to control in this type of system (see Fig. 14.2c), and vigilance is necessary to ensure that adequate air quality is maintained at all times. An idealized sublevel caving ventilation system is shown in Fig. 14.5.

Block Caving

Disregarding the distinction between panel caving and block caving, and between slusher-drift systems and grizzly-drift systems, block caving mines can be considered as being developed in either of two ways: (1) the grizzly or slusher drifts and the haulage drift are actually separate levels, or (2) the grizzly or slusher drifts are simply sublevels accessed only from the primary haulage level.

In the first type of system, which is the more commonly used today, ventilation air is drawn from the intake shaft on the grizzly or slusher level,

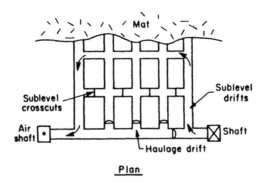

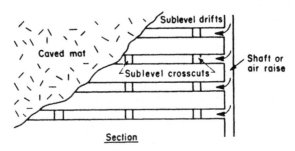

FIGURE 14.5 Ventilation for sublevel caving.

usually by a booster fan, and directed through the various grizzly or slusher drifts by control brattices or auxiliary fans (Fig. 14.6). The air then downcasts to the haulage level and into the exhaust through ventilation raises.

The large amount of muck moving on the grizzly levels causes dust problems that cannot be fully mitigated by wetting down, as the dust is generated in the movement of the ore falling from the fingers onto the grizzlies. Large quantities of air are necessary to keep the dust within acceptable limits. Filtration of the air before downcasting to the haulage level is a desirable though uncommon practice. Short-circuiting of the air through open ore-passes is also a problem. This could be minimized by maintaining some ore at all times in the orepasses, again an uncommon practice. Small exhaust fans at the exhaust ends of the grizzly drifts to maintain air moving in the right direction are better; however, this might result in some recirculation from the haulage level back up through the orepasses.

A more ideal system is to ventilate haulage levels and grizzly levels separately, with both levels exhausting directly to the exhaust shaft or a third exhaust sublevel. Thus the two levels would be maintained at the same pressure, and little air movement through the orepasses between levels would occur. This would require a much larger quantity of ventilation air and might not be considered economically feasible. A third alternative would be to maintain a third, separate sublevel to receive exhaust air.

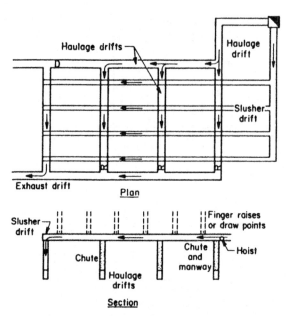

FIGURE 14.6 Ventilation of block caving mines with separate grizzly and slusher levels (plan view).

In the second type of system, air must be brought through the haulage level and upcast to the grizzly or slusher level through the manway. The best arrangement for a slusher-type system is to direct the air over the slusher and down across the drawpoints to the end of the slusher drift (Fig. 14.7). The air from all the slusher drifts in the area then is collected by a sublevel drift connecting the ends of the slusher drifts in the area and exhausted either to a special main-level exhaust drift or is filtered and exhausted to the haulage level. If grizzly drifts are used in this system, the layout is the same as for slusher drifts, except that more air is needed to dilute the dust being generated at the various drawpoints where personnel are working. Caving undercuts are of short-term duration and can easily be ventilated with small auxiliary fans located on the grizzly level and flexible tubing directing air to the undercut working areas.

Room and Pillar or Stope and Pillar Mining

Perhaps the most commonly used underground metal and nonmetal mining method for flat-lying deposits is open stoping, with pillar support on either regular or irregular spacings. The Viburnum lead mines in Missouri are excellent examples of this method. In plan view, metal mines in this category often appear similar to room and pillar coal mines, and the ventilation systems in general are similar. Thus the ventilation system principles discussed in Chapter 13 are especially applicable here.

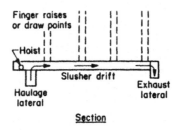

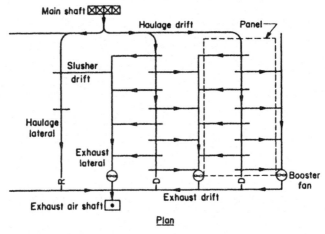

FIGURE 14.7 Ventilation of block caving mines having grizzly and slusher sublevels (section view).

In practice, however, some unique ventilation design factors are encountered in stope and pillar metal mines. First, fans need not be located exclusively on the surface, giving greater flexibility in providing additional ventilation to specific workplaces. A second feature is the extensive use of diesel equipment in stope and pillar mines. The installed diesel horsepower is probably greater in such mines than for any other mining method. Finally, the size of openings excavated in stope and pillar mines can be so large that the construction of ventilation stoppings often becomes impractical. Rooms in the Missouri lead mines, for example, can be 60 ft (18 m) high and 40 ft (12 m) wide. In a mine of this type, the enormous quantity of air present serves as a large reservoir to dilute noxious exhaust gases and combat oxygen depletion. Some buildup of diesel exhaust (layering) can be detected during an operating shift, but such concentrations rarely exceed acceptable levels.

14.4 CASE STUDIES OF VENTILATION SYSTEMS

The ventilation systems described in this section are typical of those used by modern mines and represent the best layouts under the stated conditions

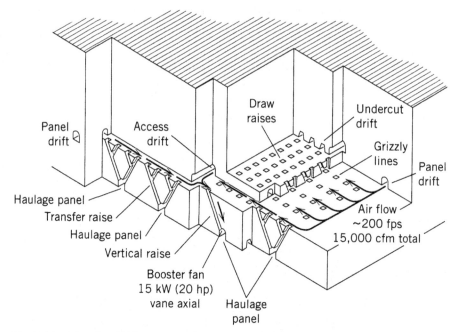

FIGURE 14.8   Ventilation layout of San Manuel mine (Johnson, 1992a). (*Conversion factors:* 1 fpm = 0.0051 m/s, 1 cfm = 0.472 × 10⁻³ m³/s.)

and constraints (Johnson, 1992a). Additional ventilation data on these mines and others can be found in O'Neil and Johnson (1982).

Block Caving Mines

San Manuel Mine   This mine is developed by driving pairs of grizzly and haulage levels 60 ft (18.3 m) apart; there is a 300-ft (91.5-m) elevation difference between level pairs. Intake is by a pair of personnel–materials shafts on the east end of the mine; these intake shafts are connected to all levels. Exhaust is via the four hoisting shafts on the west end of the mine; the exhaust shafts are connected only to the haulage levels. Motivation for airflow is by main intake fans near the intake shafts on the grizzly levels.

Air leaving the intake fan on each grizzly level is coursed along the north periphery of the ore body and enters through panel drifts driven into the ore body and along the strike (Fig. 14.8). The air then is directed through the grizzly lines to 20-hp (15-kW) exhaust fans, which are located in raises connecting the grizzly level to the haulage level; these raises hole into intermediate drifts that are driven between panel drifts. Additional boosting for ventilation air through individual grizzly lines is provided by compressed-air movers that are used only when muck is running; inactive grizzly lines are bratticed

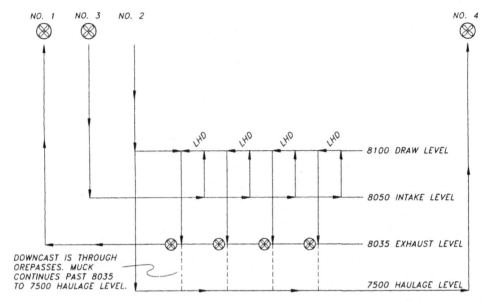

NO. 1   NO. 3   NO. 2   NO. 4

8100 DRAW LEVEL

8050 INTAKE LEVEL

8035 EXHAUST LEVEL

DOWNCAST IS THROUGH
OREPASSES. MUCK
CONTINUES PAST 8035
TO 7500 HAULAGE LEVEL.

7500 HAULAGE LEVEL

FIGURE 14.9 Ventilation layout of Henderson mine (Knape, 1985).

off. A quantity of 3000–6000 cfm (1.4–2.8 m/s) is provided for each grizzly line being worked.

In areas where load-haul-dumps (LHDs) are used, grizzly lines are replaced by draw drifts that access the caving ore via chambers driven at the sides of the draw drifts. Here one exhaust–raise fan to the haulage level is provided for each draw drift, and jet fans are used for moving the air rather than compressed-air movers.

Henderson Mine   Air is downcast through No. 3 shaft and directed across the 8050 intake level to the production area (Fig. 14.9). Air is then upcast to individual draw drifts on the 8100 level, where LHDs are in use. Fresh air travels past the LHD and downcasts the orepass in which the LHD is dumping and is drawn off the orepass on the 8035 exhaust level by booster fans; the muck continues down the orepass to the 7500 haulage level. Air travels across the 8035 level to exhaust at No. 1 shaft. Main intake fans are located at the collar of No. 3 shaft and main exhaust fans at the collar of No. 1.

The 7500 haulage level is ventilated by a separate intake shaft, No. 2, and a separate exhaust shaft, No. 4. Motivation for this system is by exhaust fans at the collar of No. 4.

Climax Mine   This was the older type of block cave that did not have separate connections to the main openings for the draw-control and haulage

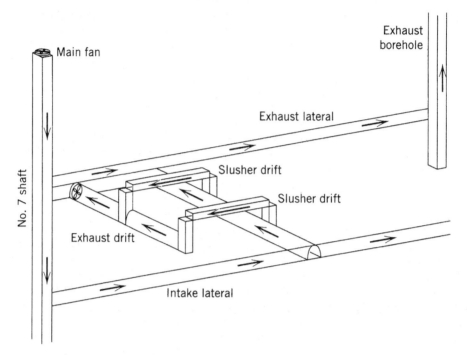

FIGURE 14.10   Ventilation layout of Climax mine (Johnson, 1992a).

levels. In the latter days, production was from the 600 level, which was below surface level, and from the Storke level, which was accessed by an adit. Air was downcast through No. 7 shaft via intake fans located at the collar, then traveled through a ventilation intake lateral and into the haulage drifts (Fig. 14.10).

Air was upcast to each scraper drift from the haulage drift; scraper drifts were driven just above and at right angles to the haulage drifts. Air traveled to the end of the scraper drifts and downcast to exhaust drifts, one of which was driven between each pair of haulage drifts and parallel to them; motivation was provided by exhaust fans in the exhaust drifts. Air was then gathered into exhaust laterals and conducted to raises that exhausted up into the open pit. There were also some exhaust fans in the main exhaust laterals. Underground operations at Climax were suspended in 1987.

Stope Mines

Homestake Mine   The mine is ventilated by three relatively independent circuits (Fig. 14.11). For the upper part, No. 2 shaft is the intake; there is an intake fan at the collar of this shaft, and exhaust is via raises that break into an open-cut mine. The middepth part of the mine intakes via No. 5 shaft

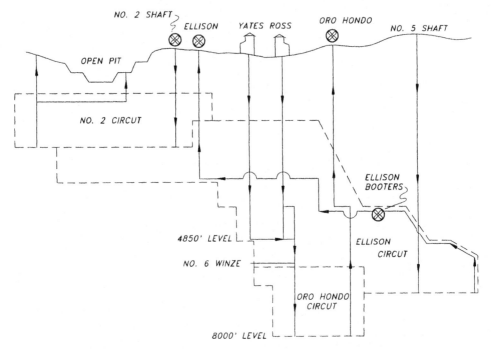

FIGURE 14.11 Ventilation layout of Homestake mine (Marks et al., 1987).

with exhaust up Ellison shaft; fans are underground boosters with an exhaust fan at the collar of the Ellison. The deepest part of the mine is designated the Oro Hondo circuit; intake is via Ross shaft, and exhaust is up Oro Hondo shaft, with main-air motivation supplied by an exhaust fan at the collar of the Oro Hondo.

Most mining nowadays is with overhand cut and fill or vertical crater retreat (VCR) stoping systems. In the cut and fill system, air is directed into the stope with auxiliary fans from the drift on the bottom level of the stope; air is often blown through a cooler before entering the stope. Exhaust is through ventilation boreholes that are sunk from the upper main level of the stope along the strike of the stope in its early days of development. A quantity of 25,000 cfm (11.8 m³/s) is provided for each cut and fill stope that uses extensive diesel equipment; smaller stopes that employ only compressed air machinery need 7000 cfm (3.3 m³/s).

In the VCR method (see Fig. 14.12), it is necessary to provide ventilation for the top sill of the stope, from whence downholes are drilled to the bottom sill and loaded, and to the bottom sill of the stope where the muck is picked up; both these sills are driven by crosscutting from main level drifts. In the top sill, air is directed through an auxiliary fan to the face through vent tubing. In the bottom sill, ventilation is flow-through since there are crosscuts

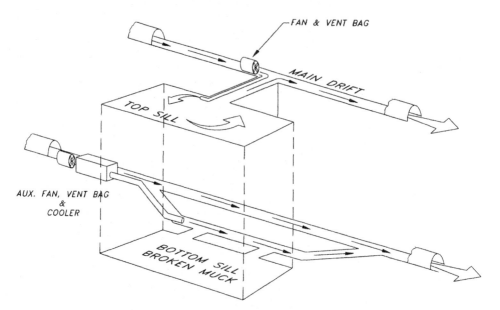

FIGURE 14.12 VCR stope ventilation at Homestake mine. (After Lapp in-Randolph, 1987).

in either end of the sill to the drift; air is directed through an auxiliary fan and vent tubing to the muck bays that connect into the broken muck area; coolers are often used in line with the auxiliary fans. An airflow of 7000 cfm (3.3 m³/s) is allowed for top sills, and 9000–15,000 cfm (9.9–7.1 m³/s) for bottom sills.

Sunshine Mine Air intakes the Jewell shaft and is divided into east and west ventilation districts. Initial motivation power is provided by underground boosters near the Jewell (Fig. 14.13). Air is then coursed through a series of winzes and raises to be distributed to the workings. Main exhaust is up the Bighole shaft, which is furnished with surface exhaust fans. Some supplementary exhaust is through the Sunshine tunnel and Silver Summit shaft.

Mount Isa Mine The overall ventilation layout of this mine is complex. Only stope ventilation is described here. Footwall and hanging wall drifts are driven on each sublevel around the periphery of the ore body; these drifts either connect directly to shafts or to inclines between main sublevels. Stopes do not extend along the full width of the ore body, so several crosscuts connect the drifts, and air is supplied off these crosscuts to individual stopes (Fig. 14.14).

The stoping method used is sublevel stoping with ring drilling. Initial de-

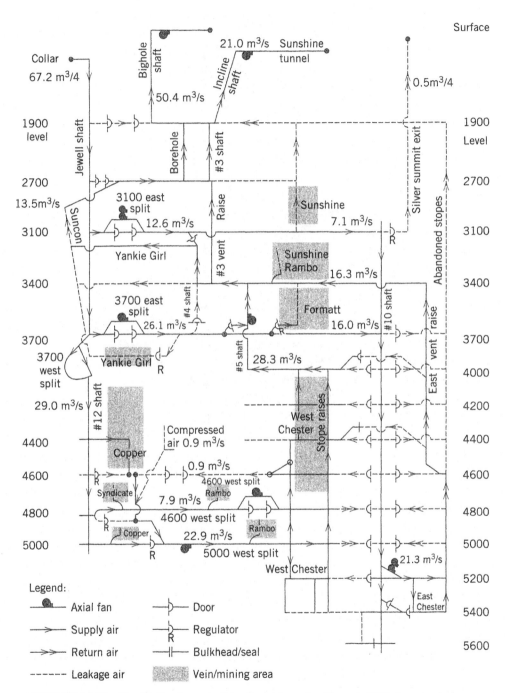

FIGURE 14.13 Ventilation layout of the Sunshine mine (Jurani, 1987; by permission of SME). (*Conversion factor:* 1 cfm = 0.472 D 10⁻³ m³/s.)

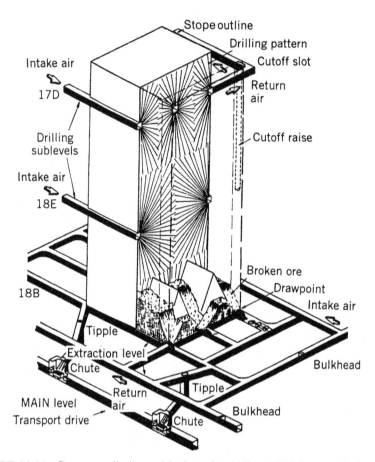

FIGURE 14.14   Stope ventilation at Mt. Isa mine (Allen, 1976; by permission of the Mine Ventilation Society of South Africa, Johannesburg).

velopment is to bore a cutoff raise in the middle of the stope and slash this out to the stope width. Sublevels are driven along the stope length to intersect the cutoff raise. The raise connects to an exhaust horizon above the top of the stope and to a fill conveyor level above the exhaust horizon. Air is supplied to the sublevels either directly via footwall pressure shafts or via access ramps and fresh-air raises. Air courses through sublevels and exhausts up the stope cutoff raise through an exhaust fan on the exhaust horizon. In stope development, auxiliary fans supply the faces of the sublevels before they connect to the cutoff raise; sometimes these fans continue to be used once hole-through to the cutoff raise is effected. About 15,000 cfm (7 m³/s) of air per sublevel is needed.

Magma Mine   Ventilation is by downcasting No. 9 shaft, the main person-nel–materials hoisting shaft, and upcasting No. 6 shaft, which is located west of the active workings. Supplemental exhaust is supplied through a series of raises that connect to No. 4 shaft in the old western portion of the mine. Three of the shafts in the western portion supply intake ventilation for the older, inactive part of the mine. Motive power is provided by intake fans near No. 9 shaft that blow through cooling coils, and by main-exhaust fans at the collars of both exhaust shafts. Some large underground exhaust booster fans are also used.

The ventilation pattern is to intake alternate levels from No. 9, upcast and downcast the air to the levels between the intake levels through mining raises, and then direct the air to the exhaust shafts. This is the ideal situation, although not always possible.

The mining method employed is undercut and fill stoping, principally in a replacement deposit that dips at 30°. A raise was driven between levels and two stopes started off each raise. One stope was just below the top sill; the other, about halfway between levels. Stoping proceeds downward, with cuts on each sublevel being filled before the cut on the sublevel below is begun. Extraction of the ore on each sublevel is by an initial sublevel crosscut being driven along with the footwall, then by parallel cuts (termed *panels*) being driven across the dip of the ore from the crosscut to the hanging wall. About three panels are driven at a time; the panels are then sand-filled, and new panels driven. This sequence continues until the sublevel is mined out. Ventilation is supplied by drawing air off the raise and directing it to the faces of the various panels with auxiliary fans, then exhausting up another raise to the level above (Fig. 14.15). The exhaust raise is often furnished with an exhaust fan at the exhaust level. A quantity of 6000 cfm (2.8 m³/s) per active panel in each stope is allowed.

14.5 DESIGN OF VENTILATION SYSTEMS

Major and Minor Features

In planning the layout of a ventilation system and determining the proper method of ventilation for a prospective mine or changes and extensions to an existing mine, attention must be given to many factors. These factors affect both the major and minor features of the ventilation system. *Major features* of the system are concerned with the fan, the main openings and airways, and the primary circuits. *Minor features* involve the details of distributing air to individual working places. McElroy (1935) summarizes the desirable major features to include in a mine ventilation system as follows:

1. Utilize haulage or hoisting openings as intakes.
2. Utilize all available openings and connections to the surface in transmitting air.
3. Utilize ascensional ventilation, that is, direct intake air to the lowest

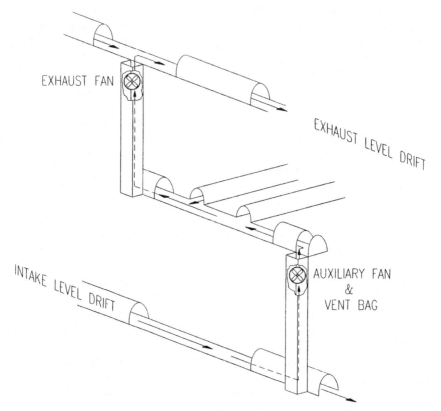

FIGURE 14.15 Stope ventilation at Magma mine (Johnson, 1992a).

active level in the mine, allowing it to course through workings as it ascends.

4. Limit distance of air travel to a minimum.

5. Maintain a balance of resistance between main intake and return airways.

6. Reduce control devices in main operating openings to a minimum.

7. Avoid leakage and recirculation.

8. Circulate air from active zones to caved ground.

9. Split air as often as needed and as close to the fan as a possible (but limit in a hot mine, where velocity must be maintained for cooling effect).

10. Consider surface location and exhausting arrangement for at least some main fans.

11. Avoid interventilation of adjacent mines, but utilize them for escape routes.

12. Use centrally located shaft(s) as intake(s) and peripheral shafts as exhausts.
13. Avoid transmitting air in opposite directions (bidirectional flow) in adjoining airways (shafts, entries).

The minor features of a ventilation system are reflected in the details of planning ventilation in conjunction with mine operation. They are influenced by a number of factors, which are listed by McElroy (1935).

1. Degree of concentration of workplaces
2. Position of active workings
3. Ground condition and methods of support required
4. Position of local sources of heat, cold, gas, or dust
5. Position of sealed areas or fire zones
6. Condition of abandoned workings
7. Precooling requirements of hot zones in deep mines

In general, as illustrated in Chapter 13, planning to ensure adequate amounts of fresh air at all workplaces can best be done by observing the following principles:

1. Attempt to coordinate mining and ventilation systems, rather than superimposing a ventilation system on an existing, rigid, or predetermined mining method (Section 13.4).
2. Utilize the primary development openings as main arteries (airways) in the ventilation system and other development openings as secondary airways.
3. Restrict the number of workplaces on one split of air, striving to have a single section or stope ventilated by one split; aside from providing fresh air in all workings, splitting ensures that disruption of ventilation in one has no effect on others.

Design Procedure

Design of a ventilation system consists of the following steps:

1. From the proposed mine plan, lay out a system of air courses in accordance with the principles listed above.
2. Determine air-quantity requirements. The limiting design criterion here may be imposed by any of the following:
   a. *Legal Quantity Requirements.* Quantity requirements are likely to prevail in mines using diesel equipment or where methane gas exists, as

federal requirements here are quite restrictive. State standards may also apply in some cases (Section 14.2).

b. *Legal Quality Requirements*. Here sufficient air must be provided to prevent contaminants from exceeding certain levels. This is always the controlling factor in uranium mining, for example. Again, both federal and state regulations should be consulted (Section 14.2).

c. *Empirical Guidelines*. In metal mining, the controlling air-quantity requirement is often due to locally established guidelines for improving worker comfort and productivity. Some rules of thumb in this category are provided below.

3. Tentatively select main- and booster-fan positions.
4. Calculate resistances in the various parts of the circuits, and calculate head and power requirements for each fan. Computer analysis is a great aid here.
5. Select a number of options, revise requirements, run economic analyses of the various options, and redesign the system if necessary.
6. Evaluate each option thoroughly from the viewpoint of safety. The response of each alternative system to simulated mine fires should be analyzed carefully. Accessibility and reliability of system components—as well as the ability of the system to deliver adequate fresh air to underground employees under the widest variety of conditions—are vitally important.
7. Decide on a final plan.

As mines are developed over a period of many years and, even after being fully developed, are constantly expanding into one area and being phased out of others, ventilation requirements and fan heads change. Therefore, in ventilation-system design, the engineer must consider the phase of mine development, starting with sinking of the first shaft, driving the initial levels, connecting the levels to the other shafts, and so forth, and the future needs. For example, a fan used during development may be too low a quantity and too high a head to be efficient in the later stages, although it may be used later in combination with other fans. Conversely, fans that move large quantities at low pressures and that are needed during production might simply stall out with very small quantities of air being moved if an attempt is made to put them into service too early in the mine's life.

Furthermore, the exact production capabilities and reserves of a mine prior to development can only be estimated, and the planned ventilation system will have to be changed as the mine progresses. For example, projected raises may not be driven or used fans may be bought instead of the new ones requested. The mine head is probably the most difficult factor to estimate realistically in that the head and quantity realized may not be exactly those as planned. For example, the projected mine requirements of, say, 250,000 cfm (118 m^3/s) at 4.0 in. water (995 Pa) may actually turn out to be 225,000 cfm (106 m^3/s) at 7.0 in. water (1742 Pa). Therefore, in mine layout and ventilation equipment selection, sufficient flexibility must be provided to meet unanticipated deviations from the plan.

Ventilation Design Rules of Thumb

The following rules of thumb are useful for determining air quantities where other considerations do not govern (O'Neil and Johnson, 1982).

1. *Drift Velocities*. Air velocity in main-level drifts should be 200–600 fpm (1.0–3.0 m/s), with 400 fpm (2.0 m/s) a good average. Where drifts open up into repair galleries or other large openings, 50 fpm (0.25 m/s) air velocity is adequate.

2. *Stope Velocities*. Air in open stopes should move at about 400 fpm (2.0 m/s), unless excessive cold or heat is a problem. Reduce the velocity to 100 fpm (0.5 m/s) for cold air, and increase to 600 fpm (3.0 m/s) or above for hot air.

3. *Development-Working Velocities*. When dead-end face ventilation is needed, vent tubing should be carried 15 ft (4.6) behind the face. Air emanating from the tubing should be of quantity such that the same speeds are felt as given above. Since the air disperses rapidly as it leaves the tubing and velocity is reduced, air velocity in front of the tubing is not the same as air velocity in the tubing (Section 11.3). The following data have been experimentally determined for 16-in. (406-mm) tubing terminating 15 ft (4.6 m) from a drift face:

Air Quantity		Face Velocity	
cfm	(m³/s)	fpm	(m/s)
900–1500	(0.42–0.71)	300–400	(1.52–2.03)
1600–2200	(0.76–1.04)	500–800	(2.54–4.06)
2300–7000	(1.09–3.30)	800–1000	(4.06–5.08)

4. *Radon-Gas Dilution*. Radon gas is generally a problem in uranium mines, although potentially hazardous concentrations have been detected in other mines as well. For example copper mines in the Bisbee, AZ, area, have encountered radon gas. The source was probably underground water moving through radioactive materials but was never definitely known. In order to maintain working levels (WLs) of radon daughters (the currently accepted criteria for evaluation of radiation hazards) below acceptable limits, it is necessary to know the emission rate of radon gas from the rock. There are methods of predicting radon-gas emissions using theoretical formulas (Bates and Edwards, 1980). Empirical data from mines operating in similar geologic formations are also useful. The Quirke mine of Rio Algom in Elliot

Lake, Ontario, uses a quantity of 150 cfm (0.07 m³/s) per ton broken per day (Bell and Black, 1979), but some American producers use 350 cfm (0.17 m³/s) per ton broken per day.

General Layout of Systems

In any stoping system, it is desirable to have a ventilation intake and a ventilation exhaust at opposite ends of the stope so that auxiliary fans and the attendant tubing may be eliminated. This, however, is seldom possible because of the mining method.

Many mines working vertical veins continue to practice ascensional ventilation. This system usually supplies adequate air and should be considered if contamination from dust or gas in the workings is not a problem. However, in the event of a fire in the lower levels, where work is often concentrated, an ascensional system may contaminate the entire mine with lethal gases; whereas in a descensional system, the gases would have been exhausted immediately. Naturally, the situation is just the reverse with a fire on the upper levels.

The generally recommended practice is to put underground booster fans on as many levels as necessary, drawing air directly off the intake shaft. Air is then directed through the workings with or without the use of auxiliary fans and passes into the exhaust shaft. Positive-pressure systems are recommended in all metal mines with either booster fans at the various levels or a single intake fan on the surface where this is considered more practical. Exhaust fans can be used as a supplement to keep the air moving in the right direction and to supply the necessary additional head, but they should not be used as the only air movers because splitting is difficult to control efficiently.

14.7 DESIGN EXAMPLE

A steeply dipping copper ore body Fig. 14.16 is to be mined by the overhand cut and fill method using LHD machinery in the stopes. Each stope is to be 200 ft (61 m) long, developed in pairs, with a 3-ft (0.9-m) round airway on the ends of each stope pair, and a 3-ft (0.9-m) orepass–ventilation borehole in the center. A 100-ft (30-m) pillar is to be left between each stope pair. Main-level haulage will be by battery-powered vehicles. Production is designed for 14 active stopes.

Drifts are timbered and driven 10 by 10 ft (3.0 × 3.0 m). The intake-hoisting shaft is 20 ft (6.1 m) in diameter and concrete-lined. Similar shafts of this type have been found to have a resistance $R = 0.01 \times 10^{-10}$ in·min²/ft⁶ per 100 ft (3.665×10^{-3} N·s²/m⁸ per 100 m). The exhaust shaft is to be round, concrete-lined, 8 ft (2.4 m) in diameter, and with no furnishings except a small cage guide for emergency service and shaft inspection.

Design a ventilation system and specify regulators and fans suitable for

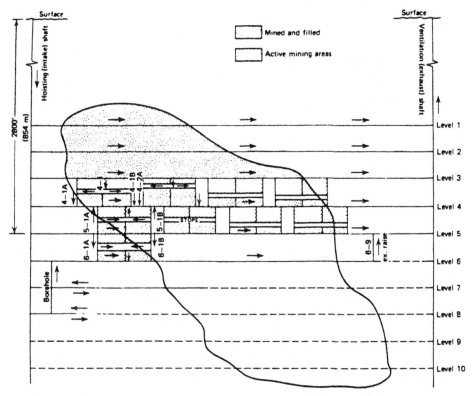

FIGURE 14.16 Simplified ventilation scheme for cut and fill mine (see Example 14.1).

this mine. Assume that shaft collars are both at sea level, and ignore air specific-weight changes, shock losses, and velocity heads.

Solution

A period about halfway through the life of the mine will be used as the point at which this problem will be solved, shown in Fig. 14.16. The ore has been mined out between levels 1 and 3; levels 3–5 are in full production, and level 6 is being driven toward the exhaust shaft past the last exhaust borehole, which has just been cut. Stope pair 6A–6B is being started. Levels 7 and 8 are being driven, and levels 9 and 10 have not been started. Levels 1 and 2, although no longer active, are still maintained.

A review of Example 6.7 would be helpful before proceeding. Since diesel exhaust is contaminating the air, none of the air that has passed through the stopes will be reused. A ventilation system using alternate levels as intakes with the intermediate levels as exhausts is selected. Levels 3 and 5 are intakes, and levels 4 and 6 exhausts. A small amount of air to ventilate the exhaust levels from the shaft to the stoping areas is necessary as well.

Step 1. Lay out the Ventilation Pattern   Level 5 will upcast through the stopes via the supply raise at the end of each stope pair and exhaust through the center boreholes. Level 3 will downcast through the center borehole and exhaust through the supply raises down to level 4. These are shown in Fig. 14.17.

Step 2. Calculate Needed Air Quantities   The required air quantities in the workplaces are determined by application of various criteria. One such criterion, for example, would be for the diesel LHD equipment. By consulting USBM Schedule 24, it is found that Approval No. 24-187 covers these specific units and calls for a minimum of 5600 cfm per unit (Anon., 1980). For an additional margin of safety, use 7000 cfm per unit, or 14,000 cfm per stope pair. Other design criteria are a minimum of 100 fpm air velocity in fresh-air drifts and 200 fpm in exhaust drifts.

The following table shows how individual air-quantity requirements were selected for the mine. The magnitude and direction of these flows have been plotted on Fig. 14.17.

Workplace	Controlling Criterion for Velocity	Air Quantity, cfm
Level 1	100 fpm minimum	10,000
Level 2	100 fpm minimum	10,000
Level 3	Diesel equipment in stopes—200 fpm minimum	76,000
Level 4	200 fpm minimum	20,000
Level 5	Diesel equipment in stopes—200 fpm minimum	90,000
Level 6	200 fpm minimum	10,000[a]
Level 7	100 fpm minimum	10,000
Level 8	100 fpm minimum	10,000
		Total 236,000
		[Will round off to 250,000 cfm (118 m³/s)]

[a] Requirement on level 6 is 20,000 cfm, which can be entirely supplied from levels 7 and 8. Nonetheless, 10,000 cfm is provided as an additional safety factor.

Step 3. Calculate Airway Head Losses and Fan Heads   In many actual metal mine ventilation systems, the number of interconnecting and overlapping circuits makes the determination of the mine static head a difficult problem. For example, 28 different airflow paths exist between the intake and exhaust shafts in this problem.

The task is to find the free split, and in this case, this is best accomplished by eliminating several possible splits by visual examination of the data. High airflow quantities are noted on levels 4 and 5, so that the highest-pressure

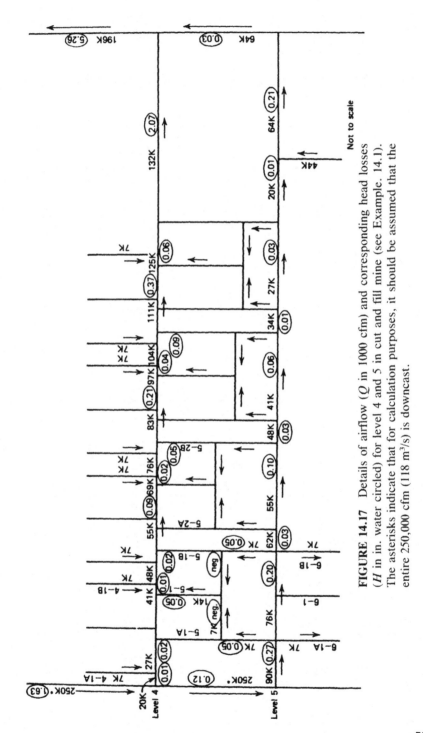

FIGURE 14.17 Details of airflow (Q in 1000 cfm) and corresponding head losses (H in in. water circled) for level 4 and 5 in cut and fill mine (see Example. 14.1). The asterisks indicate that for calculation purposes, it should be assumed that the entire 250,000 cfm (118 m³/s) is downcast.

551

drop (or head loss) branch is probably located in this area. Therefore, calculate head losses along these levels in the shaft and through the stopes.
(a)   Find the head loss in the intake shaft, using Eq. 7.7.

$$R = 0.01 \times 10^{-10} \text{ in.·min.}^2/\text{ft}^6 \text{ per 100 ft}$$

$$H_l = RQ^2$$

Although we could calculate head losses for each segment going down the shaft, the shaft resistance between levels is so small that we can assume without introducing significant error that the entire 250,000 cfm is downcast to either level 4 or level 5:

$$H_l = (0.01)(10^{-10})(250,000)^2 = 1.75 \text{ in. water (level 5)}$$

Similarly,

$$H_l = 1.625 \text{ in. water (level 4)}$$

(b)   Find the losses across levels 4 and 5, using the following K values:

$K \times 10^{10}$ lb·min^2/ft^4	Airway
110	Main levels—timbered drift
20	Access raises—bored
100	Stopes

The following head losses are calculated using Eq. 5.20:

	Level 4				Level 5		
Airway	L ft	Q cfm	H_l in. water	Airway	L ft	Q cfm	H_l in. wate
Shaft to				Shaft to			
4-1A raise	200	20,000	0.01	5-1A raise	400	90,000	0.27
4-1A to 5-1	350	27,000	0.02	5-1A to 5-1B	400	76,000	0.20
5-1 to 4-1B	50	41,000	0.01	5-1B to 5-2A	100	62,000	0.03
4-1B to 4-2A	100	48,000	0.02	5-2A to 5-2B	400	55,000	0.10
4-2A to 5-2	350	55,000	0.09	5-2B to 5-3A	100	48,000	0.02
5-2 to 4-2B	50	69,000	0.02	5-3A to 5-3B	400	41,000	0.06
4-2B to 4-3A	100	76,000	0.05	5-3B to 5-4A	100	34,000	0.01
4-3A to 5-3	350	83,000	0.21	5-4A to 5-4B	400	27,000	0.03
5-3 to 4-3B	50	97,000	0.04	5-4B to 6-9	400	20,000	0.01
4-3B to 4-4A	100	104,000	0.09	raise			
4-4A to 5-4	350	111,000	0.37	6-9 raise to	800	64,000	0.28
5-4 to 4-4B	50	125,000	0.06	exhaust			
4-4B to exhaust shaft	1400	132,000	2.07	shaft			

Average head losses for airflow through the stopes are as follows:

Airway	L, ft	Q, cfm	H_l, in. water
Access raise	100	7,000	0.05
Stope	100	7,000	negl.
Fill raise	100	14,000	0.05

(c) Find head losses in exhaust shaft. Assume an 8-ft diameter exhaust shaft with $K = 10 \times 10^{-10}$ lb·min²/lb⁴, yielding an R value of 0.038×10^{-10} in·min²/ft⁶ per 100 ft. Thus head losses are

Airway	L, ft	Q, cfm	H_l, in. water
Level 5–level 4	200	64,000	0.03
Level 4–level 3	200	196,000	0.29
Level 3–level 2	200	216,000	0.35
Level 2–level 1	200	226,000	0.39
Level 1–surface	2000	250,000	<u>4.75</u>
			5.81 in.

Some of these values are plotted on schematics of levels 4 and 5 in Fig. 14.17.

Step 4. Determine the Free Split By calculating the cumulative head loss for the various paths across levels 4 and 5, one can identify the highest head loss branch as

Intake → level 5 → raise 5-2B → level 4 → exhaust shaft

This path has a total head loss of

Intake shaft	1.75 in. water
Level 5	0.60
Between levels	0.10
Level 4	2.91
Exhaust shaft	<u>5.81</u>

$$H_l = 11.17 \text{ in. water } (2.792 \text{ kPa})$$

To obtain the prescribed airflow through every airway, only the free split (highest head-loss branch) will require no regulation.

Step 5. Specify Fans Now that the mine static head has been determined to be 11.17 in. water (say, 11 in. water), the fans must be specified. It is decided to use one booster fan in an intake position on each level and a single exhaust fan on the exhaust shaft. To maintain positive pressure on the working areas, the booster fans could be selected to operate against 5.5 in. water, and the exhaust fan would be specified to deliver 250,000 cfm at 5.5 in. water. With this configuration, failure of the exhaust fan will still result in a fair amount of ventilation in the mine.

Because the pressure difference between intake and exhaust levels is so slight (0.1 in. water), fans producing the same pressure should be selected for each level. Thus fan requirements are as follows:

| Location | Q | | H_s | |
	cfm	(m³/s)	in. water	(kPa)
Level 1	10,000	(4.72)	5.5	(1.369)
Level 2	10,000	(4.72)	5.5	(1.369)
Level 3	76,000	(35.87)	5.5	(1.369)
Level 4	20,000	(9.44)	5.5	(1.369)
Level 5	90,000	(42.48)	5.5	(1.369)
Level 6	10,000	(4.72)	5.5	(1.369)
Exhaust shaft	250,000	(117.98)	5.5	(1.369)

The fans for levels 3 and 5 can probably be of the same type with different blade pitch settings. Levels 1, 2, and 6 can also use the same type. Levels 7 and 8 at this time need auxiliary fans blowing through tubing, with another auxiliary fan on level 6 blowing air to their intakes. This is an auxiliary-fan problem and is not treated here.

The fan selection procedure illustrated in Section 13.15 can be used here to select the main fan. The relative large head loss in the exhaust shaft suggests that perhaps a larger-diameter shaft could be justified. By reviewing the principles contained in Chapter 12 and current shaft-sinking or raise-boring costs, the reader can determine the most economical size for the exhaust shaft.

PROBLEMS

14.1   Lay out the ventilation system for a mine of five active levels being worked entirely by sublevel stoping. Three shafts are available for ventilation use. Assume a steeply pitching, medium-width vein deposit.

14.2   Lay out the ventilation system for a block caving mine consisting of one haulage level and one grizzly level. The grizzly drifts are on 20-ft (6.1-m) centers and perpendicular to the haulages, which are on 50-ft (15.24-m) centers. The mine is developed through a service shaft on the north end of the 2000-ft (610-m) long by 400-ft (122-m) wide ore body, an upcast hoisting shaft on the south end, and fringe drifts

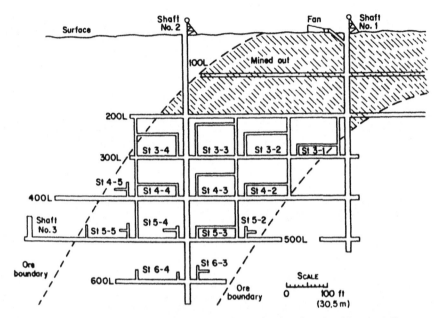

FIGURE 14.18 Vertical section of metal mine (see Problem. 14.3).

on the grizzly levels. Mining is done by starting at the north end and progressing steadily southward in a panel-type system across the width of the ore body. Twenty grizzly drifts at a time are being mined, with six drawpoints at a time being pulled in each grizzly drift. Dust generation is 0.5 mg/min at each drawpoint. Design a ventilation system, and show air quantities and distribution. Maintain dust level at no greater than 0.8 mg/m^3.

14.3 Given the metal mining plan shown in Fig. 14.18, lay out the ventilation system so that each workplace is on a separate split of air.

14.4 You have been asked to increase ventilation in the mine shown in Fig. 14.19 to accommodate 200,000 cfm (94.4 m^3/s), as the mine now employs diesels and this increased quantity is now necessary. The curve of the fan now in use is shown in Fig. 14.20 and is working approximately on the 15 setting with a 50-hp (37-kW) motor. Another of these fans is currently available on the property, and two other fans with the curves shown in Fig. 14.21 and equipped with 200-hp (150-kW) motors have been located. Determine whether these three other fans in conjunction with the one now in service can be used for the new required air quantity. New placement of all fans, operating characteristics, and any modifications to the fans such as new motors should also be specified.

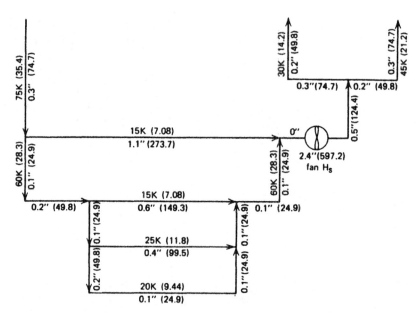

FIGURE 14.19   Ventilation schematic for Problem 14.4. Airflows in 1000 cfm (m³/s) and head losses in in. water (Pa).

14.5   Section I of the operating mine shown in Fig. 14.22 is being shut down, and the newly discovered section V is being opened. Determine an economic way to ventilate the new section. Only 10,000 cfm (4.72 m³/s) of air will now be required in the old section I. Section V will require about 30,000 cfm (14.2 m³/s), and friction losses can be roughly estimated from losses under similar circumstances in other airways in Fig. 14.22.

14.6   (a)   Calculate the mine and fan quantities and static heads in the exhaust ventilation system shown in Fig. 14.23. Quantities of air to be provided the various branches of the circuit and pressure drops between points in the network are indicated. Determine all missing quantities for individual segments of airway. Locate regulators where needed and calculate the amount of regulation at each.

    (b)   Calculate the mine and fan static heads and quantity assuming that a blower fan was used instead of two exhaust fans.

    (c)   Which arrangement, considering only H–Q–P_m requirements, is preferable?

14.7   Calculate the mine quantity and heads (static, velocity, and total) to four decimal places; the fan quantity, heads, and horsepower; and the location and pressure drop of regulators for the mine of Problem 14.3 (see Fig. 14.18), given the following conditions:

    (a)   Dimensions and friction factors for workings:

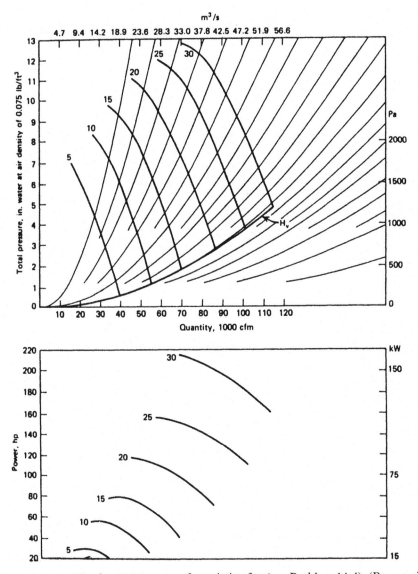

FIGURE 14.20 Performance curves for existing fan (see Problem 14.4). (By permission from Howden Group America, Nashua, NH. Copyright 1966.)

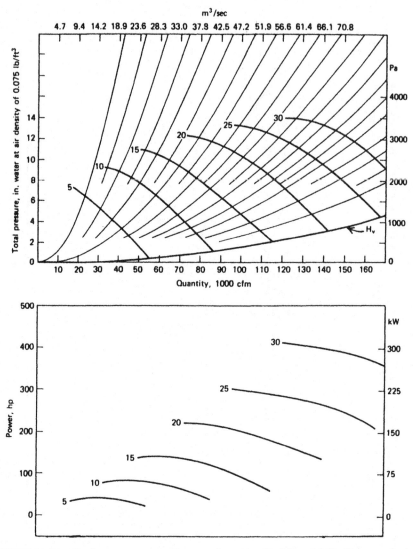

FIGURE 14.21 Performance curve for new fans (see Problem 14.4). (By permission from Howden Group America Nashua, NH. Copyright 1966.)

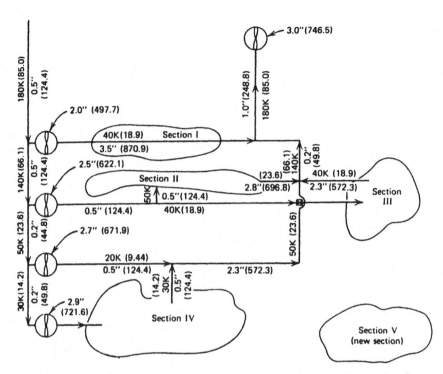

FIGURE 14.22 Schematic of mine in Problem 14.5. Airflows in 1000 cfm (m³/s) and head losses in in. water (Pa).

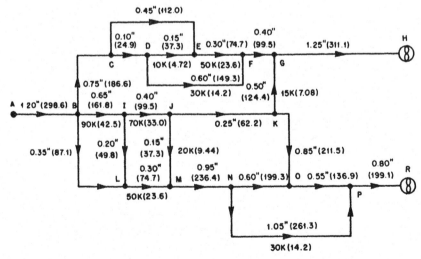

FIGURE 14.23 Schematic of mine ventilation system (see Problem 14.6). Airflows in 1000 cfm (m³/s) and head losses in in. water (Pa).

| | Size | | K | |
Working	ft	(m)	lb·min²/ft⁴	(kg/m³)
Shafts	20 × 10	(6.1 × 3.0)	25 × 10⁻¹⁰	(0.0047)
Drifts	8 × 10	(2.4 × 3.0)	100 × 10⁻¹⁰	(0.0186)
Raises	5 × 5	(1.5 × 1.5)	100 × 10⁻¹⁰	(0.0186)
Stopes	5 × 30	(1.5 × 9.1)	100 × 10⁻¹⁰	(0.0186)
Subcrosscuts	5 × 7	(1.5 × 2.1)	100 × 10⁻¹⁰	(0.0186)
Fan duct, diameter	6	(1.8)	10 × 10⁻¹⁰	(0.0019)

(b)   Airflow requirements:

Minimum velocity at face   50 fpm (0.24 m³/s)
Minimum quantity at face   100 cfm/person (0.047 m³/s·person)

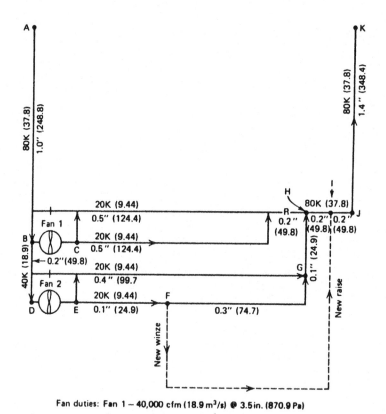

Fan duties: Fan 1 – 40,000 cfm (18.9 m³/s) @ 3.5 in. (870.9 Pa)
Fan 2 – 40,000 cfm (18.9 m³/s) @ 3.5 in. (870.9 Pa)

FIGURE 14.24   Ventilation schematic for Problem 14.8. Airflows in 1000 cfm (m³/s) and head losses in in. water (Pa).

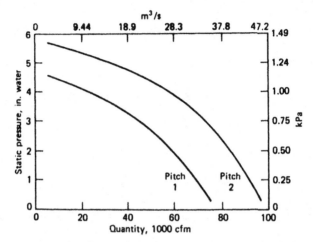

FIGURE 14.25 Performance curves for fans used in Problem 14.8.

(c) Auxiliary ventilation: Allow 25% leakage, and assume blower supplies all pressure needed to force air to the face through vent pipe and return it through the mine opening.
(d) Evasé duct: angle 6°, height 15 ft (4.6 m).
(e) Fan: axial flow, assume operating efficiency 75%; select the most suitable fan from a manufacturer's catalog.

Sketch the system characteristic curves and pressure gradients at the fan.

14.8 The mine shown in Fig. 14.24 is being deepened by means of a new winze and a planned exhaust raise. Exiting quantities and heads are given. The new circuit will require 40,000 cfm (18.9 m³/s), resulting in $H_l = 0.2$ in. water (49.9 Pa). Thus total mine quantity will increase to 120,000 cfm (56.6 m³/s), and the system of fans will therefore need to be revised. Fan curves for the units currently in service are shown in Fig. 14.25. A third fan of this type is available, but a fourth fan will probably be required. Design a ventilation system for the new total quantity, utilizing the three available fans and a fourth new fan. Specify the characteristics and placement of all fans and regulators. Calculate all motor sizes needed, assuming 85% fan efficiency.

CHAPTER 15

Control of Mine Fires and Explosions

Disaster-level explosions and fires have been reduced significantly over the last few decades because of expanded knowledge, strengthened safety regulations, more effective training, improved industry attitude, and better vigilance by mine workers. However, the continued occurrences of mine fires and explosions throughout the world demonstrate that engineering efforts for their control must be intensified. The mine ventilation engineer is clearly in an important position to affect the incidence of mine fires and explosions. The primary objective of the mine ventilation engineer should always be to eliminate fires and explosions altogether, a challenging and difficult goal, but one that must be relentlessly pursued.

In this chapter, the causes and prevention of fires and explosions and the part that ventilation plays in this process are emphasized. In addition, the response to these events and the importance of ventilation in escaping from a fire or explosion, in sealing a mine, and in recovering a sealed mine are discussed.

15.1 PREVENTION STRATEGIES

To prevent fires and explosions, it is necessary to understand the nature of their initiation and to identify the causes. For several main categories of fires and explosions in mines, the primary causes are identified in Table 15.1. On the basis of this knowledge, the elimination of these problems can be attempted by overcoming the conditions that permit their initiation or removing the causes. The following sections provide additional insights into the causes and the means to overcome them.

Mine Fires

Because of the nature of mines, a fire in a mine is potentially a disaster. As a matter of record, 57 of the 98 industrial fires resulting in 50 or more deaths during the years 1900–1990 in the United States have occurred in mines (Pomroy, 1990). This results from the confined space in mines and the fact that the mine must maintain a constant supply of ventilation air. Thus ventila-

TABLE 15.1 Primary Causes of Underground Mine Fires and Explosions in the United States

Category of Incident	Years Covered	Reference	Causes
Underground coal mine fires	1978–1992	Pomroy and Carigiet (1995)	Electrical (41%) Frictional sources (23%) Welding/cutting (19%) Spontaneous combustion (15%)
Underground metal and nonmetal mine fires	1950–1984	Butani and Pomroy (1987)	Electrical (40%) Welding/cutting (16%) Engine heat (13%) Spontaneous combustion (11%)
Underground coal gas or dust explosions	1969–1980	Richmond et al. (1983)	Frictional sources (81%) Welding/cutting (9%) Electrical (5%) Explosives (3%)

tion can carry combustion products and play a major part in the occurrences of fatal fires and in their prevention.

From the standpoint of the level of knowledge, the initiation of fires is a well-understood area of science. Three specific precursors of fire initiation are required:

1. Fuel
2. Oxygen
3. Heat

These three items are often referred to as the *fire triangle*. An understanding of these three requirements is all that is required to prevent fires, but control complications may be present. In a coal mine, for example, both oxygen and fuel are always present when the mine is operated. The only major point of control is the availability of heat sources. Therefore, controlling sources of heat whose temperature is sufficient to ignite the fuel is the principal method of preventing coal mine fires.

The causes of fires in mines has changed over time as equipment, mining methods, and power sources change. A study of the causes also identifies methods of preventing them. Table 15.1 contains some of the statistical analysis of reportable fires (those of 30 min duration after discovery or causing some injury) by Pomroy and Carigiet (1995). The prevalent ignition sources identified in underground coal mine fires were electrical (41% of fires for which the ignition source was known), friction sources (23%), welding and cutting operations (19%), and spontaneous combustion (15%). These sources indicate some of the methods of controlling coal mine fires and also some of the difficulties of achieving zero fires in an operating mine.

Of the four major sources of fires, welding and cutting operations are the easiest to manage as they are always performed by mine workers who can ensure that the heat source is used only when the fuel can be controlled or only when an accidental ignition can be immediately extinguished. These types of fires are normally a result of human failure to prepare and protect the work site.

Electrical and frictional ignition sources are somewhat harder to control. These fires generally occur in a variety of locations in the mines when electrical equipment malfunctions, trolley wires cause arcing, or friction generates a great deal of heat on belt conveyors. The location of the initial ignition generally determines whether the fire can be easily extinguished by mine personnel. Fires that are remote from the working sections are much harder to detect and control. However, they can in some cases be controlled or extinguished by fire-suppression equipment located at key locations on the belt conveyor or other critical points in the mine openings. In addition, monitoring stations can also provide early detection and quick intervention of electrical and frictional fires.

Spontaneous combustion fires in coal mines are the result of oxidation of coal under conditions that allow the heat of oxidation to be partially retained in the coal until the self-ignition temperature is achieved. These fires generally occur in gobs or old sections of mines in high-pyrite, low-rank coal seams. They can also occur in pillars or floor material in some mines. Because of the nature of spontaneous combustion, fires are difficult to prevent in coal seams where geologic conditions favorable to spontaneous combustion are found. Sealing of old mine sections and gob areas is the primary method of controlling these fires. In addition, the maintenance of adequate mine ventilation to dissipate heat of oxidation in older unsealed openings minimizes the incidences of ignition.

In underground metal and nonmetal mines, sources of ignition are somewhat different than in coal mines (see Table 15.1 for a comparison). The principal ignition sources for the years from 1950 to 1984 in the United States (Butani and Pomroy, 1987) were electrical (40%), welding or cutting (16%), and engine heat (13%). For metal mines, fire-prevention efforts can be oriented toward fire-suppression systems on equipment and safer and more reliable electrical systems. In addition, safer welding and cutting practices can greatly reduce the number of fires.

Fatalities that occur from mine fires have been decreasing steadily over time. These deaths usually occur as a result of carbon monoxide or other toxic gases given off during the course of the fire or due to asphyxiation resulting from low oxygen content. While the ventilation system can help in preventing fires, it should also be considered once a fire starts. This is a result of the fact that the ventilation airflow can carry toxic gases to other areas, increasing the number of workers exposed to gaseous products of the fires. For this reason, the coal mine ventilation plan should provide escapeways that can be easily accessed from either the intake or return airways of a mine in order to allow the miners a means of egress from the mine in airways that are not subjected to gases from a fire zone. This is the basis of any plan that utilizes two escapeways from the working sections of the mine.

The cardinal rule for eliminating deaths from mine fires is to prevent fires from occurring. The second important principle is to provide for early detection so that most fires can be quickly extinguished. The third most important safety factor is to design the ventilation system so that escape is possible in the event of an uncontrollable fire.

Mine Explosions

Most explosions in mines occur in underground coal mines and are the result of either strata gases gathering in parts of the mine or dusts being suspended in the airstream by some disturbance. The requirements that must exist for an explosion are the following:

1. Fuel
2. Oxygen
3. Heat
4. Suspension or mixture of the fuel and oxygen
5. A degree of confinement

This is sometimes called the *explosion pentagon* (Stephan, 1990), which is comparable to the fire triangle. All five factors at favorable levels must be present for an explosion. The first four factors alone can produce combustion of the fuel, but confinement is necessary for an explosion to occur. Without confinement, methane will burn without an explosion. The five conditions for a dust explosion can sometimes occur if a small gas ignition raises float coal dust from underground mine surfaces. Thus rock dusting on a regular basis is the normal preventive for dust explosions because the rock dust absorbs heat from the combustion of coal particles and prevents the transfer of combustion from particle to particle in a dust cloud (i.e., it takes away the degree of confinement necessary for an explosion).

Coupled with sufficient rock dusting, the primary method of eliminating mine explosions is control of conditions that initiate explosions. Statistics from previous explosions are helpful to mine management and engineers in this endeavor. Table 15.1 indicates the four primary causes for underground explosions, based on experience in the United States during 1969–1980. By far the biggest causes of explosions (81%) are frictional sources of heat such as sparks caused by the cutting picks on mining machinery, sparks due to roof falls, and frictional heating that can occur on belt conveyors and other machinery. Frictional sources can be reduced significantly through inspection and monitoring of the equipment, but they cannot be entirely eliminated. The reason is that frictional sparks at the working face or from roof falls cannot be totally controlled. It is therefore imperative that concentrations of mine methane in working faces be controlled to prevent explosions due to frictional sources.

Other major causes of gas and dust explosions are welding and cutting operations (9%), electrical sources (5%), and explosives (3%). Two of these causes are easier to control (welding/cutting and explosives) because they should be used only under conditions that will not result in an explosion. This type of explosion is almost always caused by human failure. Explosions due to electrical sources are somewhat different. They can often be caused by sparks from trolley wires or electrical faults in equipment. To combat these fires, the electrical system must be installed using all known safety precautions and be maintained in good condition. In addition, the electrical system and sources of mine gases must be isolated as much as possible.

Current mining regulations are normally written to attempt to eliminate mine gas and dust explosions in several ways. First, explosive contaminants (gases or dusts) are targeted for control such that they are always below their explosive limit. In addition, heat sources in a mine are controlled to keep ignitions from occurring even if explosive conditions exist. For example, only permissible equipment is used at the face, and other heat sources may be used only if a gas check has been conducted. Finally, rock dusting is used in an attempt to quench dust explosions. Only an aggressive campaign aimed at all elements of the explosion pentagon can further reduce the number of explosions.

15.2   MINE MONITORING

Monitoring and detection of mine fires and other environmental variables are an important aspect of a mine loss-prevention effort. Statistics indicate that over 70% of fires detected during the first 15 min cause little or no damage to the mine (Pomroy, 1990). In addition, fires that are quickly detected cause fewer injuries and deaths than do other fires. A monitoring and detection system is therefore an important part of any concerted effort to reduce the number of mine fires, particularly in coal mines.

The strategies and technology of mine monitoring systems are outlined in Section 6.8 and in other publications (Kohler, 1992; Bandopadhyay, 1992). Monitoring is an area of technology that has been rapidly changing and improving as new electronics and computer technology are implemented. The technology is being rapidly advanced with the use of programmable logic controllers (PLCs), helping to make the systems more versatile and better designed for each mine's application. Many mine monitoring systems were first applied to the mine environmental conditions underground to control methane, carbon monoxide, carbon dioxide, air velocity, and other indicator variables of the mine environment. However, capability exists in the newer systems to monitor production, environmental, materials handling, materials storage, electrical power system, and maintenance variables. Thus a mine-wide monitoring system is often employed rather than a system dedicated solely to the mine environment.

The detection of a fire in its early stages may be the most important task of the minewide monitoring system. To accomplish this task, a number of types of sensors are available. Three main principles of operation are employed to detect fires: optical field of view, thermal, and products of combustion (Conti and Litton, 1992). While all of these types of sensors can detect a fire adequately at times, they each have advantages and disadvantages. Two problems that are often encountered are an excessive number of false alarms and difficulty of discriminating an actual fire from the exhaust of diesel engines. These problems can normally be solved by a correct choice of the fire sensors, proper placement of the sensors in the mine layout, and optimal setting of the alarm level (Conti and Litton, 1992; Morrow and Litton, 1992).

15.3 RESPONSES TO FIRES AND EXPLOSIONS

When a fire or explosion occurs in a mine, all normal production activity ceases immediately, and the attention of all personnel is directed toward ending this threat to life and property. If the incident is a major one and is considered a threat to personnel, a set of four emergency phases is initiated to counteract the threat and to bring about a restoration of a safe mine environment. The four emergency phases are

1. Escape of the miners
2. Rescue of trapped miners
3. Recovery of bodies
4. Recovery of the mine

A well-planned and coordinated ventilation and escapeway system is crucial at this point. In addition, a good communication system, knowledge of the mine ventilation system, and a thorough training regimen for all personnel positively affect the success of the emergency response. In the following are discussed some of the requirements and resources necessary to overcome the results of a fire or explosion.

Mine Communications

When a fire or explosion occurs in a mine, one of the immediate concerns is the need to communicate with all the personnel in the mine. A variety of methods is used. In most U.S. coal mines, a paging-type telephone system is used to communicate with each section. This system is useful in an emergency situation but is often out of the hearing range of personnel in the section. As a result, emergency notification may be supplemented by turning off the electric power to the mine. Individual sections then contact someone on the mine staff to determine the nature of the shutdown, and an evacuation can be initiated. This normally allows nearly all mine personnel to be notified of the emergency.

In noncoal mines, the communication system used for normal communication may not extend into every workplace. A stench warning system may be used instead. Using this system, a strong-smelling chemical is distributed throughout the mine using the ventilating air or the compressed air system as the distribution medium. Ethyl mercaptan or tetrahydrothiaphene are some possibilities that provide an unmistakable signal. However, this system suffers from slow warning speeds if ventilation velocities are low or if their path through the mine is long. Under these conditions, a stench warning system with multiple points of injection is recommended (McPherson, 1993, p. 879).

Ventilation Management

If a fire or explosion occurs in a mine, management of the ventilation system becomes crucial to the escape of personnel and to the safety of rescue personnel who may enter the mine atmosphere. Ventilation management normally becomes the responsibility of the disaster management team. A command center is set up to control the ventilation, attempt to aid escaping miners, and direct the rescue personnel. The mine ventilation engineer ordinarily becomes a member of or an advisor to the disaster management team. This person's job is to lend expertise to the decisionmaking group and to provide a perspective on the role of the ventilation in the escape and rescue efforts.

Conventional wisdom in the control of the mine fans and ventilation (Mitchell, 1990b) is to allow the fans and ventilation system to operate without change unless overwhelming evidence exists to indicate that mine employees are endangered by such a policy. Several reasons justify this approach. The first is that personnel are taught to escape through previously established escapeways with established airflow directions. Changing fan or air directions would therefore confuse and endanger mine personnel. Second, changes in the airflows can send explosive gas mixtures into a fire zone. This can result in a gas explosion that will only worsen the situation.

During an emergency such as a fire or explosion, modern technology can be a tremendous help, both in terms of preparation for a fire or explosion and in the management of the mine ventilation system during the course of the event. Mine fire simulators can provide invaluable assistance both before and during a mine fire. Use of a simulator may be necessary because the heat from a fire can alter the ventilation airflows in terms of both magnitude and direction. By carefully studying fire scenarios before they occur, mine management can often determine the optimum escape routes for a fire with a known location.

A number of mine-fire simulation programs are available to allow for the analysis of fluid flows during a mine fire. The U.S. Bureau of Mines (USBM) has supported the development of a mine-fire program known as MFIRE (Chang et al., 1990). This program is now available in a personal computer format and allows a mine ventilation network solution for any set of given conditions, including those existing during a fire. Another mine-fire simulator called PCVENT is also available for use on personal computers (Wolski, 1991). Both of the programs are available to the public; however, they should be used only if an accurate ventilation survey has recently been conducted, and if proper preplanning and training has been completed.

Firefighting

The occurrence of a small fire in a mine may not be a serious problem if the fire can be quickly extinguished. For this reason, many locations in both metal and coal mines are equipped with automatic or manual water-deluge or foam-generating fire-suppression devices designed to quickly extinguish a fire. Belt conveyors, mobile equipment, fueling stations, and large surface mining equipment are often so equipped. The strategy is to quickly extinguish a fire through proper placement and utilization of a fire-suppression device.

In mining sections, a specified set of firefighting resources is normally required by federal law. In each coal mine section, the following are required for most mines:

1. Two portable fire extinguishers
2. 240 lb (109 kg) of rock dust
3. Waterlines into the section or two portable water or chemical cars
4. A portable foam-generating machine or a portable high-pressure rock-dusting machine and 60 bags of rock dust

In addition, portable fire extinguishers are also required on each piece of mobile equipment, at electrical installations, at oil storage locations, and at other crucial points in the mine. For noncoal mines, the firfighting equipment is not as specifically outlined. Instead, the *Code of Federal Regulations* specifies that each mine have on-site firefighting equipment to extinguish fires of any class that would occur because of the hazards present (Anon., 1995b).

If a fire occurs in a mine opening where no fire-supression equipment is located, then a firefighting effort must be initiated. Basic strategy in this type of an effort has been outlined by Dougherty (1969) and by McPherson (1993). In an underground mine, the first duty of the person who discovers the fire is to immediately notify all personnel downwind of the fire of its existence.

This being accomplished, all personnel can then retreat to the fresh-air side of the fire and proceed in an attempt to extinguish it, using the firefighting equipment provided in the mine section. For small fires, removing power from the fire source (if applicable) and utilizing dry-type extinguishers would normally be the best and most expedient procedure.

If the fire is too large for fighting with a handheld extinguisher, the mine water supply lines are then used for firefighting. Current federal law requires that a water line and related fire-extinguishing equipment be available on each coal mine section. This equipment will be useful if the dangers of a methane ignition are not present and if the spread of the fire is not too great. Specially trained fire brigades are often used in mines to fight any fire of significance. These personnel are trained in firefighting strategies, in recognizing dangerous situations, and in assessment of conditions that dictate when the personnel should evacuate the mine rather than continuing the firefighting effort.

Escape Procedures

Whenever a fire or an explosion occurs, the escape of mine personnel is of paramount importance. Three strategies of escape may be utilized:

1. Utilize the escapeways, donning self-protection equipment where needed.
2. Barricade a nonaffected part of the mine by trapping a supply of uncontaminated air.
3. Move to a refuge chamber.

In the United States, refuge chambers are not utilized on a routine basis. This leaves immediate escape and barricading as strategies in most mines. Because of current regulations that require a self-contained self-rescuer for all personnel in coal mines and a filter self-rescuer in noncoal mines, the best strategy is normally to execute an immediate escape, using procedures that are rehearsed on a regular basis. If escapeways are logically laid out and personnel are well trained, an escape will normally be the most reliable strategy of survival. However, hands-on education of all personnel on a regular basis is essential.

Barricading in the mine is a strategy that has been used with some success historically. Dougherty (1969) outlined the basic strategy of barricading: Choose a sizable section of mine openings containing uncontaminated air that has a minimum number of openings to seal. The miners may have little choice as to section, but the sealing-off points may be a decision of great importance. By short-circuiting the intake air and quickly erecting temporary stoppings to isolate the section, they can often provide a sizable volume of air to sustain themselves until rescue is achieved by rescue personnel. Under regulations currently enforced in the United States, each coal mine section must have a supply of materials to construct temporary stoppings for the purpose of barricading. However, with better escapeways and self-rescuers, immediate escape is currently the preferred procedure.

Mine Rescue

The rescue of personnel is always the primary concern in the aftermath of a fire or an explosion. The check-in system of the mine provides the means of determining how many personnel remain in the mine after the expected escape time has been exceeded. When it is determined that some personnel are not able to escape immediately after a fire or explosion, the rescue effort begins. The rescue attempt is also managed by the disaster management team.

Mine rescue attempts normally involve entry into the mine by mine rescue squads or the drilling of escape boreholes into mine workings suspected to be the location of survivors. All mines in the United States are required to maintain mine rescue squads if they exceed a certain number of employees; others are required to have a mine rescue squad under contract. Mine rescue squads are an historical tradition in mining. Readiness of mine rescue squads is tested in mine rescue and first-aid contests in mining regions, and a prize-winning mine rescue team is a source of pride to a mine and its personnel. In some countries, rescue teams are maintained at central rescue stations and travel where needed at a moment's notice.

The operations performed by a mine rescue team are outlined in some detail in texts by Strang and Mackenzie-Wood (1990) and Ramlu (1991). The team normally consists of at least six experienced miners who are physically fit, properly trained as a team, and expert in the application of first aid to injured personnel. The teams are arranged in a hierarchy with a captain, a second in command, and so on. A team is generally equipped with oxygen breathing apparatus for each team member capable of supplying oxygen for 4, a lifeline for connecting all personnel when traveling in smoke, a means of communicating with its base, a reserve breathing apparatus, gas-measuring equipment, first-aid supplies, and a stretcher if applicable.

A mine rescue team is trained to establish and work from a fresh-air base in the mine. On a six-person rescue team, five miners normally perform the mine exploration while the sixth remains at the fresh-air base. The base is set up at a location in the mine where the ventilation is strong enough to keep all combustion gases swept away. Several rescue teams may operate from the base, with one team at a time traveling inby for limited periods to search for survivors, to locate and fight fires, or to recover the mine. Communications with the base must be maintained at all times or the rescue team excursion must be terminated. The fresh-air base may be advanced into the mine in safe increments through the erection of temporary stoppings. This must be performed systematically and only if the rescue team is relatively certain of the safety of advance. Under conditions where serious hazards to the rescue team are encountered, rescue efforts must be suspended (Ramlu, 1991, p. 355).

When the traditional means of rescue by rescue personnel is untenable, other means may be employed. One procedure that has been successful in a number of countries is to drill large-diameter boreholes into mine openings for escape purposes. To effect successful use of this system, trapped miners must be able to signal the surface by seismic means (Durkin and Greenfield, 1981). In the United States, the standard procedure is for the miners to rest in the mine opening until three shots are heard (signaling the readiness of a seismic locational array on the surface). They then pound on a roof bolt or other hard surface 10 times at 15-min intervals until a 5-shot sequence is heard (signaling a successful effort at location). To effect a successful rescue by means of drilled boreholes, the Mine Safety and Health Administration maintains a seismic location detection system in a ready condition to be flown to the site of a mine disaster.

Mine Sealing

When a mine fire is out of control, is too remote, or cannot be fought without endangering personnel, it may be necessary to seal the mine or part of the mine. Sealing requires that all oxygen be cut off from the fire zone and that the seals remain intact and effective if a gas explosion occurs behind the seals. Current federal coal mine regulations require that such seals be able to withstand a 20-psig (138-kPa) pressure without structural damage or loss of sealing capability. In addition, the seals must be coated with materials with a flame spread index of 25 or less, according to ASTM Standard E162-87.

The standard type of mine seal used in the coal industry is 16 in. (0.4 m) thick and made of solid concrete blocks with a center pilaster. The seal is keyed into the mine opening using channels cut into the floor and ribs. The blocks are laid in an interlocking pattern and mortared to provide the necessary strength to resist explosion forces. Additional details of the seal design are provided by Weiss et al. (1993).

The building of thick-wall, solid-block seals is a labor-intensive and time-consuming activity. As a result, recent tests (Greninger et al., 1991; Weiss et al., 1993) have shown that alternative designs using crib blocks and rock dust, cementitious foam, or low-density foam blocks can be used to meet the requirements of federal regulations. However, not all designs of these seals have been judged to be adequate. It is, therefore, important to refer to the two references above for designs that will provide a seal that meets both the structural integrity and leakage specifications of the regulations.

Two additional concerns when sealing a mine fire should be addressed. The first is the disposition of ventilation of a fire zone during the building of the seals. Forbes and Grove (1948), Dougherty (1969), and Strang and

Mackenzie-Wood (1990) discuss this subject in some detail. The basic danger that is encountered is the buildup of methane or gases of distillation in the fire zone as the ventilation is being reduced. This can result in an explosive mixture behind the seals. To combat this hazard, the general procedure is to enclose a significant area of the mine around the fire zone using temporary seals with cutoff of the ventilation conducted quickly and efficiently. The mine is then evacuated until the gases behind the temporary seals are stabilized. At this point, the building of permanent seals can be completed with a greatly reduced danger of an explosion. McPherson (1993, pp. 849–850) recommends access tubes built into the seals for reduction of the explosion hazard during sealing. These tubes provide for maintaining some ventilation during the sealing process and access to the sealed area at a later time.

The second concern in erecting permanent seals is the provision of sampling lines for monitoring of the gas chemistry behind the seals. In addition, a tubing for measurement of pressure difference is normally provided. The sampling tubes must be extended a considerable distance inby the seal to obtain reliable samples. Dougherty (1969) suggests 40–50 ft (12–15 m), while McPherson (1993) suggests 30 m (98 ft).

Monitoring Behind the Seals

Once a mine or an area of the mine has been sealed, the process of monitoring the atmosphere behind the seals can begin. The ability to draw samples and measure head behind the seals is facilitated if small-diameter tubing and a manometer are installed through the seals. In other cases, it may be necessary to draw samples through boreholes from the surface to determine the state of the atmosphere. It is important to recognize that proper procedures must be followed to ensure that the samples taken are valid representations of the gaseous atmosphere behind the seals. A variety of problems can be overcome if the sampling rules outlined in Mitchell and Burns (1979), Mitchell (1990b), and Ramlu (1991) are followed.

The samples taken from behind the seal are normally analyzed for N_2, O_2, CO, CO_2, CH_4, H_2, Ar, C_2H_6 (ethane), C_2H_2 (acetylene), and C_2H_4 (ethylene). Each gas is normally analyzed by gas chromatograph directly rather than by subtraction. An exception to this rule is sometimes made for Ar, which can be determined by calculation if time is critical. It is also important to calibrate the gas chromatograph daily to ensure accurate analyses.

The interpretation of the results is the most crucial part of the monitoring process. A great deal of knowledge from past mine fires can be used to help in the assessment of the state of the fire and when it would be possible to attempt a recovery of the sealed area. Many "fire ratios" have been used over the years to assess the atmosphere behind the seals. Eight of these ratios are listed in the publication by Ren and Richards (1993). The ones most commonly used in recent decades are Graham's ratio (Vutukuri and

Lama, 1986, pp. 292–293) and Trickett's ratio (Jones and Trickett, 1954); both are dimensionless. In addition, an explosibility diagram should also be used to insure that the atmosphere within the sealed volume will not become explosive when ventilation is restored.

Graham's ratio, also called the *index of carbon monoxide* (ICO), can be calculated using the equation,

$$ICO = \frac{100 \ CO}{0.265(N_2 + Ar) - O_2} \tag{15.1}$$

where CO, N_2, Ar, and O_2 represent the volume percentage of the respective gases in the atmosphere. This relationship is often applied with the argon term omitted from the equation. Graham's ratio indicates the existence of a fire or increasing temperatures if it is rising over time. The value of ICO usually rises after a mine is sealed and will remain high until the available oxygen is consumed and the temperatures fall to near-ambient mine levels (Timko and Derick, 1991, p. 11). Values of ICO > 0.5 normally indicate that a fire is active, while values <0.2 indicate that the fire is under control. In mines where the formation will not readily absorb CO, a level, stable value for ICO may be more indicative of a fire under control than using an absolute value.

Trickett's ratio (TR), often called the *Jones–Trickett ratio,* is used in a similar manner. It is calculated from the expression

$$TR = \frac{CO_2 + 0.75 \ CO - 0.25 \ H_2}{0.265(N_2 + Ar) - O_2} \tag{15.2}$$

where CO_2, CO, H_2, N_2, Ar, and O_2 are again the volume percentages of the respective gases in the mine environment. Some references use Eq. 15.2 without the Ar term. The value of TR has more utility than the ICO value, as it can also be an indication of which fuels are being combusted. However, it is best applied if the gases in the equation originate from the fire alone. Carbon dioxide often is derived from the strata and from the heating of rock dust. This will cause erroneous results. Consequently, Trickett's ratio is often used only to verify other results (Urosek, 1995). TR ranges associated with various possibilities are as follows:

TR Value	Implications
<0.4	Fire probably out
0.4–0.5	Methane probably burning
0.5–1.0	Coal, oil, or conveyor belting probably burning
0.8–1.6	Wood probably burning
>1.6	Not normally possible in a coal mine fire; sample is suspect

Interpretation of TR and ICO values cannot be made based on a single sample. The long-term trend in the data must be analyzed instead. One advantage of the TR value is that it is not affected significantly in the case of dilution by air from nonburning regions of the mine. Thus it can be applied to samples taken at an exhaust shaft location.

The third analysis that is often performed on a sealed mine atmosphere is an investigation of the explosibility of the atmosphere, both by itself and when mixed with ventilation air. To perform this operation, explosibility diagrams are ordinarily used. A number of them are available in the literature. The *USBM explosibility diagram* (Zabetakis et al., 1959) is often used in the United States as a measure of the condition of a sealed mine environment. The diagram, as shown in Fig. 15.1, is used by determining the *effective*

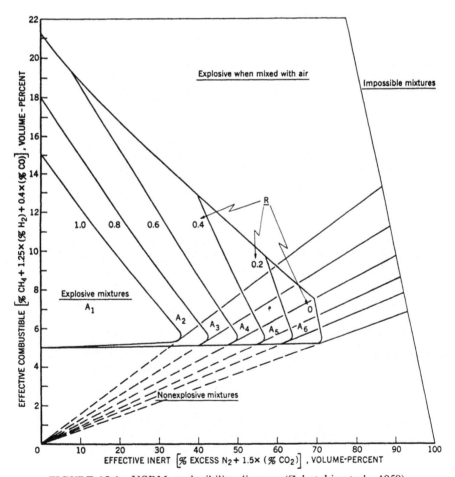

FIGURE 15.1 USBM explosibility diagram (Zabetakis et al., 1959).

combustible, the *effective inert*, and the *ratio of the methane to total combustible* in the atmosphere. These parameters are calculated using the equations that follow.

Effective combustible in %:

$$E_c = CH_4 + 1.25 \, H_2 + 0.4 \, CO \tag{15.3}$$

Effective inert in %:

$$E_i = N_2 - 3.8 \, O_2 + 1.5 \, CO_2 \tag{15.4}$$

Ratio of methane to total combustible (dimensionless):

$$R_{mc} = \frac{CH_4}{CH_4 + H_2 + CO} \tag{15.5}$$

In addition, an alternate and more accurate form of Eq. 15.3 (Urosek, 1995) is

$$E_c = CH_4 + 1.25 \, H_2 + 0.4 \, CO + 0.6 \, C_2H_6 + 0.54 \, C_2H_4 \tag{15.6}$$

Equations 15.3 and 15.6 are not valid for CO greater than 3% and H_2 greater than 5%. The use of the explosibility diagram is demonstrated in Examples 15.1 and 15.2.

Example 15.1   A large coal mine section has been sealed for a number of weeks. The gas analysis behind the seals is as follows:

Carbon dioxide	9.5%	Carbon monoxide	5.3%
Oxygen	2.9%	Hydrogen	5.2%
Nitrogen	71.5%	Methane	5.5%

Determine the ICO value, the TR value, the position of the atmosphere on the explosibility diagram, and the implications of the results.

Solution: Summing the gaseous components gives a value of 99.9%, indicating that all other gases are negligible in their percentage. The calculations can proceed assuming that all other gases have 0% content in the atmosphere:

ICO value (Eq. 15.1):

$$ICO = \frac{100(5.3)}{0.265(71.5 + 0.0) - 2.9} = 33.0$$

This indicates an active fire still exists behind the seals.

TR value (Eq. 15.2):

$$TR = \frac{9.5 + 0.75(5.3) - 0.25(5.2)}{0.265(71.5 + 0.0) - 2.9} = 0.904$$

This indicates that coal, wood, oil, or conveyor belting is burning. Explosibility diagram parameters (Eqs. 15.3–15.5):

$$E_c = 5.5 + 1.25(5.2) + 0.4(5.3) = 14.12\%$$

$$E_i = 71.5 - (3.8)(2.9) + (1.5)(9.5) = 74.73\%$$

$$R_{mc} = \frac{5.5}{5.5 + 5.2 + 5.3} = 0.34$$

These three values identify a point on the explosibility diagram that indicates that the atmosphere behind the seals is explosive if mixed with air.

Example 15.2 A large coal mine has been sealed for a number of months. Its atmosphere, as measured on a gas chromatograph, is determined to be as follows:

Hydrogen	0%	Carbon monoxide	0.0196%
Carbon dioxide	2.43%	Methane	10.03%
Oxygen	7.16%	Ethane	0.14%
Nitrogen	80.22%	Other hydrocarbons	0%

(a) Determine the ICO value, the TR value, and the position on the explosibility diagram.
(b) State the implication of each of these calculations.
(c) Provide a conclusion as to whether the mine should be reopened.

Solution: The calculations should begin with an analysis of the gas percentages. The sum of the values is 99.999%, which gives confidence in the individual values. The remaining calculations are then carried out as follows.
(a) Using Eq. 15.1,

$$ICO = \frac{100(0.0196)}{0.265(80.22 + 0.0) - 7.16} = 0.139$$

Using Eq. 15.2,

$$TR = \frac{2.43 + 0.75(0.0196) - 0.25(0.0)}{0.265(80.22 + 0.0) - 7.16} = 0.173$$

Using Eq. 15.6 to obtain the value of the effective combustible,

$$E_c = 10.03 + (1.25)(0) + (0.4)(0.0196) + (0.6)(0.14) + (0.54)(0)$$

$$= 10.12\%$$

Using Eq. 15.4,

$$E_i = 80.22 - (3.8)(7.16) + (1.5)(2.43) = 56.66\%$$

Using Eq. 15.5,

$$R_{mc} = \frac{10.03}{10.03 + 0 + 0.0196} = 0.998$$

(b) The implication of the ICO value is that the fire is probably under control. The value of Trickett's ratio also indicates that the fire is most likely under control. The location on the explosibility diagram, however, places the mine atmosphere in the area that indicates that the atmosphere is explosive if mixed with air.

(c) The mine should not be opened immediately for two reasons. First, the low TR and ICO values must be proved accurate through continued analysis and interpretation of the trends. Second, the location on the explosibility diagram indicates the mixture is potentially explosive if mixed with air. If the mine is kept sealed, more of the remaining oxygen may be consumed, reducing the explosibility of the mine atmosphere.

The methods used here to evaluate the mine atmosphere are only a few of the many that can be employed. A number of additional atmospheric indices have also been used, both alone and in conjunction with others. Some of the more widely used methods include the Ellicott diagram (Strang and Mackenzie-Wood, 1990, p. 345), the index of carbon dioxide (Timko and Derick, 1991, p. 13), the Coward diagram (Ren and Richards, 1993, p. 584), the Drekopf method (Ramlu, 1991, pp. 257–269), and Willett's ratio (Banerjee, 1985, p. 75).

Use of Inert Gases

The use of inert gases in an attempt to control mine fires, often termed *inertization,* is intended to reduce the percentage of oxygen in the mine atmosphere. The introduction of inert gas to flush a fire zone while maintaining normal ventilation is generally termed *local inertization;* filling a sealed area with inert gas is called *area* or *zone inertization* (Ramlu, 1991, pp. 66–67). The technique has been used in eastern Europe, India, and the United States fairly frequently. Inert gases may be helpful in firefighting activities, in gob fire extinguishing, and in recovering sealed mine areas by more quickly controlling combustion.

The most common gases used for inertization are nitrogen and carbon dioxide. Mitchell (1990b, pp. 133–167) outlines the properties of each and the methods of using them in fire-control activities. Nitrogen is the most commonly used gas for major mine fires. It is close to air in specific weight and thus mixes well with air in a fire zone. In addition, it will not dissolve in water and is not absorbed by coal or coke on the mine surfaces. One ton (907 kg) of liquid nitrogen produces about 27,600 ft^3 (782 m^3) of gas at 1 atm (101.3 kPa) and 60°F (15.6°C). This high volume of gas per ton is one of its advantages. Carbon dioxide has a specific weight of 0.1234 lb/ft^3 (1.98 kg/m^3), making it considerably denser than air. Thus it is quite useful in fighting fires that are lower in elevation than the point of injection. In addition, it will flow to points of lower elevation even when the ventilation air is moving in the opposite direction. One ton of liquid carbon dioxide expands into 17,200 ft^3 (487 m^3) of gas at 1 atm (101.3 kPa) and 60°F (15.6°C). The gas will degrade rubber hoses and is soluble in water. In wet mines, some of the CO_2 will be lost as a result of wet ribs and roof or standing water in the mine openings. Details of the injection techniques for N_2 and CO_2 are available in the text by Mitchell (1990b).

One additional strategy is the use of spent combustion gases as an inerting agent. This method has been used in Poland, Russia, and Czechoslovakia. The basic system consists of a jet engine powered by kerosene and a water-injection system to cool the exhaust (Strang and Mackenzie-Wood, 1990, pp. 318–319). The jet engine also generates power to operate the pumps necessary for fueling and water cooling. This method provides the cheapest form of inert gas, based on the operating cost, but the equipment and a system for mobilizing it costs about $1 million.

Mine Recovery

The proper choice of time for the recovery of a mine following a period of isolation behind seals is the most important decision to be made in the recovery operation. In most cases, the sealed area of the mine has been monitored, and the progression of the fire ratios has been observed. When the atmosphere behind the seals has stabilized and the ratios are favorable, planning for the recovery operation can begin. However, a variety of conditions should exist before the reentry can begin. These have been outlined by Forbes and Grove (1948), Mitchell (1990b), and Ramlu (1991). Some of the more important of these are as follows:

1. The oxygen content of the sealed area should be low enough to make an explosion impossible.
2. The carbon monoxide should have nearly disappeared.

3. The temperature of the fire area should have been reduced to minimize the chance of rekindling.

In assessing the possibilities for rekindling the fire or igniting an explosive mixture of gases, several additional variables of the fire, mine, seal, and barometric conditions must be taken into account. These are outlined in some detail in the three references just cited.

When the decision has been made to attempt a recovery of the mine, the next decision will be to decide on the method of recovery. Two methods of recovery are generally used: reventilation and airlocking. Reventilation is the simplest, easiest, safest, and least expensive way to recover a sealed area (Mitchell, 1990b, p. 121). The reventilation procedure is to open a return seal, then open an intake seal. All personnel are then evacuated, the fan is started, and a waiting period is observed to allow the fan to clear the sealed area. If no rekindling occurs, the mine can then be reentered without danger to the personnel.

The disadvantage of the reventilation method is that small areas of smoldering fuel can reignite the fire. Using airlocking, these smoldering fire remains can often be extinguished by mine exploration personnel. Airlocking is a strategy for recovering the mine using a base station and standard rescue team technique. The procedure starts from the intake seal where an airlock is constructed and recovers the mine on a step-by-step basis. The airlock is used to limit the airflow into the sealed area to keep its atmosphere below the lower explosive limit. As exploration is advanced, new airlocks are built inby the fresh-air base to establish new base stations. The mine is systematically recovered in blocks using this method of advance. The reference by Mitchell (1990b) is particularly useful in setting up a strategy or plan for a recovery operation.

PROBLEMS

15.1 A mine fire has occurred, and the mine has been sealed. Analysis of the mine atmosphere inside the seals yields these results:

CO_2	11.0%	H_2	3.1%
O_2	8.1%	C_2H_4	0.2%
N_2	72.6%	C_2H_6	0.5%
CH_4	2.1%	CO	2.4%

Determine the following:
(a) The position of the mine atmosphere on the USBM explosibility diagram using Eqs. 15.3–15.5.
(b) The position of the mine atmosphere on the USBM explosibility diagram using Eqs. 15.4–15.6.

(c) Does the substitution of Eq. 15.6 for Eq. 15.3 make a big difference in the position on the explosibility diagram?

15.2 A mine has been sealed for some time after a fire. The atmosphere has stabilized and a typical analysis is

CO_2	1.0%	CH_4	3.7%
O_2	17.1%	H_2	0.0%
CO	0.03%	N_2	78.2%

Determine:
(a) Graham's ratio (ICO) and its implication,
(b) Trickett's ratio and its implication,
(c) The position on the USBM explosibility diagram and its implication.

15.3 An area of a mine has been sealed because of a fire. After some time, the atmosphere is rather stable and has the following analysis:

CO_2	15.0%	H_2	4.7 %
O_2	2.6%	N_2	74.0%
CO	1.1%	C_2H_4	0.03%
CH_4	3.8%	C_2H_6	0.27%

Use the explosibility diagram to determine whether work should go on in the mine outside the seals.

15.4 A large section of a mine has been sealed after a fire occurred. After a period of monitoring the atmosphere behind the seals, the analysis is

CO_2	10.0%	CH_4	1.1%
O_2	9.2%	H_2	3.1%
CO	2.9%	N_2	73.7%

The mine manager wishes to reopen the other parts of the mine for construction and production work while the sealed area continues to cool down and achieve a stable atmosphere. What would you recommend? What logical basis do you use for your recommendation?

PART IV

MINE AIR CONDITIONING

CHAPTER 16

Heat Sources and Effects in Mines

16.1 NEED FOR AIR CONDITIONING IN MINES

Temperature–humidity control, one of the three functions of total mine air conditioning, is essentially heat control. It consists of those processes that are designed to regulate the sensible- and/or latent-heat content of the air: heating, cooling, humidification, and dehumidification. Temperature–humidity control is akin to quality control, in that it pertains to the physical quality of the air, whereas quality control pertains to the chemical quality of the air (Section 1.2).

In keeping with current mining usage and the nomenclature adopted in this book, *air conditioning* is the term applied to temperature–humidity control processes, and most commonly to cooling and dehumidification. The usual reason for employing air conditioning in mines is for *comfort* rather than *product* or process purposes. The heat content of the mine air is maintained within limits prescribed for the comfort, safety, and working efficiency of human beings. Occasionally, product air conditioning is employed, as in coal mines where slaking of the roof in warm, moist, summer air, or in salt mines where excessive absorption of moisture by the mineral product may constitute environmental problems.

Mine air conditioning for temperature–humidity control becomes necessary when ventilation alone is inadequate to maintain acceptable atmospheric-heat standards. If required, this aspect of total air conditioning supplements rather than replaces ventilation and quality control. The number of mines and mining districts finding it necessary to condition air, although still small, has risen sharply in the last few decades. Air conditioning can be expected to play an increasingly important role in mining under the increasingly hostile environmental conditions now being encountered underground.

Cooling

As discussed in Chapter 1, the ultimate and eventual limitation in all mining is depth; and, along with pressure, the most serious problem attendant with depth is heat. Cooling and dehumidification processes have become necessary in providing bearable environmental conditions in mines of great depth [over 1 mi (1.5 km)] and in some of only moderate depth, especially those that are extensively mechanized or in zones of high wall-rock temperatures. The simplest applications are small portable units called *spot coolers,* but large central cooling plants are becoming more common.

Underground air conditioning began in Brazil in 1920 at the Morro Velho gold mine, approximately $1\frac{1}{2}$ mi (2.4 km) deep. In the United States, two copper mines, the Anaconda (at Butte, MT) since inactive, and Magma (at Superior, AZ), were the first to employ air conditioning [both at less than 1 mi (1.5 km) in depth]. More recently, the Homestake gold mine [at Lead, SD, over 8000 ft (2.4 km) deep], uranium mines in New Mexico [over 3000 ft (0.9 km) deep], and lead–zinc–silver mines in the Coeur d'Alene district of northern Idaho [the deepest is over 8000 ft (2.4 km)] have installed cooling plants. All told, deep-mining districts in over 15 countries have had to resort to cooling systems, the best known and most extensive consisting of the deep gold mines [maximum 12,000 ft (3.7 km)] of the Witwatersand, South Africa. In Europe, practically all air conditioning occurs in deep coal mines [3000–5000 ft (0.9–1.5 km)] located in several different countries. Other kinds of mines, worldwide, that make use of cooling and dehumidification include nickel, potash, and salt.

Heating

Relatively shallow mines in cold climates sometimes find it necessary to heat air being taken underground, for comfort reasons as well as for prevention of freezing in intake openings. As with cooling, heating installations range from small local units to large central systems. Mines employing heating are located in countries in high latitudes and/or at high elevations, including metal, nonmetal, and coal mines in the United States, Canada, Scandinavia, and Russia. Localized applications to warm shafts in U.S. coal mines extend as far south as Alabama.

16.2   SOURCES OF HEAT IN MINES

Because cooling and dehumidification are the most critical needs in mine air conditioning, underground heat sources have to be identified and quantified. There are nine potential sources of heat in mines, the first four of which are considered major and capable of creating intolerable environmental conditions:

1. Autocompression
2. Wall rock
3. Underground water

4. Machinery and lights
5. Human metabolism
6. Oxidation
7. Blasting
8. Rock movement
9. Pipelines

In this list and the discussion that follows, heat sources are ranked in approximate order of importance. Since any or all of them may be operative, however, it is important in mine temperature–humidity control both to understand the nature of the heat sources and to be able to calculate or estimate the magnitude of the heat flow from each, broken down into sensible and latent components. For other treatments of heat sources, see Johnson (1992b) and McPherson (1993).

Autocompression

Very similar to the way gas in a compressor reacts, air entering a mine through a shaft is compressed and heated as it flows downward. Autocompression occurs when potential energy is converted to thermal energy. If no interchange in the heat or moisture content of the air takes place in the shaft, the compression occurs adiabatically, with the attendant temperature rise following the adiabatic law:

$$\frac{T_2}{T_1} = \left(\frac{p_2}{p_1}\right)^{(\gamma - 1)/\gamma} \tag{16.1}$$

where T is absolute dry-bulb temperature, p is atmospheric pressure, γ is the ratio of the specific heats of air at constant volume and pressure, and the subscripts 1 and 2 denote initial and final conditions, respectively. Values of γ are 1.402 for dry air and 1.362 minimum for saturated; the exponent $(\gamma - 1)/\gamma$ thus assumes values of 0.287 for dry air and approximately 0.266 for saturated. Theoretical calculation of dry-bulb temperature increases due to autocompression by Eq. 16.1 is usually not warranted, however, because of the nonadiabatic airflow that often prevails in mine shafts (direct calculation of wet-bulb temperature increases is even more complicated). Pickup of both heat and moisture from the wall rock can occur. Autocompression may also be masked by the presence of other heating or cooling sources located in or near the shaft, such as air and water lines, hoisting equipment, or other machinery or electrical facilities.

Reasonable accuracy can be obtained if increases in both dry-bulb and wet-bulb temperatures are estimated by the following relationships:

$$\frac{\Delta t_d}{\Delta Z} = 5.3°F/1000 \text{ ft } (9.66°C/1000 \text{ m}) \tag{16.2}$$

$$\frac{\Delta t_w}{\Delta Z} = 2.4°F/1000 \text{ ft } (4.37°C/1000 \text{ m}) \tag{16.3}$$

where Δt_d is dry-bulb temperature increase, Δt_w is wet-bulb temperature increase, and ΔZ is elevation change, usually expressed per 1000 ft (or 1000 m). Results obtained by calculation (of Δt_d) and estimation (of Δt_d and Δt_w) are compared in Example 16.1.

Example 16.1   (a) Compute the dry-bulb temperature increase by formula, and (b) estimate both dry-bulb and wet-bulb temperature increases due to autocompression of air flowing adiabatically to the bottom of a 5000-ft (1524-m) shaft, if surface temperatures are 80°F (26.7°C) dry-bulb and 60°F (15.6°C) wet-bulb. Elevation of the shaft collar $Z = 5000$ ft (1524 m).

Solution: (a) Using Eq. 16.1, assume $(\gamma - 1)/\gamma = 0.276$, an average value, and read $p_1 = 12.22$ psi (at 5000 ft elevation) and $p_2 = 14.70$ psi (at sea level) from Appendix Table A.1:

$$T_2 = (460 + 80)\left(\frac{14.70}{12.22}\right)^{0.276} = 568.2°R$$

$$t_{d_2} = 108.2°F$$

The increase is

$$\Delta t_d = 108.2 - 80 = 28.2°F \ (15.7°C)$$

or

$$\frac{\Delta t_d}{\Delta Z} = \frac{28.2}{5000/1000} = 5.64°F/1000 \text{ ft } (10.28°C/1000 \text{ m})$$

(b) Estimate increases in dry-bulb and wet-bulb by Eqs. 16.2 and 16.3:

$$\Delta t_d = (5000)\left(\frac{5.3}{1000}\right) = 26.5°F$$

$$t_{d_2} = 80 + 26.5 = 106.5°F \ (41.4°C)$$

$$\Delta t_w = (5000) \left(\frac{2.4}{1000} \right) = 12.0°F$$

$$t_{w_2} = 60 + 12.0 = 72.0°F \ (22.2°C)$$

The value of $\Delta t_d / \Delta Z$ determined in (a) agrees reasonably well with that assumed in (b).

If the heat flow q in Btu/h (W) rather than the temperature difference is desired in cooling–dehumidification computations, enthalpies for the state points involved (1 and 2) may be calculated by psychrometric equations or read from the psychrometric chart, as explained in Chapter 2.

Autocompression—along with wall-rock heat—constitutes one of the two principal sources of heat in mines. It can, of course, be eliminated from the mine cooling load (Section 17.1) by locating the air conditioning plant underground, but heat addition to the air then occurs prior to its entry into the plant. During exhaust to the surface in the upcast shaft, the air temperature drops because of decompression by exactly the same amount it increases during autocompression, but unfortunately without benefit to the air conditioning system.

Wall Rock

Geothermal Gradient   The temperature of subsurface rock rises steadily with depth. In most climates, the so-called *virgin-rock temperature* t_r ceases to be affected by surface temperature changes and is taken as constant for reference purposes at about 50 ft (15 m) beneath the surface. It then increases with depth at approximately a uniform rate in a given locale and rock formation; the rise is termed the *geothermal gradient,* or the temperature change per unit depth $\Delta t / \Delta Z$. However, even within a single mining district, it varies and seldom can be considered a constant (Morris, 1953; Jones, 1972). The age of the rocks, their thermal properties, and their proximity to recent igneous activity or hot springs largely determine the gradient.

Although it varies in different mining districts and can be determined accurately only by field measurement, the earth's *heat flux* q/A (heat flow per unit area) ranges from 0.013 to 0.019 Btu/h·ft² (0.04–0.06 W/m²) and averages 0.016 Btu/h·ft² (0.05 W/m²). Knowing the heat flux and the thermal conductivity k (in Btu/h·ft·°F or W/m·°C) of the various rock formations encountered, the geothermal gradient in °F/100 ft (°C/100 m) for a given formation can be calculated from conductive heat-transfer theory (Whillier, 1974a):

$$q = kA \frac{\Delta t}{\Delta Z} \tag{16.4}$$

Rearranging,

$$\frac{\Delta t}{\Delta Z} = \frac{q/A}{k} \tag{16.5}$$

The overall gradient can then be determined by cumulating the gradients for the individual formations. Note in Eqs. 16.4 and 16.5 that although the heat flow varies directly with thermal conductivity, the geothermal gradient varies inversely with k.

Example 16.2   A limestone has a thermal conductivity of 1.2 Btu/h·ft·°F (2.08 W/m·°C), a relatively low value. Calculate the geothermal gradient for the formation and the virgin-rock temperature at 4500-ft (1372-m) depth, if the surface rock temperature t_0 = 70°F (21.1°C). Assume an average heat flux.

Solution: Find the geothermal gradient by Eq. 16.5:

$$\frac{\Delta t}{\Delta Z} = \frac{0.016}{1.2} \times 100 = 1.33°\text{F}/100 \text{ ft } (2.42°\text{C}/100 \text{ m})$$

which is higher than average [about 1°F/100 ft (1.8°C/100 m)]. Temperature gain is Δt = (1.33/100) × 4500 = 60.0°F (33.3°C); and at 4500 ft, t_r = t_0 + Δt = 70 + 60 = 130°F (54.4°C).

Listed in Table 16.1 are geothermal gradients representative of various world mining districts. Values of the surface rock temperature, which along

Table 16.1   Approximate Geothermal Gradients and Surface Rock Temperatures for Different Mining Districts

Mine or District	Temperature Rise		Surface Rock Temperature	
	°F/100 ft	(°C/100 m)	At 50 ft, °F	(At 15 m, °C)
Anaconda copper, Montana	2.5–3.3	(4.6–6.0)	40	(4)
Magma copper, Arizona	2.1	(3.8)	65	(18)
Coal mines, United Kingdom	1.0–2.2	(1.8–4.0)	50	(10)
Agnew nickel, Australia	0.7	(1.3)	77	(25)
Kolar gold, India	0.6	(1.1)	75	(24)
Witwatersrand gold, S. Africa	0.4–0.7	(0.7–1.3)	65	(18)
McIntyre gold, Canada	0.4	(0.7)	40	(4)

Sources: Forbes et al. (1949), Morris (1953), and unpublished data.

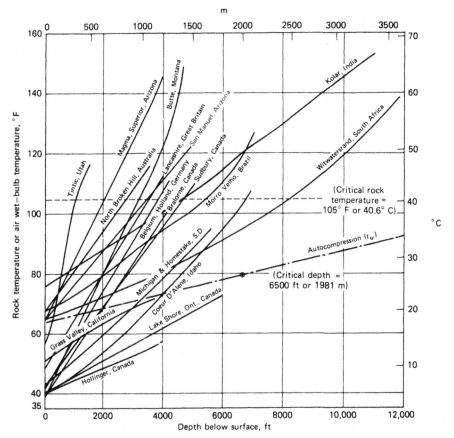

FIGURE 16.1 Average geothermal gradients of various worldwide mining districts and effects of autocompression on air wet-bulb temperature (assumed surface t_w = 64°F or 18°C, and Δt_w = 2.4°F/1000 ft or 4.4°C/1000 m), both as a function of depth below surface. Critical rock temperature and critical depth for t_w are also shown. (After Fenton, 1972; Enderlin, 1973.)

with the geothermal gradient are critical in fixing mine temperatures, are also given in the table. For example, the surface temperature is 25°F (13.9°C) higher at the Magma mine than at the Anaconda; and thus, in spite of its lower geothermal gradient, Magma must contend with higher rock temperatures underground.

Geothermal gradients for various world mining districts are plotted in Fig. 16.1. Also superimposed for comparison purposes is the autocompression effect based on wet-bulb temperature [assume surface t_w = 64°F (18°C) and $\Delta t_w / \Delta Z$ = 2.4°F/1000 ft (4.4°C/1000 m)]. Considering only autocompression, and imposing an air threshold limit value (TLV) of t_w = 79°F (26°C), it is evident from Fig. 16.1 that at an approximate depth of 6500 ft (1980 m), autocompression heating effects alone dictate that air conditioning must be installed. This so-called *critical depth* occurs because the ventilation air, heated by autocompression, has no remaining cooling potential (Pretorius et al., 1980). While these values are only illustrative, and the critical depth varies with the surface wet-bulb temperature, the relentless influence of auto-compression—independent of rock heat—is compelling.

Still, the rock mass in which the mine opening is located is almost always the major source of heat underground. The *critical temperature* of the rock (see Fig. 16.1), and its associated depth at which rock heat alone governs and air conditioning becomes essential for the hottest and poorest-ventilated workings of the mine, is considered to be about 105°F (41°C) (Whillier, 1972). It is evident from Fig. 16.1 that that temperature is reached in many mines at depths considerably less than the critical depth because of autocompression alone.

Sensible-Heat Flow   Heat transfer from the wall rock to the ventilation air in a mine opening is unsteady and complex. The mechanism is illustrated in Fig. 16.2. Initially, when the opening is being advanced, heat flow into the mine is at a very high rate. Subsequently, the layer of rock adjacent to the opening cools to essentially the temperature of the air in the opening [within 1–3°F (0.6–1.7°C)], thus retarding heat transfer. But in mechanized mining, with rapidly advancing workings, heat flow remains high at the face where the greatest number of miners is exposed.

To minimize the heat flow from rock to air, it is essential that (1) the length of airway and (2) the temperature differential between rock and air be kept to absolute minima. The design and layout of the air conditioning plant and mine plan with regard to operating temperatures and distances are thus critical. Another key factor is (3) moisture, which, if present, exacer-bates heat transfer by lowering heat-transfer resistance at the interface and lowering air dry-bulb temperature. All other considerations, such as airway size and shape, air velocity, rock surface irregularities, and nature of the air–rock interface, are secondary.

Attempts to insulate the walls of mine openings to alter the heat-transfer coefficient and reduce rock heat inflow have to date had little application or success (Hinsley, 1943/44; Whillier, 1974a). The reason has to do with resistance to heat transfer, which, as Whillier points out, is far greater within the rock itself than at the air–rock interface. It is much more beneficial to reduce the wetness of the rock surface than to insulate it.

In addition to heat flow from the wall rock, there is also a heat addition from the rock broken in mining. It may be substantial, up to 60% of the total rock heat inflow into an airway, especially in high-production bulk mining (Gracie and Matthews, 1975/76). The heat flow from broken rock can be approximated if the volumetric heat capacity of the muck and the tempera-ture difference are known (Whillier, 1974a).

Heat transfer into mine openings has been studied in the United States

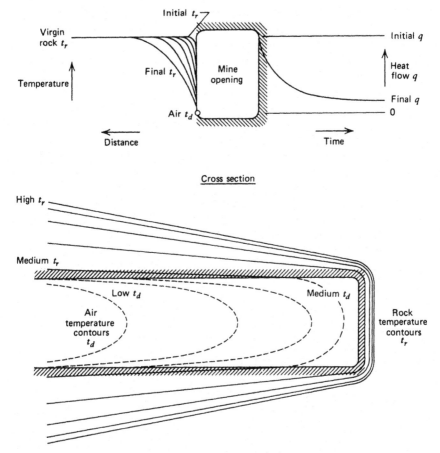

FIGURE 16.2 Idealized representation of heat flow into a mine opening from surrounding rock, together with resulting temperature distribution. Opening is being advanced to the right.

and abroad for many years (Carrier, 1938; Jordan, 1965; Whillier and Ramsden, 1975; Whittaker, 1980; McPherson, 1993). Because it is unsteady, exact mathematical solution is not possible. The irregular shape of most mine openings, the complexity of the air-rock interface, the usual presence of water, and unknown and continually changing temperature distributions in both rock and air account for the difficulty of analysis. Rock thermal properties must also be determined with some precision (Tucker, 1968; Vost, 1973). To permit solutions to be made numerically or graphically, simplifying assumptions are required, namely to employ conductive heat-transfer theory within the rock and to approximate the convective process within the airway itself (Whillier, 1974a).

To calculate wall-rock heat transfer, the virgin-rock temperature of the workings must first be known, along with the thermal properties of the rock. Virgin-rock temperature may be estimated and projected by Eq. 16.5, based on data obtained from drillhole temperature measurements (Chapter 6). Then a plot is prepared for the affected levels of the mine, similar to Fig. 16.1. Even assuming that the rock thermal conductivity is constant, the heat flow is considerably higher during the initial period after a mine opening is excavated than several years later when steady-state conditions have developed. In young mining areas, heat inflow must be computed using transient-heat-flow techniques. A graph of heat flow from wall rock to air versus time can be plotted, calculating the heat flow at various times (see Fig. 16.2, cross section). The point at which the heat flow assumes near-steady state can be determined from the graph where the curve begins to flatten out.

Solutions in use for wall-rock heat flows into mine openings derive from Fourier equations for conductive heat transfer. One of the most widely used is a numerical solution by Goch and Patterson (1940), presented in tabular form. It involves the determination of two coefficients, one for rock thermal properties and the other for rock heat flow. The resulting equation for sensible heat flow is expressed in the form

$$q = S \frac{k}{r_e} (t_r - t_d)\omega \tag{16.6}$$

where S is surface area of opening in ft^2 (m^2) $= O \cdot L$, O is perimeter in ft (m), L is length in ft (m), r_e is hydraulic radius of opening modified for roughness in ft (m), and ω is the Goch–Patterson heat flow coefficient (dimensionless). The value of hydraulic radius modified for roughness by multiplication by $\sqrt{2}$ is calculated as

$$r_e = \sqrt{2} \frac{A}{O} \tag{16.7}$$

where A is cross-sectional area in ft^2 (m^2).

To obtain the heat flow coefficient, reference is made to Table 16.2 where values of ω are presented as a function of the rock thermal properties coefficient ϵ, also dimensionless:

$$\epsilon = \frac{\alpha\tau}{r_e^2} \tag{16.8}$$

where α is rock thermal diffusivity in ft^2/h (m^2/s), and τ is time in h (s). The diffusivity, in turn, reflects other thermal properties and is found from the relation

$$\alpha = \frac{k}{w_r c_r} \tag{16.9}$$

Table 16.2 Coefficients for Calculating Heat Flow into Mine Airways[a]

ϵ	ω	ϵ	ω	ϵ	ω	ϵ	ω
0.010	6.1289	0.125	2.0571	1.60	0.8536	28	0.4317
0.011	5.8658	0.130	2.0255	1.70	0.8387	30	0.4261
0.012	5.6362	0.135	1.9957	1.80	0.8250	35	0.4140
0.013	5.4336	0.140	1.9674	1.90	0.8123	40	0.4040
0.014	5.2531	0.150	1.9152	2.00	0.8006	45	0.3955
0.015	5.0910	0.160	1.8679	2.20	0.7795	50	0.3882
0.016	4.9442	0.170	1.8248	2.40	0.7609	55	0.3818
0.017	4.8106	0.180	1.7853	2.60	0.7444	60	0.3761
0.018	4.6883	0.190	1.7489	2.80	0.7296	65	0.3710
0.019	4.5757	0.200	1.7152	3.00	0.7162	70	0.3664
0.020	4.4716	0.220	1.6548	3.20	0.7040	75	0.3622
0.022	4.2852	0.240	1.5020	3.40	0.6929	80	0.3583
0.024	4.1225	0.260	1.5554	3.60	0.6827	85	0.3548
0.026	3.9790	0.280	1.5138	3.80	0.6732	90	0.3515
0.028	3.8510	0.300	1.4763	4.00	0.6644	95	0.3484
0.030	3.7360	0.320	1.4423	4.50	0.6449	100	0.3456
0.032	3.6320	0.340	1.4113	5.00	0.6282	110	0.3404
0.034	3.5373	0.360	1.3829	5.50	0.6137	120	0.3358
0.036	3.4505	0.380	1.3567	6.00	0.6009	130	0.3316
0.038	3.3707	0.400	1.3325	6.50	0.5895	140	0.3278
0.040	3.2968	0.420	1.3100	7.00	0.5793	150	0.3244
0.042	3.2283	0.440	1.2890	7.50	0.5700	160	0.3212
0.044	3.1645	0.460	1.2694	8.00	0.5615	180	0.3157
0.046	3.1049	0.480	1.2510	8.50	0.5538	200	0.3108
0.048	3.0491	0.500	1.2336	9.00	0.5467	220	0.3065
0.050	2.9966	0.550	1.1944	9.50	0.5401	240	0.3027
0.055	2.8781	0.600	1.1600	10.00	0.5339	260	0.2993
0.060	2.7746	0.650	1.1296	11.00	0.5228	280	0.2962
0.065	2.6832	0.700	1.1025	12.00	0.5130	300	0.2933
0.070	2.6018	0.750	1.0780	13.00	0.5042	350	0.2872
0.075	2.5287	0.800	1.0558	14.00	0.4963	400	0.2820
0.080	2.4624	0.850	1.0355	15.00	0.4892	450	0.2776
0.085	2.4020	0.900	1.0169	16.00	0.4826	500	0.2738
0.090	2.3467	0.950	0.9997	17.00	0.4766	600	0.2674
0.095	2.2959	1.000	0.9838	18.00	0.4711	700	0.2622
0.100	2.2488	1.100	0.9552	19.00	0.4660	800	0.2579
0.105	2.2050	1.200	0.9301	20.00	0.4611	900	0.2542
0.110	2.1643	1.300	0.9078	22.00	0.4524	1000	0.2510
0.115	2.1263	1.400	0.8879	24.00	0.4448		
0.120	2.0906	1.500	0.8699	26.00	0.4379		

Source: Goch and Patterson (1940). By permission from South African Institute of Mining and Metallurgy, Marshalltown.

[a] Values are the same in English and SI units (dimensionless).

Table 16.3   Measured Values of Rock Thermal Properties

Location	Rock Type	Thermal Conductivity k		Thermal Diffusivity α	
		Btu/h·ft·°F	(W/m·°C)	ft²/h	(10^{-6} m²/s)
Ambrosia Lake, New Mexico	Sandstone (uranium-bearing)	1.39	(2.40)	Not given	
Broken Hill, Australia	Gneiss	1.84	(3.18)	0.064	(1.65)
	Greenstone	1.83	(3.16)	Not given	
Butte, Montana	Granite (fresh)	1.51	(2.61)	0.068	(1.75)
	Granite (silicified)	1.97	(3.41)	Not given	
	Rhyolite	1.03	(1.78)	Not given	
	Copper ore	2.46–2.84	(4.25–4.91)	Not given	
Coal measures, Great Britain	Mudstone	0.40	(0.69)	0.013	(0.34)
	Siltstone	0.60	(1.04)	0.010	(0.26)
	Shale	1.15	(1.99)	Not given	
Grass Valley, California	Granodiorite	1.84	(3.18)	Not given	
Moffat Tunnel, Adams, Colorado	Granite	1.89	(3.27)	Not given	
Mount Isa, Australia	Chloritic siltstone	3.40	(5.88)	Not given	
	Quartzite	4.03	(6.96)	Not given	
	Dolomite shale (fractured)	1.27	(2.20)	Not given	
Superior, Arizona	Dacite	1.18	(2.04)	0.039	(1.01)
	Conglomerate	1.66	(2.87)	0.042	(1.08)
	Limestone	1.82	(3.15)	0.049	(1.26)
Witwatersrand, South Africa	Quartzite	3.15	(5.45)	0.0253	(0.65)

where w_r is rock specific weight in lb/ft³ (kg/m³), and c_r is rock specific heat in Btu/lb·°F (kJ/kg·°C).

In computing wall-rock heat flow, thermal properties of the rock are determined by laboratory or field tests or approximated from Table 16.3. After calculating the diffusivity and the coefficient ϵ, the corresponding value of ω is selected by interpolation from Table 16.2. The heat flow is then calculated by Eq. 16.6. An example illustrates the procedure.

Example 16.3   Calculate the rock heat flow in a 100-ft (30-m) segment of drift being driven in Superior, AZ limestone, given the following conditions:

Virgin-rock temperature	140°F (60°C)
Drift size	10 × 12 ft (3.0 × 3.7 m)
Entering air temperatures	dry-bulb 75°F (23.9°C)
	wet-bulb 70°F (21.1°C)
Barometric pressure	30 in. Hg (762 mm Hg)
Time period	7 mo.

Solution: From Table 16.3, read values of thermal properties of rock: $k = $ 182 Btu/h·ft·°F and $\alpha = 0.049$ ft²/h. Using Eq. 16.7,

$$r_e = \sqrt{2}\,\frac{(10 \times 12)}{(2 \times 10 + 2 \times 12)} = 3.86 \text{ ft}$$

Using Eq. 16.8,

$$\epsilon = \frac{(0.049)(7 \times 30 \times 24)}{(3.86)^2} = 16.57$$

From Table 16.2, interpolate

$$\omega = 0.4792$$

Using Eq. 16.6,

$$q = (44 \times 100)\left(\frac{1.82}{3.86}\right)(140 - 75)(0.4792) = 64{,}620 \text{ Btu/h (18.9 KW)}$$

If it is necessary to establish the change in air temperature Δt in a mine opening, the dry-bulb temperature, t_{d_1} at the end of a segment of drift L can be found as follows:

$$t_{d_1} = t_{d_0} + \Delta t = t_{d_0} + \frac{vq}{60\,c_v Q} = t_{d_0} + \frac{q}{Gc_v} \qquad (16.10)$$

where v is specific volume of dry air in ft³/lb (m³/kg), Q is quantity of airflow in cfm (m³/s), G is weight flow rate of air in lb/h (kg/s), and c_v is specific heat of moist air in Btu/lb·°F (kJ/kg·°C). The specific heat is determined from the psychrometric relation

$$c_v = 0.24 + 0.45W \qquad (16.11)$$

$$c_v = 1.005 + 1.884W \qquad (16.11a)$$

where W is specific humidity in lb water vapor per lb dry air (kg/kg). A continuation of Example 16.3 demonstrates this application.

Example 16.4 Find the dry-bulb and wet-bulb temperatures 100 ft (30 m) downstream ($= L$) from the entrance to the drift, described in Example 16.3, given $Q = 40{,}000$ cfm (18.88 m³/s).

Solution: Refer to the psychrometric chart in Fig. 2.2 to find specific humidity and specific volume. For $t_{d_0} = 75°F$ and $t_{w_0} = 70°F$, read $W = 0.01455$ lb/lb and $v = 14.79$ ft^3/lb. Using Eq. 16.11,

$$c_v = 0.24 + 0.45 (0.01455) = 0.2465 \text{ Btu/lb·°F}$$

Using Eq. 16.10,

$$t_{d_1} = 75 + \frac{(14.79)(64,620)}{(60)(0.2465)(40,000)} = 75 + 1.6 = 76.6°F \ (24.8°C)$$

To find t_{w_1}, refer to Fig. 2.2 and, assuming no change in moisture content or latent heat, read

$$t_{w_1} = 70.4°F \ (21.3°C)$$

To compensate for significant surface-moisture effects on wall-rock heat transfer, the Goch-Patterson method has been modified by others (Starfield, 1966; Van der Walt and Whillier, 1979). Because of their complexity, these methods are not utilized here. As an alternative to calculation, Johnson (1992b) proposes using an approximation that increases the air dry-bulb temperature gain by 0.2°F per 100 ft (0.36°/100 m) over the temperature gain in a dry drift, if the water is not too hot. If the water is very hot, an increase of 2°F per 100 ft (3.65°C/100 m) up to a maximum dry-bulb temperature of 110°F (43.3°C) is sufficiently close for estimation purposes. The air wet-bulb temperature is then assumed to be within 2°F (1.1°C) of the dry-bulb temperature at all times in very wet drifts.

Underground Water

Two different water sources are encountered underground: groundwater and mine water. All groundwater, but especially that from hot fissures and natural rock reservoirs, is a prolific source of heat in mine workings (Enderlin, 1973). Since water and heat both are derived from the surrounding rock or geothermal sources, the water temperature will approach or even exceed that of the rock. The water transfers its heat to the mine air mainly by evaporation, increasing the latent heat of the air. Evaporation in airways should, and can, be minimized by grouting the rock, isolating the water inflow, and constructing covered ditches or laying pipe to transport all large flows of water. It is particularly crucial to maintain separation of conditioned air and groundwater if the water temperature is high or exceeds the air wet-bulb temperature. An additional hazard associated with very hot water in mines at temperatures over approximately 125°F (52°C) is burns to the body.

Equally large amounts of latent heat may be added to the mine air through the evaporation of service water supplied for drilling and wetting down and of drainage water from filling operations, so-called mine water, which also is heated by the rock (Sharp, 1967). As a result, careful control over the open-channel flow of waste service water as well as groundwater is necessary in hot mines. On the other hand, the chilling of service water to supply partial air conditioning in working places is an alternative now receiving wide application in mines (Section 17.5).

The gain in latent heat content of the air due to evaporation of moisture from a free-water surface in a mine opening follows a psychrometric process. The evaporation rate is proportional to the difference in vapor pressures or temperatures of the air and water, the air velocity, and the water surface area. Depending on the difference in temperatures, the air may gain (or lose) sensible as well as latent heat.

The total heat gain in the mine air from hot underground water in open-channel flow can be calculated from the relation (McPherson, 1993)

$$q = G_w c_w (t_1 - t_2) \tag{16.12}$$

where G_w is weight flow rate of water in lb/h (kg/s), c_w is specific heat of water = 1 Btu/lb·°F (4.187 kJ/kg·°C) at standard conditions, and t_1 and t_2 are water temperatures in °F (°C) at points of emission and exit from the mine airway, respectively.

Example 16.5   A mine produces 200 gpm (12.62 L/s) of underground water on the deepest level. The water enters the workings at 110°F (43.3°C) and exits at 98°F (36.7°C). Determine the total heat gain to the ventilation system.

Solution: Convert water flow rate from volume to weight basis:

$$G_w = (200)(60) \left(\frac{62.4}{7.48} \right) = 100,000 \text{ lb/h}$$

Using Eq. 16.12,

$$q = (100,000)(1)(110 - 98) = 1,200,000 \text{ Btu/h} \ (352 \text{ kW})$$

To determine the distribution of latent heat q_L and sensible heat q_S transmitted from the mine water to the ventilating air, see the solution developed by McPherson (1993). For estimating purposes, at least 90% may be considered latent heat. Thus in the example preceding, the approximate distribution would be

$$q_L \simeq (1{,}200{,}000)(0.90) \simeq 1{,}080{,}000 \text{ Btu/h (317 kW)}$$

$$q_S \simeq (1{,}200{,}000)(0.10) \simeq 120{,}000 \text{ Btu/h (35 kW)}$$

Machinery and Lights

Nearly all the energy consumption of machinery underground adds heat to the mine air, since the power losses and most of the work done are converted directly to heat or indirectly through friction. This is true of electrical, compressed-air, or internal-combustion (diesel) machinery, although compressed-air machines also exhibit a local cooling effect at discharge. Only that portion of the load of materials-handling equipment used to elevate and raise the potential energy of broken ore or coal accomplishes useful work, and it is sufficiently minor to be neglected. Hence, in practice, the entire energy consumption of underground machinery in the air conditioned zone of the mine is assumed to contribute to the sensible-heat gain, often a very sizable fraction of the mine cooling load, especially in highly mechanized operations (Williams, 1960b).

Calculation of the sensible-heat gain from machinery proceeds from a knowledge of the total power connected in the mine and the load factor. If a single workplace or portion of the mine is under consideration, then only the heat contributed by machinery located there need be considered. Fans should be handled separately, depending on their location with respect to the cooling zone of the mine. If the equipment is diesel-powered, then the heat-gain calculation must allow for the combustion of the fuel, commonly assumed to be 90% of the heat value of the fuel.

Example 16.6 Calculate the heat liberated by machinery (a) in an entire mine having a connected underground electrical load of 2400 kW and a load factor of 70% and (b) in a section of the mine continuously using the following equipment: 8 compressed-air rock drills at 10 hp (7.4 kW) each, 1 electric loader at 150 hp (112 kW), 1 electric hoist at 75 hp (56 kW), and 1 electric locomotive at 125 hp (93 kW).

Solution: (a) For the mine, find the total operating load and convert to heat liberated per hour:

$$q = (2400)(0.70)(3412 \text{ Btu/kWh})$$

$$= 5{,}732{,}000 \text{ Btu/h (1680 kW)}$$

(b) For the section, find the operating load and heat generated:

$$\text{Load} = (8)(10) + 150 + 75 + 125 = 430 \text{ hp}$$

$$q = (430)(2544 \text{ Btu/hp·h}) = 1{,}094{,}000 \text{ Btu/h (321 kW)}$$

Example 16.7 Recalculate Example 16.6(b) for a diesel rather than an electric locomotive. Fuel consumption of the diesel is 8 gal/h (30.3 L/h); heat value of the diesel fuel is 140,000 Btu/gal (38,985 kJ/L).

Solution:

Electrical load = (8)(10) + 150 + 75 = 305 kW
Electrical q = (305)(2544) = 775,920 Btu/h (227 kW)
Diesel q = (8)(140,000)(0.90) = 1,008,000 Btu/h (295 kW)
Total q = 775,920 + 1,008,000 = 1,783,920 Btu/h (523 kW)

If any of the stationary machinery in a mine (e.g., a main fan, pump, or compressor) can be placed at the surface, or if any of the section equipment (e.g., a fan) can be located in an exhaust airway, then the sensible-heat gain in the workplaces can be proportionally reduced. The savings may be substantial: For example, Whillier et al. (1969) estimated that the dry-bulb temperature rise across an axial-flow fan is 0.8°F per 1 in. water (1.8°C/Pa) rated head and 0.4°F per 1 in. water (0.9°C/Pa) across a centrifugal fan.

The entire electrical energy of lights is converted into sensible heat, which, although small, adds to the cooling load of the mine air. Both the central lighting system and individual cap lamps should be considered.

Example 16.8 Calculate the sensible heat liberated by lights: central underground system 30 kW, and 250 cap lamps at 5 W each.

Solution:

$$q = \left(30 + \frac{5 \times 250}{1000}\right)(3412) = 106,600 \text{ Btu/h (31.2 kW)}$$

Other Heat Sources

The five remaining heat sources in mines are grouped here summarily because they are (1) less important quantitatively and (2) can only be approximated numerically (McPherson, 1993, p. 566). Sometimes the most valid approach is to conduct a heat balance of the mine, subtracting all the known heat sources from the measured heat gain in order to isolate the unknown source (Section 17.4).

Human Metabolism The body's waste heat is continually being rejected by heat-transfer processes. The result is an increase in both the sensible- and latent-heat contents of the mine air, in small to moderate amounts. Calculation of the gains requires knowledge of the miners' activity and attire and the ambient conditions, as is demonstrated in Section 16.5.

Oxidation   Oxidation processes involving the mineral, backfill, and timber in mines contribute heat to the mine air. In coal mines, under unusual circumstances, it is reported that this source may constitute 80% of the total load (Forbes et al., 1949). In ore mines having a high sulfide content, the addition of heat may also be considerable. Particularly in filled stopes where timber, setting cement, and/or sulfides are present, heat liberation may be intensive enough to cause spontaneous combustion and to constitute a fire hazard, raising the air temperature by many degrees. Spontaneous combustion and fires, when they occur underground, generate inordinate amounts of heat and, if not extinguished or controlled promptly, have disastrous consequences on the mine atmosphere and human life (see Chapter 15).

At present, there is no effective way to calculate the amount of heat produced by oxidation processes. In some high-sulfide ore and coal mines, this can be significant, and in such cases, the mine cooling load has to be increased by an appropriate estimated amount.

Blasting   Since over 50% and perhaps 90% of the energy released by the detonation of a high explosive is liberated in the form of heat, blasting on occasion can be a significant heat source. If the blasting is done on-shift, especially in development openings, the ventilation heat gain locally is considerable. The exact amount of heat added to the mine air cannot be calculated directly, since most is probably absorbed by the rock (Hemp and Deglon, 1980). Fenton (1972) has developed a method to approximate the heat gain. Blasting off-shift or during lunch hour is an effective way of reducing or eliminating this factor from consideration in exposure of miners to heat gains in workplaces.

Rock Movement   Movement of ground due to geologic causes or mining subsidence is another heat source that is difficult to quantify. Caving or collapse of waste or ore in stopes or abandoned areas is the most common cause of heat liberation due to ground movement. The problem can be approached if the mass, distance, and friction factor involved are known, but an exact calculation is not possible. It is doubtful that even 1% of the heat generated by rock movement actually enters the airstream, since much is absorbed in the rock mass itself, as in blasting; hence it is usually negligible.

Pipelines   Lines carrying drainage water are frequently hotter than the mine air and thus contribute some heat. Drainage water is usually the only warm water that is found in underground pipelines; service water and sand-fill-slurry water are normally at or slightly below drift air temperature. Chilled-water lines or service-water lines from cold-water sources will be below the air temperature of the workings and will remove heat from the air, which is normally considered a waste of cooling. In all cases where underground water pipelines operate at significant temperature differences from the mine air, insulation is required.

Heat-transfer calculations for pipelines are difficult, and gains or losses are generally neglected (Whillier, 1974a). Compressed-air lines are generally at about the same temperature as the air in the workings, as the compressor aftercoolers reduce the temperature to ambient, and the compressed-air temperature equalizes with the various ambient temperatures as it moves through the mine.

Energy Losses in Airflow

The flow of air through mine openings is accompanied by head losses and the dissipation of energy as friction and shock losses are overcome. This energy loss might be expected to generate heat, but such is not the case (Whillier, 1974b). Rather, from the general energy equation (5.1), it can be demonstrated that the change in enthalpy and temperature must be zero; only the entropy increases as the flow potential decreases. Thus airflow head losses are not reflected in any addition of heat to the mine air.

16.3 PHYSIOLOGICAL EFFECTS OF HEAT AND HUMIDITY

Body Adaptation to Heat and Cold

The human body employs a remarkable thermostatic control system for regulating body heat and holding body temperature nearly constant at 98.6°F (37.0°C). This mechanism maintains heat-loss and heat-gain processes in balance to prevent harmful thermal effects (heat strain) to the body.

Control for the body's heat regulation is provided by the nervous system. When sensory nerves near the surface of the skin detect a change in environmental conditions or body temperature, they act like a thermostat and transmit an impulse to sympathetic nerves controlling the functioning of the circulatory system and the sweat glands in the body. The body responds by dissipating more or less heat, which is conveyed by the bloodstream from the deep tissues (body core) to the skin for dissipation. Core and skin temperatures usually must differ by 2°F (1°C) or more for effective heat transfer to occur.

Metabolism is the source of all heat produced within the body (muscle flexure is a negligible contributor). Oxidation and other chemical processes continually generate heat while digesting food and oxygenating blood. This is termed *basal metabolism* and is measured as the amount of heat liberated by a person at rest in a comfortable environment [about 400 Btu/h (115 W)]. In addition, waste heat [400–2000 Btu/h (115–585 W), or more] is produced during physical exertion. A mechanism of the body that may accumulate or give up heat as needed to maintain the body temperature constant is termed *storage*. At low ambient temperatures, storage gives up heat to keep the body warm, and at high temperatures, takes up heat to keep the body cool. If overloaded, the storage function eventually breaks down, permitting the body temperature to change and possible heat strain and damage to occur.

At work, the human body is a heat engine, and not a very efficient one (perhaps 0–25%, and on average 10%, productive work is accomplished). The waste heat it produces and heat in the ambient air may be exchanged by conventional heat-transfer processes: *convection, radiation,* and/or *evaporation* (of body fluids, mostly sweat). Heat flows from the warmer mass to the cooler mass, and transfer by any or all of these processes is possible.

A basic heat-balance equation, developed by the American Society of Heating, Refrigerating & Air Conditioning Engineers (ASHRAE), may be written for the human body to express heat changes, setting the heat losses ($+$) and gains ($-$) equal to metabolism less mechanical work accomplished (Anon., 1993):

$$M - W_k = q_e + q_r + q_c + q_s \qquad (16.13)$$

where M is metabolism, W_k is work, q_e is evaporation, q_r is radiation, q_c is convection, and q_s is storage, all measured in units of Btu/h (W). Work is positive when accomplished by the body, such as climbing a ladder, and existent in the form of potential energy; it must be subtracted from metabolism to find net body-heat production (work is negative when potential energy is added, such as walking down steps). Symbolically, the heat balance for the human body can be represented as in Fig. 16.3.

Heat losses and gains from the human body have been measured using a room-sized calorimeter, with $q_r + q_c$ combined (Anon., 1993). Experimental results over a wide range of dry-bulb temperatures with little air movement and normal moisture content (about 45% relative humidity) are plotted in Fig. 16.4 for a clothed male at rest. Note that the neutral point for storage ($q_s = 0$) occurs at approximately 83°F (28°C). It is possible to write the heat-balance equation using numerical values taken from this graph, and to determine the heat gain to the atmosphere.

Example 16.9 On the basis of Eq. 16.13, write the heat–balance equation at 50°F (10°C) dry-bulb temperature for a clothed miner at rest, indicating correct algebraic signs and reading heat flows from Fig. 16.4. What is the net heat flow?

Solution:

$$M - W_k = q_e + q_r + q_c + q_s$$
$$420 = 60 + 650 - 290$$
$$420 = 420$$

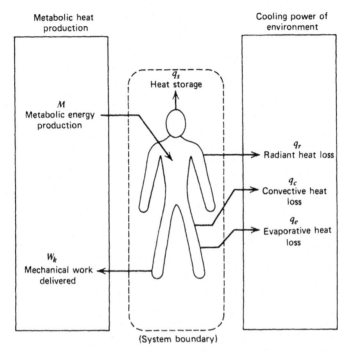

FIGURE 16.3 Heat balance of the human body. (After Stewart, 1980. By permission of publisher, SME-AIME, New York. Copyright 1980.)

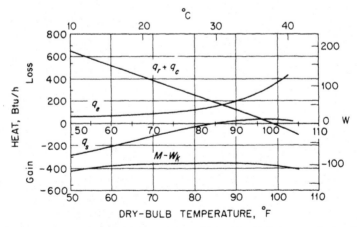

FIGURE 16.4 Heat gains and losses for a clothed male at rest, relative humidity 45%, little air motion. (After Anon., 1993.)

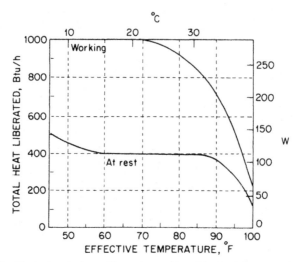

FIGURE 16.5 Variation in heat liberation with effective temperature of person at rest or working. (After Anon., 1993.)

The heat liberation to the atmosphere is 420 Btu/h (123 W), with a negative storage function (heat gain).

Additional data are necessary to make more extensive use of the basic heat-balance equation (16.13). Two other ASHRAE graphs are helpful. In Fig 16.5, the heat liberation from a male at rest or at work is plotted against effective temperature, a convenient thermal-stress index, which is discussed in Section 16.4. The percentage of sensible heat in the total heat liberated (the ratio of $q_r + q_c$ to the total) and the percentage of latent heat (the ratio of q_e to the total) are shown in Fig. 16.6 at various dry-bulb temperatures. In this graph, note the strong predominance of q_e over $q_r + q_c$ at the high temperatures usually encountered in mining. Data from these figures are useful in calculating additions to the mine cooling load due to human metabolism (Section 16.5).

Heat Stress and Heat Strain

The measure of the body's response and adjustment to environmental and metabolic *heat stress* is termed *heat strain*. When heat stress rises, the body attempts to maintain thermal equilibrium through the heat-transfer mechanisms pictured in Fig. 16.3. Heat strains or adjustments within the body result when the storage function q_s is overloaded (Misaqi et al., 1976).

There are three main body vital signs or criteria that are indicative of the degree to which the body is able to cope with heat stress: body core temperature, heart rate, and sweat rate; the most reliable and critical of these

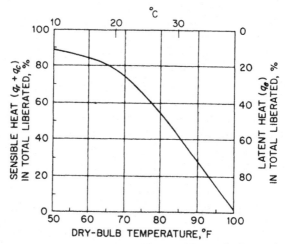

FIGURE 16.6 Proportion of body heat liberated by radiation and convection and by evaporation. (After Carrier, 1950. By permission from *Heating/Piping/Air Conditioning,* Chicago.)

is core temperature. A rise or fall of a few degrees in core temperature is the body's early warning of heat distress, with a recommended upper limit of 100.4°F (38.0°C) in this country (by the American Conference of Governmental Industrial Hygienists, ACGIH) for work in hot environments (Anon., 1996). Heart rate is also a reliable indicator of heat strain but is less sensitive at low heat stress. Perhaps the most widely used of the three criteria is sweat rate, because it is easily observed and rapidly responsive to mild heat conditions. It is, however, less reliable and may decrease at high heat stress. An important factor in assessing heat stress limits is the differing tolerance of adult male and female workers; it reportedly is less for women, because their body temperature may rise 2–3°F (1–2°C) more than men's before they begin to perspire (Anon., 1980).

To assist the body's heat regulating mechanism in maintaining thermal equilibrium in hot workplaces such as mines, there are several important environmental measures to employ.

1. *Clothing.* Wear light, minimal clothing to aid heat loss through radiation and convection; it should also be absorbent and of loose weave (to allow passage of air) to facilitate evaporation of sweat.

2. *Air Motion.* Provide adequate ventilation to enhance heat loss, particularly by evaporation; but avoid excessive velocities, especially if air temperature is above body temperature, because of heat gain due to $q_r + q_c$.

3. *Acclimatization.* Acclimatize miners to hot, humid environments over a period of 5–8 (or more) days so their bodies can adjust to the extreme

conditions (Wyndham, 1961, 1974; Martinson, 1977). Work activity is increased about 10% per day, starting from 50% of normal activity.

4. *Others.* Alternate periods of work and rest, provide cool (not cold) drinking water, supplement salt and vitamin C intake.

Notwithstanding all the body's built-in safeguards and the preventive measures that can be taken against heat stress, illnesses in excessively hot mine environments still occur frequently. Although varying in their symptoms, effects, and treatment, there are several recognized forms of heat illness that mine personnel must be trained to recognize and treat, here listed in order of increasing seriousness (Harrington and Davenport, 1941; Misaqi et al., 1976):

1. Body dehydration, salt loss
2. Heat rash over large areas of body (blisters, infection)
3. Heat cramps (spasms, nausea, dizziness)
4. Heat exhaustion (faintness, prostration, stupor)
5. Heat retention (rise in body temperature, headache, irrational behavior)
6. Heat stroke [body temperature rise over 104°F (40°C), halt in perspiration, coma, eventual death unless treated immediately]

Human Work Capacity and Safety

Certain associated, undesirable effects occur indirectly in working environments as a result of excessively high temperatures and humidities. Because of the inability of some workers to produce at their normal rate in hot, humid surroundings, work efficiency decreases. There is also an attendant increase in carelessness and accidents, because workers are less alert. These effects are costly in terms of safety, morale, production, and dollars.

Work Capacity   It can be demonstrated empirically that, in a hot environment, the capacity of a person to do work is limited by and somewhat less numerically than the cooling power of the air in which that person is working, when both are expressed in comparable units (Anon., 1993). *Cooling power* is a measure of the ability of the air to remove metabolic heat, in heat flow units of Btu/h (W). It is usually measured by an instrument such as the kata thermometer (Sections 6.3, 16.4).

A more rational analytic approach to cooling-power limitation of work output proceeds from the fundamental heat-balance relation for the human body (Eq. 16.13), deriving theoretical equations for the three heat losses, q_e, q_r, and q_c (Whillier and Mitchell, 1968; Stewart, 1980; Anon., 1993). From these studies, the principle of a near-proportional cooling-power–work-capacity relationship is apparent. Also other work has led to

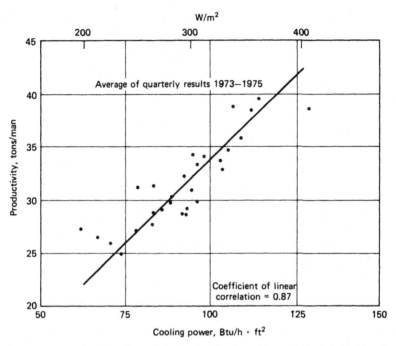

FIGURE 16.7   Relation of cooling power (in kata thermometer units) and miner productivity in South African gold mine stopes. (After Howes, 1978. By permission from *Mine Ventilation Society of South Africa,* Johannesburg.)

the development of relationships between productivity and the thermal characteristics of the work environment (Wyndham, 1974; Howes, 1978). Experimental confirmation in South African gold mines has been obtained showing a strong, direct correlation between miner productivity and air cooling power, as evident in Fig. 16.7.

Safety   There is similarly a strong correlation between cooling power and human safety. Accident rates of persons working in hot, humid conditions (such as mines) are decidedly higher than rates in normal environments. In tests conducted in British mines, the lowest accident rates have been related to men working at temperatures below 70°F (21°C), the highest to temperatures of 80°F (27°C) and above (Harrington and Davenport, 1941; Bedford and Tredre, 1955).

Studies at the Ooregum mine in India confirmed that the accident rate fluctuated seasonally with the weather but improved dramatically when temperatures were lowered continuously after air conditioning. An improvement in safety is nearly always associated with improvement of temperature-humidity conditions in workplaces.

Table 16.4   Heat Regulations in West German Coal Mines Prescribing Work and Rest Periods

Temperature °F (°C), Dry-Bulb or Effective	Work and Rest Periods, min				
	Total Shift Time	Maximum Time at Face	Statutory Rest Time	Additional Break, Union Agreement	Maximum Working Time at Face
<82.4 dry (28)	480	480	30	0	450
82.4 dry-84.2 effective (28–29)	420	360	30	0	330
84.2–86.0 effective (29–30)	420	300	30	10	260
86.0–87.8 effective (30–31)	420	300	30	15	255
87.8–89.6 effective (31–32)	420	300	30	20	250
>89.6 effective (32)	Work prohibited				

Source: Hamm (1980). By permission of publisher, SME-AIME, New York. Copyright 1980.

16.4   HEAT INDEXES AND STANDARDS

By 1960, laws or prohibitions against excessive heat stress in mines had been enacted by five countries: Belgium, Germany, The Netherlands, New Zealand, and Poland. The first was Germany (1905), which passed legislation limiting the air dry-bulb temperature to 82.4°F (28°C) or reducing the shift length; regulations presently in force there are shown in Table 16.4 (Hamm, 1980). The list of legislating countries has been extended and now includes Russia [limiting air temperature 78.8°F (26°C)] and the United Kingdom [limiting effective temperature 82°F (28°C), or compensating wages]. In Australia, some of the states have enacted legislation similar to that in Germany [the TLV in New South Wales is 80°F (27°C) wet-bulb temperature].

Currently, there are no legislated heat standards in U.S. mines, although there are recommended TLVs based on wet-bulb-globe temperature readings established by ACGIH (Anon., 1996). Individual mining companies on their own initiative, however, have adopted heat standards. One western U.S. metal mine, which has been operating in hot conditions for over 50 years, uses a "goal" effective temperature of 70°F (21°C); the limit is 80°F (27°C).

Heat indexes in current use in mining are classified as direct, empirical, and rationally derived (based on heat-transfer theory) (Anon., 1993). Since direct methods retain little favor, and rational methods are in the development stage, the following emphasizes empirical indexes.

Direct Indexes

So-called direct indexes are based on a single psychrometric measurement to establish air heat standards. Features: ease of measurement, broad instrument use, ignore other factors, obsolete as accepted standard.

Dry-Bulb Temperature (t_d)   Its familiarity as a scale and ease of measurement have much to commend dry-bulb temperature as a direct standard. Unfortunately, it lacks faithfulness to the human subject as an index of heat stress and today is rarely employed alone. Preferred standard: ≤82°F (28°C).

Wet-Bulb Temperature (t_w)   The single psychrometric property most relevant as a direct heat-stress indicator, wet-bulb temperature has enjoyed considerable use in mining. In hot, humid working environments, t_w governs, because human comfort almost always depends on evaporation of sweat and the rate at which moisture can be taken up by the atmosphere. Relatively, a change of 1°F (0.6°C) in wet-bulb temperature produces the same physiological sensation as a 10°F (5.6°C) change in dry-bulb temperature (Wyndham et al., 1965). In place of t_d or t_w, the temperature "spread" $t_d - t_w$ has been used as a standard; at least 3–5°F (2–3°C) is considered desirable. It has significance only when the dry-bulb or wet-bulb temperature is also specified. Preferred standard (Fig. 16.8): ≤80°F (27°C); limit ≤90°F (32°C).

Air Velocity (***V***)   Next to wet-bulb temperature, air velocity is the most important direct environmental factor in determining human comfort. A velocity increase, proportionally, is about as effective as a wet-bulb temperature decrease, especially at low V and t_d under 100°F (38°C) (Wyndham, 1961; Mitchell and Whillier, 1971). Sometimes employed as a comfort standard, velocity is preferably coupled with another indicator such as the wet-bulb temperature. Preferred standard (Fig. 16.8): ≥200–500 fpm (1.0–2.5 m/s); limit ≤500–600 fpm (2.5–3.0 m/s) to prevent airborne dust problem.

Empirical Indexes

On the basis of experimental rather than theoretical justification, empirical indexes were the earliest criteria employed for heat-stress standards in mines and still the most widely used. Features: based on comprehensive factors, generally good correlation with heat stress, lack rational basis.

Effective Temperature (t_e)   Effective temperature is the most popular empirical heat-stress index and is still commonly used in the United States and United Kingdom, both in surface industry and mining. Derived by ASHRAE in laboratory studies in 1923, t_e refers the combined effect of temperature, humidity, and velocity of the air to a single, empirical temperature scale reflecting equal sensations of warmth or cold (Anon., 1993). It has certain deficiencies, for example, underemphasizing the effect of humidity and velocity in hot, humid environments, a serious shortcoming in mines; and it cannot with any confidence be correlated with kata cooling power (Hinsley, 1943/44; Cook, 1977; Stewart, 1980). Nonetheless, wide experience has been built up in the United States with effective temperature as an index. A graph is necessary to find t_e from the three determining psychrometric readings,

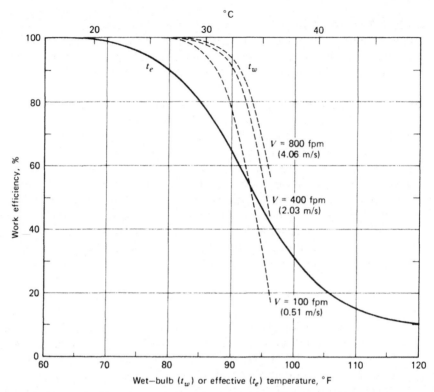

FIGURE 16.8 Effect of wet-bulb and effective temperatures on work efficiency. Results for t_w at three air velocities are based on acclimatized mine workers. Results for t_e are based on industrial workers. (After Wyndham, 1961, and Carrier, 1950. By permission from South African Institute of Mining and Metallurgy, Marshalltown; and *Heating/Piping/Air Conditioning*, Chicago.)

t_d, t_w, and V (Fig. 16.9). Preferred standard (Fig. 16.8): $\leq 80°F$ (27°C); limit $\leq 90°F$ (32°C).

Kata Thermometer (h_c) The kata thermometer was very likely the first instrument designed to measure air cooling power h_c. It has had most use in South Africa and is still employed there to some extent. Use of the kata to measure h_c was discussed in Section 6.3; Figs. 6.4 and 6.5 convert kata readings to cooling power. The wet-kata thermometer simulates the behavior of the human body in a hot, humid environment; how well is not widely agreed upon (Mitchell and Whillier, 1971; Lambrechts, 1972; Young et al., 1978). The kata's major defect is its overresponse to air movement. Hinsley (1943/44) concluded that effective temperature is "a far more satisfactory measure of environmental warmth than is the wet-kata cooling power." Preferred standard (Table 16.5): 4.5–12.0 mcal/cm²·s.

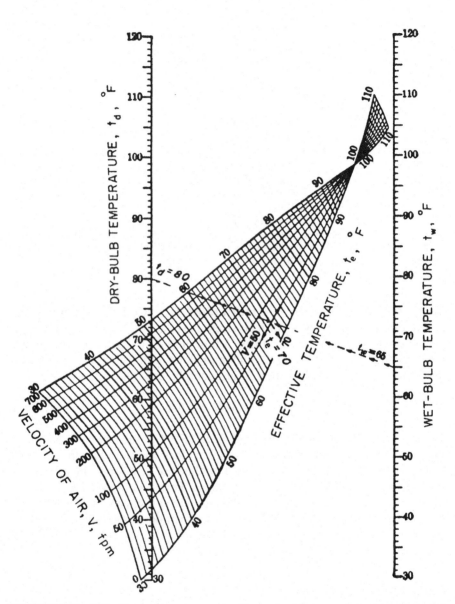

FIGURE 16.9 Effective temperature scale. Basic chart for men stripped to the waist and at rest or doing light work. Example: given t_d = 80°F (27°C), t_w = 65°F (18°C), V = 50 fpm (0.25 m/s); read t_e = 70°F (21°C). Conversion factors: °C = $\frac{5}{9}$ (°F − 32), 1 fpm = 0.00508 m/s. (After McElroy, 1935.)

Table 16.5 Cooling-Power Standards for Acclimatized Miners

Work Level	Required Cooling Power			Minimum Wet Kata, mcal/cm²·s
	W/m^2	$(Btu/h \cdot ft^2)$	(Btu/h^a)	
Light	115	(36)	(720)	4.5
Moderate	180	(57)	(1140)	7.0
Heavy	280	(89)	(1780)	12.0

Mining Tasks Corresponding to Work Levels		
Light	Moderate	Heavy
Operating hoist	Building stoppings	Shoveling rock
Sweeping	Operating box holes	Pushing mine cars
Fitting	Erecting roof supports	
	Drilling	
	Supervising	

Source: Stewart (1980). By permission of publisher, SME-AIME, New York. Copyright 1980.
[a] Assumes body surface area $\sim$ 20 ft² (1.86 m²).

***Wet-Bulb-Globe Temperature Index*(WBGT)** ACGIH advocates an empirical threshold limit value for heat stress based on the wet-bulb-globe temperature index (Misaqi et al., 1976; Anon., 1996). The technique in mines employs two thermometers: wet-bulb and globe, a black hollow sphere. Instrument readings are correlated to an index by the following relation:

$$WBGT = 0.7t_w + 0.3t_g \qquad (16.14)$$

where t_g is globe thermometer reading in °F (°C). Inclusion of the globe thermometer makes allowance for body heat loss due to radiation, an effect neglected in other direct and empirical methods of heat-stress measurement (but of questionable importance in mines). Permissible heat-exposure indexes or TLVs for acclimatized, lightly clothed workers engaged at different metabolic heat rates over an 8-h shift have been developed by ACGIH and are shown in Fig. 16.10. Preferred standard (Fig. 16.10): ≤79°F (26°C) WBGT.

Rationally Derived Index

***Specific Cooling Power*(h_{SCP})** Efforts recently have been directed toward the development of a rational heat-stress index, expressed as specific cooling power h_{SCP} in Btu/h and based on heat-transfer theory (Mitchell and Whillier, 1971; Stewart and Wyndham, 1975). There is currently no instrument available to measure h_{SCP} directly. The only one that provides even an approximation is the wet-kata thermometer, whose h_c reading can be

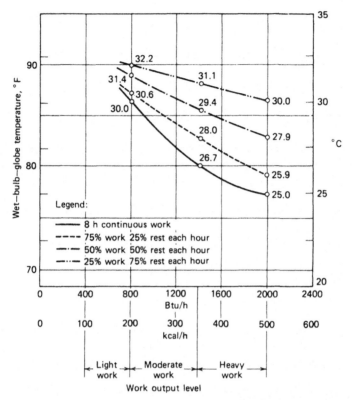

FIGURE 16.10 Permissible heat-exposure TLVs recommended for acclimatized workers over an 8-h shift, based on the wet-bulb-globe temperature index (WBGT). (After Anon., 1996. From American Conference of Governmental Industrial Hygienists, Inc., (ACGIH®): *1996 Threshold Limit Values (TLVs®) for Chemical Substances and Physical Agents and Biological Exposure Indices (BEIs®).* Reprinted with permission.)

correlated with h_{SCP}. Normally a graph is used to plot values of cooling power and safe working levels against wet-bulb temperatures and air velocities on which are superimposed wet-kata readings. The h_{SCP} values in Table 16.5, termed *required cooling power,* can be used to establish standards for underground conditions (Stewart, 1980). Preferred standard (Table 16.5): $\geq$1780 Btu/h (280 W/m^2).

Comparison of Indexes

Much remains to be done before an acceptable heat-stress indicator or index receives widespread adoption in the mining world. None currently in use satisfies completely the requirements of rationality and of faithful, reliable, and simple measurement, although specific cooling power shows promise for the future. Presently, because of its long use and wide application throughout U.S. industry, effective temperature remains the favored index in this country in mining.

16.5 CALCULATION OF MINE HEAT FROM HUMAN METABOLISM

Although not considered a major heat source in mines, human metabolism contributes significant heat to the ventilating air. The amount can be estimated, using the data presented in Sections 16.3 and 16.4.

Example 16.10   Given the following, calculate the amount of heat (q, q_S, q_L) added to the air by metabolism: 800 miners, desired work efficiency η_w = 80%, air V = 50 fpm (0.25 m/s), air temperature spread $t_d - t_w$ = 4°F (2.2°C). Base on effective temperature as a heat index.

Solution:

From Fig. 16.8, for η_w = 80%, read t_e = 85°F (29.4°C)
From Fig. 16.9, read t_d = 89°F (31.7°C) and t_w = 85°F (29.4°C)
Using Fig. 16.5, read q = 820 Btu/h·person (240 W/person)
Using Fig. 16.6, read ratio $q_S{:}q$ = 30%
Calculate the heat addition:
    Latent heat q_L = (0.70)(820)(800) = 459,200 Btu/h (135 kW)
    Sensible heat q_S = (0.30)(820)(800) = 196,800 Btu/h (57.7 kW)
    Total heat q = 656,000 Btu/h (192 kW)

PROBLEMS

16.1 (a) Using an average value for $(\gamma - 1)/\gamma$ in Eq. 16.1, calculate the rise in dry-bulb temperature of air in a shaft due to autocompression between 1000 ft (305 m) elevation and sea level. Assume sea-level temperature 70°F (21°C) and barometer 760 in. water (29.92 in. Hg), and find the dry-bulb temperature at 1000 ft (305 m). Repeat for a shaft whose collar is at 5000-ft (1524-m) elevation and bottom at sea level.

   (b) Assuming that the wet-bulb temperature initially is 60°F (16°C), estimate the wet-bulb increase and the resulting wet-bulb temperature for both cases in (a).

16.2 Recalculate the dry-bulb temperature gain in the 5000-ft (1524-m) shaft in Problem 16.1 for 1000-ft (305-m) increments in elevation. Use an average value of the exponent $(\gamma - 1)/\gamma$. Compare the temperatures at the shaft collar by two methods.

16.3 The heat flux in a mining district has been determined to average 0.016 Btu/h·ft² (0.05 W/m²). Two rock formations occur, overlying volcanics 1200 ft (366 m) thick with a thermal conductivity of 1.8 Btu/h·ft·°F (3.12 W/m·°C) and underlying quartzite 3000 ft (914 m) thick with a conductivity of 3.5 Btu/h·ft·°F (6.06 W/m·°C). Calculate the geothermal gradient in °F per 100 ft (°C per 100 m) for each formation and overall, and sketch the composite gradient. Also compute the maximum rock temperature at depth, if the surface temperature is 60°F (15.6°C).

16.4 Determine the heat flow in Btu/h (W), per unit length and total, in an excavation 12 × 30 ft (3.7 × 9.1 m) in cross section and 300 ft (91.4 m) long, given that the virgin-rock temperature is 125°F (51.7°C) and the dry-bulb temperature of the entering air is 82°F (27.8°C). The working is in gneiss and has been open 30 days.

16.5 In Problem 16.4, calculate the dry-bulb temperature increase of the air per unit length of excavation and at the face, if the weight flow rate of air is 95,000 lb/h (12 kg/s). Compare with results from Example 16.3, and account for the difference.

16.6 Determine the total heat gain from 150 gpm (9.46 L/s) of underground water flowing on a mine level. Water temperatures are 130°F (54°C) in and 105°F (41°C) out. Estimate the latent and sensible components.

16.7 In a mine workplace, the following electrical equipment is operating, rated as indicated: continuous miner at 225 hp (168 kW), two shuttle cars at 40 hp (29.8 kW) each, four drills at 7 1/2 hp (5.6 kW) each, and one face fan at 25 hp (18.7 kW). Compute the heat gain in Btu/h (W), assuming continuous operation throughout the shift.

16.8 The connected electrical machinery load in a mine is 5250 kW, and the load factor is 55%. Find the heat gain.

16.9 Compare the amount of heat liberated by two mine locomotives, one electric and the other diesel. Both are rated at 100 hp (75 kW).

16.10 Determine the heat gain from mine lighting in a stope with floodlights rated at 1500 W and 12 cap lamps rated at 5 W each.

16.11 Using Fig. 16.4, write the heat-balance equation for a miner working at 60°F (16°C) dry-bulb temperature. Repeat, at 100°F (38°C) dry-bulb.

16.12 Twenty miners work in a stope 10 × 15 ft (3.0 × 4.6 m) in cross section with 70,000 cfm (33.0 m³/s) of air flowing at 94°F (34°C) dry-bulb temperature. The atmospheric environment is to be maintained at a cooling power sufficient to permit moderate work activity at an efficiency of 75%.

(a) Determine the required design conditions (t_e, t_w) from Fig. 16.8, based on effective temperature.

(b) Based on wet-bulb temperature, what are the design conditions (t_e, t_w)? Use Fig. 16.8.

(c) What is the wet-kata cooling power for the design conditions in (a)? See Fig. 6.4. Using Table 16.5, are the conditions adequate for "moderate" work activity? Estimate the specific cooling power from the table.

(d) For (a) and (c), what is the wet-bulb-globe temperature index (see Fig. 16.10)?

CHAPTER 17

Mine Air Conditioning Systems

Air conditioning systems used in mines utilize various temperature–humidity control processes. These, in turn, are based on changes in the psychrometric properties of air, discussed in Chapter 2, called *psychrometric processes.* The present chapter covers psychrometric processes, heat-transfer processes, and air cooling and heating plants utilized in mining.

17.1 PSYCHROMETRIC PROCESSES

All psychrometric processes for temperature–humidity control involve flowing air. Conditioning of a given mass of air moving at an assumed linear rate is classified as a steady-state process (Section 5.1), and all the conditioning processes dealt with here fall in this category.

Because changes in heat content and temperature are both involved, the specific weight of the air being conditioned is not constant, and the quantity or volume flow rate Q is variable and must be related to barometric conditions. Hence flow rates are generally expressed on a *weight flow rate* basis as G in units of lb/h (kg/s) of dry air, which in a given process is the only constant. The relation at any state point between Q in cfm (m³/s) and G is

$$G = \frac{60Q}{v} \tag{17.1}$$

$$G = \frac{Q}{v} \tag{17.1a}$$

where v is specific volume in ft³/lb (m³/kg).

In an air conditioning or heat-transfer process, the *total heat change q* in Btu/h (kW) is calculated from G and the change in enthalpy h in Btu/lb (kJ/kg) as follows:

$$q = G(h_2 - h_1) \tag{17.2}$$

619

where the subscripts 1 and 2 denote initial and final conditions, respectively. The *weight flow rate of water vapor* G_w in lb/h (kg/s) being added or removed in a process is found similarly, based on the change in specific humidity W in lb/lb (kg/kg):

$$G_w = G(W_2 - W_1) \qquad (17.3)$$

The total heat change q that a given amount of air G undergoes in a temperature–humidity control process is composed of a *sensible component* q_S and a *latent component* q_L:

$$q = q_S + q_L \qquad (17.4)$$

A change in dry-bulb temperature represents a change in sensible heat while a change in specific humidity represents a change in latent heat (the wet-bulb changes in both cases). The ratio of the change in sensible heat to the change in total heat (i.e., the process slope) is termed the *sensible-heat factor F*

$$F = \frac{q_S}{q} \qquad (17.5)$$

expressed as a decimal. Its use is demonstrated in subsequent examples and problems. F appears as a scale along one side of the psychrometric chart (Fig. 2.2 for normal conditions); the process slope is found by connecting F on the scale with the scale origin. For example, in Fig. 2.2, the origin is at $t_d = 80°$ (26.7°C) and $\phi = 50\%$. (Note in the charts appearing in Appendix A, however, that Fig. A.4 has no F scale.)

By their nature, all temperature–humidity control processes in air conditioning can be represented on the psychrometric chart. This is of tremendous assistance in the solution of problems. Not only can the process be visualized better, but heat and humidity changes can be determined quickly from the chart. The line connecting two state points (drawn as a straight line, although the process may not actually proceed linearly) represents the process. The position of the line indicates the nature of the process and whether the change in heat content is latent, sensible, or a combination. *Simple processes* proceed vertically or horizontally. For example, as shown in Fig. 17.1, a horizontal line indicates a change in sensible heat: If it proceeds from right to left, the process is cooling; if from left to right, it is heating. A vertical line indicates a change in latent heat: If it proceeds upward, the process is humidification; if downward, it is dehumidification (these are drawn dotted, since the processes are usually not straight lines). *Combination processes* (e.g., cooling and dehumidification) run diagonally, as shown.

Basic processes of importance in mine air conditioning (Fig. 17.2), those

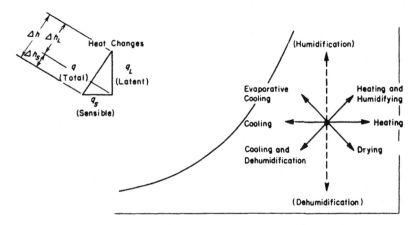

FIGURE 17.1   Nature of thermodynamic changes on a psychrometric chart.

fundamental to all temperature–humidity control, are discussed first, followed by applied processes. Sample calculations are included. Further details are available in the ASHRAE *Handbook of Fundamentals* (Anon., 1993).

Basic Processes

Heating   This process—increasing the sensible heat (dry-bulb temperature) of air without changing the latent heat (specific humidity)—is represented

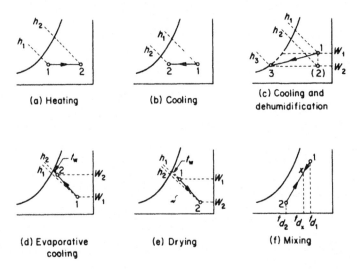

FIGURE 17.2   Representation of basic air conditioning processes on a psychrometric chart showing heat and moisture changes.

by a horizontal line running away from the saturation curve, as shown in Fig. 17.2a. Heating of air is generally accomplished by convection with hot-air furnaces or steam coils. Referring to Eq. 17.2, the heat-change relation is

$$q = q_s = G(h_2 - h_1) \qquad (17.6)$$

Notice that the dry- and wet-bulb temperatures, relative humidity, and enthalpy change, whereas the specific humidity and dew-point temperature remain constant.

Example 17.1   A quantity of 50,000 cfm (23.60 m³/s) of air at sea level is to be heated from 35°F (1.7°C) dry-bulb and 30°F (-1.1°C) wet-bulb to 60°F (15.6°C) dry-bulb. Determine the amount of heat required for the process.

Solution: Using Fig. 2.2, locate the state points and lay out the process line for sensible heating (see Fig. 17.2a). Point 2 is located at $t_d = 60$°F (15.6°C) and $t_w = 44$°F (6.7°C). At point 1, $t_d = 35$°F (1.7°C) and $t_w = 30$°F (-1.1°C), read $v = 12.51$ ft³/lb (0.7810 m³/kg).
  Using Eq. 17.1,

$$G = \frac{(60)\,(50,000)}{12.51} = 239,800 \text{ lb/h (30.21 kg/s)}$$

From the chart, read the enthalpies:

$$h_1 = 10.91 + 0.15 = 11.06 \text{ Btu/lb (25.73 kJ/kg)}$$
$$h_2 = 17.16 - 0.04 = 17.12 \text{ Btu/lb (39.82 kJ/kg)}$$

Calculate heat addition by Eq. 17.2 or 17.6:

$$q_s = (239,800)\,(17.12 - 11.06) = 1,453,000 \text{ Btu/h (426 kW)}$$

Cooling   Cooling is a process that decreases the sensible heat of air without changing the latent heat. It also is represented by a horizontal line, running toward (but not intersecting) the saturation curve (Fig. 17.2b). The process takes place by convection and conduction when air is brought into contact with cooling coils (or by convection and evaporating when cold water is sprayed into the air). The heat-change relation (from Eq. 17.2) is

$$q = q_s = G(h_1 - h_2) \qquad (17.7)$$

Again the dry-bulb and wet-bulb temperatures, relative humidity, and enthalpy change, whereas the specific humidity and dew-point temperature remain constant.

Example 17.2 Determine the change in heat content when air at sea level is cooled from 98°F (36.7°C) dry-bulb and 63°F (17.2°C) wet-bulb to 50°F (10.0°C) dry-bulb.

Solution: Locate the state points—point 2: $t_d = 50°F$ and $t_w = 43°F$ (10.0°C and 6.1°C)—then lay out the process line for sensible cooling (see Fig. 17.2b), and determine the enthalpies:

$$h_1 = 28.59 - 0.25 = 28.34 \text{ Btu/lb (65.92 kJ/kg)}$$

$$h_2 = 16.65 - 0.02 = 16.63 \text{ Btu/lb (38.68 kJ/kg)}$$

Calculate heat removal by Eq. 17.7 for $G = 1$ lb/h (0.126×10^{-3} kg/s):

$$q_S = (1)(28.34 - 16.63) = 11.71 \text{ Btu/h (3.43 W)}$$

Cooling and Dehumidification This process, a combination one—decreasing both the sensible and latent heat and lowering the moisture content—is represented along a diagonal line sloping downward to the left Fig. 17.2c); more likely, it proceeds horizontally to the saturation curve and then follows the curve down to the final state point. It may not end on the saturation curve, however, since heat exchangers are not perfectly efficient, and the distance away from the curve is referred to as the *bypass factor*. As with cooling, this process is accomplished with cooling coils or sprays. The heat-change and moisture-addition relations, referring to Eqs. 17.2 and 17.3, are as follows:

$$q_S = G(h_2 - h_3) \tag{17.8}$$

$$q_L = G(h_1 - h_2) \tag{17.9}$$

$$q = q_S + q_L = G(h_1 - h_3) \tag{17.10}$$

$$G_w = G(W_1 - W_3) \tag{17.11}$$

Notice in Fig. 17.2c that the change in latent heat can be represented by the vertical component 1–2 of the process line 1–3 and the change in sensible heat, by the horizontal component 2–3. The reduction in latent heat occurs because of condensation of moisture from the air; the latent heat of vaporization given up by the water is transferred to the coils or sprays. The process is accompanied by a decrease in value of all the psychrometric properties, except relative humidity, which increases.

Example 17.3 Determine the changes in heat and moisture contents when air at sea level is cooled and dehumidified from 90°F (32.2°C) dry-bulb and 80°F (26.7°C) wet-bulb to 70°F (21.1°C) saturated.

Solution: Proceed as in Examples 17.1 and 17.2, referring to Fig. 17.2c.

$$h_1 = 43.70 - 0.12 = 43.58 \text{ Btu/lb (101.37 kJ/kg)}$$
$$W_1 = 0.0199 \text{ lb/lb (kg/kg)}$$
$$h_2 = 34.09 \text{ Btu/lb (79.29 kJ/kg)}$$
$$W_2 = 0.0159 \text{ lb/lb (kg/kg)}$$

Calculate q and G_w by Eqs. 17.10 and 17.11, for $G = 1$ lb/h (0.126×10^{-3} kg/s):

$$q = (1)(43.58 - 34.09) = 9.49 \text{ Btu/h (2.78 W)}$$
$$G_w = (1)(0.0199 - 0.0159) = 0.0040 \text{ lb/h } (0.504 \times 10^{-6} \text{ kg/s})$$

Evaporative Cooling Evaporative cooling, another combined process, is the conversion of sensible heat to latent heat with addition of moisture and practically no change in total heat content of air. Graphically (Fig. 17.2d), the cooling curve proceeds upward to the left along a diagonal line parallel with the wet-bulb temperature lines. Evaporative cooling (the "cooling" is a misnomer since a small net increase in heat content occurs) takes place when water at the wet-bulb temperature of the air is sprayed in the air. As sensible heat of the entering air vaporizes the water, the dry-bulb temperature of the air decreases. The sensible heat used to vaporize the water enters the air as latent heat contained in the added moisture. Since sensible heat is converted to latent in evaporative cooling and no heat is added or removed, it is an adiabatic process. If it proceeds to an intersection with the saturation curve, the process is termed *adiabatic saturation*. The heat balance, expressed in Btu/lb (kJ/kg) dry air, is as follows:

$$h_2 = h_1 + h_f(W_2 - W_1) \tag{17.12}$$

where $h_f = t_w - 32$, the (sensible) heat of the liquid added in evaporation. The heat- and moisture-change relations from Eqs. 17.2 and 17.3, therefore, are

$$q = 0 \tag{17.13}$$

$$G_w = G(W_2 - W_1) \tag{17.14}$$

The wet-bulb temperature remains constant, whereas the dry-bulb decreases and the dew point, enthalpy, specific humidity, and relative humidity increase.

Example 17.4 Write the heat balance and determine the changes in heat and moisture contents when air at sea level is evaporatively cooled from 85°F (29.4°C) dry-bulb and 70°F (21.1°C) wet-bulb to saturation.

Solution: Read properties at state points 1 and 2 for the process shown in Fig. 17.2d:

$$h_1 = 34.10 - 0.14 = 33.96 \text{ Btu/lb (78.99 kJ/kg)}$$

$$W_1 = 0.0123 \text{ lb/lb (kg/kg)}$$

$$h_2 = 34.10 \text{ Btu/lb (79.32 kJ/kg)}$$

$$W_2 = 0.0159 \text{ lb/lb (kg/kg)}$$

Using Eq. 17.12, calculate $h_f = 70 - 32 = 38$ Btu/lb (88.39 kJ/kg), and write the heat balance:

$$34.10 = 33.96 + 38 (0.0159 - 0.0123) = 33.96 + 0.14$$

$$34.10 = 34.10$$

Calculate q and G_w by Eqs. 17.13 and 17.14 for $G = 1$ lb/h (0.126×10^{-3} kg/s):

$$q = 0$$

$$G_w = (1) (0.0159 - 0.0123) = 0.0036 \text{ lb/h } (0.454 \times 10^{-6} \text{ kg/s})$$

Drying The drying process, a combination one, involves adsorption or absorption of water vapor from air by chemical dehydration. A suitable chemical reagent (e.g., silica gel or lithium, calcium, or magnesium chloride) possessing the ability to take on moisture is brought in contact with air to be dried. The latent heat liberated in the condensation of moisture is added to the air in the form of sensible heat, and the amounts are approximately equal. As indicated in Fig. 17.2e, the process follows closely a wet-bulb temperature line; the slope is greater or less depending on whether heat is liberated or absorbed during drying. Applications in mining have been limited to local comfort-cooling installations of small scale and product (mineral) drying. There are also applications in commercial underground storage. For calculations involving drying, see ASHRAE handbook, *Heating, Ventilating and Air Conditioning Systems and Equipment* (Anon., 1992b).

Mixing This combined process involves the mixing of air at two different state points to obtain air at an intermediate state point. The location of the new state point, which determines the properties of air in the resulting mixture, is a function of the flow rates of the original masses of air. Mixing of fresh and recirculated air is employed frequently to obtain air of the desired properties in industrial air conditioning, and because it may be necessary in mines. To find the new state point x, the original state points 1 and 2 are connected by a straight line and a distance scaled from one proportional to the weight flow rate of the other (Fig. 17.2*f*). Mathematically, the interpolation can be performed using temperature, humidity, or enthalpy. For example, using dry-bulb temperatures, which is the easiest procedure, the following relation is obtained:

$$t_{d_x} = t_{d_2} + \frac{G_1}{G_T}(t_{d_1} - t_{d_2}) \qquad (17.15)$$

where $G_T = G_1 + G_2$ in lb/h (kg/s). If the process follows a near-vertical line, wet-bulb temperatures or specific humidities may be used for scaling. Obviously, all properties of the air will change unless the process proceeds by coincidence along a coordinate line.

Example 17.5 Determine the state point resulting from mixing at sea level (1) 40,000 lb/h (5.04 kg/s) of air at 95°F (35.0°C) dry-bulb and 78°F (25.6°C) wet-bulb, and (2) 20,000 lb/h (2.52 kg/s) of air at 70°F (21.1°C) dry-bulb and 62°F (16.7°C) wet-bulb.

Solution: Lay out the process as in Fig. 17.2*f*. Scale point x on chart 20,000/60,000 = 1/3 of distance from point 1 to point 2. Or calculate t_{d_x} by Eq. 17.15:

$$t_{d_x} = 70 + \left(\frac{40,000}{60,000}\right)(95 - 70) = 70 + (0.67)(25) = 86.7°F\ (30.5°C)$$

and read from chart $t_{w_x} = 73.3°F$ (23.0°C).

Applied Processes

The most frequently used basic air conditioning process in mines is cooling and dehumidification. Generally, it is employed alone for temperature–humidity control, but it can be used supplemental to processes to achieve additional environmental objectives underground, as is frequently the case in surface industrial air conditioning. These are referred to as *applied processes*.

In order to conduct air conditioning calculations and design, it is customary to employ the psychrometric chart to represent not only the conditioning process itself but also the psychrometric changes undergone by the air as it passes through the workplaces that are being cooled and dehumidified. Calculations involve (1) location of the state points of the air as thermodynamic changes occur during air conditioning, (2) plotting all conditioning processes on the psychrometric chart, (3) determination of the various psychrometric properties of the air at all state points, and (4) computation of the flow rates for air and water and the total heat changes that occur in the various processes. The psychrometric data thus obtained find use in the subsequent design of the air cooling plant and selection of equipment.

Two-Line Process The treatment process of cooling and dehumidifying the mine air together with the conditioning process of heating and humidification that the air undergoes in the workplaces are represented by a block diagram and on the psychrometric chart in Fig. 17.3. They constitute a *two-line process,* the commonest form of air conditioning and the most used underground. For convenience, they are represented as though the cooling plant and workplace are contiguous, but they need not be. The heat exchange unit utilized is a (1) cooling coil or (2) spray.

In process (1), outside air enters the conditioner at state point O. It passes through the coil, giving up heat to the chilled water and moisture. The coil temperature C, termed the *apparatus dew point,* will be on the saturation curve only if the coil bypass factor is zero, but it always lies near the curve. Assuming the fan is present simply to provide motion and does not contribute any heat to the process, air leaves the conditioner and enters the space (workplaces) at condition C. There the air absorbs both sensible and latent heat and is discharged from the space at condition S. The fan can be located anywhere in the system, and it is sometimes placed at the coil discharge C to permit it to handle the cleanest and least quantity of air (smallest Q for a given G). If the fan horsepower is large, heat addition becomes the prime consideration, and the fan should be located before the coil at O or—better—in the exhaust airway at S, where its heat can be dissipated outside the

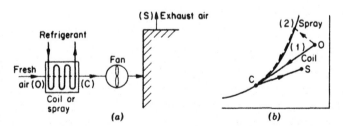

FIGURE 17.3 Two-line mine air conditioning process: (*a*) process flowsheet; (*b*) process on psychrometric chart. Either a (1) coil or (2) spray is used for cooling and dehumidification.

conditioned-air circuit. Because of difficulties with leakage and control of circulation, however, location of the fan in an exhaust position may not be practical.

If a spray instead of a coil is used for cooling the air, as with path (2) in Fig. 17.3, the process accomplishes the same overall objective, but by different means, and the air is nearly always saturated as it leaves the conditioner. Sprays and coils are covered in Section 17.3.

Supplemental Processes   Although the two-line process is adequate for most underground cooling, *supplemental processes* are in common use for surface industrial applications and may have utility in mines, probably for large centralized systems. There are several reasons for employing processes that condition beyond basic cooling and dehumidification: (1) the undesirability of discharging air from the conditioner at saturated conditions, (2) the need to recirculate some air because of limited availability, and (3) the improved operating economy of the system that may be afforded by additional conditioning. Supplemental processes that warrant consideration in mining are mixing, precooling, and reheating.

Recirculation and Reheating   A cooling coil or spray performs its joint functions of cooling and dehumidification satisfactorily, but it discharges air that is saturated, uncomfortable, and less effective in absorbing heat from sweating miners. Also the two-line process makes no provision for recirculating air, sometimes a necessity when flow volume is limited or the fresh-air temperature substantially exceeds that of the exhaust air.

Recirculation is employed to reduce the load on the coil, since the outside air often has a higher heat content than the exhaust air from the space. In surface air conditioning, recirculation is limited by building or health codes requiring a prescribed minimum quantity of fresh air per person [usually 20 cfm (0.0094 m³/s)]; underground, it is governed by air-quality considerations (Chapter 11). Reheating is introduced (by electric resistors or by steam or hot-water coils) to restore sufficient heat to the cooled and dehumidified air to reduce its relative humidity to a comfortable range (35–65%)—in effect, reheating brings the air away from the saturation line and saturated conditions. Sometimes this is accomplished without reheating by reducing the number of banks in the cooling coil, thus increasing the bypass factor of the coil and discharging air that is not saturated. The same effect is also produced by mixing untreated air with conditioned air before admitting it to the space. Coil bypass is most effective when the coil process line is relatively flat.

Generally, the maximum dry-bulb temperature differential in the workplace that can be tolerated for comfort is 20–25°F (1–14°C). Surface air conditioning can usually keep within this limit by providing ample volumes of air, but it may be sacrificed of necessity in mines because of the difficulty of providing sufficient air and controlling the extreme conditions encountered underground. The space load line is typically quite flat in industrial condition-ing, indicating a high ratio of sensible to latent heat; in mining, it is relatively steep because of the high latent load.

Reheating, mixing, or bypassing following cooling is useful to consider in centralized plants in hot, humid mines where the working atmosphere is saturated, but it is not necessary in exceedingly hot or dry mines because the saturated air leaving the conditioner is warmed considerably and its relative humidity reduced as it flows through hot airways to the workplaces. Recircu-lation should be used sparingly if exhaust air picks up excessive heat and moisture in long return airways.

Precooling and Reheating To cope more efficiently with extreme envi-ronmental conditions, industrial air conditioning often utilizes a supplemen-tal system that includes a precooler operating in closed circuit with a re-heater. The process is similar to the previous one, supplemented by the addition of a precooler. This is more complex but improves the efficiency of conditioning by reducing the coil load, and when operated in closed circuit with a reheater, the precooler removes from the air approximately the same amount of heat returned later in reheating. For all practical purposes, the heat exchange between the two units balances:

$$\text{Precooler } q = h_1 - h_2 = \text{reheater } q = h_3 - h_4 \qquad (17.16)$$

and the only energy consumed in the circuit is used in pumping a heat-transfer medium, usually water, between the reheater and the precooler.

Calculations Involving Cooling-Dehumidification Systems

Certain terms need definition that are employed in calculations involving temperature–humidity control systems. *Mine cooling load* is the total amount of heat, sensible and latent, in Btu/h that must be removed by the air in the workplaces to be cooled. Calculation of the cooling load is explained in Section 17.4. *Coil load* is the total amount of heat in Btu/h at the apparatus dew point that must be removed from the air by the cooling coils (or sprays). *Refrigeration* q_R is expressed in units of tons of refrigeration, equal to the coil cooling load divided by 12,000 (a ton of ice in melting in 24 h liberates heat at the rate of 200 Btu/min or 12,000 Btu/h (3.517 kW). *Space* (*workplace*) *design conditions* are the temperature–humidity conditions required in the space to be conditioned, as represented by state point S. The sensible-heat factor F determines the slope of the process line through point S. *Outside design conditions* refer to the state point O of outside air.

Example 17.6 Given the following design conditions:

Surface (outside) air $t_d = 100°F$ (37.8°C), $t_w = 83°F$ (28.3°C),

$$p_b = 14.7 \text{ psi } (101.35 \text{ Pa})$$

Stope (space) air $t_d = 85°F$ (29.4°C), $t_w = 75°F$ (23.9°C)

Mine cooling load $q_S = 140,000$ Btu/h (41.03 kW),

$$q_L = 100,000 \text{ Btu/h } (29.31 \text{ kW})$$

Determine the air conditioning requirements of the system, utilizing a two-line process, and locate all state points. Compare fan quantities for different locations.

Solution: Lay out the process on the normal-temperature psychrometric chart (see Fig. 17.4), plotting state points O and S. Find the total cooling load and sensible-heat factor (Eqs. 17.4 and 17.5):

$$q = q_S + q_L = 240,000 \text{ Btu/h } (70.34 \text{ kW})$$

$$F = \frac{140,000}{240,000} = 0.583$$

Assuming zero coil bypass, read state point C, the apparatus dew point, at intersection of F line with saturation curve (plot from S):

$$t_{dp} = 64.7°F \ (18.2°C)$$

Find air requirements of system. Weight flow rate required (Eq. 17.2, rearranged):

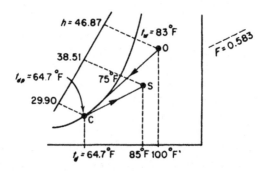

FIGURE 17.4 Two-line process of Example 17.6.

$$G = \frac{240,000}{38.51 - 29.90} = 27,870 \text{ lb/h } (3.51 \text{ kg/s})$$

Quantity of fresh air (at O) (Eq. 17.1):

$$Q = \frac{(14.56) (27,870)}{60} = 6760 \text{ cfm } (3.19 \text{ m}^3/\text{s})$$

Quantity of conditioned air (at C):

$$Q = \frac{(13.50) (27,870)}{60} = 6270 \text{ cfm } (2.96 \text{ m}^3/\text{s})$$

Quantity of exhaust air (at S):

$$Q = \frac{(14.10) (27,870)}{60} = 6550 \text{ cfm } (3.09 \text{ m}^3/\text{s})$$

To handle the least air, locate fan at C; fan $Q = 6270$ cfm (2.96 m³/s). Assuming that heat gain is detrimental, locate fan at S; fan $Q = 6550$ cfm (3.09 w³/s):

Coil load (Eq. 17.10):

$$q = (27,870) (46.87 - 29.90) = 472,950 \text{ Btu/h } (138.61 \text{ kW})$$

Refrigeration:

$$q_R = \frac{472,950}{12,000} = 39.4 \text{ tons}$$

Amount of moisture removed (Eq. 17.11):

$$G_w = (27,870) (0.0206 - 0.0132) = 206 \text{ lb/h or } 0.4 \text{ gpm } (0.026 \text{ kg/s})$$

Calculations for air conditioning systems using supplemental processes proceed in a similar manner.

17.2 CHILLED-WATER SOURCES

When cooling and dehumidification is the central process of a mine air conditioning system, two alternative methods are utilized to produce chilled fluids (usually water) for heat transfer underground. They are refrigeration and evaporative cooling by cooling towers.

As in mine ventilation, the mining engineer is not called on to design air conditioning equipment but to be knowledgeable of its operation and selection. For more comprehensive treatment of design calculations of chilled-water sources, reference can be made to an air conditioning text or manual (e.g., McQuiston and Parker, 1982; ASHRAE handbooks, Anon., 1992b, 1993). Also see expanded coverage of these topics in the second edition of this book (Hartman et al., 1982).

Refrigeration

The mechanical process of absorption of heat from one location and its transfer to and rejection at another place is termed *refrigeration*. It is the prevailing means used in mines to produce chilled water, except for the infrequent application of cooling towers in hot, dry climates. In the most commonly used form of refrigeration, the working medium or refrigerant alternates between the liquid and vapor phases. Hence such processes are called *change of state* or *vapor refrigeration*. Household refrigerators and most commercial units employ the vapor cycle. Other less frequently used systems utilize air, gas, or steam cycles, but only vapor refrigeration to date has had application in mines.

A refrigeration system consists of a cycle of four basic processes circulating a *refrigerant,* the heat-transfer medium. The purpose of the refrigerant is to absorb heat from a "source" (*evaporator*) and discharge it through a "sink" (*condenser*). Some type of vapor pump must be located between the source and sink so that the energy absorbed by the refrigerant in the evaporator may be transferred to the condenser for discharge; it takes the form of a *compressor*. The final component of a vapor refrigeration system is an *expansion valve,* used to control flow rate and permit cooling of the refrigerant in its return to the evaporator. Thus vapor refrigeration is essentially a compression system, involving heat exchange through a change of state of the refrigerant from liquid to gas and then back to liquid.

A vapor refrigeration cycle in schematic is represented by the block diagram of Fig. 17.5. The path of flow of the refrigerant and the changes it undergoes can be traced through the system as follows:

1. *Evaporator:* Refrigerant "boils" (evaporates), changing state from predominantly liquid to gas and absorbing heat from substance to be cooled, with no change in temperature.
2. *Compressor:* In the vapor state, refrigerant flows to compressor, where work is done in compressing it.
3. *Condenser:* Vapor condenses to liquid again, giving up heat without a temperature change.
4. *Expansion valve:* Temperature and pressure of liquid drops during expansion, as refrigerant completes cycle.

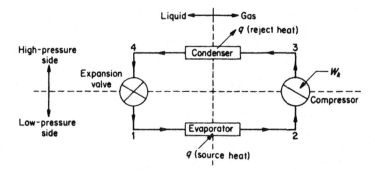

FIGURE 17.5 Schematic and block diagram of a vapor refrigeration system.

The state of the refrigerant, liquid or gas, is indicated in the schematic, referenced to a vertical axis. Notice that the changes of state of the refrigerant are actually controlled by varying the pressure; this leads to the designations *low-pressure side* and *high-pressure side* of the system, as shown, with reference to a horizontal axis.

Because of the changes of state and thermodynamic changes undergone by the refrigerant in a vapor refrigeration cycle, the representation of refrigeration processes is best accomplished on state diagrams of the refrigerant. A Mollier diagram (p–h coordinates) is most generally used for this purpose, because both enthalpies h, used in determining heat changes, and pressures p can be read directly (Fig 17.6). It has the same application for refrigeration processes that the psychrometric chart has for psychrometric processes.

A typical refrigeration machine using a vapor cycle appears in Fig. 17.7. The plant shown employs a centrifugal compressor, the most common type, although reciprocating and rotary units are also used. Centrifugal compressors are more versatile and capable of handling large volumes of gas. In mines, they are driven electrically, operate at efficiencies of 70–80%, and

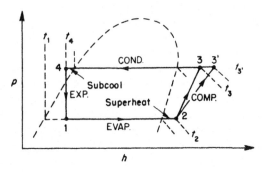

FIGURE 17.6 Modified vapor refrigeration cycle plotted on a Mollier diagram (p–h coordinates).

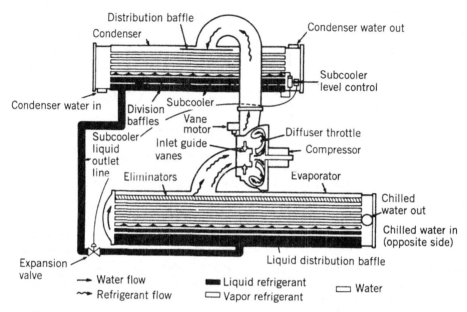

FIGURE 17.7 Diagrammatic sketch of a single-stage centrifugal refrigeration machine. (By permission from the Carrier Corp., New York.)

are capable of duties of 100–3000 tons refrigeration per unit (Burrows, 1974). Multistage compressors are employed when large pressure and temperature differences must be maintained (excessively high compression ratios cause low volumetric efficiencies). The evaporator and condenser are nearly identical in appearance and have a shell-and-tube construction. In both cases, the refrigerant flows over the outside of the tubes. In the evaporator, chilled water or brine (acting as a secondary refrigerant or intermediate heat-transfer medium) flows in the tubes, while in the condenser, water or another coolant used to reject heat to the atmosphere in cooling towers or spray ponds occupies the tubes. Expansion is accomplished by automatic throttling valves of various designs; one valve is required for each stage of compression.

A variety of fluids, both liquids and gases, with good heat-transfer properties are used as refrigerants. Although ammonia is low in cost, it can be employed only in surface plants where leaks are not hazardous. For underground installations, the freon refrigerants [chlorofluorocarbons (CFCs)] have been most commonly used but are being phased out because they pose an environmental threat to the earth's ozone layer. Besides safety and cost, other considerations in selecting a refrigerant are volumetric requirements, operating pressures, power requirements, stability, and corrosive properties. In air conditioning design, the information of greatest importance about a refrigerant and its performance is obtainable from its Mollier diagram, which provides temperature, entropy, and specific volume as well as $p–h$ data.

Design of a mine refrigeration plant commences with specification of the cooling load and selection of plant type, size (tons of refrigeration and chilled-water requirements), and location (surface or underground). After choosing a suitable refrigerant, the cycle temperatures are established. The evaporator temperature is based on chilled-water requirements: a desired flow at a desired temperature. The condenser temperature is dependent on the type of cooling and fluid available. For acceptable plant efficiency, the difference between the two temperatures should be no greater than necessary. Once these operating conditions are known, the cycle can be laid out on a Mollier diagram for that particular refrigerant (Fig. 17.6). Pressures and enthalpies are then readily determined and the remaining operating characteristics (flow rate of refrigerant, size of compressor) calculated.

Cooling Towers

A *cooling tower* is a heat-exchange device that cools liquids, usually water, by a combination of heat and mass transfer. Although cooling towers are principally used to dissipate heat from refrigerator condenser water, either surface or underground, their unique application in mine air conditioning is to produce chilled water (or brine) on the surface for use in cooling air directly in underground heat exchangers. This air conditioning application originated in the deep copper mines of Butte, MT and, although not widely applicable, has potential to supplement or replace refrigeration for mine cooling and dehumidification, especially in hot, dry climates, at a considerable cost saving.

The principal means of heat transfer involved in cooling towers is evaporative cooling, a psychrometric process discussed in Section 17.1. Evaporation of a small portion of the water being sprayed or percolated in the air of the tower removes sensible heat from the remaining water. The increase in heat content of the air is of no consequence since the air is exhausted to the atmosphere on the surface or to the upcast shaft underground. Depending on the relative temperatures of air and water, sensible cooling of the water and latent heating plus sensible heating or cooling of the air occurs.

The theoretical limit to which water can be cooled by spraying it into air is the wet-bulb temperature of the air. In practice, the water temperature can be lowered to within a few degrees of t_w. Although manageable in cooling condenser water, such a limit imposes a climatic restriction on the application of evaporative cooling to produce chilled water to those locations in which the outside wet-bulb temperature is comparatively low. Mines in the western part of the United States, where dry weather is prevalent, are situated in an ideal climate for this application of cooling towers.

To lower the limiting temperature to which chilled water can be cooled in a surface tower, an ingenious addition to the system has been made at Butte (Richardson, 1943, 1950). Before the air enters the cooling tower, it is precooled by a portion of the tower water diverted for that purpose. The result is to lower both the dry- and wet-bulb temperatures of the air entering the tower proper from the precooler. Hence, with precooling, the limit to which the water can be cooled is the dew-point temperature of the air. This permits more effective utilization of the capacity of a given tower for chilled-water production. If the cooling capacity of the water is still inadequate for underground air conditioning, then the water temperature may be further reduced with supplemental refrigeration, as practiced at the Magma mine in Superior, Arizona (Augustadt, 1951; Dobbs and Johnson, 1979).

A schematic of a surface cooling tower equipped with a precooler and intended to produce chilled water is shown in Fig. 17.8. The process as it affects the air can be represented on a psychrometric chart also (Fig. 17.9), the letters corresponding to those in Fig. 17.8. From point O to point P, the air is precooled before entering the tower. It then is evaporatively cooled from P to E as it rises in the tower, evaporating some of the spray water and cooling the remainder. This process generally reaches saturation. With continued, prolonged contact between air and water, some sensible-heat transfer occurs by convection between E and T; the water is further cooled, and the air heated and humidified. The air is discharged, saturated, at T. The line O to T symbolizes the overall process line; while the heat exchange cools the water, it produces a net heat gain in the air ($h_T - h_O$), shown in Fig. 17.9 with its sensible and latent-heat components (notice in this case that there is a conversion of some q_S to q_L). In a conventional cooling tower

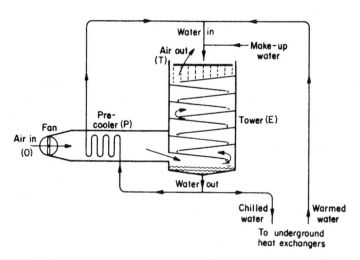

FIGURE 17.8 Schematic of a cooling tower. System shown includes a precooler, appropriate for a surface installation preparing chilled water for underground heat exchangers. (Modeled after an installation at Anaconda Copper Mines in Butte, MT.)

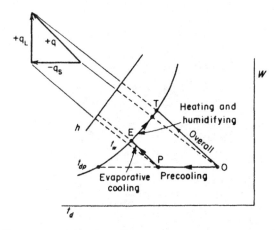

FIGURE 17.9 Representation of cooling tower process on a psychrometric chart. Refer to Fig. 17.8 for schematic.

not equipped with a precooler, the sensible cooling step from O to P does not occur, and the air often gains both sensible and latent heat.

The cutaway view of Fig. 17.10 illustrates the construction of a surface cooling tower equipped with a precooler. It is a mechanical-draft tower, with fans located both in blowing and exhaust positions. Natural draft may also be used, although underground towers are always mechanical draft. Water is sprayed into the tower at the top, and air enters at the bottom. The two streams pass each other and exchange heat in a counterflow arrangement. Intimate contact between the air and water is ensured by providing a system of either baffles (inclined wooden decks or vertical corrugated-metal sheets) or fill (film or splash type). The air is discharged at the top, and the chilled water collects in a sump at the foot of the tower. Part of the water is diverted to precooling coils, and the remainder is taken underground to heat exchangers conditioning the mine air. The underground water is returned in closed circuit to the top of the tower. Makeup water is added to replace the small amount of moisture lost in evaporation (about 1–3% for each 10°F or 5.6°C change in water temperature).

Conventional surface towers for cooling condenser water from refrigeration units lack the precooler but otherwise differ only in minor details. Underground cooling towers are also used and are constructed in existing or special excavations in solid rock adjacent to exhaust airways or upcast shafts. Often lacking packing, they resemble spray chambers and are arranged vertically or horizontally; if vertical, they have heights of 100–200 ft (30–60 m).

In place of cooling towers, spray ponds are sometimes used on the surface to dissipate heat from condenser water. Underground, the only alternatives to cooling towers are disposable mine water, if cool enough (Carrier, 1938),

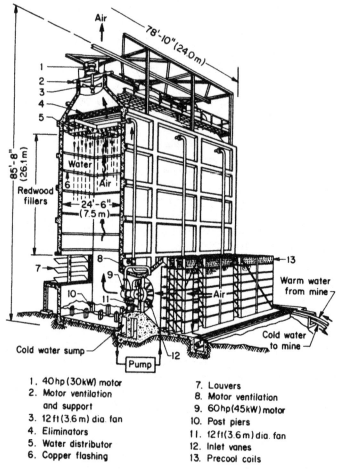

FIGURE 17.10 Diagrammatic sketch of a mechanical-draft cooling tower, with precooler for mine application. (After Augustadt, 1951. By permission from *Engineering and Mining Journal*, Chicago, IL.)

or recirculated chilled water that services the condenser subsequent to the cooling coils and dissipates its heat at the surface (Augustadt, 1951).

Analysis and design of cooling towers is complicated, because no direct solution of operating conditions is possible. Again, as a rule, mining engineers are involved in the selection of equipment, not its design, and hence need mainly an understanding of the process and the approach to calculations. Cooling tower design remains empirical and based on performance curves or parameters. With the progress made recently in design theory and analysis, however, largely by South African engineers, calculations and equipment selection have moved from a trial-and-error to more of a rational

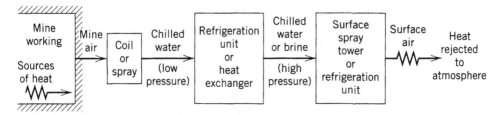

FIGURE 17.11 Schematic of heat-transfer circuits utilized in a conventional mine air conditioning system.

basis. The heat-flow balance procedure developed by Whillier (1977) is generally employed.

17.3 OTHER HEAT-TRANSFER PROCESSES

Recalling that the prime objective of the mine air conditioning system is to remove heat from the working faces and reject it at the surface, provision must be made for certain intermediate steps of heat transfer that occur between the underground and surface atmospheres. In the conventional system, heat transfer takes place between the mine air and chilled water in coils or sprays (Fig. 17.3). The low-pressure chilled water, in turn, originates in an underground refrigeration plant or heat exchanger serviced by a high-pressure fluid cooled in a surface refrigeration plant or cooling tower. A flow sheet of the several heat-transfer and heat-exchange circuits is pictured in Fig. 17.11.

Coils function by cooling, usually coupled with dehumidification; sprays first saturate by evaporative cooling and then cool and dehumidify. Observe that the overall process line (shown dashed in Fig. 17.12) is the same for both coils and sprays. In a coil, heat is transferred by conduction and convection,

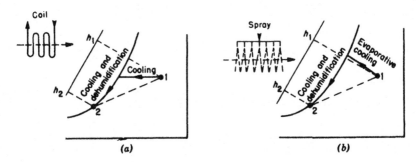

FIGURE 17.12 Comparison of air conditioning by (a) coil and (b) spray, with processes shown on the psychrometric chart.

whereas in a spray, it occurs by convection and evaporation. The same amount of heat ($q = h_1 - h_2$) is removed in either process, theoretically, if saturation is reached, and if water at constant temperature is supplied. In practice, as evidenced by the coil bypass factor, neither is perfectly efficient in heat transfer (80–98%), although coils generally outperform sprays in this regard.

Similar to cooling tower analysis, the analysis here also requires that a heat balance be employed in calculating coil or spray requirements. Equating the heat flows of air and water and neglecting evaporation effects (the error is normally less than 5%),

$$\text{Air } q = \text{water } q$$

$$G \,\Delta h = G_w c \,\Delta t \tag{17.17}$$

where c is specific heat of water $= 1$ Btu/lb·°F (4.19 kJ/kg·°C), Δt is temperature difference of water entering and leaving, and Δh is enthalpy difference of air entering and leaving the tower. Empirical data and charts from the manufacturers are required to specify operating temperatures and other performance conditions. An example demonstrates the use of a heat balance when temperatures are known and only one flow rate must be determined (if one or more temperatures are unknown, then a trial-and-error procedure is followed).

Example 17.7   A heat-transfer unit (coil or spray) operates at the following temperatures:

Water, in	60°F (15.6°C)
Water, out	72°F (22.2°C)
Air in, dry-bulb	85°F (29.4°C)
Air in, wet-bulb	75°F (23.9°C)
Air out, saturated	65°F (18.3°C)

If 50,000 cfm (23.6 m³/s) of air is circulated through the conditioner, how much water is required, disregarding evaporation in the spray? What refrigeration is accomplished? Barometric pressure is 29.92 in. Hg (101.03 kPa).

Solution: Plot the process (as it affects the air) on a psychrometric chart and read properties. Determine weight flow rate of air (Eq. 17.1):

$$G = \frac{(60)(50{,}000)}{14.08} = 213{,}100 \text{ lb/h}$$

Calculate change in heat content of air (Eq. 17.10):

$$q = (213,100)\ (38.50 - 30.05) = 1,800,700 \text{ Btu/h}$$

Amount of refrigeration

$$q_R = \frac{1,800,700}{12,000} = 150 \text{ tons}$$

Assuming that the amount of heat lost by the air is equal to the amount of heat gained by the water, then the weight flow rate of water required is found by Eq. 17.17:

$$G_w = \frac{1,800,700}{(1)(72 - 60)} = 150,100 \text{ lb/h (18.9 kg/s)}$$

Volume flow rate of water:

$$Q_w = \left(\frac{150,100}{60 \text{ min/h}}\right) \left(\frac{7.48 \text{ gal/ft}^3}{62.4 \text{ lb/ft}^3}\right) = 300 \text{ gpm (19.0 L/s)}$$

Coils

The *cooling coil*, constructed with metal tubing and fins, is an extended-surface heat exchanger. In contrast with sprays used for heat transfer, coils make possible a closed and balanced cooling-water circuit, with its attendant advantages of balanced pumping heat, absence of water-level control devices, elimination of flooding danger in multilevel operation, and dual use for summer cooling and winter heating. On the other hand, sprays are simpler and cheaper to construct and install, foul and corrode less, usually require less space, have far lower resistance to airflow, and perform some air cleaning as well as cooling. The choice between the two considers these factors in the application and location in which they are called to operate.

Numerous types and styles of coil arrangements and fin styles are used commercially. They are usually arranged in banks, a typical coil standing 18 tubes high and 6 tubes deep, connected to provide 6–36 parallel water circuits. As with cooling towers and other heat-transfer units, the relative flow direction of the water and air (or second liquid)—counterflow, cross flow, and parallel flow—is probably the most important design consideration in coils. Counterflow is preferred, because it provides the highest possible mean temperature difference.

Coils operating with a bypass factor sacrifice cooling temperature but discharge air that is unsaturated, a highly desirable feature (see Section 17.1). The usual heat balance can be employed to determine chilled-water requirements or the amount of air that can be treated in a coil. Example 17.8 concerns a coil with a bypass factor.

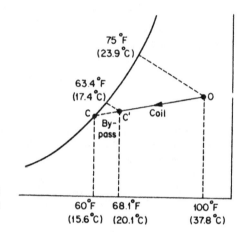

FIGURE 17.13 Heat-transfer process of a coil with bypass. See Example 17.8.

Example 17.8 A cooling coil followed by a fan operates at a temperature of 60°F (15.6°C) with a 20% bypass factor. Assuming that 1500 gpm (94.63 L/s) of chilled water is supplied the coil at a temperature differential of 15°F (8.3°C), calculate the quantity of air at 100°F, 75°F (37.8°C, 23.9°C), dry-bulb and wet-bulb, that can be conditioned.

Solution: Lay out the process on a psychrometric chart Fig. 17.13). Air is cooled and dehumidified from state point O (100°F, 75°F) along a straight line connecting O with state point C, the coil temperature (60°F). State point C′, the condition at which the air leaves the conditioner, is located one-fifth (bypass factor = 0.20) of the distance from C to O; t_d = 68.1°F, t_w = 63.4°F. Read enthalpies at points O and C′: h_1 = 38.36, h_2 = 28.87 Btu/lb. Convert water volume to weight flow rate:

$$G_w = (1500) \frac{(60)(62.4)}{7.48} = 750{,}800 \text{ lb/h}$$

Calculate the amount of heat that can be absorbed by the water, using Eq. 17.17:

$$q = (1)\,(750{,}800)\,(15) = 11{,}262{,}000 \text{ Btu/h}$$

This is equal to the amount of heat removed from the air. Find the weight flow rate of air by Eq. 17.10:

$$G = \frac{11{,}262{,}000}{38.36 - 28.87} = 1{,}186{,}700 \text{ lb/h}$$

At point O, v = 14.39 ft³/lb. Find Q by Eq. 17.1:

$$Q = \frac{(14.39)(1,186,700)}{60} = 284,600 \text{ cfm } (134.3 \text{ m}^3/\text{s})$$

Sprays

In contrast with a coil, a *spray* is a direct-contact heat-transfer device. It consists of a chamber or enclosure (the walls of the mine opening, if installed underground), a spray-nozzle system, a tank or sump for collection of used spray water, and usually an eliminator at the discharge to remove entrained water droplets from the air. Two types of spray equipment are used in mine air conditioning: (1) spray chambers, or air washers, and (2) cooling towers, a special form previously discussed. So-called evaporative coolers, which are wetted-media air coolers, used in building air conditioning, have had little application in mines because of higher first cost and maintenance. Only basic spray devices are dealt with here.

Unlike the situation with cooling towers and coils, there is little standardization in spray chambers. A few mines construct their own. Manufacturers publish tables giving specifications and performance of their sprays. Air velocity and quantity, spray density, water pressure and volume, and other design factors must be considered in each application.

In predicting spray-chamber performance, a rigorous analysis is neither possible nor warranted [for a design procedure, see Whillier (1974a)]. Performance objectives in cooling with spray chambers vary widely. Generally, the spread between air dry- and wet-bulb temperatures at discharge is 1°F (0.6°C) or less. The rise in water temperature is usually between 5 and 12°F (2.8 and 6.7°C). Depending on spray operating conditions, the spread in exit-air wet-bulb and water temperatures varies from about 7–15°F (3.9–8.3°C).

Shell-and-Tube Heat Exchangers

Heat-transfer units of this type are most commonly used in chemical plants and oil refineries when it is desired to cool one liquid while heating the other but to maintain the two liquids separate. The principle of operation is the same as for cooling coils in which only sensible cooling occurs, and the same heat-transfer equations apply. The purpose of *shell-and-tube heat exchangers* in mine cooling systems, however, is to utilize chilled water prepared at the surface to cool deep underground workings without having to resort to high-pressure piping throughout the system or pressure-reduction devices to overcome the attendant high static head.

The construction of this type of heat exchanger resembles that of a refrigeration evaporator or condenser (Fig. 17.7). In design, the objective is generally to produce a desired flow volume of chilled water on the shell side at as low an outlet temperature as possible. To achieve optimal cooling, the cold-water approach between tube side and shell side should be as close as possible. Although the approach can be effectively zero in exchangers with large surface areas, mine installations involving high-pressure and low-pressure lines and very large flow volumes are only capable of achieving an approach in the range of 3–7°F (1.7–3.9°C).

17.4   MINE COOLING LOAD

The design of systems for the cooling and dehumidification of mine air is contingent on first determining the mine cooling load. Cooling load, as defined in Section 17.1, is the total amount of sensible and latent heat required to be removed from a space to maintain desired design conditions. For operating mines, it is possible to measure the actual amount of heat generated in underground operings by observing temperature changes in a known weight flow rate of mine air, although if several heat-generating processes are operating simultaneously, separation of the processes is usually difficult. In an actual underground heat analysis, the dynamic nature of the mine airflows, diurnal and seasonal inlet air-temperature changes, groundwater inflow and evaporation fluctuation, variations in wall-rock thermal conductivity and other thermal properties, aging of mine airways, nonadiabatic compression in shafts, and the mobility and changing nature of electromechanical and other heat sources add to the uncertainties and complications of the solution.

When plans are being developed to deepen or laterally extend operating mines, the additional amount of heat produced may be calculated or extrapolated, on the basis of previous experience. When studies involving the planning of entirely new projected mines are being conducted, however, a detailed analysis of all heat sources is required, including a breakdown of the sensible and latent components. These calculations are complicated by the fact that heat flows from wall rock, underground water, and human metabolism are dependent on the temperature and humidity of the ventilating air. Other heat sources are independent of air temperature: autocompression, machinery and lights, oxidation, blasting, rock movement, and pipelines. The interacting relationships among the various heat sources and installed cooling plants, their variability and complexity, emphasize the distinct advantage of utilizing computer simulation to model the mine air conditioning system, calculate heat sources, predict ambient air temperatures, calculate cooling load requirements, and locate mine air cooling plants for maximum positional efficiency.

The quantitative results of a study of heat sources and cooling loads in seven underground mines in the United States and Canada are depicted in Fig. 17.14. Several of the mines employ air conditioning, and all are considered "hot." A heat balance of the results appears in Fig. 17.15, in which heat sources are compared with heat losses and averaged on a percentage basis. The order of importance of the heat sources corresponds in general to

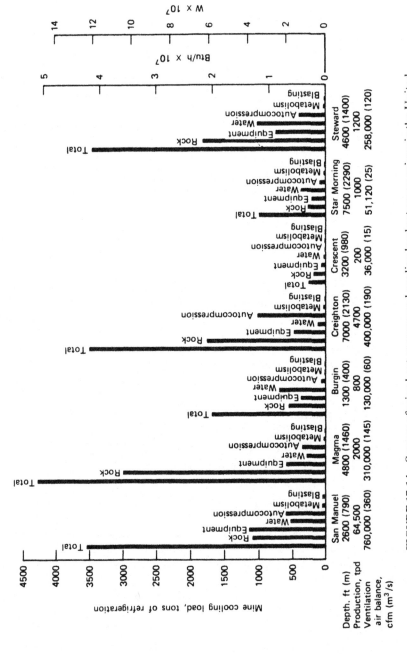

FIGURE 17.14 Survey of mine heat sources and cooling loads at seven mines in the United States and Canada. (After Enderlin, 1973.)

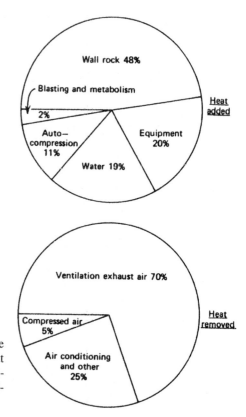

FIGURE 17.15 Average heat balance for hot mines of Fig. 17.14. Heat sources are shown above and heat-removal mechanisms below. (After Enderlin, 1973.)

the rank order in Section 16.2; notice that the wall rock and autocompression account on average for about 60 % of the heat liberated (Fig. 17.15). The peak cooling load, 4250 tons refrigeration (51,000,000 Btu/h), was measured at the Magma copper mine in Arizona (Fig. 17.14).

Cooling Load Determination

In designing the air conditioning system for a new mine, the cooling load is determined by calculating or estimating the magnitude of each operative heat source, using the methods developed in Sections 16.2 and 16.5, and then by summation calculating the sensible and latent components of the total cooling load. To demonstrate the procedure, a design example is solved using the results already obtained from the individual heat-source calculations.

Example 17.9 Air conditioning is being planned for a newly developed mine. Given conditions and calculation results relevant to cooling and dehumidification may be found in the examples of Chapter 16. Heat gains are reiterated here. Determine the mine cooling load and its sensible and latent components.

Solution:

1. Autocompression (Example 16.1). Rather than calculate a heat gain, the state point of the air delivered to the bottom of the 5000-ft (1524-m) shaft was estimated:

$$t_d = 106.5°F \ (41.4°C)$$
$$t_w = 72.0°F \ (22.2°C)$$

   Air entered the top of the shaft at $t_d = 80°F$ (26.7°C) and $t_w = 60°F$ (15.6°C).
2. Wall rock (Example 16.3). Assume 6 drifts of 1000 ft (300 m) each:

$$q_s = (6) \left(\frac{1000}{100} \right) (64,620) = 3,877,200 \text{ Btu/h } (1136 \text{ kW})$$

3. Underground water (Example 16.5):

$$q_s = 1,200,000 \text{ Btu/h } (35 \text{ kW})$$
$$q_L = 1,080,000 \text{ Btu/h } (317 \text{ kW})$$

4. Machinery and lights (Examples 16.6, 16.8):

$$\text{Machinery } q_s = 1,094,000 \text{ Btu/h } (321 \text{ kW})$$
$$\text{Lights } q_s = 106,600 \text{ Btu/h } (31.2 \text{ kW})$$

5. Human metabolism (Example 16.10):

$$q_s = 196,800 \text{ Btu/h } (57.7 \text{ kW})$$
$$q_L = 459,200 \text{ Btu/h } (135 \text{ kW})$$

6. Other heat sources: considered negligible in this example.
7. Cooling load summation:

$$q_S = 5,394,600 \text{ Btu/h } (1,580,995 \text{ kW})$$

$$q_L = \underline{1,539,200} \text{ Btu/h } (451,093 \text{ kW})$$

$$q = 6,933,800 \text{ Btu/h } (2,032,089 \text{ kW})$$

8. Required refrigeration:

$$q_R = \frac{6,933,800}{12,000} = 578 \text{ tons}$$

Once the cooling load is calculated, determination of the remaining cooling plant specifications proceeds as outlined previously in this chapter, with the selection of suitable psychrometric processes, their layout with state points on the psychrometric chart, and the designation of a suitable chilled-water source, underground heat exchanger (if needed), and heat-transfer devices at the working faces. The final step is the design of the overall air cooling system, including plant type, location, size, and cost, discussed next. Complete design problems are found in Johnson (1992b) and McPherson (1993).

17.5 MINE COOLING PLANTS

The goal of mine air conditioning is to improve air cooling power. Cooling of some type is necessary when ventilation alone is insufficient to maintain adequate comfort conditions in deeper, hotter workings.

Van der Walt and Whillier (1979) group engineering requirements for mine cooling installations into five categories, each of which must be analyzed thoroughly when making plant design decisions:

1. Physiological parameters
2. Surface meteorological parameters
3. Geologic parameters
4. Production and mining parameters
5. Ventilation parameters

Next, an appropriate strategy is established that helps to determine when each subsequent cooling phase, based on depth and rock temperature, should

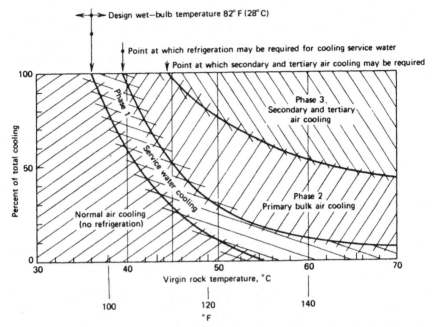

FIGURE 17.16 Strategy for air conditioning South African gold mines in three phases, according to depth and rock temperature. (After Van der Walt and Whillier, 1980. By permission from AIME, New York. Copyright 1980.)

be introduced. Current South African practice holds that when air conditioning is necessary, it should be introduced in three phases (Fig 17.16). *Phase 1* involves the use of refrigeration to cool the mine service water to the lowest practical temperature. *Phase 2* involves cooling the incoming ventilation air in bulk to counteract the effects of humid summertime air, autocompression, and sources of heat in the shaft system. *Phase 3* involves recooling the ventilation air from time to time in the stopes by the use of water cooling coils. Details are not strictly applicable to all mines because of varying conditions, of course, but are a good illustration of the application of years of experience in one large group of similar mines.

Among major design decisions involving the cooling plant are these (Johnson, 1992b):

1. *Method of Rejecting Plant Heat Underground.* There are three choices: (a) using mine drainage water as condenser water, (b) using exhaust air to cool condenser water in underground cooling towers, and (c) using chilled water prepared on the surface to cool mine air in heat transfer units. Method (a) is limited to small systems, (b) is popular in deep South African gold mines circulating large quantities of unsaturated air, and (c) is now the most widely used method, here and abroad.

2. *Method of Cooling Mine Air.* Again there are several alternatives: (a) evaporative cooling, (b) coils and sprays, and (c) direct expansion cooling coils integrated into the refrigeration unit. Method (a) is limited to dry surface air, (b) has general application, and (c) is restricted to small plants.

3. *Method of Handling Underground Water.* High static heads on chilled-water systems pose a problem in deep mines. If pipeline pressures in conventional units are unacceptable, underground heat exchangers are usually installed, and low-pressure water is circulated in the mine. Mine sumps or turbines offer alternatives, or heavy-duty pipe may be installed.

Classification of Cooling Plants

The following outline classifies the commonly used types of mine cooling plants or systems, mainly on the basis of location, secondarily on place of heat rejection, and finally on method of cooling.

Surface cooling plant (air cooling and heat rejection at surface)
  1. Refrigeration
  2. Evaporative cooling in cooling towers
Underground cooling plant (air cooling and heat rejection underground)
  1. Surface water with rejection to drainage water
  2. Refrigeration units with heat rejection to drainage water
  3. Refrigeration with heat rejection to underground towers
  4. Cooling of service water
Combination cooling plant (air cooling underground and heat rejection at surface)
  1. Direct evaporative cooling of water with towers
  2. Indirect cooling with refrigeration
  3. Combinations of direct and indirect cooling
  4. Combinations of underground chillers

Surface Cooling Plant   This was the original type of cooling employed at Brazil's Morro Velho and South Africa's Robinson Deep mines (Carrier, 1938), then falling into disfavor, and now finding new acceptance (Johnson, 1992b; O'Neil and Johnson, 1982). In this system, all of the downcast air is passed through a cooling plant at the top of the downcast shaft, and the heat is rejected directly to the atmosphere. Since the air usually has to travel a considerable distance to the workings, and since rock heat is usually the principal heat source, the air suffers considerable reheating before it arrives at the workings where it is needed. This type of system is normally practical only when the major source of heat is in the surface air itself as it enters the shaft, such as exists in very hot climates during the summer months. It is not practical where the surface air and the underground rock are both major heat sources. A significant advantage is the avoidance of heat exchangers underground and chilled water lines in the shaft. The cooling method used with this type of plant is usually refrigeration, although evaporative cooling of the air at the surface might be feasible in extremely dry climates.

Underground Cooling Plant   This is the plant location formerly favored in South Africa and employed to a lesser extent elsewhere.

1. *Surface Water with Heat Rejection to Drainage Water.* An unusual situation, this method necessitates a large supply of cold surface water such as from a nearby lake or stream. The water is taken underground and used in heat exchangers to cool air, and then the water is discharged into the drainage system to be pumped to the surface. A few Idaho mines with abundant surface water have used it (Bossard and Stout, 1973).

2. *Refrigeration Units with Heat Rejection to Drainage Water.* In these systems, refrigeration chillers sited underground are used to cool air in one of two ways. In one system, the refrigerant is used to cool air directly by means of expansion coils, a basis for most portable self-contained units called spot coolers. In another system, mine water is admitted to the condensers of central refrigeration plants, and the chillers of these plants are used to cool water that is then circulated to heat exchangers or spray ponds at strategic locations. The condensers are cooled by mine water, which is subsequently discharged into the mine drainage water. This arrangement requires an ample supply of mine water and thus is not applicable on a large-scale basis for mines located in relatively dry regions. However, if it is necessary to pump large quantities of water from the mine anyway, the water can first be utilized as condenser water for the refrigeration units. Even water that is fairly warm to start with, up to 120°F (40°C), can be utilized in this manner if the refrigeration machinery and refrigerant are correctly selected.

3. *Refrigeration with Heat Rejection to Underground Cooling Towers.* This is the system most commonly used in the South African gold mines (Grave, 1974; Whillier, 1977). The condenser water, rather than being rejected, is circulated through cooling towers located at the entrances to exhaust shafts. Large amounts of upcast air must be available, since the air is usually near saturation and consequently cannot absorb much water in the evaporative cooling of the condenser water.

4. *Cooling of Service Water.* This is a recently developed concept that has been widely adopted in South Africa as standard practice to supplement, but generally not to wholly replace, conventional air cooling systems (Van der Walt and Whillier, 1979, 1980). In this method, heat transfer occurs at the face as chilled service water for dust control is applied to newly blasted rock and exposed wall rock. For effective heat removal, the incoming water temperature must be less than 60°F (15°C), which may require insulating the

pipes. The water flow rate to be supplied is calculated by heat transfer equations (O'Neil and Johnson, 1982; McPherson, 1993).

Combination Cooling Plant   This is the commonly used system in American and German mines (Bossard and Stout, 1973; Hamm, 1980; Van der Walt et al., 1996) and is gaining more acceptance in South Africa.

1. *Direct Evaporative Cooling of Water with Towers.* In this system, heated water from underground cooling plants is taken to the surface, cooled in a conventional cooling tower, and returned to the mine. Because of the limitation that the water can be cooled only to the wet-bulb temperature of the outside air (without air precooling), this system is restricted to relatively dry, cool climates. Use is more common in combination with refrigeration, as discussed below (combination of direct and indirect cooling). The Magma copper mine at Superior, AZ (Dobbs and Johnson, 1979) and some German mines (Hamm, 1980) use this system in winter.

2. *Indirect Cooling with Refrigeration.* Here the mine cooling water is cooled by refrigeration, and the water circulating between surface and underground does not contact the outside air. The water can actually be chilled below 32°F (0°C) if brine is used or an antifreeze reagent such as ethylene glycol is added. The Schlägel–Eisen colliery in Germany employs this practice (Hamm, 1980). The disadvantage of this system over the previous one is that all the cooling capacity must be obtained by expensive refrigeration rather than by inexpensive cooling in a spray tower. Magma's San Manuel, AZ mine uses this system but has since supplemented the refrigeration unit with a surface cooling tower for summer use (O'Neil and Johnson, 1982).

3. *Combinations of Direct and Indirect Cooling.* In this system, water from the mine is first cooled in towers, then further cooled in refrigeration machinery before going underground. In winter, the tower alone is often sufficient; in summer, the surface air wet-bulb temperatures sometimes are higher than that of the water from the mine, so the towers are bypassed. This is the type of system used at the Magma mine at Superior (Dobbs and Johnson, 1979).

4. *Combinations with Underground Chillers.* Rather than the chilled water being brought to its ultimate low temperature on the surface, the water is sometimes passed through heat exchangers in combination with underground refrigeration units.

Comparison of Cooling Plant Features

As discussed, the location and arrangement of the cooling plant are probably the critical decisions to be made in designing a mine air conditioning system.

TABLE 17.1 Comparison of Locations for Mine Cooling Plants of the Refrigeration Type

Factor	Surface Plant	Underground Plant	Combination Plant
Capital cost	Same	Same	Considerably higher
Operating cost	Same	Same	Same
Installation	Simple; does not interfere with underground work	Underground excavation required	Much shaft work for pipelines
Heat disposal	Simple	Difficult, unless mine water plentiful	Simple
Water disposal	Simple	Additional expense in dry mining	
Supplying treated air to workings	May be complicated	Simple	Simple
Heat gain in transit	Large, due to autocompression	Small	Small
Refrigerant	Can use ammonia	Costly; must be safe	Can use ammonia
Natural ventilation	Increases	Little effect	Little effect
Supplying fresh air to conditioner	Simple	Difficult; small shafts may limit supply and necessitate recirculation	
Effect of temperature difference on personnel	Little	Great	Great
Intermittent operation	Unsuitable	Economically advisable only with short distribution drifts, few workings	

Source: Spalding (1949).

Table 17.1 summarizes the advantages and disadvantages of the three possible locations of refrigeration plants.

The final consideration in selecting the system, of course, is cost. Comparative costs—capital, operating (power), and maintenance—are provided in Table 17.2. Actual costs are not given, since these soon become obsolete. Capital costs are total, whereas operating and maintenance costs are expressed on an annual basis. Depreciation costs and costs of pumping service and/or drainage water are not included.

Underground Area Cooling versus Localized Cooling

Some hot mines utilize centralized underground plants for control or bulk cooling of air by chilled-water plants; that is, one plant cools a large area of the mine. One such system is shown in Fig. 17.17, typical of the plants used in Butte. However, when working headings are distant or scattered, this arrangement results in low positional efficiency. The air is heated sensibly in transit and in many cases also picks up moisture, which severely limits

TABLE 17.2   **Comparative Costs of Mine Air Cooling Plants**[a]

Type of Plant	Total Capital	Annual Operating	Annual Maintenance
Heat rejection underground			
Refrigeration units with heat rejection to Drainage water			
Direct air cooling	127	39	21
Water chilling	33	36	1
Refrigeration units with rejection to underground towers	37	39	1
Underground chilling of service water	15	29	1
Heat rejection at surface			
Direct evaporative cooling of water with towers			
With water–water heat exchangers	91	15	4
With underground turbine	65	19	1
Cooling of water with surface refrigeration units and condenser water tower			
With water–water heat exchangers	106	39	4
With underground turbine	81	43	2

Source: O'Neil and Johnson (1982).

[a] Costs are relative only. Capital costs are total, whereas operating and maintenance are expressed on an annual basis.

the air cooling power in the production area. Aside from possible positional disadvantages, area plants perform very well in cooling air. Experience with underground area cooling plants indicates that the maximum effective distance from plant to workplaces is about 1500 ft (450 m). The distance may be decreased when exigencies of mine layout require recirculation of used air from stoping. The optimal location of the plants should be determined from the completed mining plan.

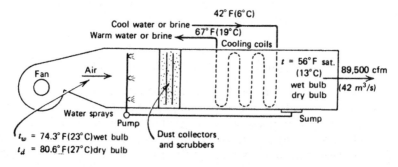

FIGURE 17.17   Schematic of underground area cooling plant used at Butte, MT. Capacity: 455 tons refrigeration. (After O'Neil and Johnson, 1982.)

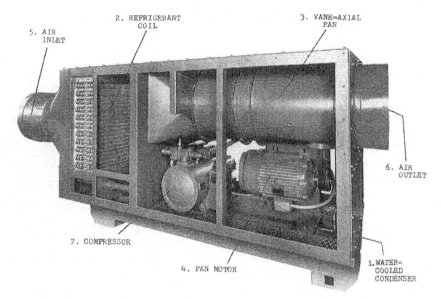

5. AIR
INLET

2. REFRIGERANT
COIL

3. VANE-AXIAL
FAN

6. AIR
OUTLET

7. COMPRESSOR

4. FAN MOTOR

1. WATER-
COOLED
CONDENSER

FIGURE 17.18 Commercially available, 40-ton spot cooler mounting expansion coil, compressor, water-cooled condenser, and fan. Airflow 12,000 cfm (5.7 m³/s), air cooling 90–60°F (32–16°C), water 80°F (27°C). (By permission from Yale, Inc., Minneapolis.)

If positional efficiency of the air conditioning plant becomes a problem, it is preferable to pump the chilled water to localized cooling plants, called *spot coolers,* close to individual production units or headings Fig. 17.18. There are several designs of localized cooling plants (Marks, 1980). These relatively light, semiportable units are attached to the normal auxiliary ventilation system, requiring no extra excavation. The fan gives good service downstream (although there is a temperature rise in the air as it passes through the fan), and no separate spray and dust collection system is required. The source of chilled water is a refrigeration unit, utilizing potable water in the condenser and discharging to the mine-water pumping system.

Spot coolers avoid excessive exposure of cooled air to hot wall rock, as the air is not delivered great distances along an airway by an auxiliary ventilation system. They are normally located within 200–300 ft (60–90 m) of the workplace. Dramatic evidence of the high positional efficiencies of spot cooling is manifest in two ways: (1) low effective temperatures in the workplace and (2) reduced overall air cooling plant requirements.

Alternative Cooling Systems

Mine air conditioning by conventional cooling processes in extreme cases may not be cost-effective or achieve satisfactory results. Alternative methods, some new or experimental, are being utilized or show promise for the future.

Microclimate Cooling   If customary mine cooling systems are unable to provide adequately for the individual worker, or when heat is an isolated problem not justifying the expense of large centralized cooling plants, management may have to adopt unconventional, localized methods of air conditioning. Space-age developments have demonstrated the feasibility of cooling and dehumidifying the space immediately adjacent to workers rather than their entire ambient surroundings or a portion of the mine itself. Such methods are termed *microclimate cooling* and have had limited use in mines, especially for isolated tasks of temporary nature such as shaft sinking (O'Neil and Johnson, 1982).

Air cooling can be provided using the *vortex tube,* a form of local pneumatic refrigeration. This system is small and compact and can be used by individuals wearing a heat-protective helmet, jacket, or complete suit, or the unit can be used for enclosures such as hoist cabs or shops. A compressed-air hose must be attached to a vortex tube, which separates the air into a cold-air fraction and a hot-air fraction. The cold air is directed into the jacket or other area to be cooled, thus providing direct spot cooling. The hot-air fraction is discharged directly to the mine atmosphere. Disadvantages to this system are the expense of compressed air, nuisance of the trailing hose where used in a helmet or suit, encumbrance of the helmet or suit, and presence of oil mist in the compressed air.

Liquid-conditioned suits, modified from the cooling suits used in the Apollo space program, are also available. In some configurations, the suits have to be connected to a cold-water source by means of a trailing hose and possess the same disadvantages of encumbrance, nuisance of the trailing hose, and lack of durability. In another version, there is no trailing hose, and heat is rejected to contained ice.

Another microclimate unit consists of a sleeveless vest jacket lined with about 10 lb (4.6 kg) of ice in removable bags. The ice is contained in a tough, plastic liner that is worn over a woolen vest next to the worker's body. Tests indicate that 10 lb (4.6 kg) of ice undergoing a phase and temperature change to warm water will protect a person working moderately hard for 4 h in conditions of severe heat and humidity. Two prefrozen jackets are required per shift. With this system, the workers are completely mobile, and the cost of the vest jacket is relatively low.

Cooling from Natural Ice Underground   In a rather unique system at one of the large underground nickel mines in the Sudbury basin of Canada, the extreme temperature variation between winter and summer is used advantageously in providing natural air conditioning at high latitudes (Stachulak, 1989). Air entering the mine in winter passes through two large abandoned stopes near the surface; water is sprayed into the air and freezes, in turn warming and humidifying the air for heating (Section 17.6). The heat flow is reversed in summer as the ice that has been formed in winter melts and cools the incoming air. With 400,000 cfm (190 m³/s) of air circulating, the discharge temperature year-round is 32–35°F (0–2°C); the equivalent refrigeration effect is 1500–2000 tons (5280–7030 kW). This system, unfortunately, cannot be utilized at mines located in temperate climates but could be applicable at high elevations.

In recent years, in South Africa, ice in water slurries has been used in underground heat exchangers to chill water circulated to air cooling stations (Sheer et al., 1985; Shone and Sheer, 1988).

Potential Applications Some alternative cooling methods utilized in other industries may have potential in mines (O'Neil and Johnson, 1982). They include

1. Air-compression—air-motor system used in aircraft
2. Vacuum cooler with a high-capacity evacuator (other than a steam jet, probably unsuited underground)
3. Synthetic liquid air
4. Heat pump
5. Thermoelectric cooling used in spacecraft and submarines

17.6 MINE HEATING SYSTEMS

Design Consideration

Although less frequently needed in mines, heating plants have application in cold climates where air temperatures drop below freezing a considerable number of times during the year, sometimes for prolonged periods. In most instances, however, heating is required not for human comfort but for protection of equipment and prevention of damage to intake airways (shafts, slopes, drifts, etc.) Types of damage may include freezing of water lines, buildup of ice on the walls of the opening, and cyclical freezing and thawing of concrete shaft linings (McNider, 1980). Raising the temperature of the intake air to 34°F (1°C) is usually adequate to prevent shaft or equipment damage. For comfort purposes, heating subfreezing intake air to 40 or 45°F (4 or 7°C) generally suffices. Heating is required only for the duration of the subfreezing temperatures; further, design temperatures should be based on average and not extreme low values. The plant can be planned to process all or a portion of the intake air (bypass and mixing may be preferable if there are flow-rate limitations in the heating equipment but higher temperatures are feasible) and to operate only during working shifts.

In designing the heating plant for a mine, the process or processes are represented on the psychrometric chart. The simplest plant involves heating only (Fig. 17.2*a*), or it may include mixing as well (Fig. 17.2*f*). For pertinent calculations of the heating load and plant size and layout of the processes, see Examples 17.1 and 17.5; for additional details, see McPherson (1993).

Heating Methods

Intake air entering the mine may be heated in any of several ways, listed in approximate order of decreasing operating cost and frequency of application (Johnson et al., 1982):

1. Direct heating with furnaces
2. Indirect heating with furnaces
3. Utilization of surface waste heat in heat exchangers
4. Heat recovery from warm exhaust air using heat exchangers
5. Liberation of heat from freezing water in ice chambers

Direct Heating (Fig. 17.19a)   The simplest and oldest mine heating plant, still in common use, is direct heating with a furnace. Outside intake air is passed through the furnace directly, as in domestic and industrial hot-air heating. Since the exhaust products are often added to the discharge air, heat-transfer efficiency is high, and essentially no heat is lost. Carbon monoxide may form in the exhaust, however, and proper combustion adjustment is critical and a toxic-gas monitor mandatory. Coal, oil, and gas are the usual

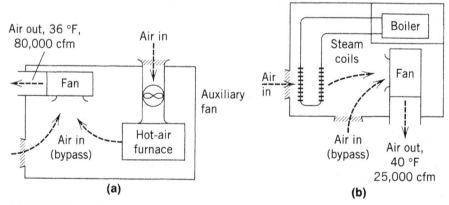

FIGURE 17.19   Furnace heating plants: (*a*) direct; (*b*) indirect (After Lewis, 1951.)

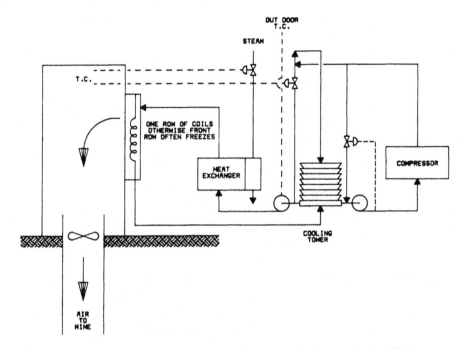

FIGURE 17.20 Surface waste heating. (After Johnson et al., 1982.)

fuels; they are injected through nozzles and burned in flame jets. Where electricity is cheap, electric resistors are used for direct heating in place of a furnace, eliminating any exhaust contamination. In these systems, thermostatic control provides even heat and efficient operation.

*Indirect Heating (**Fig. 17.19b**)*   This plant is a refinement of the first, using the furnace to warm the heat-transfer fluid (water or glycol–water) that in turn heats the intake air in a heat exchanger. A boiler may also be used to prepare steam as the fluid. No combustion products enter the airstream, although there is some penalty to heating efficiency. Furnace, fuel, and controls are the same as in the direct method.

*Surface Waste Heat (**Fig. 17.20**)*   A variety of sources of waste heat typically are available in the surface plant adjacent to the main fresh air intake at most mines (Johnson, 1992b). Rather than construct a separate heating plant if warming of the ventilating air becomes necessary, it may be possible to harness one or more of the waste heat sources using a heat exchanger to transfer heat into the air. Promising sources are steam plant boilers, cooling water from air compressors or electrical generators, mineral processing, and geothermal energy supplied by heat pumps.

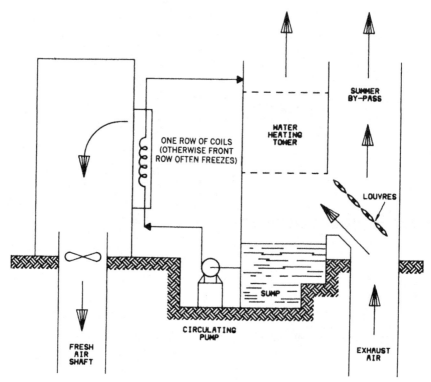

FIGURE 17.21   Heating from exhaust air. (After Johnson et al., 1982.)

***Heat from Exhaust Air* (*Fig. 17.21*)**   This is an ideal arrangement when the portals of the mine intake and exhaust airways, usually downcast and upcast shafts, are in reasonably close proximity (McPherson, 1993). A cooling tower is constructed over the exhaust opening that heats water for transmission to a heat exchanger located over the intake opening. The exhaust air dry-bulb temperature should be at least 70°F (21°C) for effective warming of the intake air. If the heat gain is insufficient, another heat source may be used in conjunction with the heat-recovery method or some air recirculation may be incorporated (Hall et al., 1989b).

Ice Chambers   A novel heating method is utilized at a Canadian mine, the obverse of the one described previously for cooling in Section 17.5 (Johnson et al., 1982). In winter, water is sprayed into empty stopes near the surface, and air is drawn through the stopes. The latent heat of freezing the water together with the sensible heat of the water supplies heat to the air. The ice so formed can then be used to cool air in the summer and the water pumped to the surface.

PROBLEMS

17.1 Air at sea level is to be heated from $t_d = 25°F$ ($-3.9°C$) and $t_w = 20°F$ ($-6.7°C$) to $t_d = 40°F$ ($4.4°C$). Determine the amount of heat required, assuming that 25,000 cfm (11.80 m^3/s) of air is flowing.

17.2 Air at sea level is to be cooled and dehumidified from $t_d = 110°F$ ($43.3°C$) and $t_w = 105°F$ ($40.6°C$) to $60°F$ ($15.6°C$) saturated. Determine the amounts of heat and moisture removed, assuming that 40,000 cfm (18.88 m^3/s) of air is flowing.

17.3 Air at sea level is evaporatively cooled to saturation. Original conditions are $t_d = 100°F$ ($37.8°C$) and $t_w = 80°F$ ($26.7°C$). Write the heat balance and determine the changes in heat and moisture contents.

17.4 Air at $105°F$ ($40.6°C$) saturated and flowing at the rate of 30,000 cfm (14.16 m^3/s) is mixed with 55,000 cfm (25.96 m^3/s) of air at $t_d = 130°F$ ($54.4°C$) and $t_w = 90°F$ ($32.2°C$). If the barometric pressure is 29.92 in. Hg (101.03 kPa), what is the resulting state point?

17.5 An underground workplace at sea level is to be maintained at design conditions of $85°F$ ($29.4°C$) dry-bulb and $80°F$ ($26.7°C$) wet-bulb. The cooling load is 800,000 Btu/h (234 kW), with a sensible-heat factor of 0.37. Surface air is delivered underground to a fan at $95°F$ ($35°C$) dry-bulb and $85°F$ ($29.4°C$) wet-bulb. Determine the following mine cooling requirements:
   (a) Layout of process on psychrometric chart, indicating all pertinent temperatures, enthalpies, humidities, and specific volumes
   (b) Apparatus dew point, assuming zero bypass factor
   (c) Weight flow rate and quantity of air at fan, located at coil entrance
   (d) Coil load, refrigeration, and water rejected
   (e) Cooling of work place (tons refrigeration) and entering air quantity and specific weight.

17.6 Rework Problem 17.5 assuming surface air conditions (t_d, t_w) are $90°F$, $75°F$ ($32.2°C$, $23.9°C$), at 5000 ft (1524 m) elevation. Consider autocompression effects.

17.7 Air is to be conditioned for cooling underground workplaces with a cooling load of 1,800,000 Btu/h (528 kW) sensible and 2,200,000 Btu/h (645 kW) latent. Design temperatures (t_d, t_w) are $103°F$, $79°F$ ($39.4°C$, $26.1°C$) for surface fresh air and $95°F$, $84°F$ ($35.0°C$, $28.9°C$) in the stopes. The conditioner, a cooling coil, is located at sea level at the portal of an adit leading to the workplaces. No recirculated air is employed. Design the simplest possible air conditioning process to supply air for these requirements. The following information is required:

(a)   Labeled sketch of processes on psychrometric chart, indicating temperatures, enthalpies, and specific humidities.
(b)   Apparatus dew point.
(c)   Weight flow rate of air required.
(d)   Quantity rating of fan on fresh air and air specific weight.
(e)   Quantity rating of fan on conditioned air and air specific weight.
(f)   Coil load, refrigeration, and water removed.

17.8   Assuming that the mine of Problem 17.7 is located at 2500 ft (762 m) elevation, calculate the air conditioning requirements. Change surface air temperatures to 98°F, 79°F (36.7°C, 26.1°C) dry-bulb, wet-bulb.

17.9   An underground heat exchanger (elevation sea level) operates with air in at 96°F, 90°F (35.6°C, 32.2°C) dry-bulb, wet-bulb, and out at 62°F (16.7°C) dry-bulb and 95% relative humidity. The temperature spread of the chilled water is 15°F (8.3°C). If 500 gpm (31.8 L/s) of water is supplied the coil, how much air (cfm or m³/s) can be conditioned?

17.10   Design a heat exchanger to supply the cooling requirements of Problem 17.5. Specify water temperatures and other design conditions for a coil that meet good performance standards. Select a coil configuration that is suitable.

17.11   For Problem 17.5, select a spray heat-transfer unit to meet the cooling requirements. Use empirical guidelines to specify operating conditions.

17.12   Calculate the heat to be removed from the air entering the excavation in Problem 16.4, assuming that the air at the far end of the excavation is not to exceed 80°F (29.4°C). Also calculate the amount of 45°F (7.2°C) cooling water needed if the heated water is not to exceed 70°F (21.1°C).

17.13   Recalculate Problem 17.12 given that the excavation contains the equipment of Problem 16.7 and the lighting of Problem 16.10.

17.14   Determine (by psychrometric chart) the amount of heat required to raise the temperature of 250,000 cfm of air (118 m³/s) from 10 to 34°F (−12 to 1°C) at an elevation of 5000 ft (1524 m) above sea level. Employ an equal quantity of exhaust air for heat exchange, available at 75°F (24°C) saturated. What will be the exhaust air temperature at discharge?

Reference Tables and Figures

TABLE A.1 Barometric Pressure, Temperature, and Air Specific Weight at Different Altitudes

Altitude Above or Below Sea Level, Z ft	Barometric Pressure p_b psi	Barometric Pressure p_b in. mercury	At Constant $t = 70°F$ Relative Air Specific Weight	At Constant $t = 70°F$ Air Specific Weight, w lb/ft³	At Varying t and Z Air Temperature, °F	At Varying t and Z Air Specific Weight, lb/ft³
−1000	15.23	31.02	1.037	0.0778	73.8	0.0771
−500	14.94	30.47	1.018	0.0764	71.9	0.0761
0	14.70	29.92	1.000	0.0750	70.0	0.0750
500	14.42	29.38	0.981	0.0736	68.1	0.0740
1,000	14.16	28.86	0.964	0.0723	66.1	0.0730
1,500	13.91	28.33	0.947	0.0710	64.2	0.0719
2,000	13.66	27.82	0.930	0.0698	62.3	0.0709
2,500	13.41	27.31	0.913	0.0685	60.4	0.0698
3,000	13.16	26.81	0.896	0.0672	58.4	0.0687
3,500	12.92	26.32	0.880	0.0660	56.5	0.0676
4,000	12.68	25.84	0.864	0.0648	54.6	0.0666
4,500	12.45	25.36	0.848	0.0636	52.6	0.0657
5,000	12.22	24.89	0.832	0.0624	50.7	0.0648
5,500	11.99	24.43	0.816	0.0612	48.8	0.0638
6,000	11.77	23.98	0.799	0.0599	46.9	0.6628
6,500	11.55	23.53	0.786	0.0590	45.0	0.0619
7,000	11.33	23.09	0.774	0.0580	43.0	0.0610
7,500	11.12	22.65	0.758	0.0568	41.0	0.0600
8,000	10.91	22.22	0.739	0.0554	39.0	0.0590
8,500	10.70	21.80	0.728	0.0546	37.1	0.0581
9,000	10.50	21.38	0.715	0.0536	35.2	0.0573
9,500	10.30	20.98	0.701	0.0526	33.3	0.0564
10,000	10.10	20.58	0.687	0.0515	31.3	0.0555
10,500	9.90	20.18	0.674	0.0506	29.4	0.0546
11,000	9.71	19.75	0.661	0.0496	27.5	0.0538
11,500	9.52	19.40	0.648	0.0486	25.5	0.0529
12,000	9.34	19.03	0.636	0.0477	23.6	0.0521
12,500	9.15	18.65	0.624	0.0468	21.6	0.0513
13,000	8.97	18.29	0.611	0.0458	19.7	0.0505
13,500	8.80	17.93	0.599	0.0449	17.7	0.0496
14,000	8.62	17.57	0.587	0.0440	15.8	0.0488
14,500	8.45	17.22	0.576	0.0432	13.9	0.0480
15,000	8.28	16.88	0.564	0.0423	12.0	0.0473

Source: Madison (1949, pp. 28–29). By permission from Buffalo Forge Co., Buffalo, NY.

Conversion factors: 1 ft = 0.3048 m, 1 psi = 6.8948 kPa, 1 lb/ft³ = 1.6018 kg/m³, °F = $\frac{9}{5}$°C + 32.

TABLE A.2 Psychrometric Data for Air-Water-Vapor Mixtures

Temperature, °F t	Saturation Pressure of Water and Steam, in. Hg p_s	Properties of Water" and Steam — Enthalpy		Specific Volume of Sat. Steam, ft³/lb v_g	Properties of Dry Air at a Pressure of 29.921 in. Hg abs.		Properties of Mixture of Dry Air and Sat. Steam at a Total Pressure of 29.921 in. Hg abs.		
		Saturated Water," Btu/lb h_f	Saturated Steam, Btu/lb h_g		True Specific Volume, ft³/lb v_a	Enthalpy, Btu/lb h_a	Volume of Mixture per lb of Dry Air, ft³ v_s	Enthalpy of Mixture per lb of Dry Air, Btu h_s	Specific Humidity Grains per lb of Dry Air W_s
1	$3.966\,(10)^{-2}$	−158.5	1061.5	14,080	11.604	0.24	11.62	1.12	5.777
2	$4.178\,(10)^{-2}$	−158.0	1062.0	13,400	11.630	0.48	11.65	1.40	6.084
3	$4.400\,(10)^{-2}$	−157.6	1062.4	12,750	11.655	0.72	11.56	1.68	6.348
4	$4.633\,(10)^{-2}$	−157.1	1062.8	12,140	11.680	0.96	11.70	1.98	6.745
5	$4.878\,(10)^{-2}$	−156.6	1063.3	11,550	11.706	1.20	11.72	2.28	7.106
6	$5.134\,(10)^{-2}$	−156.1	1063.7	11,000	11.731	1.44	11.75	2.58	7.478
7	$5.402\,(10)^{-2}$	−155.7	1064.2	10,480	11.756	1.68	11.78	2.88	7.867
8	$5.683\,(10)^{-2}$	−155.2	1064.6	9979	11.782	1.92	11.80	3.18	8.280
9	$5.977\,(10)^{-2}$	−154.7	1065.1	9507	11.807	2.16	11.83	3.49	8.711
10	$6.286\,(10)^{-2}$	−154.2	1065.5	9060	11.832	2.40	11.86	3.80	9.161
11	$6.608\,(10)^{-2}$	−153.7	1065.9	8636	11.857	2.64	11.88	4.11	9.633
12	$6.946\,(10)^{-2}$	−153.3	1066.4	8234	11.883	2.88	11.91	4.43	10.13
13	$7.300\,(10)^{-2}$	−152.8	1066.8	7851	11.918	3.12	11.94	4.75	10.64
14	$7.669\,(10)^{-2}$	−152.3	1067.3	7489	11.933	3.36	11.96	5.07	10.64
15	$8.056\,(10)^{-2}$	−151.8	1067.7	7144	11.959	3.60	11.99	5.40	11.18
16	$8.461\,(10)^{-2}$	−151.3	1068.1	6817	11.984	3.84	12.02	5.73	11.73
17	$8.884\,(10)^{-2}$	−150.8	1068.6	6505	12.009	4.08	12.05	6.06	12.34
18	$8.884\,(10)^{-2}$	−150.8	1068.6	6210	12.035	4.32	12.07	6.40	12.96
19	$9.789\,(10)^{-2}$	−149.8	1069.5	5929	12.060	4.56	12.10	6.75	13.51
20	0.1027	−149.4	1069.9	5662	12.085	4.81	12.13	7.10	14.28
21	0.1078	−148.9	1070.3	5408	12.110	5.05	12.15	7.45	14.99
22	0.1130	−148.4	1070.8	5166	12.136	5.29	12.18	7.81	15.73
23	0.1186	−147.9	1071.2	4936	12.161	5.53	12.21	8.18	16.51
24	0.1243	−147.4	1071.7	4717	12.186	5.77	12.24	8.55	17.32

25	0.1303	−146.9	1072.1	4509	12.211	6.01	12.27	8.92	19.04
26	0.1366	−146.4	1072.5	4311	12.237	6.25	12.29	9.31	19.96
27	0.1431	−145.9	1073.0	4122	12.262	6.49	12.32	9.70	20.92
28	0.1500	−145.4	1073.4	3943	12.287	6.73	12.35	10.09	21.92
29	0.1571	−144.9	1073.8	3771	12.313	6.97	12.38	10.49	22.98
30	0.1645	−144.4	1074.3	3608	12.338	7.21	12.41	10.90	24.07
31	0.1723	−143.9	1074.7	3453	12.363	7.45	12.43	11.32	25.21
32	0.1803	−143.4	1075.2	3305	12.389	7.69	12.46	11.75	26.40
33	0.1878	1.0	1075.6	3180	12.414	7.93	12.49	12.16	27.49
34	0.1955	2.0	1076.0	3062	12.439	8.17	12.52	12.57	28.63
35	0.2034	3.0	1076.5	2948	12.464	8.41	12.55	13.00	29.80
36	0.2117	4.0	1076.9	2839	12.490	8.65	12.58	13.42	31.02
37	0.2202	5.0	1077.4	2734	12.515	8.89	12.61	13.86	32.28
38	0.2290	6.0	1077.8	2634	12.540	9.13	12.64	14.30	33.58
39	0.2382	7.0	1078.2	2538	12.565	9.37	12.67	14.75	34.94
40	0.2477	8.0	1078.7	2445	12.591	9.61	12.70	15.21	36.34
41	0.2575	9.0	1079.1	2357	12.616	9.85	12.73	15.68	37.80
42	0.2676	10.1	1079.5	2272	12.641	10.09	12.76	16.16	39.30
43	0.2781	11.1	1080.0	2190	12.667	10.34	12.79	16.64	40.86
44	0.2890	12.1	1080.4	2112	12.692	10.58	12.82	17.13	42.47
45	0.3002	13.1	1080.9	2037	12.717	10.82	12.85	17.63	44.14
46	0.3119	14.1	1081.3	1965	12.742	11.06	12.88	18.13	45.86
47	0.3239	15.1	1081.7	1896	12.768	11.30	12.91	18.66	47.65
48	0.3363	16.1	1082.2	1829	12.793	11.54	12.94	19.19	49.51
49	0.3491	17.1	1082.6	1766	12.818	11.78	12.97	19.73	51.42
50	0.3624	18.1	1083.1	1704	12.844	12.02	13.00	20.28	53.40
51	0.3761	19.1	1083.5	1645	12.869	12.26	13.03	20.84	55.44
52	0.3903	20.1	1083.9	1589	12.894	12.50	13.06	21.41	57.56
53	0.4049	21.1	1084.4	1534	12.919	12.74	13.10	21.99	59.75
54	0.4200	22.1	1084.8	1482	12.945	12.98	13.13	22.59	62.01
55	0.4356	23.1	1085.2	1431	12.970	13.22	13.16	23.20	64.36
56	0.4518	24.1	1085.7	1383	12.995	13.46	13.19	23.82	66.78

(continued)

| Temperature, °F t | Saturation Pressure of Water and Steam, in. Hg p_s | Properties of Water^a and Steam | | | Properties of Dry Air at a Pressure of 29.921 in. Hg abs. | | Properties of Mixture of Dry Air and Sat. Steam at a Total Pressure of 29.921 in. Hg abs. | | |
| | | Enthalpy | | Specific Volume of Sat. Steam, ft^3/lb v_g | True Specific Volume, ft^3/lb v_a | Enthalpy, Btu/lb h_a | Volume of Mixture per lb of Dry Air, ft^3 v_s | Enthalpy of Mixture per lb of Dry Air, Btu h_s | Specific Humidity Grains per lb of Dry Air W_s |
		Saturated Water,^{a} Btu/lb h_f	Saturated Steam, Btu/lb h_g						
57	0.4684	25.1	1086.1	1336	13.020	13.70	13.23	24.45	69.28
58	0.4856	26.1	1086.5	1292	13.046	13.94	13.26	25.10	71.86
59	0.5033	27.1	1087.0	1249	13.071	14.18	13.29	25.76	74.54
60	0.5216	28.1	1087.4	1207	13.096	14.42	13.33	26.43	77.29
61	0.5405	29.1	1087.9	1167	13.122	14.66	13.36	27.11	80.14
62	0.5599	30.1	1088.3	1129	13.147	14.90	13.40	27.82	83.09
63	0.5800	31.1	1088.7	1092	13.172	15.14	13.43	28.54	86.14
64	0.6007	32.1	1089.2	1056	13.197	15.38	13.47	29.27	89.27
65	0.6221	33.1	1089.6	1022	13.223	15.62	13.50	30.03	92.51
66	0.6441	34.1	1090.0	988.6	13.248	15.85	13.54	30.79	95.86
67	0.6668	35.1	1090.5	956.8	13.273	16.10	13.58	31.58	99.32
68	0.6902	36.1	1090.9	926.1	13.298	16.35	13.61	32.38	102.9
69	0.7143	37.1	1091.3	896.5	13.324	16.59	13.65	33.20	106.6
70	0.7392	38.1	1091.8	868.0	13.349	16.83	13.69	34.04	110.4
71	0.7648	39.1	1092.2	840.5	13.374	17.07	13.72	34.90	114.3
72	0.7911	40.1	1092.6	814.0	13.399	17.31	13.76	35.79	118.4
73	0.8183	41.1	1093.1	788.4	13.425	17.55	13.80	36.69	122.6
74	0.8463	42.1	1093.5	763.8	13.450	17.79	13.84	37.61	126.9
75	0.8751	43.1	1093.9	740.0	13.475	18.03	13.88	38.55	131.3
76	0.9047	44.1	1094.4	717.0	13.501	18.27	13.92	39.52	135.9
77	0.9352	45.1	1094.8	694.9	13.526	18.51	13.96	40.51	140.6
78	0.9667	46.1	1095.2	673.5	13.551	18.75	14.00	41.52	145.5
79	0.9990	47.1	1095.7	652.9	13.576	18.99	14.04	42.56	150.6

80	1.0323	48.1	1096.1	633.0	13.602	19.23	14.09	43.63	155.8
81	1.0665	49.1	1096.6	613.8	13.627	19.47	14.13	44.72	161.2
82	1.1017	50.1	1097.0	595.3	13.652	19.71	14.17	45.84	166.7
83	1.1380	51.1	1097.4	577.4	13.678	19.95	14.22	46.98	172.4
84	1.1752	52.1	1097.8	560.1	13.703	20.19	14.26	48.16	178.3
85	1.2136	53.1	1098.3	543.3	13.738	20.43	14.31	49.36	184.4
86	1.2530	54.0	1098.7	527.2	13.753	20.67	14.35	50.59	190.6
87	1.2935	55.0	1099.1	511.6	13.778	20.91	14.40	51.86	197.0
88	1.3351	56.0	1099.6	496.5	13.804	21.15	14.45	53.14	203.7
89	1.3779	57.0	1100.0	482.0	13.829	21.39	14.50	54.48	210.6
90	1.4219	58.0	1100.4	467.9	13.854	21.64	14.55	55.85	217.6
91	1.4671	59.0	1100.9	454.3	13.880	21.88	14.60	57.25	224.9
92	1.5136	60.0	1101.3	441.1	13.905	22.12	14.65	58.69	232.4
93	1.5613	61.0	1101.7	428.4	13.930	22.36	14.70	60.16	240.1
94	1.6103	62.0	1102.2	416.1	13.955	22.60	14.75	61.67	248.1
95	1.6607	63.0	1102.6	404.2	13.981	22.84	14.80	63.12	256.4
96	1.7124	64.0	1103.0	392.7	14.006	23.08	14.86	64.81	264.8
97	1.7655	65.0	1103.4	381.5	14.031	23.32	14.91	66.20	273.6
98	1.8200	66.0	1103.9	370.7	14.057	23.56	14.97	68.53	282.6
99	1.8759	67.0	1104.3	360.3	14.082	23.80	15.02	70.03	293.0
100	1.9334	68.0	1104.7	350.2	14.107	24.04	15.08	71.62	301.5
101	1.9923	69.0	1105.2	340.4	14.132	24.28	15.14	73.44	311.3
102	2.0529	70.0	1105.6	331.0	14.157	24.52	15.20	75.31	321.5
103	2.1149	71.0	1106.0	321.8	14.183	24.76	15.26	77.22	332.0
104	2.1786	72.0	1106.4	313.0	14.208	25.00	15.32	79.19	342.8
105	2.2440	73.0	1106.9	304.4	14.233	25.24	15.39	81.21	353.9
106	2.3110	74.0	1107.3	296.0	14.259	25.48	15.45	83.29	365.4
107	2.3798	75.0	1107.7	288.0	14.284	25.72	15.52	85.42	377.2
108	2.4503	76.0	1108.2	280.2	14.309	25.96	15.59	87.62	389.4
109	2.5226	77.0	1108.6	272.6	14.334	26.20	15.65	89.87	402.0
110	2.5968	78.0	1109.0	265.3	14.360	26.45	15.72	92.19	414.9
111	2.6728	79.0	1109.4	258.2	14.385	26.69	15.80	94.58	428.3

(continued)

TABLE A.2—(continued)

Temperature, °F t	Saturation Pressure of Water and Steam, in. Hg p_s	Properties of Water[a] and Steam — Enthalpy			Properties of Dry Air at a Pressure of 29.921 in. Hg abs.		Properties of Mixture of Dry Air and Sat. Steam at a Total Pressure of 29.921 in. Hg abs.		
		Saturated Water,[a] Btu/lb h_f	Saturated Steam, Btu/lb h_g	Specific Volume of Sat. Steam, ft³/lb v_g	True Specific Volume, ft³/lb v_a	Enthalpy, Btu/lb h_a	Volume of Mixture per lb of Dry Air, ft³ v_s	Enthalpy of Mixture per lb of Dry Air, Btu h_s	Specific Humidity Grains per lb of Dry Air W_s
112	2.7507	80.0	1109.9	251.3	14.410	26.93	15.87	97.03	442.1
113	2.8306	81.0	1110.3	244.6	14.435	27.17	15.94	99.55	456.3
114	2.9125	82.0	1110.7	238.1	14.461	27.41	16.02	102.16	471.0
115	2.9963	83.0	1111.1	231.8	14.486	27.65	16.10	104.81	486.1
116	3.0823	84.0	1111.6	225.8	14.511	27.89	16.18	107.55	501.6
117	3.1703	85.0	1112.0	219.9	14.537	28.13	16.26	110.38	517.7
118	3.2606	86.0	1112.4	214.1	14.562	28.37	16.34	113.29	534.3
119	3.3530	87.0	1112.8	208.6	14.587	28.61	16.43	116.28	551.4
120	3.4477	88.0	1113.3	203.2	14.612	28.85	16.51	119.36	569.0
121	3.5446	89.0	1113.7	197.9	14.637	29.09	16.60	122.52	587.2
122	3.6439	90.0	1114.1	192.9	14.663	29.33	16.70	125.79	606.0
123	3.7455	91.0	1114.5	188.0	14.688	29.57	16.79	129.15	625.3
124	3.8496	92.0	1114.9	183.2	14.713	29.82	16.89	132.61	645.3
125	3.9561	93.0	1115.4	178.5	14.739	30.06	16.98	136.17	665.9
126	4.0651	94.0	1115.8	174.0	14.764	30.30	17.08	139.88	687.2
127	4.1768	95.0	1116.2	169.6	14.789	30.54	17.19	143.64	709.2

128	4.2910	96.0	1116.6	165.4	14.814	30.78	17.29	147.54	731.9
129	4.4078	97.0	1117.0	161.3	14.839	31.02	17.40	151.57	755.4
130	4.5274	98.0	1117.5	157.3	14.865	31.26	17.52	155.72	779.6
131	4.6498	99.0	1117.9	153.4	14.890	31.50	17.63	160.00	804.6
132	4.7750	100.0	1118.3	149.6	14.915	31.74	17.75	164.43	830.5
133	4.9030	101.0	1118.7	145.9	14.941	31.98	17.87	168.98	857.2
134	5.0340	102.0	1119.2	142.4	14.966	32.22	17.99	173.69	884.8
135	5.1679	103.0	1119.6	138.9	14.991	32.46	18.12	178.54	913.3
136	5.3049	104.0	1120.0	135.5	15.016	32.70	18.25	183.57	942.8
137	5.4450	105.0	1120.4	132.2	15.043	32.94	18.39	188.75	973.4
138	5.5881	106.0	1120.8	129.1	15.067	33.18	18.53	194.09	1000.0
139	5.7345	107.0	1121.2	126.0	15.092	33.43	18.67	199.64	1038
140	5.8842	108.0	1121.7	123.0	15.117	33.67	18.82	205.34	1071
141	6.0371	109.0	1122.1	120.0	15.143	33.91	18.97	211.27	1106
142	6.1934	110.0	1122.5	117.2	15.168	34.15	19.13	217.39	1143
143	6.3532	111.0	1122.9	114.4	15.193	34.39	19.29	223.70	1180
144	6.5164	112.0	1123.3	111.7	15.218	34.63	19.45	230.28	1219
145	6.6832	113.0	1123.7	109.1	15.244	34.87	19.62	236.94	1259
146	6.8536	114.0	1124.1	106.6	15.269	35.11	19.81	244.06	1301
147	7.0277	115.0	1124.6	104.1	15.294	35.35	19.99	251.34	1344
148	7.2056	116.0	1125.0	101.7	15.319	35.59	20.18	258.88	1389
149	7.3872	117.0	1125.4	99.32	15.345	35.83	20.37	266.71	1436
150	7.5727	118.0	1125.8	97.04	15.370	36.07	20.58	274.34	1485

Source: Goff and Gratch (1946, p. 95). Reprinted by permission of the American Society of Heating, Refrigeration, & Air Conditioning, Inc., Atlanta, GA.

[a] For temperatures of 1 to 32°F, the data are for ice instead of water.

Conversion factors: °F = $\frac{9}{5}$ °C + 32, 1 in. Hg = 3.3768 kPa, 1 Btu/lb = 2.326 kJ/kg, 1 ft³/lb = 0.06243 m³/kg, 1 grain/lb = 1.429 × 10⁻⁴ kg

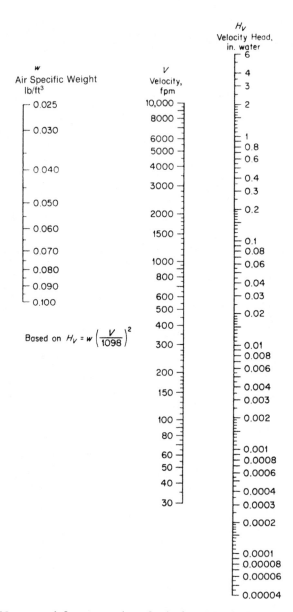

FIGURE A.1 Nomograph for conversion of velocity and velocity head. (*Conversion factors*: 1 lb/ft³ = 16.018 kg/m³, 1 fpm = 0.005080 m/s, 1 in. water = 248.84 Pa.) (After Baumeister, 1935.)

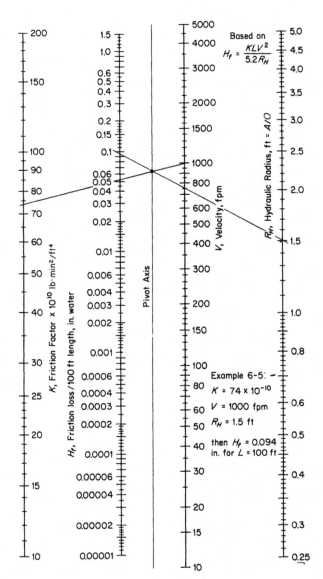

FIGURE A.2 Nomograph for determination of friction loss in mine airways. (*Conversion factors*: 1 lb·min²/ft⁴ = 1.855 × 10⁶ kg/m³, 1 in. water = 248.84 Pa, 1 fpm = 0.005080 m/s, 1 ft = 0.3048 m.) (After Hartman, 1954. By permission from *Eng. Mng. J.*, Chicago, IL.)

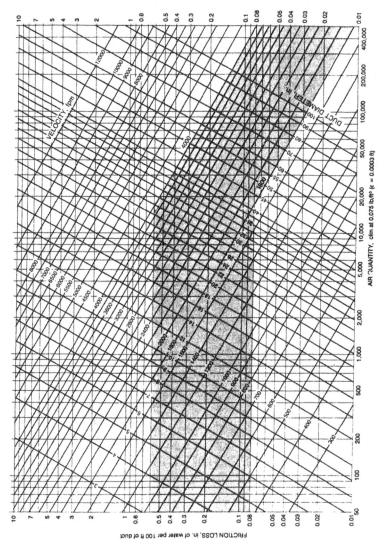

FIGURE A.3 Graph for determination of friction loss in ventilation pipe or tubing. Use without correction for new, steel pipe; see correction table, Section 5.5, for other pipe or tubing. Based on standard air specific weight. The shaded area of the graph contains the recommended velocities for efficient use of steel ventilation pipe. (*Conversion factors:* 1 cfm = 4.7195 × 10⁻⁴ m³/s, 1 fpm = 0.005080 m/s, 1 in. = 25.4 mm, 1 in. water = 248.84 Pa., 1 ft = 0.3048 m.) (After Anon., 1993. Reprinted with permission of the American Society of Heating, Refrigerating, & Air Conditioning, Inc., Atlanta, GA.)

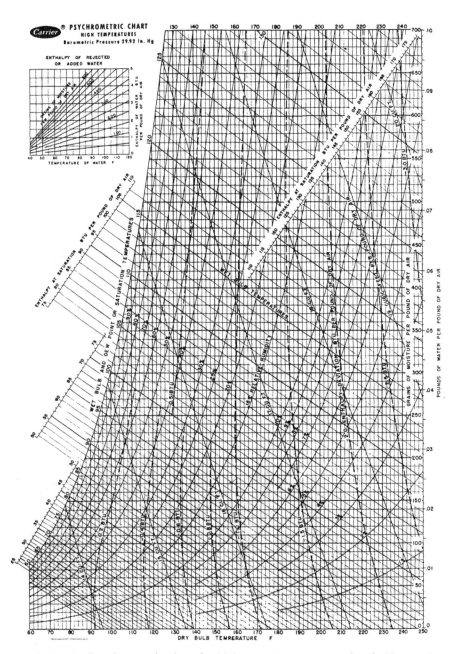

FIGURE A.4 Psychrometric chart for high temperatures at sea level. (*Conversion factors:* $°F = \frac{9}{5}°C + 32$, 1 grain/lb $= 1.429 \times 10^{-4}$ kg/kg, 1 Btu/lb $= 2.326$ kJ/kg, 1 ft³/lb $= 0.06243$ m³/kg, 1 in. Hg $= 3.3768$ kPa.) (By permission from the Carrier Corp. Copyright 1946, Carrier Corp., Syracuse, NY.)

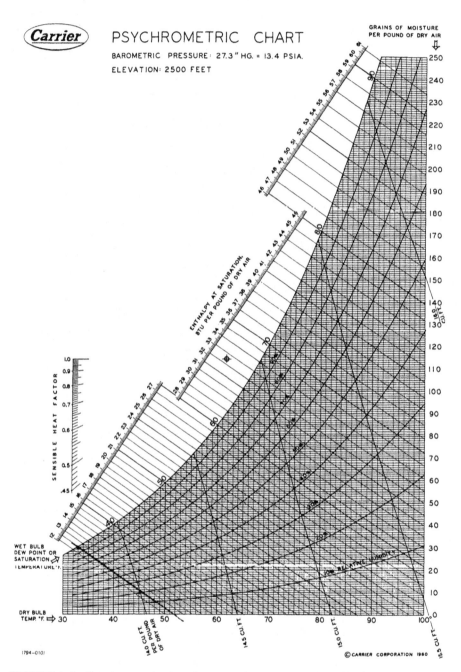

FIGURE A.5 Psychrometric chart for normal temperatures at 2500-ft (762-m) elevation. (*Conversion factors:* $°F = \frac{9}{5}°C + 32$, 1 grain/lb = 1.429×10^{-4} kg/kg, 1 Btu/lb = 2.326 kJ/kg, 1 ft^3/lb = 0.06243 m^3/kg, 1 in. Hg = 3.3768 kPa.) (By permission from the Carrier Corp. Copyright 1960, Carrier Corp., Syracuse, NY.)

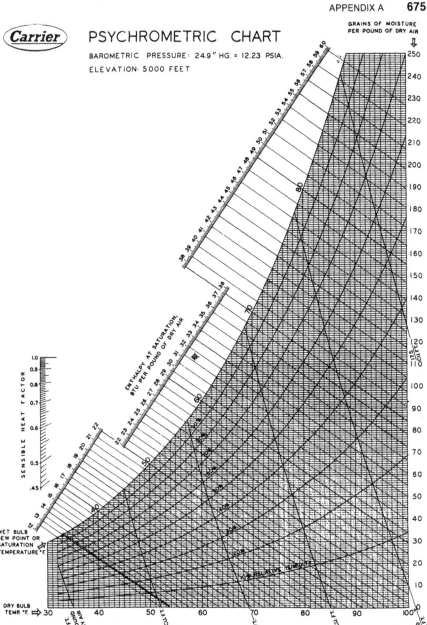

FIGURE A.6 Psychrometric chart for normal temperatures at 5000-ft (1524-m) elevation. (*Conversion factors:* °F = $\frac{9}{5}$°C + 32, 1 grain/lb = 1.429 × 10^{-4} kg/kg, 1 Btu/lb = 2.326 kJ/kg, 1 ft³/lb = 0.06243 m³/kg, 1 in. Hg = 3.3768 kPa.) (By permission from the Carrier Corp. Copyright 1960, Carrier Corp., Syracuse, NY.)

SI Units in Mine Ventilation

The International System of Units (often called the SI system) is used throughout most of the world. While the SI system is taught in college-level engineering programs in the United States and is slowly gaining acceptance in industry, the English system of units remains the standard for most mine ventilation calculations. Accordingly, this text is written using the English system as the primary system of units. However, several steps have been taken to make the book more useful abroad and to help educate U.S. ventilation engineers in the SI system:

1. SI units are provided in parentheses in the descriptive material, the examples, and in the problems at the end of the chapters.
2. The List of Mathematical Symbols at the front of the book is provided with the proper units to be used in both systems.
3. Equations, where appropriate, are provided in both the English and SI forms. For example, the English units form of Eq. 5.18 is supplemented by Eq. 5.18a, the correct form of the equation for SI units.
4. A number of example problems are worked in SI units.
5. This Appendix was provided to aid those who are interested in applying SI units to ventilation problems.

While the English system of units remains the primary system in this text, the SI system has several advantages that should be recognized. The sections that follow outline the basics and advantages of the SI system and provide a table of conversion factors to convert from one system to the other.

B.1 BASICS OF THE SI SYSTEM

The International System of Units was named and adopted by the Eleventh General Conference on Weights and Measures, which was convened in Paris in 1960. It was intended that this system become the standard for the world, and so the system was carefully designed to be as simple and logical as

TABLE B.1   Units Used in the SI System

Quantity	Unit	Symbol
Base Units		
Length	meter	m
Mass	kilogram	kg
Time	second	s
Electric current	ampere	A
Thermodynamic temperature	kelvin	K
Amount of substance	mole	mol
Luminous intensity	candela	cd
Supplementary Units		
Plane angle	radian	rad
Solid angle	steradian	sr

Quantity	Unit	Symbol	Definition
Derived Units			
Area	hectare	ha	$1 \text{ ha} = 1 \text{ hm}^2$
Energy	kilowatt-hour	kWh	$1 \text{ kWh} = 3.6 \text{ MJ}$
Exposure	roentgen	R	$1 \text{ R} = 2.58 \times 10^{-4} \text{ C/kg}$
Mass	metric ton	t	$1 \text{ t} = 1000 \text{ kg}$
Plane angle	degree	°	$1° = (\pi/180) \text{ rad}$
	minute	′	$1' = (\pi/10,800) \text{ rad}$
	second	″	$1'' = (\pi/648,000) \text{ rad}$
Pressure	bar	bar	$1 \text{ bar} = 10^5 \text{ Pa}$
Temperature	degree Celsius	°C	$1°\text{C} = 1 \text{ K}$ (in temperature interval only)
Time	minute	min	$1 \text{ min} = 60 \text{ s}$
	hour	h	$1 \text{ h} = 3.6 \text{ ks}$
	day	day	$1 \text{ day} = 86.4 \text{ ks}$
Volume	liter	L	$1 \text{ L} = 10^{-3} \text{ m}^3$

possible. To fully appreciate the working of the SI system, one should understand its basics and the advantages of its use.

The units used in the SI system are divided into three types: base units, supplementary units, and derived units. These are outlined in Table B.1. The *base units* of the system are a set of seven independent quantities that form the basis for the measurement of most physical quantities. Although most physical quantities can be measured by means of the base units and their combinations, plane and solid angles require the use of the *supplementary units*. These two units can also be considered as base or derived units.

Derived units are any units that are formed by the combination of base and supplementary units. A derived unit normally results when two or more

TABLE B.2   **Additional Units Used in the SI System**

Quantity	Unit	Symbol	Formula
Frequency (of a periodic) phenomenon	hertz	Hz	1/s
Force	newton	N	$kg \cdot m/s^2$
Pressure, stress	pascal	Pa	N/m^2
Energy, work, quantity of heat	joule	J	$N \cdot m$
Power, radiant flux	watt	W	J/s
Quantity of electricity, electric charge	coulomb	C	$A \cdot s$
Electric potential, potential difference, electromotive force	volt	V	W/A
Capacitance	farad	F	C/V
Electric resistance	ohm	Ω	V/A
Conductance	siemens	S	A/V
Magnetic flux	weber	Wb	$V \cdot s$
Magnetic flux density	tesla	T	Wb/m^2
Inductance	henry	H	Wb/A
Luminous flux	lumen	lm	$cd \cdot sr$
Illuminance	lux	lx	lm/m^2
Activity (of radionuclides)	becquerel	Bq	1/s
Absorbed dose	gray	Gy	J/kg

base or supplementary units are multiplied or divided, resulting in a compound unit. A good example of this process is the means for providing a unit of force. As we can see from the list of base and supplementary units, no primary unit of force exists. Instead, force is defined as the product of mass and acceleration. The unit of mass (kg) multiplied by the units of acceleration (m/s^2) results in units of $kg \cdot m/s^2$. Thus the unit of force in the SI system is the $kg \cdot m/s^2$. This is a derived unit that is commonly used and thus is given a special name: the newton (N). A group of these derived units that are commonly used have been provided with special names; an approved list appears in Table B.1.

Several additional units in common use that are not strict combinations of the base units have been adopted for use in the SI system. These are generally permitted where long-standing use has made them common. The units here that are pertinent to mining and ventilation are summarized in Table B.2.

B.2   ADVANTAGES OF THE SI SYSTEM

The primary advantage of the SI system is that it is more logically consistent or coherent than any other system of units. This is readily appreciated from a comparison of various systems. Although some American engineers find the system difficult to employ, it is most likely unfamiliarity that creates a problem and not the system itself.

One of the advantages of SI usage is the smaller number of units in use. The only primary unit of length is the meter, and all other units of length are formed using prefixes. In the English system, a large number of units of length have been used including inches, feet, rods, perches, chains, fathoms, leagues, miles, and light years. Because such a large number of arbitrarily adopted units have existed, the conversion from one unit to another is troublesome. The SI system has solved this problem by providing a decimal system of prefixes to accommodate any measure of length.

Another advantage of the SI system is the lack of any units that have double meanings, such as the unit pound in the English system. Because the pound was used to mean both the pound mass (lbm) and the pound force (lbf), this has always been a source of confusion in calculations and in interpreting data. The SI system of using the kg as the only acceptable unit of mass and the newton (N) as the only acceptable unit of force provides the ultimate solution to this problem.

The facility of using the SI system in calculations is one property that becomes obvious in ventilation. It is often true that the relationship between units in the SI system is simpler than that in the English system. In particular, some of the calculations normally performed in ventilation problems are simplified by the absence of arbitrary conversion factors in the equations. This simplifies both the calculations and the determination of the correct units to be assigned to the answer.

The final advantage of the SI system is that it is now the standard system of units used in world trade. In 1975, when the United Kingdom adopted the SI system and abandoned the old British Imperial system of units, the United States remained the only major country in the world that did not use the SI system. Thus the use of SI as the universal system of units appears inevitable.

B.3 TABLES OF CONVERSION FACTORS

Table B.3 contains useful conversion factors for the units normally encountered in ventilation calculations. It provides the multiplication factors for both English to SI units and the reverse conversion. For more encompassing conversion tables, standard references (Anon., 1992c; Wandmacher and Johnson, 1995) may be consulted.

TABLE B.3 Conversion Factors for Common Ventilation Units

Units	To Convert from	To	Multiply By	To Convert from	To	Multiply By
Area (A)	in.^{2}	mm^2	645.16	mm^2	in.^{2}	1.545×10^{-3}
	ft^2	m^2	0.0929	m^2	ft^2	10.764
	yd^2	m^2	0.8361	m^2	yd^2	1.196
Barometric pressure (p_b)	mm Hg	Pa	133.32	Pa	mm Hg	7.50×10^{-3}
	atm	kPa	101.32	kPa	atm	9.869×10^{-3}
Cooling power (h_c)	Btu/h·ft^2	W/m^2	3.155	W/m^2	Btu/h·ft^2	0.3170
Dust concentration (B)	mppcf	(ppcc)^a	35.3	(ppcc)^a	mppcf	0.0283
Energy	Btu	kJ	1.05506	kJ	Btu	0.9478
	(cal)^a	J	4.1868	J	(cal)^a	0.2388
	kWh	kJ	3600.0	kJ	kWh	0.2778×10^{-3}
Enthalpy (h)	Btu/lb	kJ/kg	2.326	kJ(kg)	Btu/lb	0.4299
Entropy (s)	Btu/lb·°F	kJ/kg·°C	4.1868	kJ/kg·°C	Btu/lb·°F	0.2388
Force (F)	lb(force)	N	4.4482	N	lb(force)	0.2248
Friction factor (K)	lb·min^2/ft^4	kg/m^3	1.855×10^6	kg/m^3	lb·min^2/ft^4	0.5391×10^{-6}
Head or pressure head (H)^b	in. water	Pa	248.84	Pa	in. water	4.0186×10^{-3}
	in. Hg	kPa	3.3768	kPa	in. Hg	0.2961
Heat flow (q) or work rate (W_k)	Btu/h	W	0.29307	W	Btu/h	3.414
	(kcal/h)^a	W	1.162	W	(kcal/h)^a	0.8606
Heat transfer ($U \cdot A$)	Btu/h·°F	W/°C	0.5275	W/°C	Btu/h·°F	1.895
Length (L)	in.	mm	25.4	mm	in.	39.37×10^{-3}
	in.	cm^a	2.54	cm^a	in.	0.3937
	ft	m	0.3048	m	ft	3.281
	mi	km	1.6093	km	mi	0.6214
Mass (m)	lb (mass)	kg	0.4536	kg	lb (mass)	2.204
Mass density (ρ)	lb·s^2/ft^4	kg/m^3	515.38	kg/m^3	lb·s^2/ft^4	1.940×10^{-3}
Power (P_a or P_m)	hp	kW	0.7457	kW	hp	1.341
Pressure (p)	psi	kPa	6.8948	kPa	psi	0.1450
Quantity or volume flow rate (Q)	cfm	m^3/s	0.47195×10^{-3}	m^3/s	cfm	2118.9
	cfm	L/s	0.47195	L/s	cfm	2.1189

Quantity	English unit	SI unit	Factor	SI unit	English unit	Factor
Refrigeration (q_R)	cfm	$(cm^3/s)^a$	471.95	$(cm^3/s)^a$	cfm	2.1189×10^{-3}
Resistance (R)	ton	kW	3.5168	kW	ton	0.2843
	in.·min²/ft⁶	N·s²/m⁸	1.117×10^9	N·s²/m⁸	in.·min²/ft⁶	0.8953×10^{-9}
Specific heat (c)	Btu/lb·°F	kJ/kg·°C	4.1868	kJ/kg·°C	Btu/lb·°F	0.2388
Specific humidity (W)	lb vapor/lb dry air	kg vapor/kg dry air	1.000	kg vapor/kg dry air	lb vapor/lb dry air	1.000
Specific volume (v)	ft³/lb	m³/kg	0.06243	m³/kg	ft³/lb	16.018
Specific weight (w)	lb/ft³	kg/m³	16.018	kg/m³	lb/ft³	0.06243
Thermal conductance (C_r) or Thermal emissivity (ϵ) or Coefficient of heat transfer (U)	Btu/h·ft²·°F	W/m²·°C	5.6783	W/m²·°C	Btu/h·ft²·°F	0.1761
Thermal conductivity (k)	Btu/h·ft·°F	W/m·°C	1.7307	W/m·°C	Btu/h·ft·°F	0.5778
Thermal diffusivity (α)	ft²/h	m²/s	25.806×10^{-6}	m²/s	ft²/h	38.75×10^3
Velocity (V)	fpm	m/s	0.00508	m/s	ft/min	196.85
Viscosity, absolute (μ)	lb·s/ft²	Pa·s	47.880	Pa·s	lb·s/ft²	0.02089
Viscosity, kinematic (v)	ft²/s	m²/s	0.0929	m²/s	ft²/s	10.764
Volume flow rate of liquid (Q_w)	gpm	m³/s	63.09×10^{-6}	m³/s	gpm	15.85×10^3
	gpm	L/s	0.06309	L/s	gpm	15.850
Volume (Y)	in.³	mm³	16.387×10^3	mm³	in.³	61.02×10^{-6}
	ft³	m³	0.028317	m³	ft³	35.315
	yd³	m³	0.76455	m³	yd³	1.308
Weight flow rate (G)	lb/h	kg/s	0.1260×10^{-3}	kg/s	lb/h	7936.7

ᵃ Former metric units *not* approved in SI usage.

ᵇ Note that present practice in most parts of the world is to use Pa or kPa for heads as well as pressures; mm water or mm Hg is falling into disuse.

Equivalence of constants:

Acceleration due to gravity $32.174 \text{ ft/s}^2 = 9.80665 \text{ m/s}^2$

Gas constant $53.35 \text{ ft·lb/lb (mass)·°R} = 287.045 \text{ J/kg·K}$

Temperature conversion equations:

$$°C = \tfrac{5}{9}(°F - 32)$$
$$°F = \tfrac{9}{5}°C + 32$$
$$K = °C + 273.15$$
$$°R = °F + 459.67$$

Laboratory Experiments

Laboratory work is an integral part of an engineer's education. Ventilation is so critical to the health, safety, and productivity of the miners and to the mining environment, a good background in both the theory and the practice of mine ventilation is an essential requirement of a modern mining engineering education. Laboratory experiments are an important component of the pedagogical process in fulfilling this requirement.

The design of laboratory experiments for a semester-long undergraduate course in mine ventilation course must achieve several goals. As a minimum, from the specifics to the general, these are

1. To enhance knowledge of physical phenomena such as psychometric properties and basic psychrometric processes
2. To become familiar with modern measurement methods and equipment for such purposes as mine air contaminants and airflow parameters
3. To verify the behavior of ventilation systems and subsystems under different constraints such as mine circuits in series and parallel, and to investigate the fan laws
4. To develop an ability to design experiments and to analyze experimental results
5. To develop the ability to report laboratory results in a concise, but complete, manner

The overall objective is to instill the importance of laboratory and field experimentation in obtaining data (1) for design and development during the planning stages of a ventilation project and (2) for monitoring and control during the operational phase of the mine.

C.1 LABORATORY EQUIPMENT NEEDS

University facilities available for conducting mine ventilation laboratory experiments are quite varied. In terms of a permanent facility duplicating the mine environment, some mining schools have ready access to experimental mines. In some schools, simulated mine environments with tunnels, ventilation instrumentation and controls, fans, and a monitoring system are permanently housed. In other schools, large systems of ducting and fans are available.

As a minimum, the ventilation laboratory must have enough equipment to achieve the goals listed above. The references to manufacturers in the following list are for identification purposes only and are not to be construed as a recommendation of specific equipment:

1. An airflow demonstration set, such as the one available from Airflow Development Ltd., for demonstrating basic principles of fluid flow and flow measurement techniques. Sufficient spare pipes of different cross sections and extra fans are recommended so as to be able to create series and parallel airways as well as series and parallel fan operations.

2. A laboratory unit such as the Hilton Air Conditioning Laboratory unit to demonstrate psychrometric properties and processes, and to study the energy transfer that occurs in hot and deep mines and refrigeration plants.

3. A mercury barometer capable of reading to 0.01 in. Hg.

4. An adjustable inclined-tube manometer.

5. Magnehelic differential pressure gages for different ranges (e.g., 0–0.5, 0–1.0, and 0–2.00 in. water).

6. Assorted pitot tubes.

7. Several altimeters, capable of reading pressures to the nearest ft of elevation.

8. Several vane anemometers. Some of these can be of the digital type.

9. Smoke tubes.

10. A sling psychrometer, a stationary hygrometer, and kata thermometers.

11. Tachometer and industrial power analyzer for fan experiments.

12. Instruments for air-quality sampling and monitoring.

There is a wide variety of air-quality sampling and monitoring instruments (item 12) to choose from. A minimum collection in the laboratory for gas detection should be flame safety lamps, aspirator bulb with stain tubes for a number of gases commonly found in mines, and portable electronic instruments for detecting methane, carbon monoxide, oxygen deficiency, and other conditions, along with cylinders of calibrated gases. Instruments for mine aerosol sampling should include personal gravimetric dust samplers, Mining Research Establishment (MRE) gravimetric dust samplers, and real-time aerosol samplers such as RAM-1 or DataRAM. Where samples have to be weighed for concentration calculation, then a high-precision balance should be available.

Technical literature on the various equipment, including any operational instructions and maintenance guidelines, should be maintained in a secure but accessible place in the laboratory. Some equipment such as vane anemometers may require calibration after several semesters of use or when damaged. When funds are available to acquire additional equipment for the laboratory, it is necessary to strike a careful balance between two choices of the most modern equipment available in the market and the most commonly used equipment in practice.

Laboratory facilities must be complemented with a computational facility where the students have access to personal computers; computers networked to larger systems; mine-ventilation-specific software such as pressure survey, heat transfer, and ventilation network programs, and general-purpose software including statistical, graphic, and word-processing packages.

C.2 OUTLINE OF EXPERIMENTS

The laboratory experiments must closely follow the lectures. Students should have learned the theoretical and practical aspects of mine ventilation and be ready to explore their validation and application through the laboratory experiments. In a semester-long course, the laboratory experiments may commence in the third week of the course and continue to the end of the course, allowing time for up to 12 experiments. Each experiment may last from 120 to 150 min. When the lectures follow the outline of this text, the following sequence of experiments is recommended:

Experiment 1:   Introduction to permanent laboratory facilities and available instrumentation; discussion of care of instruments, laboratory safety requirements, individual contributions in team experiments, report requirements, and report format.

Experiment 2:   Mine air-quality measurements with appropriate gases such as methane, oxygen, and carbon monoxide (Chapter 3).

Experiment 3:   Mine air-particulate measurements with DataRAM and, where possible, with personal samplers (Chapter 4).

Experiment 4:   Air specific-weight determination from barometric pressure and wet- and dry-bulb temperature readings; static-, velocity-, and total-head measurements; calculation of head losses and verification of Bernoulli's general equation for fluid flow (Chapters 5 and 6).

Experiment 5:   Friction factor and shock-loss determinations (Chapter 5).

Experiment 6:   Direct (manometer) survey of a mine circuit (Chapter 6).

Experiment 7:   Indirect (altimeter) survey of a mine circuit; use of pressure survey calculation programs (Chapter 6).

Experiment 8:   Equivalent resistance of series and parallel circuits; analysis of fan heads; quantity divider rule (Chapter 7).

Experiment 9:   Verification of the fan laws (speed); characteristics of fans in combination (series–parallel) (Chapter 9).

Experiment 10:   Determination of the operating point of a fan (fan head and quantity) and the fan (H–Q) characteristic; determination of the mine characteristic (Chapter 10).

Experiment 11:   Demonstration of psychrometric properties and air conditioning processes in mines (Chapters 2 and 16).

Experiment 12:   Computer-oriented mine ventilation network analysis, including deletion and addition of ventilation splits, new shafts, and new fans (Chapters 7, 13, and 14).

C.3 LABORATORY REPORTS

The writing of a good report of the experiment is as important as the conduct of the experiment itself. While individual preferences for report format and writing style should be respected, the need for embodying key technical concepts and concise technical writing in a logical format must be stressed. Even when an experiment is conducted by a team of students, each student must write a separate report. All graphs and charts must be drawn to scale, and all units must be clearly specified. The report should preferably be typed. The report must be evaluated for the validity of the experimental results as well as for the manner of presentation. Firm deadlines should be established for submission of the reports.

As a minimum, the body of the report should contain descriptions, data, and results under the following headings:

1. Purpose and scope of the experiment
2. Theory, hypothesis, and applicable formulas and equations (with units and any references)
3. List of instruments (with model numbers)
4. Experimental procedure
5. Experimental data (with units) in tabular form with section headings
6. Data analysis and results with tables, graphs, and charts as applicable, including a discussion of sources of errors
7. Conclusions regarding the fulfillment of the purpose and scope of the experiment and the hypothesis studied
8. Any other attachments (e.g., appendix of relevant material)

Computer Applications and Software

Computers have become an essential tool for mine ventilation engineers in performing their duties. Computers are of three types: mainframes, workstations, or personal computers. The trend at the current time is toward performing most ventilation engineering activities on personal computers, but monitoring, control, and other applications require larger computer systems. Furthermore, the computers used in the monitoring, control, and mine planning areas may be linked with other computers in a local area network (LAN) or a wide area network (WAN). The purpose of this appendix is to discuss briefly the scope of mine ventilation applications, to provide some of the history of ventilation network analysis methods, and to outline some of the computer software programs that the engineer can use to perform many of the engineering calculations related to ventilation analysis.

D.1 SCOPE OF COMPUTER APPLICATIONS

Most of the computer applications in mine ventilation and environmental analysis fall into the following areas (Ramani, 1982; Bandopadhyay, 1992):

1. Monitoring and control
2. Data storage and analysis
3. Atmospheric-component analysis
4. Ventilation network analysis

Other capabilities including the use of computer-assisted design (CAD) systems, multimedia graphic outputs, and applications of expert systems and artificial intelligence are also important in mine ventilation computer analyses.

Monitoring and control operations in mines are rapidly becoming commonplace (see Section 6.8). As shown in Fig. 6.22, monitoring and control of the mine atmosphere is generally a part of a larger system that may also provide for monitoring of belts, pumps, bins, fans, power-distribution equipment, and preparation plant facilities. In addition, it is not unusual for a mine monitoring system to be used for control of the ventilation system and a methane drainage system simultaneously. Furthermore, it can be used with an expert system or optimization software to monitor, analyze, and optimize the ventilation system as shown in Fig. 6.23.

The second major area of computer application is in the analysis of ventilation system data. Many of the continuing tasks of the mine ventilation engineer are performed using the computer. These include the reduction of data from periodic mine pressure surveys (see Section. 6.7 for the parameters measured and calculated), the determination of airway friction and resistance factors, and the calculation of cooling loads in hot mines. Computer programs for calculating pressure losses from altimeter surveys are useful adjuncts to generate inputs for ventilation network analysis (Harris et al., 1973; Anderson and Nugent, 1977; Luxbacher et al., 1977; Hadden and Smith, 1987).

Atmospheric-component analysis, the third principal computer application area, includes analysis of methane flows, heat generation and dissipation, dust generation and control, diesel exhaust gases and particulates, radon emissions, and other environmental parameters that will affect the quality, comfort, level of safety, and healthfulness of the mine environment. Further discussion of these topics is presented in Chapter 3 on gases and Chapter 4 on aerosols.

Computer-oriented network analysis is essential for analysis of the current or future ventilation networks in terms of the air velocities and quantities at key locations, the head losses in the various airways, and the requirements of the fans. Because many of these problems are related to mine planning, it is advantageous to have a network program interfaced with the mine planning software system, particularly under a CAD umbrella. This would ease the production of graphic output from the ventilation program. The ability to produce single-line network diagrams with control of scale and orientation is a valuable auxiliary capability for any network software package.

D.2 NETWORK ANALYSIS SOFTWARE

Mathematical aspects of mine ventilation networks have been provided in Chapter 7. Many of the current procedures are based on a generalized iterative solution method for complex networks dealing with fluid flows that was published by Hardy Cross (1936). This procedure was then adopted by Scott and Hinsley (1951) for use in solving ventilation networks. Adaptation of computers to the Hardy Cross algorithm for ventilation networks was successfully used in a number of places throughout the world in the 1960s.

At Penn State University, a succession of FORTRAN programs oriented toward ventilation network solution was developed during the 1960s and 1970s. The first program was written and published by Hartman and Trafton (1963); the second, by Wang and Hartman (1967). The 1967 version of the network solution program was widely distributed and used. Some further developments were also outlined by Wang and Saperstein in 1970. Additional advancements in the capabilities of network software were then produced when the U.S. Bureau of Mines supported research oriented toward improving safety in coal mine ventilation systems. These developments were described by Stefanko and Ramani (1972), Stefanko et al. (1977), and Didyk et al. (1977).

While computer methods were being developed at Penn State, others were establishing their own programs. Two of the early programs were outlined by Hashimoto (1961) and McPherson (1966). Others developments were oriented toward the analysis of heat and humidity problems or toward the simulation of ventilation when a mine fire is present in the mine airways. Simulation of mine fires and their effects on mine ventilation systems were undertaken in projects conducted over several decades at Michigan Tech (Chang and Greuer, 1987; Chang et al., 1990). Other researchers in Poland, South Africa, and Japan contributed greatly to the knowledge required for the modeling of mine atmospheric conditions.

The availability of mine ventilation simulation programs makes the job of the today's ventilation engineer much easier as digital computers can eliminate the burden associated with data analysis, network generation, and ventilation map production. Using computers for computation in this way, the engineer can concentrate on design aspects and optimization of the ventilation system. The Penn State ventilation programs, originally written in FORTRAN, have been reprogrammed into other languages and supplemented with additional input and output options over the years. A variety of versions of that program are still being used, particularly in the coal industry. One version, called MineVent, is available through Ohio Automation, Inc. (Athens, OH). This ventilation analyzer runs as a module under AutoCAD in the Microsoft Windows environment. The module forms part of a mine planning software package called ICAMPS. The value of the Windows environment and the capability of AutoCAD are evident in this package.

A second program that is widely used is the VnetPC network analyzer available from Mine Ventilation Services, Inc. (Fresno, CA). This program also runs under Windows, possesses a convenient data input scheme, and has the capability of producing network diagrams from the raw data. The Windows environment is also used in the ventilation program known as P.C. Vent (FABWOL, Inc., Socorro, NM). This program provides basic network analysis with the additional capability of being able to simulate a fire on a real-time basis. In addition, the package provides a graphic display of the network in which the spread of contamination during a fire can be observed during a simulation.

The thermal energy within the rock and groundwater surrounding a mine will have a significant effect on the flow of air through the mine due to the net conversion of heat to flow energy within the mine airways (see Section. 5.8). ENVIRON (Von Glehn et al., 1987) is a computer program developed at the Chamber of Mines in South Africa that allows a mine to be modeled in terms of heat sources and sinks, and at the same time allows the airflow distribution throughout the mine to be simulated. The program consists of two modules, HEATFLOW and VENTFLOW, the former calculating the heat loads on the ventilating air, and the latter calculating the air distribution on the basis of the predicted temperatures from HEATFLOW; the two modules are accessed by ENVIRON iteratively until the solution converges. The modules can, however, be executed as two independent programs. Another program is CLIMSIM (Mine Ventilation Services, Inc., Fresno, CA). This program uses numerical procedures to determine the net transfer of energy between the moving air mass and the surrounding mine airways and to predict the psychometric properties of the air at various points within the mine. It provides graphic output of the important thermodynamic and heat stress values and can be used to evaluate the need for air conditioning within the mine openings. Capabilities to design a mine ventilation and air conditioning network are also inherent in the software package known as MIVENA (Sasaki et al., 1996). The software implements the nodal pressure technique and the skyline-modified Choleski decomposition scheme to provide a network solution that satisfies both airflow and thermodynamic balances. The program works only in the Windows NT system, but the package has advanced three-dimensional graphics and CAD capabilities.

The MFIRE program developed by Michigan Technological University simulates the response of a mine ventilation system to various parameters, including the interaction of the ventilation system to a fire located within the mine airways. The MFIRE software, which runs under DOS, is currently available from the National Institute for Occupational Safety and Health (NIOSH).

This list of available programs is certainly not comprehensive but is indicative of the many ventilation-oriented computer programs that are available, both in the public domain and from software vendors. Description of all the programs and their characteristics is beyond the scope of this appendix. With a diligent search, ventilation engineers should readily be able to locate hardware and software resources that meet their requirements.

References

Agricola, G., 1556, *De Re Metallica,* H. C. and L. H. Hoover, translators, Dover, New York, 1950.

Allan, A. L., Hollwey, J. R., and Maynes, J. H. B., 1968, *Practical Field Surveying and Computations,* American Elsevier, New York.

Allen, R. F., 1976, "Ventilation of the 1100 Ore Body at the Mount Isa Mine," *Proc. Intl. Mine Vent. Cong.,* R. Hemp and F. H. Lancaster, eds., Mine Ventilation Society of South Africa, Johannesburg.

Ambs, J. L., and Hillman, T. L., 1992, "Disposable and Reusable Diesel Exhaust Filters," *USBM I.C.,* No. 9324, pp. 67–73.

Anderson, T. C., and Nugent, J. W., 1977, "Mine Pressure Differentials with a Programmable Hand Calculator," *MESA I.R.,* No. 1050, 29 pp.

Anon., 1955, "Planning the Ventilation of New or Reorganized Collieries," *National Coal Board, Inf. Bull.,* No. 55/153, London.

Anon., 1956, "Coal Age Mining Guidebook," *Coal Age,* Vol. 61, No. 7, July, p. 65.

Anon., 1971, "Ventilation Planning as a Prerequisite for Winning Higher Outputs," *Mng. Eng.,* Vol. 30, Part 12, pp. 796–811.

Anon., 1972a, "Criteria for a Recommended Standard, Occupational Exposure to Carbon Monoxide," U.S. Dept. Health, Education, and Welfare, National Institute for Occupational Safety and Health, Washington.

Anon., 1972b, *Routine Mine Ventilation Measurements,* Chamber of Mines of South Africa, Johannesburg.

Anon., 1972d, *Threshold Limit Values for Chemical Substances and Physical Agents and Biological Exposure Indices,* American Conference of Governmental Industrial Hygienists, Cincinnati (updated annually).

Anon., 1976, *Mining Code, State of Arizona,* rev. ed., Arizona State Mine Inspector's Office, Phoenix.

Anon., 1978, *Steam,* 39th ed., Babcock & Wilcox Co., New York.

Anon., 1979, "J & L Looks at Longwall Ventilation Problems," *Coal Age,* Vol. 84, No. 7, July, p. 130.

Anon., 1980, "Woman vs. Man in the Sporting Field," *Morning Herald,* Sydney, N.S.W., Australia, June 7, pp. 13–14.

Anon., 1983, "New Machine-Mounted Dust Collector System for Continuous Mining Machines," *USBM Tech. News,* No. 178, Aug.

Anon., 1984, "Infrared Determination of Respirable Coal Mine Dust," Method No. P7, Mine Safety and Health Administration.

Anon., 1985a, "Laboratory Methods of Testing Fans for Rating," *AMCA Standard 210–85,* Air Movement and Control Assn., Inc., Arlington Heights, IL.

Anon., 1985b, "Novel Use of Ventilation Tubing Bypasses Dust Away from Returnside Roof Bolting Operations," *USBM Tech. News,* No. 235, Dec.

Anon., 1987, *IARC Monographs on the Evaluation of the Carcinogenic Risk of Chemicals to Humans: Silica and Some Silicates,* Vol. 42, World Health Organization, International Agency for Research on Cancer, Lyons, France.

Anon., 1988, "Carcinogenic Effects of Exposure to Diesel Exhaust," U.S. Dept. Health and Human Services, National Institute for Occupational Safety and Health, Current Intelligence Bull. 50, Aug.

Anon., 1989a, "Approval Requirements for Diesel-Powered Machines and Approval, Exposure Monitoring, and Safety Requirements for the Use of Diesel-Powered Equipment in Underground Coal Mines; Proposed Rules," U.S. Dept. Labor, *Federal Register,* Vol. 54, No. 191, Washington, DC, pp. 40950–40997.

Anon., 1989b, "Air Quality, Chemical Substances, and Respiratory Protection Standards," U.S. Dept. Labor, MSHA, *Federal Register,* Vol. 54, No. 166, pp. 35760–35852.

Anon., 1992a, "Permissible Exposure Limit for Diesel Particulate; Proposed Rule," U.S. Dept. Labor, *Federal Register,* Vol. 57, No. 3, Washington, pp. 500–503.

Anon., 1992b, *Heating, Ventilating and Air Conditioning Systems and Equipment,* I-P Edition, 37th ed., American Society of Heating, Refrigerating & Air Conditioning Engineers, Atlanta (updated periodically).

Anon., 1992c, "American National Standard for Metric Practice," ANSI/IEEE Standard 268–82, American National Standards Institute, New York.

Anon., 1993, *ASHRAE Handbook of Fundamentals,* I-P Edition, American Society of Heating, Refrigerating & Air Conditioning Engineers, Atlanta (updated periodically).

Anon., 1995a, *Code of Federal Regulations,* Title 30, Mineral Resources, U.S. Govt. Printing Office, Washington, July 1 (updated annually).

Anon., 1995b, personal communication, ABC Industries, Inc., Warsaw, IN.

Anon., 1995c, 30 *CFR* Parts 70 and 75, Safety Standards for Underground Coal Mine Ventilation; *Federal Register,* Vol. 57, No. 95, pp. 20868–20929.

Anon., 1996, *1995–1996 Threshold Limit Values for Chemical Substances and Physical Agents and Biological Exposure Indices,* American Conference of Governmental Industrial Hygienists, Cincinnati (updated annually).

Artz, R. T., 1951, "Some Practical Aspects of Coal-Mine Ventilation," *USBM Handbook,* 45 pp.

Attfield, M., Reger, R., and Glenn, R., 1984a, "The Incidence and Progression of Pneumoconiosis Over Nine Years in U.S. Coal Mines: I. Principal Findings," *Am. J. Indust. Med.,* Vol. 6, pp. 407–415.

Attfield, M., Reger, R., and Glenn, R., 1984b, "The Incidence and Progression of Pneumoconiosis Over Nine Years in U.S. Coal Mines: II. Relationship with Dust Exposure and Other Potential Causative Factors," *Am. J. Indust. Med.,* Vol. 6, pp. 417–425.

Augustadt, G. L., 1951, "How Magma's New Cooling Plant Combats the Mine Heat Problem," *Eng. Mng. J.,* Vol. 152, No. 9, Sept., pp. 72–79.

Aul, G., and Ray, R., Jr., 1991, "Optimizing Methane Drainage Systems to Reduce Mine Ventilation Requirements," *Proc. 5th U.S. Mine Vent. Symp.*, Y. J. Wang, ed., SME, Littleton, CO., pp. 638–646.

Bakke, P., Leach, S. J., and Slack, A., 1964, "Some Theoretical and Experimental Observations on the Recirculation of Mine Ventilation," *Colliery Eng.* (London), No. 141, Nov., pp. 471–477.

Bandopadhyay, S., 1992, "Computer Applications in Mine Ventilation and the Environment," Chap. 11.10 in *SME Mining Engineering Handbook*, 2nd ed., H. L. Hartman, sr. ed., SME, Littleton, CO, pp. 1139–1153.

Banerjee, S. C., 1985, *Spontaneous Combustion of Coal and Mine Fires*, Balkema, Rotterdam.

Barenbrug, A. W. T., 1965, *Psychrometry and Psychrometric Charts*, Transvaal and Orange Free State Chamber of Mines, Johannesburg (also in SI units, 1974).

Bates, R. C., and Edwards, J. C., 1980, "Mathematical Modeling of Time–Dependent Radon Flux Problems," *Proc. 2nd Intl. Mine Vent. Cong.*, P. Mousset-Jones, ed., SME-AIME, New York, pp. 412–419.

Baumeister, T., 1935, *Fans*, McGraw-Hill, New York.

Bedford, T., and Tredre, B. E., 1955, "Heat Stress in Industry," *Colliery Guard.*, Vol. 191, No. 4933, Sept. 15, pp. 317–323.

Bell, A. R., and Black, K. P., 1979, "Ventilation Planning for the Reactivation of Rio Algom's Panel Mines," *J. Mine Vent. Soc. S. Afr.* (Johannesburg), Vol. 32, No. 8, pp. 145–160.

Belton, A. E., 1962, "Mining Projections in Northern West Virginia," *Trans. SME-AIME*, Vol. 223, pp. 329–335.

Berger, L. B., and Schrenk, H. H., 1948, "Sampling and Analysis of Mine Atmospheres," *USBM M.C.*, No. 34.

Berry, C. H., 1963, *Flow and Fan*, 2nd ed., Industrial Press, New York.

Bexon, I., and Pargeter, D., 1976, "Practical Aspects of Computers in Mine Monitoring and Control Systems," *Proc. 3rd WVU Conf. Coal Mine Electrotechnology*, West Va. Univ., Morgantown, WV, pp. 22-1–22-16.

Bhaskar, R., and Gong, R., 1992, "Effect of Foam Surfactants on Quartz and Dust Levels in Continuous Miner Sections," *Mng. Eng.*, Vol. 44, No. 9, Sept., pp. 1164–1168.

Bhaskar, R., and Ramani, R. V., 1986, "Behavior of Dust Clouds in Mine Airways," *Trans. SME*, Vol. 280, pp. 2051–2060.

Bhaskar, R., and Ramani, R. V., 1988, "Dust Flows in Mine Airways: A Comparison of Experimental Results and Mathematical Predictions," *Trans. SME*, Vol. 284, pp. 1859–1864.

Bickel, K. L., Thomson, P., and Hillman, T. L., 1992, "In-Service Performance of Catalyzed Ceramic Wall-Flow Diesel Particulate Filters," *USBM I.C.*, No. 9324, pp. 74–81.

Biswas, N., 1966, "The Effect of Periodic Variations in Temperature and Humidity of Air Due to Seasonal and Diurnal Changes on Climate in Underground Roadways," *Mng. Minl. Eng.* (London), Vol. 2, No. 6, June, pp. 219–225.

Bolz, R. E., and Tuve, G. L., eds., 1973, *Handbook of Tables for Applied Engineering Science*, 2nd ed., CRC Press, Cleveland.

Boshkov, S., and Wane, M. T., 1955, "Errors in Underground Air Measurements," *Mng. Eng.,* Vol. 7, No. 11, Nov., pp. 1047–1053.

Bossard, F. C., and Stout, K. S., 1973, "Underground Mine Air Cooling Practices," *USBM O.F.R.,* Research Contract Final Report, No. GO 122137, June.

Brightwell, J., Foillet, X., Cassano-Zoppi, A.-L., Gatz, R., and Duchosal, F., 1986, "Neoplastic and Functional Changes in Rodents after Chronic Inhalation of Engine Exhaust Emissions," *Carcinogenic and Mutagenic Effects of Diesel Engine Exhaust: Proc. Symp. Toxicological Effects of Emissions from Diesel Engines,* N. Ishinishi et al., eds., Elsevier, New York, pp. 471–487.

Bromilow, J. G., 1957/58, "Descensional and Homotropal Ventilation," *Trans. Inst. Mng. Engr.* (London), Vol. 117, pp. 441–460.

Bruce, W. E., 1986, "Natural Draft: Its Measurement and Modeling in Underground Mine Ventilation Systems," *MSHA I.R.,* No. 1183.

Bruzewski, R. G., and Aughenbaugh, N. E., 1977, "Effects of Weather on Mine Air," *Mng. Cong. J.,* Vol. 63, No. 9, Sept., pp. 23–25.

Bryom, R. D., 1957, "Defense Against CO—the Silent Killer," *Eng. Mng. J.,* Vol. 158, No. 10, Oct., pp. 88–89.

Burrows, J. H. J., 1974, "Refrigeration," *The Ventilation of South African Gold Mines,* J. H. J. Burrows, ed., *Mine Ventilation Society of South Africa,* Johannesburg, pp. 180–215.

Burrows, J. H. J., and Roberts, B. G., 1980. "Managing and Manning Ventilation Departments of South African Gold Mines," *Proc. 2nd Intl. Mine Vent. Cong.,* P. Mousset-Jones, ed., SME-AIME, New York, pp. 42–50.

Butani, S. J., and Pomroy, W. H., 1987, "A Statistical Analysis of Metal and Nonmetal Mine Fire Incidents in the United States from 1950 to 1984," *USBM I.C.,* No. 9132.

Campbell, J. A. L., 1987, "The Recirculation Hoax," *Proc. 3rd U.S. Mine Vent. Symp.,* J. M. Mutmansky, ed., SME, Littleton, CO, pp. 24–30.

Campbell, J. C., 1988, "The Campbell Flooded-Bed Scrubber System," *Respirable Dust in the Mineral Industries: Health Effects, Characterization and Control,* R. L. Frantz and R. V. Ramani, eds., Penn. State Univ., University Park, PA, pp. 307–311.

Cantrell, B. K., and Rubow, K. L., 1991, "Development of Personal Diesel Aerosol Sampler Design and Performance Criteria," *Mng. Eng.,* Vol. 43, No. 2, Feb., pp. 232–236.

Carrier, W. H., 1938, "Air Cooling in the Gold Mines on the Rand," *AIME Mng. Tech.,* T.P. 970, Sept.; also *Trans. AIME,* Vol. 141, 1940, pp. 176–287.

Carrier, W. H., 1950, "Principles of Air Conditioning," *Heating Piping Air Cond.,* Vol. 22, No. 8, Aug., p. 108.

Cashdollar, K. L., and Hertzberg, M., eds., 1987, *Industrial Dust Explosions,* ASTM Spec. Tech. Pub. 958, American Society for Testing and Materials, Philadelphia.

Cervik, J., 1969, "Behavior of Coal-Gas Reservoirs," *USBM Tech. Prog. Rept.,* No. 10, April.

Cervik, J., 1977, "Water Infusion for Dust Control," *USBM I.C.,* No. 8753, pp. 63–77.

Chang, X., and Greuer, R. E., 1987, "A Mathematical Model for Mine Fires," *Proc. 3rd U.S. Mine Vent. Symp.*, J. M. Mutmansky, ed., SME, Littleton, CO, pp. 453–461.

Chang, X., Laage, L. W., and Greuer, R. E., 1990, "A User's Manual for MFIRE: A Computer Simulation Program for Mine Ventilation and Fire Modeling," *USBM I.C.*, No. 9245.

Christensen, H. E., and Luginbyhl, T. T., eds., 1975, "Registry of Toxic Effects of Chemical Substances," U.S. Dept. Health, Education, and Welfare, National Institute for Occupational Safety and Health, Washington, p. 798.

Cole, D. E., 1984, "Longwall Dust Control Respirators," *Proc. Coal Mine Dust Conf.*, S. S. Peng, ed., West Va. Univ., Morgantown, WV, pp. 61–64.

Colinet, J. F., McClelland, J. J., Erhard, L. A., and Jankowski, R. A., 1990, "Laboratory Evaluation of Quartz Dust Capture of Irrigated-Filter Collection Systems for Continuous Miners," *USBM R.I.*, No. 9313.

Conti, R. S., and Hertzberg, M., 1987, "Thermal Autoignition Temperatures from the 1.2L Furnace and Their Use in Evaluating the Explosion Potential of Dusts," *Industrial Dust Explosions*, K. L. Cashdollar and M. Hertzberg, eds., ASTM Pub. STP958, American Society for Testing and Materials, Philadelphia, pp. 45–59.

Conti, R. S., and Litton, C. D., 1992, "Response of Underground Fire Sensors: An Evaluation," *USBM R.I.*, No. 9412.

Cook, N. G. W., 1977, "Ventilation Techniques Developed in South Africa," *Mng. Cong. J.*, Vol. 63, No. 10, Oct., pp. 40–44.

Courtney, W. G., and Cheng, L., 1977, "Control of Respirable Dust by Improved Water Sprays," *USBM I.C.*, No. 8753, pp. 92–108.

Coward, H. F., and Jones, G. W., 1952, "Limits of Flammability of Gases and Vapors," *USBM Bull.*, No. 503.

Crank, J., 1975, *The Mathematics of Diffusion*, Claredon Press, Oxford.

Cross, H., 1936, *Analysis of Flow in Networks of Conduits or Conductors*, Bull. 286, Engineering Experiment Station, Univ. Illinois, Urbana, IL.

Cross, J., and Farrer, D., 1982, *Dust Explosions*, Plenum Press, New York.

Dahl, H. D., 1976, "The Impact of Mine Development Lead Times on New Coal Mining Ventures," *Proc. 1976 Rapid Excavation and Tunneling Conf.*, SME-AIME, New York, pp. 3–12.

Dalzell, R. W., 1972, "Longwall Ventilation Systems," *Mng. Cong. J.*, Vol. 58, No. 3, pp. 53–60.

Dement, J. A., Merchant, J. A., and Green, F. H. Y., 1986, "Asbestosis," *Occupational Respiratory Diseases*, J. A. Merchant, ed., National Institute for Occupational Safety and Health, U.S. Govt. Printing Office, Washington.

Deul, M., and Kim, A. G., 1986, "Methane Control Research: Summary of Results, 1964–80," *USBM Bull.*, No. 687.

Diamond, W. P., Murrie, G. W., and McCulloch, C. M., 1976, "Methane Gas Content of the Mary Lee Group of Coalbeds, Jefferson, Tuscaloosa, and Walker Counties, Ala.," *USBM R.I.*, No. 8117.

Didyk, M., Ramani, R. V., Stefanko, R., and Luxbacher, G. W., 1977, "Advancement of Mine Ventilation Network Analysis from Art to Science," Vol. 3, PB 290 193, National Technical Information Service, Springfield, VA, 196 pp.

Dilworth, E. L., 1939, "Ventilating Fans for Mines," *Can. Mng. J.* (Westmount, Que.), Vol. 60, No. 3, pp. 136–139.

Divers, E. F., and Cecala, A. B., 1990, "Dust Control in Coal Preparation and Mineral Processing Plants," *USBM I.C.,* No. 9248.

Divers, E. F., and Janosik, J. J., 1978, "Comparison of Five Types of Low-Energy Scrubbers for Dust Control," *USBM R.I.,* No. 8289.

Dobbs, D., and Johnson, B., 1979, "Ventilation and Cooling at the Magma Mine," unpublished field trip guide for 2nd Intl. Mine Vent. Cong., Magma Copper Co., Superior, AZ.

Dorsett, J. G., Jr., and Nagy, J., 1968, "Explosibility of Chemicals, Drugs, Dyes, and Pesticides," *USBM R.I.,* No. 7132.

Dougherty, J. J., 1969, "Control of Mine Fires," Mining Extension Service, West Va. Univ., Morgantown, WV.

Drummond, J. A., 1974, "The Measurement of Airflow," *The Ventilation of South African Gold Mines,* J. H. J. Burrows, ed., Mine Ventilation Society of South Africa, Johannesburg, pp. 266–301.

Dunn, M. F., Kendorski, F. S., Rahim, M. O., and Mukherjee, A., 1983, "Testing Jet Fans in Metal/Monmetal Mines with Large Cross-Sectional Airways," *USBM O.F.R.,* No. 106–84, PB 84-196393, National Technical Information Service, Springfield, VA, 132 pp.

Durkin, J., and Greenfield, R. J., 1981, "Evaluation of the Seismic System for Locating Trapped Miners," *USBM, R.I.,* No. 8567.

Enderlin, W., 1973, *Evaluating Underground Heat Sources in Deep Mines,* M.S. thesis, Mont. Coll. Minl. Sci. Tech., Butte, MT; also *USBM O.F.R.,* Research Contract Report, No. GO 12137, June.

Ettinger, I. L., and Sulla, M. B., 1964, "The Gas-Capacity of the Brown Coals and Liberation of Gas in the Collieries of the Moscow Coal Fields," *Met. Mng.* (Moscow, Izvestia Akademi Nauk SSR), Vol. 5.

Falkie, T. V., 1958, "A Study of the Effects of Reynolds Number on Friction Factor in Mine Airways," unpublished report, Penn. State Univ., University Park, PA.

Fenton, J. L., 1972, "Survey of Underground Mine Heat Sources," M.S. thesis, Mont. Coll. Minl. Sci. Tech., Butte, MT; also *USBM O.F.R.,* Research Contract Report, No. GO 12137, June.

Ferber, B. I., and Wieser, A. H., 1972, "Instruments for Detecting Gas in Underground Mines and Tunnels," *USBM I.C.,* No. 8548.

Field, P., 1982, *Dust Explosions,* Elsevier, Amsterdam.

Fink, Z. J., and Adler, D. T., 1975, "Continuous Monitoring System for Mine Gas Concentrations Using Tube Bundles," *USBM I.C.,* No. 8060.

First, M. W., 1989, "Air Sampling and Analysis for Contaminants: An Overview," *Air Sampling Instruments,* S. V. Hering, ed., American Conference of Governmental Industrial Hygienists, Cincinnati, pp. 1–19.

Forbes, J. J., Davenport, S. J., and Morgis, G., 1949, "Review of Literature on Conditioning Air for Advancement of Health and Safety in Mines, Part III. Methods of Controlling the Chemical and Physical Qualities of Underground Air," *USBM I.C.,* No. 7528, Oct.

Forbes, J. J., and Grove, G. W. 1948, "Procedure in Sealing and Unsealing Mine Fires and in Recovery Operations Following Mine Explosions," *USBM M.C.,* No. 36.

Forbes, J. J., and Grove, G. W., 1954, "Mine Gases and Methods for Detecting Them," *USBM M.C., No. 33.*

Fuller, J. L., 1989, "An Overview of Longwall Ventilation System Design," Preprint No. 89-103, SME Annual Meeting.

Glenn, R. E., and Craft, B. F., 1986, "Air Sampling for Particulates," *Occupational Respiratory Diseases,* J. A. Merchant, ed., American Conference of Governmental Industrial Hygienists, Cincinnati, pp. 69–87.

Goch, D. C., and Patterson, H. S., 1940, "The Heat Flow into Tunnels," *J. Chem., Met. Mng. Soc. S. Afr.* (Johannesburg), Vol. 41, No. 3, pp. 117–128.

Goff, J. A., and Gratch, S., 1946, "Low–Pressure Properties of Water in the Range − 160 to 212°F," *Trans. Am. Soc. Heating Vent. Eng.,* Vol. 52, p. 95.

Goodfellow, H. D., 1985, *Advanced Design of Ventilation Systems for Contaminant Control,* Elsevier, Amsterdam.

Goodman, G. V. R., Taylor, C. D., and Thimons, E. D., 1992, "Jet Fan Ventilation in Very Deep Cuts—Preliminary Analysis," *USBM R.I.,* No. 9399.

Gracie, A., and Matthews, R., 1975/76, "Strata, Machinery and Coal in Transit—Their Respective Roles as Heat Sources," *Mng. Engr.* (London), Vol. 135, No. 178, Dec.-Jan., pp. 181–188.

Graham, J. B., 1975, "Fan Selection and Installation," *Trans. Am. Soc. Heating Refrig. Air Cond. Eng.,* Vol. 81, p. 488.

Grant, E. L., and Ireson, W. G., 1970, *Principles of Engineering Economy,* 5th ed., Ronald-Wiley, New York.

Grave, D. F. H., 1974, "Main and Auxiliary Ventilation Practice," *The Ventilation of South African Gold Mines,* J. H. J. Burrows, ed., Mine Ventilation Society of South Africa, Johannesburg, pp. 398–424.

Gregory, C. E., 1980, *A Concise History of Mining,* Pergamon, Elmsford, NY.

Greninger, N. B., Weiss, E. S., Luzik, S. J., and Stephan, C. R., 1991, "Evaluation of Solid Block and Cementitious Foam Seals," *USBM R.I.,* No. 9382.

Hadden, J. D., and Smith, R. L., 1987, "A Coal Mine Ventilation Survey Method," *MSHA I.R.,* No. 1162, 62 pp.

Hall, A. E., McLaine, D. M., Botsford, J. D., and Rohr, S., 1989a, "Sulfide Dust Explosion Studies at H-W Mine of Westmin Resources Limited," *Proc. 4th U.S. Mine Vent. Symp.,* M. J. McPherson, ed., SME, Littleton, CO, pp. 532–539.

Hall, A. E., McLaine, D. M., and Hardcastle, S. G., 1989b, "The Use of Controlled Recirculation to Reduce Winter Heating Costs in Canada," *Proc. 4th Intl. Mine Vent. Cong.,* S. Gillies, ed., Australasia Institute of Mining & Metallurgy, Melbourne, Australia, pp. 301–307.

Hall, C. J., 1967, *Airflow in Mines,* C. J. Hall & Assoc., Houghton, MI.

Hall, C. J., 1981. *Mine Ventilation Engineering,* SME-AIME, New York.

Hamm, E., 1980, "Central Refrigerating Plants for Air Conditioning in the Mines of Ruhrkohle AG," *Proc. 2nd Intl. Mine Vent. Cong.,* P. Mousset-Jones, ed., SME-AIME, New York, pp. 635–648.

Haney, R. A., 1980, "Characteristics and Calibration of Air Velocity Measuring Instruments," *Proc. 2nd Intl. Mine Vent. Cong.,* P. Mousset-Jones, ed., SME-AIME, New York, pp. 25–30.

Haney, R. A., 1995, "Trends in Implementation of Longwall Dust Controls," *Proc. 7th U.S. Mine Vent. Symp.,* A. Wala, ed., SME, Littleton, CO, pp. 311–318.

Hardcastle, S. G., Grenier, M. A., and Butler, K. C., 1993, "Electronic Anemometry—Recommended Instruments and Methods for Routine Airflow Measurements," *Proc. 6th U.S. Mine Vent. Symp.,* R. Bhaskar, ed., SME, Littleton, CO, pp. 571–576.

Hargrave, A. J., 1973, "Planning and Operation of Gaseous Mines," *Can. Mng. Met. Bull.* (Montreal), Vol. 66, No. 731, March, pp. 110–128.

Harrington, D., and Davenport, S. J., 1941, "Review of Literature on Conditioning Air for Advancement of Health and Safety in Mines. Part II, Need for Air Conditioning Indicated by Physical Quality of Underground Air," *USBM I.C.,* No. 7182.

Harris, E. J., Dalzell, R. W., Kline, R. J., and Miller, E. J., 1973, "A Method for Calculating Mine Ventilation Pressure Losses Using Computers and Desktop Calculators," *USBM I.C.,* No. 8594, 22 pp.

Hartman, H. L., 1954, "Rapid Graphic Solution of Air Friction Loss," *Eng. Mng. J.,* Vol. 155, No. 7, July, pp. 100–101.

Hartman, H. L., 1960, "Pressure Changes at Splits and Junctions in Mine Ventilation Circuits," *Trans. AIME,* Vol. 217, pp. 163–170.

Hartman, H. L., 1961, *Mine Ventilation and Air Conditioning,* Ronald, New York.

Hartman, H. L., 1962, "Determining Ventilation Requirements for Continuous Miners," *Mng. Eng.,* Vol. 14, No. 3, March, pp. 58–62.

Hartman, H. L., 1968, "Environmental Control and Safety," Panel Report 2F, *Rapid Excavation—Significance, Needs, and Opportunities,* E. P. Pfleider, ed., National Academy of Sciences, Committee on Rapid Excavation, Report 1690, Washington, pp. 75–83.

Hartman, H. L., 1973, "Mine Atmospheres and Gases," Sec. 16.1 in *SME Mining Engineering Handbook,* A. G. Cummins and I. A. Given, eds., SME-AIME, New York, pp. 16.2–16.4.

Hartman, H. L., ed., 1982, "New Recognition for an Old Technical Field," Keynote Address, *Proc. 1st U.S. Mine Vent. Symp.,* SME-AIME, Littleton, CO, pp. 3–5.

Hartman, H. L., Mutmansky, J. M., and Wang, Y. J., eds., 1982, *Mine Ventilation and Air Conditioning,* 2nd ed., Wiley, New York.

Hartman, H. L., and Trafton, B. O., 1963, "Digital Computer May Find New Uses in Determining Mine Ventilation Networks," *Mng. Eng.,* Vol. 15, No. 9, Oct., pp. 39–42.

Hartmann, I., 1948a, "Recent Research on the Explosibility of Dust Dispersions," *Ind. Eng. Chem.,* Vol. 40, April, p. 754.

Hartmann, I., 1948b, "Explosion and Fire Hazards of Combustible Dusts," *Industrial Hygiene and Toxicology,* Interscience, New York, pp. 446–449.

Hashimoto, B., 1961, "Analysis of Mine Ventilation Distribution Networks by Digital Computers," Waseda University, *Bulletin of the Science and Engineering Research Laboratory* (Japan), No. 17, pp. 18–29.

Heinrich, U., Muhle, H., Takenaka, S., Ernst, H., Fuhst, R., Mohr, U., Pott, F., and Stober, W., 1986, "Chronic Effects on the Respiratory Tract of Hamsters, Mice, and Rats after Long-Term Inhalation of High Concentrations of Filtered and Unfiltered Diesel Engine Emissions," *J. Appl. Toxicol.*, Vol. 6, No. 6, pp. 383–395.

Hemp, R., 1982, "Pressure Surveys," Chap. 6 in *Environmental Engineering in South African Mines*, Mine Ventilation Society of South Africa, pp. 123–153.

Hemp, R., and Deglon, P., 1980, "A Heat Balance in a Section of a Mine," *Proc. 2nd Intl. Mine Vent. Cong.*, P. Mousset-Jones, ed., SME-AIME, New York, pp. 523–533.

Hering, S. V., 1989, "Inertial and Gravitational Collectors," *Air Sampling Instruments*, S. V. Hering, ed., 19, *American Conference of Governmental Industrial Hygienists*, Cincinnati, pp. 337–385.

Hertzberg, M., and Cashdollar, K. L., 1987, "Introduction to Dust Explosions," *Industrial Dust Explosions*, K. L. Cashdollar and M. Hertzberg, eds., ASTM Pub. STP958, American Society for Testing and Materials, Philadelphia, pp. 5–32.

Hess, S. L., 1959, *Introduction to Theoretical Meteorology*, Holt, Rinehart & Winston, New York.

Higgins, J., and Shuttleworth, S., 1958, "A Tracer Gas Technique for the Measurement of Airflow in a Heading," *Colliery Eng.* (London), Vol. 35, Nov., pp. 483–491.

Hinds, W. C., 1982, *Aerosol Technology, Properties, Behavior, and Measurement of Airborne Particles*, Wiley, New York.

Hinsley, F. B., 1943/44, "Some Aspects of Deep-Mine Ventilation," *Trans. Inst. Mng. Engr.* (London), Vol. 103, pp. 567–590.

Hinsley, F. B., 1948/49, "The Determination of Pressure Losses in Mine Shafts," *Trans. Inst. Mng. Engr.* (London), Vol. 108, pp. 614–639.

Hinsley, F. B., 1950/51, "Natural and Mechanical Ventilation of Mines," *Trans. Inst. Mng. Engr.* (London), Vol. 110, pp. 651–691.

Hinsley, F. B., 1965, "New Method of Calculating the Natural Ventilating Pressure in Mines," *Colliery Guard.* (London), Vol. 211, No. 5454, Oct. 29, pp. 653–656.

Hodgson, R. A., 1979, "Precision Altimeter Survey Procedures," 4th rev. ed., instruction booklet, American Paulin System, Los Angeles.

Hogan, K. B., 1993, *Anthropogenic Methane Emissions in the United States; Estimates for 1990*, U.S. Environmental Protection Agency, Report EPA 430-R-93-003, Washington, pp. 3-1–3-30.

Holaday, D. A., Rushing, D. E., Coleman, R. D., Woolrich, P. F., Kusnetz, H. L. and Bale, W. F., 1957, "Control of Radon and Daughters in Uranium Mines and Calculations on Biologic Effects," U.S. Dept. Health, Education, and Welfare, Public Health Service Pub. No. 494, Washington.

Holdsworth, J. F., Pritchard, F. W., and Walton, W. H., 1951, "Fluid Flow in Ducts with a Uniformly Distributed Leakage," *Br. J. Appl. Phys.* (London), Vol. 2, Nov., pp. 321–324.

Hormozdi, I., 1979, "A Review of the Developments in Automatic Monitoring and Control of Mine Atmospheric Environment," unpublished M. Eng. report, Mineral Engineering Management, Penn. State Univ., University Park, PA.

Houghton, A., 1991, "The Planning and Methods Used at Selby in the Design of High Productivity Systems," *Proc. Intl. Conf. Reliability, Producing and Control in Coal Mines,* Australasia Institute of Mining & Metallurgy, Melbourne, Australia, pp. 336–342.

Howes, M. J., 1978, "Development of a Functional Relationship between Productivity and the Thermal Environment," *J. Mine Vent. Soc. S. Afr.* (Johannesburg), Vol. 31, No. 2, Feb., pp. 21–38.

Huffer, W. D., 1995, personal communication, ABC Industries, Inc., Warsaw, IN.

Hunt, H. J., 1960, "The Flow of Fluids," *Mine Ventilation,* A. Roberts, ed., Cleaver-Hume Press, London, p. 11.

Ishinishi, N., Kuwabara, N., Nagase, S., Suzuki, T., Ishiwata, S., and Kohno, T., 1986, "Long-Term Inhalation Studies on Effects of Exhaust form Heavy- and Light-Duty Diesel Engines on F344 Rats," *Carcinogenic and Mutagenic Effects of Diesel Engine Exhaust: Proc. Symp. Toxicological Effects of Diesel Engines,* N. Ishinishi et al., eds., Elsevier, New York, pp. 329–348.

Iwai, K., Udagawa, T., Yamagishi, M., and Yamada, H., 1986, "Long-Term Inhalation Studies of Diesel Exhaust on F344 SPF Rats, Incidence of Lung Cancer and Lymphona," *Carcinogenic and Mutagenic Effects of Diesel Engine Exhaust: Proc. Symp. Toxiocological Effects of Diesel Engines,* N. Ishinishi et al., eds., Elsevier, New York, pp. 349–360.

Jacobson, M., 1972, "Sampling and Evaluating Respirable Coal Mine Dust," *Coal Workers' Pneumoconiosis,* M. M. Key and D. H. K. Lee, eds., *Annals of the New York Academy of Science,* Vol. 200, Dec., pp. 661–665.

Jacobsen, M., Rae, S., Walton, W. H., and Rogan, J. M., 1970, "New Dust Standards for British Coal Mines," *Nature,* Vol. 227, pp. 445–447.

Jacques, E. J. P. M., 1976, "Generalization of Cross' Iterative Method for Computing Ventilation Networks," *Proc. Intl. Mine Vent. Cong.,* R. Hemp and F. H. Lancaster, eds., Mine Ventilation Society of South Africa, Johannesburg, pp. 1–3.

Jankowski, R., 1995, personal communication, USBM, Bruceton, PA.

Jankowski, R. A., Jayaraman, N. I., and Potts, J. D., 1993, "Update on Ventilation for Longwall Mine Dust Control," *USBM I.C.,* No. 9366.

Johnson, B. R., 1982, "Mine Air Conditioning Systems," Chap. 22 in *Mine Ventilation and Air Conditioning,* 2nd ed., H. L. Hartman, J. M. Mutmansky, and Y. J. Wang, eds., Wiley, New York, pp. 653–695.

Johnson, B. R., 1992a, "Mine Ventilation Design: Metal Mine Ventilation," Sec. 11.7.1 in *SME Mining Engineering Handbook,* 2nd ed., H. L. Hartman, sr. ed., SME, Littleton, CO, pp. 1093–1106.

Johnson, B. R., 1992b, "Heat, Humidity, and Air Conditioning," Chap. 11.4 in *SME Mining Engineering Handbook,* 2nd ed., H. L. Hartman, sr. ed., SME, Littleton, CO, pp. 1028–1039.

Johnson, B. R., Bossard, F. C., and Walli, R. S., 1982, "Mine Air Heating Plants," Chap. 22 in *Mine Ventilation and Air Conditioning,* 2nd ed., H. L. Hartman, J. M. Mutmansky, and Y. J. Wang, eds., Wiley, New York, pp. 695–702.

Jones, C., 1972, "Estimating Heat and Humidity in Coal Mine Airflow," *Colliery Eng.* (London), Vol. 39, No. 462, Sept., pp. 372–376; No. 464, Oct., pp. 420–425.

Jones, J. H., and Trickett, J. C., 1954, "Some Observations on the Examination of

Gases Resulting from Explosions in Collieries," *Trans. Inst. Mng. Engr.* (London), Vol. 114, pp. 768–791.

Jones, T. M., 1985, "Firedamp Accumulations and Their Dispersal with Special Reference to Use of Controlled Recirculation Systems," *Proc. 2nd U.S. Mine Vent. Symp.,* P. Mousset-Jones, ed., Balkema, Rotterdam, pp. 239–248.

Jones, T. M., 1987, "The Application of Controlled Recirculation to Mine Ventilation Planning," Ph.D. thesis, University of Nottingham, UK.

Jordan, D. W., 1965, "Numerical Solution of Underground Heat Transfer Problems, Parts 1, 2, and 3," *Intl. J. Rock Mechanics Mng. Sci.* (Oxford), Vol. 2, No. 3, Sept., pp. 247–270; No. 4, Dec., pp. 341–387.

Jorgensen, R., ed., 1970, *Fan Engineering,* 7th ed., Buffalo Forge, Buffalo, NY.

Jorgensen, R., ed., 1983, *Fan Engineering,* 8th ed., Buffalo Forge, Buffalo, NY.

Jurani, R. F., 1987, "Optimization of Ventilation and Cooling System of Sunshine Mine's Kellogg Operation," Preprint No. 87–124, SME, Littleton, CO.

Kalasky, J. D., and Krikovic, S., 1973, "Ventilation of Pillared Areas by Bleeder Entries, Bleeder Systems or Equivalent Means," *Trans. SME-AIME,* Vol. 254, pp. 284–291.

Kane, A., 1993, "Epidemiology and Pathology of Asbestos-Related Diseases," *Health Effects of Mineral Dusts,* G. D. Guthrie, Jr., and B. T. Mossman, eds., *Reviews in Mineralogy,* Vol. 28, Mineralogical Society of America, Washington.

Kawenski, E. M., Mitchell, D. W., Bercik, G. R., and Frances, A., 1965, "Stoppings for Ventilating Coal Mines," *USBM R.I.,* No. 6710.

Kenzy, G. W., and Ramani, R. V., 1980, "An International Review of Regulations for Diesel-Powered Equipment in Underground Mines," *Proc. 2nd Intl. Vent. Cong.,* P. Mousset-Jones, ed., SME-AIME, New York, pp. 175–190.

Kharkar, R., Ramani, R. V., and Stefanko, R., 1974, "Analysis of Leakage and Friction Factors in Coal Mine Ventilation Systems," Special Research Report No. SR-99, Penn. State Univ., University Park, PA.

Kim, A. G., 1973, "The Composition of Coalbed Gas," *USBM R.I.,* No. 7762.

Kimmins, E. J., 1971, "Firedamp Drainage in the North Western Area," *Colliery Guard.* (London), Vol. 219, Annual Review Issue, pp. 39–44.

Kingery, D. S., 1960, "Introduction to Mine Ventilation Principles and Practices," *USBM Bull.* No. 589.

Kingery, D. S., and Harris, E. J., 1958, "Coal Mine Ventilation Without Doors to Control Main Air Currents," *USBM I.C.,* No. 7853.

Kingery, D. S., and Kapsch, F. A., 1959, "Airflow Changes in Multiple-Fan Systems," *USBM I.C.,* No. 7889.

Kislig, R. E., 1968, "A Method for the Direct Measurement of Pressure Losses in Mine Shafts," Research Report No. 1/68, Chamber of Mines of South Africa, Johannesburg, Jan.

Kissell, F. N., and Bielecki, R. J., 1972, "An In-Situ Diffusion Parameter for the Pittsburgh and Pocahontas No. 3 Coalbeds," *USBM R.I.,* No. 7668.

Kissell, F. N., and Bielecki, R. J., 1975, "Methane Buildup Hazards Caused by Dust Scrubber Recirculation at Coal Mine Working Faces, A Preliminary Estimate," *USBM R.I.,* No. 8015.

Kissell, F. N., McCulloch, C. M., and Elder, C. H., 1973, "The Direct Method of Determining Methane Content of Coalbeds for Ventilation Design," *USBM R.I.,* No. 7767.

Klinowski, W. G., and Kennedy, D. J., 1991, "Tracer Gas Techniques Used in Mine Ventilation," *Proc. 5th U.S. Mine Vent. Symp.,* Y. J. Wang, ed., SME, Littleton, CO, pp. 662–666.

Knape, H. W., 1985, "Ventilation Report for Henderson Mine Tours on 12/12/85," Magma Copper Co. Internal Report, San Manual, AZ, Feb. 15.

Kohler, J. L., 1992, "Monitoring, Control, and Communications," Chap. 12.6 in *SME Mining Engineering Handbook,* 2nd ed., H. L. Hartman, sr. ed., SME, Littleton, CO, pp. 1237–1251.

Krickovic, S., 1945/46, "The Need and Use of Altimeter Surveys in Coal Mine Ventilation," *Proc. Kentucky Mining Institute,* Lexington, pp. 117–134.

Kruger, D., 1994, "Identifying Opportunities for Methane Recovery at U.S. Coal Mines; Draft Profiles of Selected Gassy Underground Coal Mines," U.S. Environmental Protection Agency, Report EPA-430-R-94-012, Washington, pp. 6-1–6-8.

Lacy, W. C., and Lacy, J. C., 1992, "History of Mining," Chap. 1.1 in *SME Mining Engineering Handbook,* 2nd ed., H. L. Hartman, sr. ed., SME, Littleton, CO.

Lambrechts, J. deV., 1972, "Critical Comparison of Specific Cooling Power and the Wet Kata Thermometer in Hot Mining Environments," *J.S. Afr. Inst. Mng. Met.* (Marshalltown), Vol. 73, No. 4, Dec., pp. 169–174.

Lambrechts, J. deV., 1974, "Mine Ventilation Economics," *The Ventilation of South African Gold Mines,* J. H. J. Burrows, ed., Mine Ventilation Society of South Africa, Johannesburg, pp. 449–474.

Langmuir, I., 1918, "The Adsorption of Gases on Plane Surfaces of Glass, Mica and Platinum," *J. Am. Chem. Soc.,* Vol. 40, pp. 1361–1403.

Lapp, N. L., 1981, "Lung Disease Secondary to Inhalation of Nonfibrous Minerals," *Clin. Chest Med.,* Vol. 2, pp. 219–233.

Lappin-Randolph, J., 1987, *Small-Block VCR Mining at Homestake,* Preprint No. 87–97, SME, Littleton, CO.

Lapple, C. E., 1961, "Characteristics of Particles and Particle Dispersoids," *SRI J.,* Vol. 5, No. 3, p. 94.

Lawton, B. R., 1933, "Local Cooling Underground by Recirculation," *Trans. Inst. Mng. Eng.,* Vol. 85, pp. 63–67.

Leach, S. J., 1969, "Recirculation of Mine Ventilation Systems," *Mng. Eng.* (London), Vol. 128, Jan., pp. 227–236.

Leach, S. J., and Thompson, H., 1968, "Observations on a Methane Roof Layer at Cambrian Colliery," *Mng. Minl. Eng.* (London), Vol. 4, No. 8, pp. 35–37.

Lee, C., and Ember, G., 1946, unpublished graph.

LeRoux, W. L., 1974, "Duties and Organization of Ventilation Services," *The Ventilation of South African Gold Mines,* J. H. J. Burrows, ed., Mine Ventilation Society of South Africa, Johannesburg, pp. 1–13.

Lewis, W. E., 1951, "Ventilation-Air Heating Plants of the Menominee Range," *USBM I.C.,* No. 7596, opp. pp. 2,6.

Linch, K. D., and Cocalis, J. C., 1994, "An Emerging Issue: Silicosis Prevention in Construction," *Appl. Occup. Environ. Hyg.,* Vol. 9, No. 9, Aug., pp. 539–542.

Lioy, P. J., 1995, "Community Air Sampling Strategies," *Air Sampling Instruments,* S. V. Hering and B. S. Cohen, eds., American Conference of Governmental Industrial Hygienists, Cincinnati.

Lippman, M., 1989, "Sampling Aerosols by Filtration," *Air Sampling Instruments,* S. V. Hering, ed., American Conference of Governmental Industrial Hygienists, Cincinnati, pp. 305–336.

Luxbacher, G. W., and Ramani, R. V., 1980, "The Interrelationships Between Coal Mine Plant and Ventilation System Design," *Proc. 2nd Intl. Mine Vent. Cong.,* P. Mousset-Jones, ed., SME-AIME, New York, pp. 73–82.

Luxbacher, G. W., and Ramani, R. V., 1982, "Developing Input Data for Computer Simulation of Mine Ventilation Systems for a Pressure-Quantity Survey," *Proc. 1st U.S. Mine Vent. Symp.,* H. L. Hartman, ed., SME-AIME, Littleton, CO, pp. 95–101.

Luxbacher, G. W., Ramani, R. V., and Stefanko, R., 1977, *Advancement of Mine Ventilation Network Analysis from an Art to a Science,* Vol. 6, PB29019/AS, National Technical Information Service, Springfield, VA, 196 pp.

Madison, R. D., ed., 1949, *Fan Engineering,* 5th ed., Buffalo Forge Co., Buffalo, NY.

Mancha, R., 1942, "Effects of Underground Stopping Leakage upon Mine-Fan Performance," *Trans. AIME,* Vol. 149, p. 178.

Mancha, R., 1946, "Surveys of Underground Mine Pressure," *Trans. AIME,* Vol. 168, pp. 106–118.

Mancha, R., 1950, *Aspects of Coal-Mine Ventilation,* Joy Manufacturing Co., Pittsburgh.

Mancha, R., 1958, "Safety Problems with Multiple Fans," *Mechanization,* Vol. 22, No. 10, Oct., pp. 71–72.

Marks, J. R., 1980, "Refrigeration Economics at the Star Mine," *Proc. 2nd Intl. Mine Vent. Cong.,* P. Mousset-Jones, ed., SME-AIME, New York, pp. 649–655.

Marks, J. R., 1997, personal communication, Homestake Mining Company, Lead, SD.

Marks, J. R., Struble, G. R., and Brown, A. B., 1987, "Recent Ventilation Improvements at the Homestake Mine," *Proc. 3rd Western Regional Conf. Precious Metals, Coal, and the Environment,* Sept. 23–26.

Martinson, M. J., 1977, "Heat Stress in Witwatersrand Gold Mines," *J. Occup. Accidents* (Amsterdam), Vol. 1, No. 2, Jan., pp. 171–193.

Mateer, R. S., 1981, unpublished class notes, Univ. Kentucky, Lexington.

Mauderly, J. L., 1995, "Current Assessment of the Carcinogenic Hazard of Diesel Exhaust," *Toxicol. Environ. Chem. J.,* Vol. 49, pp. 167–180.

Mauderly, J. L., Jones, R. K., Griffith, W. C., Henderson, R. F., and McClellan, P. F., 1987, "Diesel Exhaust is a Pulmonary Carcinogen in Rats Explosed Chronically by Inhalation," *Fund. Appl. Toxicol.,* Vol. 9, pp. 208–221.

Mbuyikamba, G., Jacques, E., and Patigny, J., 1992, "Mine Air Flow Quantity Measurements: Investigations to Improve the Traverse Anemometry Method," *Proc. 5th Intl. Mine Vent. Cong.,* R. Hemp, ed., Mine Ventilation Society of South Africa, pp. 363–367.

McClelland, J. J., Organiscak, J. A., Jankowski, R. A., and Pothini, B. R., 1987, "Water Infusion for Coal Mine Dust Control: Three Case Studies," *USBM R.I.,* No. 9096.

McCulloch, C. M., and Diamond, W. P., 1976, "Inexpensive Method Helps Predict Methane Content of Coalbeds," *Coal Age,* Vol. 81, No. 6, June, pp. 102–106.

McDonald, J. C., 1973, "Cancer in Chrysotile Mines and Mills," *Biological Effects of Asbestos,* P. Bogovski et al., eds., International Agency for Research on Cancer, Lyons, France, pp. 189–196.

McElroy, G. E., 1935, "Engineering Factors in the Ventilation of Metal Mines," *USBM Bull.,* No. 385, 196 pp.

McElroy, G. E., 1943, "Air Flow at Discharge of Fan Pipe Lines in Mines," *USBM R.I.,* No. 3730.

McElroy, G. E., 1945, "Design of Injectors for Low-Pressure Airflow," *USBM T.P.,* No. 678.

McElroy, G. E., 1947, "A Mine Air Conditioning Chart," *USBM R.I.,* No. 4165.

McElroy, G. E., and Kingery, D. S., 1957, "Making Ventilation-Pressure Surveys with Altimeters," *USBM I.C.,* No. 7809.

McNider, T. E., 1980, "Shaft Heater Evaluation," *Proc. 2nd Intl. Mine Vent. Cong.,* P. Mousset-Jones, ed., SME-AIME, New York, pp. 475–494.

McPherson, M. J., 1966, "Mine Ventilation Network Problems, Solution by Digital Computer," *Colliery Guard.* (London), Vol. 209, No. 5392, pp. 253–259.

McPherson, M. J., 1971, "The Metrication and Rationalization of Mine Ventilation Calculations," *Mng. Eng.* (London), Vol. 130, No. 131, pp. 729–738.

McPherson, M. J., 1985, "The Resistance to Airflow of Mine Shafts," *Proc. 2nd U.S. Mine Vent. Symp.,* P. Mousset-Jones, ed., Balkema, Rotterdam, pp. 531–542.

McPherson, M. J., 1987, "The Resistance to Airflow on a Longwall Face," *Proc. 3rd U.S. Mine Vent. Symp.,* J. M. Mutmansky, ed., SME, Littleton, CO, pp. 465–477.

McPherson, M. J., 1993, *Subsurface Ventilation and Environmental Engineering,* Chapman & Hall, London.

McQuiston, F. C., and Parker, J. D., 1982, *Heating, Ventilating, and Air Conditioning Analysis and Design,* 2nd ed., Wiley, New York.

Meakin, W. D., 1979, "Mine Fan Selection for Dependability and Low Maintenance," Joy Manufacturing Co., New Philadelphia, OH.

Merchant, J. A., ed., 1986, *Occupational Respiratory Diseases,* National Institute for Occupational Safety and Health, Cincinnati.

Miller, E. J., 1978, "Early Warning Fire Detection Using Low Level Carbon Monoxide Monitors," *Proc. 4th WVU Conf. Coal Mine Electrotechnology,* West Va. Univ., Morgantown, WV, Aug., pp. 39-1–39-13.

Miller, E. J., and Dalzell, R. W., 1982, "Mine Gases," Chap. 3 in *Mine Ventilation and Air Conditioning,* 2nd ed., H. L. Hartman, J. M. Mutmansky, and Y. J. Wang, eds., Wiley, New York, pp. 39–68.

Miller, E. J., Turcic, P. M., and Banfield, J. L., Jr., 1980, "Equivalency Tests of Fire Detection Systems for Underground Coal Mines Using Low Level Carbon Monoxide Monitors," *Proc. 2nd Intl. Mine Vent. Cong.,* P. Mousset-Jones, ed., SME-AIME, New York, pp. 239–246.

Misaqi, F. D., Inderberg, J. G., Blumenstein, P. D., and Naiman, T., 1976, "Heat Stress in Hot U.S. Mines and Criteria for Standards for Mining in Hot Environments," *MESA I.R.*, No. 1048.

Mishra, R., Ramani, R. V., and Wang, Y. J., 1978, "Systems Concept for Coal Mine Ventilation," *Trans. SME-AIME*, Vol. 264, pp. 1569–1575.

Mitchell, D., 1990a, "Safety Considerations in the Development of Recirculation Techniques in Primary Mine Ventilation Systems," *Mng. Engr.* (London), Vol. 149, No. 340, Jan., pp. 271–275.

Mitchell, D. W., 1990b, *Mine Fires: Prevention, Detection, Fighting,* Mclean Hunter Pub., Chicago.

Mitchell, D. W., and Burns, F. A., 1979, "Interpreting the State of a Mine Fire," *MSHA R.I.*, No. 1103.

Mitchell, D., and Whillier, A., 1971, "Cooling Power of Underground Environments," *J. S. Afr. Inst. Mng. Met.* (Marshalltown), Vol. 72, No. 5, Oct., pp. 93–99.

Montgomery, W. J., 1936, *Theory and Practice of Mine Ventilation,* Jeffrey Manufacturing Co., Columbus.

Morgan, W. K. C., and Seaton, A., 1975, *Occupational Lung Diseases,* Saunders, Philadelphia.

Morris, I. H., and Hinsley, F. B., 1951, "Some Factors Affecting the Choice of Fans for Mine Ventilation," *Trans. Inst. Mng. Engr.* (London), Vol. 111, pp. 489–524.

Morris, W. J., 1953, "Temperature Problems in Mining, etc.", *Mine Quarry Eng.* (London), Vol. 19, No. 3, March, pp. 88–92; No. 4, Apr., pp. 127–132; No. 5, May, pp. 163–168; No. 6, June, pp. 193–197.

Morrow, G. S., and Litton, C. D., 1992, "In-Mine Evaluation of Smoke Detectors," *USBM I.C.*, No. 9311.

Muche, G., 1975, "Methane Desorption Within the Area of Influence of Workings," *Proc. Intl. Vent. Cong.,* R. Hemp and F. H. Lancaster, eds., Mine Ventilation Society of South Africa, Johannesburg, pp. 125–133.

Mullins, R., and Hinsley, F. B., 1958, "Measurement of Geothermal Gradients in Boreholes," *Trans. Inst. Mng. Engr.* (London), Vol. 117, pp. 380–396.

Munson, B. R., Young, D. E., and Okiishi, T. H., 1994, *Fundamentals of Fluid Mechanics,* 2nd ed., Wiley, New York.

Mutmansky, J. M., and Greer, T. K., 1984, "Statistical Determination of the Optimal Velocity in Coal Mine Entries," *Trans. SME-AIME*, Vol. 276, pp. 1939–1943.

Mutmansky, J. M., Wang, Y. J., and Hartman, H. L., 1985, "Determining the Operating Point for a Mine Fan Installation—a Comprehensive Method," *Trans. SME-AIME*, Vol. 278, pp. 1847–1853.

Nagy, J., Dorsett, H. G., and Cooper, A. R., 1965, "Explosibility of Carbonaceous Dusts," *USBM R.I.*, No. 6597.

Nagy, J., and Verakis, H. C., 1983, *Development and Control of Dust Explosions,* Marcel Dekker, New York.

Nilsson, C. E., 1995, "Ventilation-as-Required, A New Monitoring System for the Swedish Mines," *Proc. 7th U.S. Mine Vent. Symp.,* A. Wala, ed., SME, Littleton, CO, pp. 83–86.

Olson, K. S., and Veith, D. L., 1987, "Fugitive Dust Control for Haulage Roads and Tailings Basins," *USBM R.I.*, No. 9096.

O'Neil, T. J., and Johnson, B. R., 1982, "Metal Mine Ventilation Systems," Chap. 14 in *Mine Ventilation and Air Conditioning*, 2nd ed., H. L. Hartman, J. M. Mutmansky, and Y. J. Wang, eds., Wiley, New York, pp. 379–420.

Organischak, J. A., Volkwein, J. C., and Jankowski, R. A., 1983, "Reducing Longwall Tailgate Workers' Dust Exposure Using Water-Powered Scrubbers," *USBM R.I.*, No. 8780.

Osborne, W. C., 1977, *Fans*, 2nd ed., Pergamon, New York.

Ower, E., 1966, *The Measurement of Airflow*, 4th ed., Pergamon, New York.

Owili-Eger, A. S. C., 1973, "Simulation of Quantity and Quality Control in Mine Ventilation," M. S. thesis, Penn. State Univ., University Park, PA, 259 pp.

Page, S. J., 1982, "Evaluation of the Use of Foam for Dust Control on Face Drills and Crushers," *USBM R.I.*, No. 8595.

Parczewski, K., and Hinsley, F. B., 1956/57, "Hydrometry in Mines," *Trans. Inst. Mng. Eng.* (London), Vol. 116, pp. 63–83.

Peele, R., ed., 1941, *Mining Engineers' Handbook*, 3rd ed., Wiley, New York.

Pomroy, W. H., 1990, "Mining Methods and Equipment," Sec. 10, Chap. 23 in *Industrial Fire Hazards Handbook*, A. E. Cota, ed., National Fire Protection Assoc., Quincy, MA, pp. 241–282.

Pomroy, W. H., and Carigiet, A. M., 1995, "Analysis of Underground Coal Mine Fire Incidents in the United States from 1978 to 1992," *USBM I.C.*, No. 9426.

Potts, E. L. J., 1945/46, "Ventilation Surveying Techniques," *Trans. Inst. Mng. Eng.* (London), Vol. 105, pp. 631–654.

Pretorius, B. D. G., Ferguson, R. O., and Ramsden, R., 1980, "Considerations in the Selection of a Cooling System for the Deep Sub-Shaft Area of a Mine with Very High Rock Temperatures," *Proc. 2nd Intl. Mine Vent. Cong.*, P. Mousset-Jones, ed., SME-AIME, New York, pp. 666–673.

Pursall, B. R., 1960, "Ventilation Planning," Chap. 13 in *Mine Ventilation*, A. Roberts, ed., Cleaver-Hume Press, London.

Quillian, J. H., 1974, "Pressure Surveys," *The Ventilation of South African Gold Mines*, J. H. J. Burrows, ed., Mine Ventilation Society of South Africa, Johannesburg, pp. 319–341.

Rahim, M. O., Patnaik, N. K., and Banerjee, S. P., 1976, "Determination of the Frictional Coefficient of Mine Airways with Varied Lining and Support Systems," *J. Mines Metals Fuels* (Calcutta), Vol. 24, No. 7, pp. 222–229.

Ramani, R. V., 1982, "Application of Computers to Mine Ventilation," Chap. 18 in *Mine Ventilation and Air Conditioning*, 2nd ed., H. L. Hartman, J. M. Mutmansky, and Y. J. Wang, eds., Wiley, New York, pp. 517–545.

Ramani, R. V., 1988, "Design of Mine Ventilation Systems," Sec. 7.5 in *Engineering Design: Better Results through Operations Research*, R. R. Levary, ed., North-Holland Pub., New York, pp. 532–555.

Ramani, R. V., 1992a, "Mine Ventilation," Chap. 11.6 in *SME Mining Engineering Handbook*, 2nd ed., H. L. Hartman, sr. ed., SME, Littleton, CO, p. 1056.

Ramani, R. V., 1992b, "Personnel Health and Safety," Chap. 11.1 in *SME Mining*

Engineering Handbook, 2nd ed., H. L. Hartman, sr. ed., SME, Littleton, CO, pp. 995–1003.

Ramani, R. V., 1992c, "Mine Ventilation Design: Coal Mine Ventilation," Sec. 11.7.2. in *SME Mining Engineering Handbook,* 2nd ed., H. L. Hartman, sr. ed., SME, Littleton, CO, pp. 1106–1121.

Ramani, R. V., Bhaskar, R., and Frantz, R. L., 1985, "An Analysis of Decision-Making Aspects in Mine Ventilation," *Proc. 2nd U.S. Mine Vent. Symp.,* P. Mousset-Jones, ed., Balkema, Rotterdam, pp. 765–772.

Ramani, R. V., and Prasad, K. V. K., 1987, "Application of Knowledge Based Systems in Mining Engineering," *Proc. 20th Intl. APCOM Symp.,* L. Wade, ed., The South African Institute of Mining and Metallurgy, Johannesburg, pp. 167–180.

Ramlu, M. A., 1991, *Mine Disasters and Mine Rescue,* Balkema, Rotterdam.

Rao, G. V. K., Rao, B. S., and Kumar, G. V., 1989, "Controlled Recirculation of Mine Air—Technique for the Future," *J. Mines Metals Fuels* (Calcutta), Vol. 37, No. 4, Apr., pp. 152–158.

Rees, J. P., 1950, *Ventilation Calculations,* Transvaal Chamber of Mines, Johannesburg.

Reist, P. C., 1993, *Aerosol Science and Technology,* 2nd ed., McGraw-Hill, New York.

Ren, T. X., and Richards, M. J., 1993, "Diagnostics of Sealed Coal Mine Fires with a Knowledge-Based Expert System," *Proc. 6th U.S. Mine Vent. Symp.,* R. Bhaskar, ed., SME, Littleton, CO, pp. 583–588.

Richardson, A. S., 1943, "Progress in Air Conditioning for the Ventilation of the Butte Mines," *Trans. AIME,* Vol. 153, pp. 203–222.

Richardson, A. S., 1950, "A Review of Progress in the Ventilation of the Butte, Montana, District," *Q. Colo. School Mines,* Vol. 45, No. 2B, pp. 148–166.

Richmond, J. K., Price, G. C., Sapko, M. J., and Kawenski, E. M., 1983, "Historical Summary of Coal Mine Explosions in the United States, 1959–81," *USBM I.C.,* No. 8909.

Roberts, A., ed., 1960, *Mine Ventilation,* Cleaver-Hume Press, London, 363 pp.

Rock, R. L., Dalzall, R. W., and Harris, E. J., 1971, "Controlling Employee Exposure to Alpha Radiation in Underground Uranium Mines," Vol. 2, USBM, Washington, DC, 180 pp.

Rock, R. L., Svilar, G., Beckman, R. T., and Rapp, D. D., 1975, "Evaluation of Radioactive Aerosols in United States Underground Coal Mines," *MESA I.R.,* No. 1025.

Rock, R. L., and Walker, D. K., 1970, "Controlling Employee Exposure to Alpha Radiation in Underground Uranium Mines," *USBM Handbook,* Vol. 1, pp. 7–10.

Rose, H. J. M., 1980, "Management of Ventilation Departments," *Proc. 2nd Intl. Mine Vent. Cong.,* P. Mousset-Jones, ed., SME-AIME, New York, pp. 17–22.

Rouse, H., 1937, "Modern Conceptions of the Mechanics of Fluid Turbulence," *Trans. Am. Soc. Civil Eng.,* Vol. 102, p. 163.

Rubow, K. L., Cantrell, B. K., and Marple, V. A., 1990, "Measurement of Coal Dust and Diesel Exhaust Aerosols in Underground Mines," *Proc. VIIth Intl. Pneumoconiosis Conf.,* NIOSH Pub. No. 90–108, pp. 645–650.

Rustan, A., and Stöckel, I., 1980, "Review of Developments in Monitoring and Control of Mine Ventilation Systems," *Proc. 2nd Intl. Mine Vent. Cong.*, P. Mousset-Jones, ed., SME-AIME, New York, pp. 223–229.

Sasaki, K., Miyakoshi, H., and Mashiba, K., 1996, "Analytical System for Ventilation Simulators with the Skyline Nodal Pressure Method and Practical Estimate System for Underground Mine Air Conditioning," *Proc. 26th Intl. Symp. Application of Computers and Operations Research in the Mineral Industry*, R. V., Ramani, ed., SME, Littleton, CO, pp. 393–399.

Schweisheimer, W., 1952, "Ventilation in Mines Down the Centuries," *Can. Mng. J.* (Westmount, Que.,), Vol. 73, No. 4, pp. 81–82.

Scott, D. R., and Hinsley, F. B., 1951, "Ventilation Network Theory," *Colliery Eng.* (London), Vol. 28, Nos. 324, 326, 328, 334, pp. 67–71, 159–166, 229–235, 497–500, respectively.

Scott, D. R., and Hinsley, F. B., 1951/52, "The Solution of Ventilation Network Problems," *Trans. Inst. Mng. Eng.* (London), Vol. 111, pp. 347–366.

Seaton, A., 1975a, "Silicosis," *Occupational Lung Diseases*, A. Seaton and W. K. C. Morgan, eds., Saunders, Philadelphia.

Seaton, A., 1975b, "Asbestosis," *Occupational Lung Diseases*, A. Seaton and W. K. C. Morgan, eds., Saunders, Philadelphia.

Sharp, D. F., 1967, "Sources of Moisture in Mine Airflow," *Intl. J. Rock Mechanics Mng. Sci.* (Oxford), Vol. 4, No. 1, Jan., pp. 71–83.

Shaw, C. T., 1980, "A Management and Reward System for Specialist Departments in Mining Companies," *Proc. 2nd Intl. Mine Vent. Cong.*, P. Mousset-Jones, ed., SME-AIME, New York, pp. 37–41.

Sheer, T. J., Cilliers, P. F., Chaplain, E. J., and Cornela, R. M., 1985, "Some Recent Developments in the Use of Ice for Cooling Mines," *J. Mine Vent. Soc. S. Afr.* (Johannesburg), Vol. 38, No. 5, May, pp. 56–59; No. 6, June, pp. 67–68.

Shirey, G. A., Colinet, J. F., and Kost, J. A., 1985, "Dust Control Handbook for Longwall Mining Operations," *USBM Final Report*, Contract J0348000, Bituminous Coal Research, Monroeville, PA.

Shone, R. D. C., and Sheer, T. J., 1988, "An Overview of Research into the Use of Ice for Cooling Deep Mines," *Proc. 4th Intl. Mine Vent. Cong.*, S. Gillies, ed., Australasia Institute of Mining & Metallurgy, Melbourne, Australia, pp. 407–413.

Short, S. R., Parker, J. E., Musgrave, K. J., Jajosky, R. A., and Wagner, G. R., 1993, "Silicosis in Sandblasters, Reporting and Surveillance: The Utility of the NIOSH Sensor Program," Abstract, *Proc. 2nd Intl. Symp. Silica, Silicosis, and Cancer*, Western Consortium for Public Health, San Francisco, Oct. 30, p. 33.

Skochinsky, A., and Komarov, V., 1969, *Mine Ventilation*, J. S. Scott, translator, D. A. Telyakovsky, ed., MIR Publishers, Moscow.

Smith, W. A., O'Malley, J. K., and Phelps, A. H., 1974, "Reducing Blade Passage Noise in Centrifugal Fans," *Trans. Am. Soc. Heating Refrig. Air Cond. Eng.*, Vol. 80, pp. 45–52.

Solow, N. M., 1977, *Optimal Sizing of Raise-Bored Ventilation Shafts*, Preprint, SME Fall Meeting, St. Louis.

Spalding, J., 1949, *Deep Mining*, Mining Pub., London.

Stachulak, J., 1989, "Ventilation Strategy and Unique Air Conditioning at Inco

Ltd.," *Proc. 4th U.S. Mine Vent. Symp.*, M. J. McPherson, ed., SME, Littleton, CO, pp. 3–9.

Stahlhofen, W., Gebhart, J., and Heyder, J., 1980, "Experimental Determination of the Regional Deposition of Aerosol Particles in the Human Respiratory Tract," *Am. Indust. Hyg. Assoc. J.*, Vol. 41, No. 6, pp. 385–398.

Starfield, A. M., 1966, "The Computation of Temperature Increase in Wet and Dry Airways," *J. Mine Vent. Soc. S. Afr.* (Johannesburg), Vol. 19, No. 10, Oct., pp. 157–165.

Stefanko, R., 1983, "Ventilation," Chap. 4 in *Coal Mining Technology: Theory and Practice*, SME-AIME, New York, pp. 60–80.

Stefanko, R., and Ramani, R. V., 1972, "Mine Ventilation Network Analysis," *Trans. SME-AIME*, Vol. 252, pp. 382–387.

Stefanko, R., Ramani, R. V., White, R. J., and Luxbacher, G. W., 1977, "Advancement of Mine Ventilation Network Analysis from Art to Science," Vol. 5: Application of PSU/MVS to the Ventilation Planning of a New Mine, *USBM O.F.R.*, No. 123(6)-78, NTIS PB 290 196/AS.

Stephan, C. R., 1990, "Coal Dust as a Fuel for Fires and Explosions," *MSHA Report*, No. 01-066-90.

Stevenson, J. W., 1980, "Establishing a Mine Ventilation Department in an Operating Coal Company," *Proc. 2nd Intl. Mine Vent. Cong.*, P. Mousset-Jones, ed., SME-AIME, New York, pp. 23–36.

Stevenson, J. W., 1988, "Ventilation Considerations in Longwall Mining of Deep Gassy Seams in Alabama," *Longwall Thick Seam Mining*, Oxford & IBH Book Pub., New Delhi, India, pp. 163–180.

Stewart, J. M., 1980, "The Use of Heat Transfer and Limiting Physiological Criteria as a Basis for Setting Heat Stress Limits," *Proc. 2nd Intl. Mine Vent. Cong.*, P. Mousset-Jones, ed., SME-AIME, New York, pp. 556–571.

Stewart, J. M., and Wyndham, C. H., 1975, "Suggested Thermal Stress Limits for Safe Physiological Strain in Underground Environments," *J. S. Afr. Inst. Mng. Met.* (Marshalltown), Vol. 76, No. 1, Aug., pp. 334–338.

Strang, J., and Mackenzie-Wood, P., 1990, *A Manual on Mines Rescue, Safety and Gas Detection*, CSM Press, Golden, CO.

Strever, M. T., Wallace, K. G., and McDaniel, K. H., 1995, "Underground Mine Ventilation Remote Monitoring and Control System," *7th U.S. Mine Vent. Symp.*, A. Wala, ed., SME, Littleton, CO, pp. 69–74.

Suboleski, S. C., and Kalasky, J. D., 1982, "Coal Mine Ventilation Systems," Chap. 15 in *Mine Ventilation and Air Conditioning*, 2nd ed., H. L. Hartman, J. M. Mutmansky, and Y. J. Wang, eds., Wiley, New York, pp. 421–452.

Sunshine, I., 1969, *Handbook of Analytical Toxicology*, CRC Press, Boca Raton, FL.

Swift, D. L., and Lippman, M., 1989, "Electrostatic and Thermal Precipitators," *Air Sampling Instruments*, S. V. Hering, ed., American Conference of Governmental Industrial Hygienists, Cincinnati, pp. 387–403.

Swirles, J., and Hinsley, F. B., 1953/54, "The Use of Vane Anemometers in the Measurement of Airflow," *Trans. Inst. Mng. Engr.* (London), Vol. 113, pp. 898–923.

Taylor, C. D., Kovscek, P. D., and Thimons, E. D., 1986, "Dust Control on Longwall Shearers Using Water-Jet-Assisted Cutting," *USBM I.C.*, No. 9077.

Teale, R., 1958, "The Accuracy of Vane Anemometers," *Colliery Eng.* (London), Vol. 35, June, pp. 239–246.

Thakur, P. C., and Dahl, H. D., 1982, "Methane Drainage," Chap. 4 in *Mine Ventilation and Air Conditioning*, 2nd ed., H. L. Hartman, J. M. Mutmansky, and Y. J. Wang, eds., Wiley, New York, pp. 69–83.

Thakur, P. C., and Davis, J. G., 1977, "How to Plan for Methane Control in the Underground Coal Mines," *Mng. Eng.*, Vol. 29, No. 10, Oct., pp. 41–45.

Thakur, P. C., and Poundstone, W. N., 1980, "Horizontal Drilling Technology for Advance Degasification," *Mng. Eng.*, Vol. 32, No. 6, June, pp. 676–680.

Thakur, P. C., and Umphrey, R. W., 1979, "Methane Drainage and Transport in Consol Mines," paper, Pittsburgh Mining Institute, Annual Meeting, Pittsburgh, Nov. 8–9.

Thimons, E. D., and Kissell, F. N., 1974, "Tracer Gas as an Aid in Mine Ventilation Analysis," *USBM R.I.*, No. 7917.

Thimons, E. D., Vinson, R. P., and Kissell, F. N., 1979, "Forecasting Methane in Metal and Nonmetal Mines," *USBM R.I.*, No. 8392.

Thomas, V. M., and Chandler, K. W., 1978, "Monitoring and Remote Control—Progress towards Automation in British Coal Mines, 1977," *Mng. Engr.* (London), Vol. 136, No. 206, Oct., pp. 251–262.

Timko, R. J., and Derick, R. L., 1991, "Applying Atmospheric Status Equations to Data Collected from a Sealed Mine, Postfire Atmosphere," *USBM R.I.*, No. 9362.

Tucker, H. S. G., 1968, "The Thermal Conductivity of Rock," *J. Mine Vent. Soc. S. Afr.* (Johannesburg), Vol. 21, No. 10, Oct., pp. 179–181.

Urosek, J. E., 1995, personal communication, U.S. Dept. Labor, Mine Safety and Health Administration, Bruceton, PA.

Van der Walt, J., and Whillier, A., 1979, "Heat Pick-up from the Rock and the Thermal Efficiency of Production," *J. Mine Vent. Soc. S. Afr.* (Johannesburg), Vol. 32, No. 7, July, pp. 125–144.

Van der Walt, J., and Whillier, A., 1980, "Engineering of Systems for Cooling Mines," *Proc. 2nd Intl. Mine Vent. Cong.*, P. Mousset-Jones, ed., SME-AIME, New York, pp. 627–634.

Van der Walt, N. T., and Hemp, R., 1989, "Thermometry and Temperature Measurements," Chap. 17 in *Environmental Engineering in South African Mines*, Mine Ventilation Society of South Africa, Johannesburg, pp. 413–433.

Van der Walt, J., Pye, R., Picterse, H., and Dionne, L., 1996, "Ventilating and Cooling at Barrick's Meikle Mine," *Mng. Eng.*, Vol. 48, No. 4, April, pp. 36–39.

Vennard, J. K., 1940, *Elementary Fluid Mechanics*, Wiley, New York.

Vincent, J. J., 1989, *Aerosol Sampling: Science and Practice*, Wiley, New York.

Von Glehn, F. H., Wernick, B. J., Chorosz, G., and Bluhm, S. J., 1987, "ENVIRON: A Computer Program for the Simulation of Cooling and Ventilation Systems in South African Mines," *Proc. 20th Symp. Applications of Computers and Mathematics in the Mineral Industries*, L. Wade et al., eds., South African Institute of Mining and Metallurgy, Johannesburg, Vol. 1, pp. 319–330.

Vost, K. R., 1973, "In Situ Measurements of the Surface Heat Transfer Coefficient in Underground Airways," *J. S. Afr. Inst. Mng. Met.* (Marshalltown), Vol. 73, No. 8, Mar., pp. 269–272.

Vost, K. R., 1976, "InSitu Measurement of the Thermal Diffusivity of Rock around Underground Airways," *Trans. Inst. Mng. Met.* (London), Sec. A, Vol. 85, April, pp. 57–62.

Vutukuri, V. S., and Lama, R. D., 1986, *Environmental Engineering in Mines,* Cambridge Univ. Press, London.

Wala, A. M., 1991, "Studies of Friction Factors for Kentucky's Coal Mines," *Proc. 5th U.S. Mine Vent. Symp.,* Y. J. Wang, ed., SME, Littleton, CO, pp. 675–684.

Wallace, K. G., and McPherson, M. J., 1991, "Use of the Gage and Tube Method to Determine Pressure Drops in Mine Shafts," *Proc. 5th U.S. Mine Vent. Symp.,* Y. J. Wang, ed., SME, Littleton, CO, pp. 507–512.

Wallace, K. G., McPherson, M. J., Brunner, D. J., and Kissell, F. N., 1990, "Impact of Using Auxiliary Fans on Coal Mine Ventilation Efficiency and Cost," *USBM I.C.,* No. 9307.

Wallis, R. A., 1961, *Axial Flow Fans,* Academic, New York.

Wandmacher, C., and Johnson, A. I., 1995, *Metric Units in Engineering: Going SI,* rev. ed., American Society of Civil Engineers, New York.

Wang, Y. J., 1982a, "Critical Path Approach to Mine Ventilation Networks with Controlled Flow," *Trans. SME-AIME,* Vol. 272, pp. 1862–1872.

Wang, Y. J., 1982b, "Ventilation Network Theory," Chap. 17 in *Mine Ventilation and Air Conditioning,* 2nd ed., H. L. Hartman, J. M. Mutmansky, and Y. J. Wang, eds., Wiley, New York, pp. 483–516.

Wang, Y. J., 1988, "A Coding Scheme for a Graphical Solution to Mine Ventilation Networks," *Mng. Sci. Tech.,* Vol. 7, No. 1, pp. 31–43.

Wang, Y. J., 1989, "A Procedure for Solving a More Generalized System of Mine Ventilation Network Equations," *Proc. 4th Mine Vent. Symp.,* M. J. McPherson, ed., SME, Littleton, CO, pp. 419–424.

Wang, Y. J., 1990, "Solving Mine Ventilation Networks with Fixed and Nonfixed Operating Points," *Mng. Eng.,* Vol. 42, No. 12, Nov., pp. 1342–1346.

Wang, Y. J., 1992, "Characteristic Curves for Multiple-Fan Ventilation Systems," *Trans. SME,* Vol. 292, pp. 1829–1836.

Wang, Y. J., 1993, "Characteristic Curves for the Series-Parallel Ventilation Network with Multiple Fans," *Trans. SME,* Vol. 294, pp. 1821–1827.

Wang, Y. J., and Hartman, H. L., 1967, "Computer Solution of Three-Dimensional Mine Ventilation Networks with Multiple Fans and Natural Ventilation," *Intl. J. Rock Mechanics Mng. Sci.* (Oxford), Vol. 4, No. 2, Apr., pp. 129–154.

Wang, Y. J., and Mutmansky, J. M., 1982, "Application of CPM in Mine Ventilation," *Proc. 1st Mine Vent. Symp.,* H. L. Hartman, ed., SME-AIME, New York, pp. 159–168.

Wang, Y. J., Mutmansky, J. M., and Walrod, G. H., 1979, "Optimal Sizing of Conventionally-Sunk Ventilation Shafts Based Upon Capital and Operating Criteria," *Mng. Eng.,* Vol. 31, No. 1, Jan., pp. 47–54.

Wang, Y. J., and Pana, M. T., 1971, "Solving Mine Ventilation Problems by Linear Programming," SME Preprint No. 71AU132, AIME Annual Meeting, New York.

Wang, Y. J., and Saperstein, L. W., 1970, "Computer-Aided Solution of Complex Ventilation Networks," *Trans. SME-AIME,* Vol. 247, pp. 238–250.

Watson, H. A., Beatty, R. L., Bekert, A. J., and Dufresne, D. E., 1966, "Portable Methane Detectors, Effects of Gases in Mine Atmospheres," *USBM I.C.,* No. 8292.

Watts, W. F., Jr., and Parker, D. R., 1987, "Respirable Dust Levels in Coal, Metal, and Nonmetal Mines," *USBM I.C.,* No. 9125.

Weast, R. C., 1973, "Mensuration Formulas," *Standard Mathematical Tables,* 23rd ed., CRC Press, Cleveland, pp. 396–402.

Webster, I., 1970, "Asbestos Exposure in South Africa," *Proc. Intl. Conf. Pneumoconiosis,* Johannesburg, H. A. Shapiro, ed., Oxford Univ. Press, Cape Town, pp. 209–212.

Weeks, W. S., 1926, *Ventilation of Mines,* McGraw-Hill, New York.

Weeks, W. S., 1928, "The Air Current Regulator," discussion by G. E. McElroy, *Trans. AIME,* Vol. 76, pp. 142–147.

Weisenfeld, S. L., Perrota, D. M., and Abraham, J. L., 1993, "Epidemic of Accelerated Silicosis in West Texas Sandblasters," abstract, *Proc. 2nd Intl. Symp. Silica, Silicosis, and Cancer,* Western Consortium for Public Health, San Francisco, Oct. 28, p. 1.

Weiss, E. S., Greninger, N. B., Stephan, C. S., and Lipscomb, J. R., 1993, "Strength Characteristics and Air Leakage Determinations for Alternative Mine Seal Designs," *USBM R.I.,* No. 9477.

Whillier, A., 1971, "Psychrometric Charts for All Barometric Pressures," *J. Mine Vent. Soc. S. Afr.* (Johannesburg), Vol. 24, No. 9, Sept., pp. 138–144.

Whillier, A., 1972, "Heat—a Challenge in Deep-Level Mining," *J. Mine Vent. Soc. S. Afr.* (Johannesburg), Vol. 25, No. 11, Nov., pp. 205–213.

Whillier, A., 1974a, "Introduction to Steady-State Heat Transfer," *The Ventilation of South African Gold Mines,* J. H. J. Burrows, ed., Mine Ventilation Society of South Africa, Johannesburg, pp. 138–163.

Whillier, A., 1974b, "Elementary Thermodynamics," *The Ventilation of South African Gold Mines,* J. H. J. Burrows, ed., Mine Ventilation Society of South Africa, Johannesburg, pp. 164–179.

Whillier, A., 1977, "Predicting the Performance of Forced-Draught Cooling; Towers," *J. Mine Vent. Soc. S. Afr.* (Johannesburg), Vol. 30, No. 1, Jan., pp. 2–25.

Whillier, A., Macken, A. D., Chacar, A. C., and Alli, H. A., 1969, "Fan Efficiency Determination from Air Temperature Measurements," *S. Afr. Mech. Eng.* (Johannesburg), Vol. 18, No. 6, Jan., pp. 125–132.

Whillier, A., and Mitchell, D., 1968, "Prediction of Cooling Rate of the Human Body," *J. S. Afr. Inst. Mng. Met.* (Marshalltown), Vol. 69, No. 2, Sept., pp. 103–114.

Whillier, A., and Ramsden, R., 1975, "Sources of Heat in Deep Mines and the Use of Mine Service Water for Cooling," *Proc. Intl. Mine Vent. Cong.,* R. Hemp and F. H. Lancaster, eds., Mine Ventilation Society of South Africa, Johannesburg, pp. 339–346.

Whitt, T. B., 1980, "Early Warning Fire Detection System in an Underground Coal

Mine," *Proc. 5th WVU Conf. Coal Mine Electrotechnology,* West Va. Univ., Morgantown, WV, July, pp. 7-1–7-9.

Whittaker, D., 1980, "Heat Emission in Longwall Coal Mining," *Proc. 2nd Intl. Mine Vent. Cong.,* P. Mousset-Jones, ed., SME-AIME, New York, pp. 534–548.

Williams, F. T., 1960a, "The Thermodynamics of Mine Ventilation," Chap. 8 in *Mine Ventilation,* A. Roberts, ed., Cleaver-Hume Press, London, pp. 160–176.

Williams, F. T., 1960b, "The Mine Climate: Air Conditioning in Hot and Deep Mines," Chap. 7 in *Mine Ventilation,* A. Roberts, ed., Cleaver-Hume Press, London, pp. 143–146.

Williamson, J. N., 1932, "How to Make Air Surveys and Calculate Results," *Coal Age,* Vol. 37, No. 9, Sept., pp. 331–333.

Winter, K., 1975, "Extent of Gas Emission in Zones Influenced by Extraction," *Proc. 16th Intl. Conf. Mine Safety Research,* Washington, Sept., pp. V3.1–V3.17.

Wolski, J. F., 1991, "Mine Fire Real-Time Simulator Can Help in Selecting the Best Firefighting Strategies," *Proc. 5th U.S. Mine Vent. Symp.,* Y. J. Wang, ed., SME, Littleton, CO, pp. 82–87.

Wright, G. W., 1978, "The Pulmonary Effects of Inhaled Inorganic Dust," *Patty's Industrial Hygiene and Toxicology,* G. D. Clayton and F. E. Clayton, eds., Wiley-Interscience, New York, pp. 165–202.

Wyndham, C. H., 1961, "Applied Physiological Research in South African Gold Mining Industry," *Trans. 7th Commonwealth Mng. Met. Cong.,* Vol. 2, South African Institute of Mining & Metallurgy, Marshalltown, pp. 775–792.

Wyndham, C. H., 1974, "The Physiological and Psychological Effects of Heat," *The Ventilation of South African Gold Mines,* J. H. J. Burrows, ed., Mine Ventilation Society of South Africa, Johannesburg, pp. 93–137.

Wyndham, C. H., Williams, C. G., and Bredell, G. A. G., 1965, "The Physiological Effects of Different Gaps Between Wet and Dry-Bulb Temperatures at High Wet-Bulb Temperatures," *J. S. Afr. Inst. Mng. Met.,* Marshalltown, Vol. 66, No. 4, Sept., pp. 52–57.

Yevdokimov, A. G., 1969, "A Theory of the Solution of Steady State Network Problems with Special Reference to Mine Ventilation Networks," *Intl. J. Num. Meth. Eng.,* Vol. 1, No. 3, pp. 279–299.

Young, P. A., Potts, W. L. J., and Mandal, A. C., 1978, "Kata Thermometry in Relation to the Specific Cooling Power of a Mine Environment," *J. Mine Vent. Soc. S. Afr.* (Johannesburg), Vol. 31, No. 1, Jan., pp. 2–15.

Zabetakis, M. G., 1973, "Methane Control in U.S. Coal Mines—an Overview," *USBM I.C.,* No. 8621, pp. 9–17.

Zabetakis, M. G., Stahl, R. W., and Watson, H. A., 1959, "Determining the Explosibility of Mine Atmospheres," *USBM I.C.,* No. 7901.

Answers to Selected Problems

Chapter 2

2.1 t_{dp} = 84.3°F (29.1°C)
ϕ = 88.7%
W = 0.0275 lb/lb (kg/kg)
μ = 88.1%
v = 15.39 ft³/lb (0.9608 m³/kg)
h = 52.30 Btu/lb (121.60 kJ/kg)

2.3 t_w = 68.3°F (20.2°C)
ϕ = 71%
W = 0.0132 lb/lb (kg/kg)
μ = 70.2%
v = 13.75 ft³/lb (08584 m³/kg)
h = 32.54 Btu/lb (75.69 kJ/kg)

2.6 70.7 psf (3.385 kPa)

2.8 13.36 ft³/lb (0.8341 m³/kg)

2.10 Calculated, 0.0697 lb/ft³ (1.1165 kg/m³)
Table, 0.0698 lb/ft³ (1.1181 kg/m³)

Chapter 3

3.1

Seam	Y_c, ft³(m³)	n
OK	3780 (107)	0.445
VA	3530 (100)	0.424
WV	2370 (67)	0.424
UT	1520 (43)	0.520
IL	990 (28)	0.577

3.2

Seam	τ,h
Pocahontas	3.3
Pittsburgh	330

3.5 (a) 42.8 h (governs)
(b) 56.9 h

3.7 9.2 min

3.10 24,800 cfm (11.70 m³/s)

3.11 2000 cfm (0.944 m³/s)

Chapter 4

4.1 $\tau = 8.66$ h
$\tau = 216$ h

4.3 $E_{TWA} = 1.88$ mg/m³
Yearly dose = 3000 mg·h/m³

4.6 Dose = 783,000 mg·h/m³

4.8 $Q = 20.16$ m³/s (42,700 cfm)

Chapter 5

5.1 (a) 413 = 413 in. (10.49 m = 10.49 m) water
(b) 6 = 6 in. (152 mm = 152 mm) water

5.2 (a) $H_s = 5.5$ in. (140 mm) water
$H_v = 0.5$ in. (13 mm) water
$H_t = 6.0$ in. (153 mm) water
(b) $H_s = 5.5$ in. (140 mm) water
$H_v = 0.5$ in. (13 mm) water
$H_t = 6.0$ in. (153 mm) water
(c) $H_s = 5.0$ in. (127 mm) water
$H_v = 0.5$ in. (13 mm) water
$H_t = 5.5$ in. (140 mm) water

5.5 (a) $H_s = 2.172$ in. (540.5 Pa)
$H_v = 0.176$ in. (43.8 Pa)
$H_t = 2.348$ in. (584.3 Pa)

5.7 (a) 47,400 cfm (22.37 m³/s)
$H_s = 4.4477$ in. water (1106.8 Pa)
$H_v = 0.0062$ in. water (1.5 Pa)
$H_t = 4.4539$ in. water (1108.3 Pa)

5.9 $H_s = 4.7485$ in. water (1181.6 Pa)
$H_v = 0.0062$ in. water (1.5 Pa)
$H_t = 4.7547$ in. water (1183.1 Pa)
Error = 6.3%

5.10 $5.70/yr

5.11 $P_a = 33.22$ hp (24.77 kW)
$P_a = 35.47$ hp (26.45 kW)
Error $= 6.3\%$

5.14 (a) $\dfrac{1,855,355 \text{ kg/m}^3}{1.0 \text{ lb·min}^2/\text{ft}^4}$

(b) 0.0186 kg/m^3

Chapter 6

6.2 1222 fpm (6.21 m/s)
97,600 cfm (46.1 m^3/s)

6.4 $\theta = 14.5°$
Rise $= 2.28$ in. (57.9 mm)
Length $= 9.18$ in. (233.2 mm)

6.5 $\Delta Z = 123$ ft (375 m)
$H_s = 1.79$ in. water (445 Pa)

6.8 Mine $H_s = \Sigma H_l = 1.81$ in. (450 Pa)
Mine $H_v = 0.12$ in. water (30 Pa)
Mine $H_t = 1.93$ in. water (480 Pa)

Chapter 7

7.14

Branch	A–B	B–C	C–D	B–E	E–F
Q, cfm	99,571	42,557	52,557	57,014	57,014
(m^3/s)	(46.99)	(20.08)	(24.80)	(26.91)	(26.91)
R, 10^{-10} in.·min^2/ft^6	0.90	4.25	7.60	0.55	2.70
(N·s^2/m^8)	(0.101)	(0.475)	(0.849)	(0.061)	(0.302)
H^L, in. water	0.892	0.770	2.099	0.179	0.878
(Pa)	(223)	(192)	(522)	(44)	(219)
H^F, in. water					
(Pa)					

Branch	F–C	F–G	G–D	D–H
Q, cfm	10,000	47,014	47,014	99,571
(m^3/s)	(4.72)	(22.19)	(22.19)	(46.99)
R, 10^{-10} in.·min^2/ft^6	9.45	5.35	2.85	4.30
(N·s^2/m^8)	(1.056)	(0.598)	(0.318)	(0.480)
H^L, in. water	0.095	1.183	0.630	4.263
(Pa)	(24)	(294)	(157)	(1060)
H^F, in. water	0.381			8.026
(Pa)	(85)			(1997)

Chapter 8

8.2 3.92 in. water (975 Pa)
4.44 in. water (1105 Pa)
3.76 in. water (936 Pa)
3.93 in. water (978 Pa)
4.10 in. water (1020 Pa)
45,800 cfm (21.6 m³/s)
30.0 hp (22.4 kW)

8.4 Mine H_s = 1.44 in. water (358 Pa)
Mine H_v = 0.06 in. water (15 Pa)
Mine H_t = 1.50 in. water (373 Pa)
39,000 cfm (18.4 m³/s)

8.5 3.95 in. (983 Pa)
4.03 in. (1003 Pa)

Chapter 9

9.1 3.46 in. water (0.861 kPa)

9.3 7.87 in. water (1.958 kPa)

9.5 Fan rating: 23,000 cfm (10.9 m³/s)
H_s = 4.95 in. water (1.232 kPa)
H_t = 5.50 in. water (1.369 kPa)
P_{as} = 17.8 hp (13.3 kW)
P_a = 20.0 hp (14.9 kW)
P_m = 23.0 bhp (17.2 kW)

9.8 2.9 psig (20.0 kPa)
3000 cfm (1.42 m³/s)

9.9 2480 cfm (1.17 m³/s)
2150 cfm (1.01 m³/s)
86.7%
57.8 bhp (43.1 kW)

Chapter 10

10.1 Mine H_s = 2.1 in. water (523 Pa)
Mine H_v = 1.0 in. water (249 Pa)
Fan H_t = 3.1 in. water (771 Pa)
Fan H_s = 2.5 in. water (622 Pa)
Fan H_v = 0.6 in. water (149 Pa)
Fan H_t = 3.1 in. water (771 Pa)
Manometer H = 2.5 in. water (622 Pa)

10.3 622 rpm

10.5 163,000 cfm (76.9 m³/s)

10.6 1075 rpm

10.7 2.7 in. water (672 Pa)
          47,000 cfm (22.2 m³/s)

10.9 With, 1040 rpm
          Against, 800 rpm

10.10 Q_a = 280,000 cfm (132 m³/s)
           Q_b = 290,000 cfm (137 m³/s)
           Q_c = 570,000 cfm (269 m³/s)

Chapter 11

11.1 39.4 in. (9.80 kPa)
          36.0 in. (8.96 kPa)

11.3 8260 cfm (3.90 m³/s)

11.5 (a) 11,200 cfm (5.28 m³/s)
          (b) 10,900 cfm (5.14 m³/s)
          (c) 10,600 cfm (5.00 m³/s)

11.6 Without recirculation:

Point G	0%
Point C	0%
Point D	0.66%
Point F	0.66%

With recirculation:

Point G	0%
Point C	0.165%
Point D	0.66%
Point F	0.66%

Chapter 12

12.1 13.9 ft (4.24 m)
12.3 10.6× 10.6 ft (3.23 × 3.23 m)
12.6 19.8 ft (6.04 m)
12.8 4 intakes and 4 returns

Chapter 13

13.3 140,820 cfm
          Mine H_s = 3.891 in. water

H_v = 1.538 in. water
H_t = 5.429 in. water
Fan H_s = 3.027 in. water
H_v = 1.538 in. water
H_t = 5.429 in. water
P_m = 142 bhp
Regulators 0.4492, 0.9483, 2.0318, 2.2277, 2.8892, 1.5275 in. water

13.4   Fan Q = 168,000 cfm
H_s = 3.2 in. water
Regulate sections 2 and 3.

Chapter 14

14.4   Yes. Increase fan A pitch, install fan B (same type) in series, install new C and D in series.
Fan A: pitch 25, motor 130 bhp (97 kW), 80,000 cfm (37.8 m³/s) at 8.4 in. water   (2.09 kPa)
Fan B: 80,000 cfm (37.8 m³/s) at 8.4 in. water (2.09 kPa), 130 bhp (97 kW)
Fans C and D: pitch 20, 120,000 cfm (56.6 m³/s) at 8.5 in. water (2.12 kPa), 200 bhp (149 kW)

14.5   Extend exhaust raise to serve sections IV and V.
30,000 cfm (14.2 m³/s) for section V
Upper fan H_s = 1.9 in. water (473 Pa)
There are alternative solutions.

14.6   (a)   Fan H: 95,000 cfm (44.8 m³/s), 4.35 in. water (1.08 kPa)
             Fan R: 105,000 cfm (49.6 m³/s), 5.30 in. water (1.32 kPa)
       (b)   Fan A: 200,000 cfm (94.4 m³/s), 5.30 in. water (1.32 kPa)
       (c)   Exhaust

14.8   Fan 1: 40,000 cfm (18.9 m³/s), 5.0 in. water (1244 Pa) 60 bhp (44.7 kW)
Fan 2: 80,000 cfm (37.8 m³/s), 2.5 in. water (622 Pa), 40 bhp (29.8 kW)
Fan 3: 80,000 cfm (37.8 m³/s), 2.5 in. water (622 Pa), 40 bhp (29.8 kW)
Fan 4: 120,000 cfm (56.6 m³/s), 3.2 in. water (796 Pa), 75 bhp (55.9 kW)
Mine head 8.2 in. water

Chapter 15

15.2   (a) ICO = 0.828 (Unfavorable, but may be OK if the atmosphere is stable.)
       (b) TR = 0.35 (Indicates the fire may be extinguished.)
       (c) E_c = 3.71%
          E_l = 14.72%

$R_{mc} = 0.992$ (These values indicate the mine atmosphere is nonexplosive.)

15.4 $E_c = 6.91\%$
$E_l = 53.74\%$
$R_{mc} = 0.17$ (These values indicate the sealed area is in the explosive range.)

Chapter 16

16.1 (a) 5.3°F (2.9°C) rise
64.7°F (18.1°C)
(b) 25.8°F (14.3°C) rise
44.2°F (6.8°C)

16.3 0.89°F (1.62°C)
0.46°F (0.84°C)
0.58°F (1.06°C)
84.4°F (29.1°C)

16.4 1739 Btu/h·ft (1670 W/m)
521,600 Btu/h (152,900 W)

16.7 992,200 Btu/h (290,800 W)

16.10 5320 Btu/h (1560 W)

16.12 (a) 86.5°F (30.3°C)
90.0°F (32.2°C)
(b) 90.0°F
93.5°F (34.2°C)
(c) 10 mcal/cm²·s
Moderate work level
(d) 82°F (27.8°C)

Chapter 17

17.1 443,000 Btu/h (129,800 W)

17.3 43.69 = 43.46 + 0.23 Btu/lb (101.62 = 101.09 = 0.53 kJ/kg)
0.0048 lb/lb (kg/kg)

17.5 (b) 67.0°F (19.4°C)
(c) 66,720 lb/h (8.407 kg/s)
16,100 cfm (7.60 m³/s)
(d) 1,187,000 Btu/h (347,900 W)
99 tons
654 lb/h (0.0824 kg/s)
(e) 67 tons
15,100 cfm (7.13 m³/s)
0.0747 lb/ft³ (1.197 kg/m³)

17.7 (b) 70.0°F (21.1°C)
 (c) 286,000 lb/h (36.04 kg/s)
 (d) 69,500 cfm (32.80 m³/s)
      0.0699 lb/ft³ (1.120 kg/m³)
 (e) 65,400 cfm (30.87 m³/s)
      0.0741 lb/ft³ (1.187 kg/m³)
 (f) 2,360,000 Btu/h (691,600 W)
      197 tons
      0

17.9 32,100 cfm (15.15 m³/s)

17.13 1,519,000 Btu/h (445,200 W)
      137 gpm (8.70 L/s)
      356 gpm (22.61 L/s)
      8,881,800 gal/yr (33,620 m³/yr)

CPSIA information can be obtained
at www.ICGtesting.com
Printed in the USA
BVOW08*1415040218
507010BV00003B/6/P